Random Surfaces
and Quantum Gravity

NATO ASI Series

Advanced Science Institutes Series

A series presenting the results of activities sponsored by the NATO Science Committee, which aims at the dissemination of advanced scientific and technological knowledge, with a view to strengthening links between scientific communities.

The series is published by an international board of publishers in conjunction with the NATO Scientific Affairs Division

A	**Life Sciences**	Plenum Publishing Corporation
B	**Physics**	New York and London
C	**Mathematical and Physical Sciences**	Kluwer Academic Publishers
D	**Behavioral and Social Sciences**	Dordrecht, Boston, and London
E	**Applied Sciences**	
F	**Computer and Systems Sciences**	Springer-Verlag
G	**Ecological Sciences**	Berlin, Heidelberg, New York, London,
H	**Cell Biology**	Paris, Tokyo, Hong Kong, and Barcelona
I	**Global Environmental Change**	

Recent Volumes in this Series

Volume 261—Z° Physics: *Cargèse 1990*
edited by Maurice Lévy, Jean-Louis Basdevant, Maurice Jacob, David Speiser, Jacques Weyers, and Raymond Gastmans

Volume 262—Random Surfaces and Quantum Gravity
edited by Orlando Alvarez, Enzo Marinari, and Paul Windey

Volume 263—Biologically Inspired Physics
edited by L. Peliti

Volume 264—Microscopic Aspects of Nonlinearity in Condensed Matter
edited by A. R. Bishop, V. L. Pokrovsky, and V. Tognetti

Volume 265—Fundamental Aspects of Heterogeneous Catalysis
Studied by Particle Beams
edited by H. H. Brongersma and R. A. van Santen

Volume 266—Diamond and Diamond-Like Films and Coatings
edited by Robert E. Clausing, Linda L. Horton, John C. Angus, and Peter Koidl

Volume 267—Phase Transitions in Surface Films 2
edited by H. Taub, G. Torzo, H. J. Lauter, and S. C. Fain, Jr.

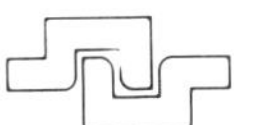

Series B: Physics

Random Surfaces and Quantum Gravity

Edited by

Orlando Alvarez

University of California
Berkeley, California

Enzo Marinari

Università di Roma II "Tor Vergata"
Rome, Italy

and

Paul Windey

Université Pierre et Marie Curie (Paris VI)
Paris, France

Plenum Press
New York and London
Published in cooperation with NATO Scientific Affairs Division

Proceedings of a NATO Advanced Research Workshop on
Random Surfaces and Quantum Gravity,
held May 27–June 2, 1990,
in Cargèse, France

Library of Congress Cataloging-in-Publication Data

NATO Advanced Research Workshop on Random Surfaces and Quantum Gravity
  (1990 : Cargèse, France)
    Random surfaces and quantum gravity / edited by Orlando Alvarez,
Enzo Marinari, and Paul Windey.
        p.   cm. -- (NATO ASI series. Series B, Physics ; vol. 262)
      "Published in cooperation with NATO Scientific Affairs Division."
      "Proceedings of a NATO Advanced Research Workshop on Random
Surfaces and Quantum Gravity, held May 27-June 2, 1990, in Cargèse,
France"--T.p. verso.
    Includes bibliographical references and index.
    ISBN 0-306-43939-5
    1. Quantum gravity--Congresses. 2. String models--Congresses.
3. Surfaces (Physics)--Congresses.   I. Alvarez, Orlando, 1953-   .
II. Marinari, Enzo. III. Windey, Paul. IV. North Atlantic Treaty
Organization. Scientific AffairsDivision. V. Title. VI. Series:
NATO ASI series. Series B, Physics ; v. 262.
QC178.N32   1990
531'.14--dc20   91-20195
                                                             CIP

ISBN 0-306-43939-5

PREFACE

The Cargese Workshop *Random Surfaces and Quantum Gravity* was held from May 27 to June 2, 1990.

Little was known about string theory in the non-perturbative regime before October 1989 when non-perturbative equations for the string partition functions were found by using methods based on the random triangulations of surfaces. This set of methods provides a description of non-critical string theory or equivalently of the coupling of matter fields to quantum gravity in two dimensions. The Cargese meeting was very successful in that it provided the first opportunity to gather most of the active workers in the field for a full week of lectures and extensive informal discussions about these exciting new developments. The main results were reviewed, recent advances were explained, new results and conjectures (which appear for the first time in these proceedings) were presented and discussed.

Among the most important topics discussed at the workshop were: The relation of KdV theory to loop equations and the Virasoro algebra, new results in Liouville field theory, effective $(1 + 1)$ dimensional theory for $2 - D$ quantum gravity coupled to $c = 1$ matter and its fermionization, proposal for a new geometrical interpretation of the string equation and possible definition of quantum Riemann surfaces, discussion of the string equation for the multi-matrix models, links with topological field theories of gravity, issues in using target space supersymmetry to define *good theories*, definition of the partition function via analytic continuation, new models of random surfaces including non-orientable surfaces, proposal of how to distinguish purely stringy effects from ordinary (field theoretic) non-perturbative effects in string theory.

The proceedings have been organized in three parts reflecting the main themes which were treated extensively during the meeting. We would like to point out that some of the contributions contain detailed reviews of the field which we hope will make this volume a useful tool even for the non specialist.

The success of the meeting was made possible thanks to the crucial help of many people to all of whom we wish to express our gratitude.

Firstly we are greatly indebted to the Nato Division for Scientific Affairs and particularly to Prof. Venturi for providing us a generous grant on very short notice and for constant help in organizing the meeting.

We are grateful to Marie-France Hanseler for her unique organizational talent and for her smiling efforts at ironing out all last minute difficulties. Without her help the meeting would have been impossible to put together.

We received the precious help of Liù Catena in editing the proceedings: without her constant attention this volume would be very different (and it would have been significantly delayed).

Last but not least, enthusiastic speakers and participants were the key ingredient for the success of the meeting: we warmly thank all of them and especially those who did not hesitate to destroy a solid reputation for writer's block by submitting their contribution ahead of the most optimistic deadline.

Orlando Alvarez

Enzo Marinari

Paul Windey

CONTENTS

2D GRAVITY AND NON PERTURBATIVE EFFECTS

DIFFUSION EQUATION, CONTINUUM LIMIT AND

UNIVERSALITY IN TWO DIMENSIONAL QUANTUM GRAVITY [1]

Orlando Alvarez

Department of Physics,
University of California,
Berkeley, CA 94720, USA

and

Paul Windey

Laboratoire de Physique Théorique et Hautes Energies
(Laboratoire associé No. 280 au CNRS)
Université Pierre et Marie Curie, Paris VI,
4 place Jussieu, F-75252 Paris Cedex 05, France

1   Introduction

In this paper we will present a brief summary of the definition of the continuum limit of the matrix model description of two dimensional gravity using a diffusion equation. This method differs from the continuum limit originally presented in [1, 2, 7] in the following way: the continuum limit in these papers is obtained by firstly computing the matrix elements of a certain operator and subsequently taking the continuum limit of the matrix elements. In this article we use a different approach. We derive an exact difference equation satisfied by the matrix elements and subsequently take the continuum limit of this equation. The continuum differential equation is then used to define the continuum limit of the original matrix elements in a totally consistent way. Both these methods lead to the same universal behavior as seen by explicit examples.

This situation is reminiscent of that encountered in the two dimensional Ising model on a rectangular lattice (see the review by Kogut [4] and references therein). On the lattice we can study the low temperature expansion of the free energy and try to understand the behavior of the series as one approaches the critical point. Alternatively, one can take the "time continuum" limit and formulate the problem as a Hamiltonian

---

[1]This work was supported in part by the Centre National de la Recherche Scientifique.

acting on a one dimensional lattice. These two approaches are very different and lead to different correlation functions at generic temperatures. In the continuum limit both theories lead to the same universal behavior but there are non-universal parts which are remnants of the original expansions. We shall see that the same is true in our approach and we shall even be able to identify the point of deviation of the two methods and the consequences.

This alternative scheme in the definition of the continuum limit leads to the same universal behavior discovered by the authors of [1, 2, 7] but we believe that it clarifies some of the issues and questions which have arisen in the subsequent literature [5, 6, 7]. In particular it leads to a simple proof of universality valid for any potential and to a direct connection to the KdV hierarchy without having to resort to special forms of the potential. Since we will present here only a sketchy account of our work we refer the reader to [8] for more details and in particular for a discussion of the case of arbitrary potentials.

We will start by stating as clearly as possible what we will call universality in this paper. Firstly we adhere to the definition of multicritical behavior given by Kazakov in [9] where he defines a critical point of order m by specifying the behavior of the free energy in the spherical limit (genus zero). The new feature introduced in references [1, 2, 7] has been called the double scaling limit. It amounts to finding a limit where all genus contribute democratically which means that one looks for a critical point where the perturbation expansion of the free energy becomes singular. We will say that universality is satisfied if all potentials which satisfy the genus zero criterion lead to identical behavior after summing over all surfaces (all genus) in the continuum limit.

In Section 2 we fix our notations and review the definition of universality on the sphere [9] and its extension to the full theory in [2, 3, 7]. In Section 3 we decribe the diffusion method to take the continuum limit and use it to prove how it leads to a universal behavior.

2   Universality in the one matrix models

The one matrix models are defined by a potential $U(\Phi)$ where Φ is a $N \times N$ hermitian matrix. The main object of interest is the behavior of the partition function

$$Z_N = \int d\Phi \, \exp(-\beta \operatorname{Tr} U(\Phi)) \tag{2.1}$$

near the critical point. The orthogonal polynomial method [10, 11] may be used to express the free energy in a convenient form. Let $P_n(\phi) = 1 \cdot \phi^n + \cdots$ be the n-th degree monic orthogonal polynomial defined with respect to the measure

$$\exp(-\beta U(\phi))d\phi \,, \tag{2.2}$$

where ϕ is a real variable. The normalization constants h_n are defined by

$$\int_{-\infty}^{+\infty} d\phi \, e^{-\beta U(\phi)} \, P_n(\phi)P_m(\phi) = h_n \delta_{nm} \,, \tag{2.3}$$

and the recursion coefficients for the orthogonal polynomials are defined by

$$\phi P_n(\phi) = P_{n+1}(\phi) + S_n P_n(\phi) + R_n P_{n-1}(\phi) \ . \tag{2.4}$$

This may be rewritten as an operator expression on the Hilbert space defined by the measure (2.2). Introduce orthonormal states $|n\rangle$ defined by

$$|n\rangle = \frac{1}{\sqrt{h_n}} \, |P_n\rangle \tag{2.5}$$

and following Gross and Migdal [3] raising and lowering operators $\exp(\pm i\widehat{\theta})$ by

$$e^{\pm i\widehat{\theta}} \, |n\rangle = |n \pm 1\rangle \ . \tag{2.6}$$

These operators are isometries but not unitary since they are not invertible. It is also convenient to introduce a number operator $\widehat{\mathbf{n}}$ defined by $\widehat{\mathbf{n}} \, |n\rangle = n \, |n\rangle$. The polynomial recursion relation now reads

$$\widehat{\phi} = \sqrt{R(\widehat{\mathbf{n}})} e^{i\widehat{\theta}} + e^{-i\widehat{\theta}} \sqrt{R(\widehat{\mathbf{n}})} + S(\widehat{\mathbf{n}}) \ , \tag{2.7}$$

where $R(\widehat{\mathbf{n}}) \, |n\rangle = R_n \, |n\rangle$ and similarly for S.

The recursion coefficients R and S are determined by the equations

$$\frac{n}{\beta} = -\frac{1}{2\beta} + \frac{1}{2} \langle n| \, V(\widehat{\phi}) \, |n\rangle \ , \tag{2.8}$$

$$0 = \langle n| \, U'(\widehat{\phi}) \, |n\rangle \ , \tag{2.9}$$

where $V(\phi) = \phi U'(\phi)$.

An exact expression for the free energy is

$$- F_N = \log Z_N = \kappa_N + N \log h_0 + \sum_{n=1}^{N-1} (N - n) \log R_n \ , \tag{2.10}$$

where κ_N is a numerical constant. In the large N limit near the critical point, the singular part of the free energy may be computed by defining a continuous variable $x = n/N$ and replacing the above sum by an integral

$$- F_N^{\text{sing}} = N^2 \int_0^1 dx \, (1 - x) \log R(x) \ , \tag{2.11}$$

where $R(x) = R_n$. The other terms in the Euler-Maclaurin formula do not contribute to the singular part of the free energy [2]. To simplify arguments we will assume that the potential U is an even function and thus the recursion coefficients S_n will vanish. Later in the article we will return to the general case.

It is useful to think of $e^{\pm i\widehat{\theta}}$ as operators which create and destroy one unit of charge. If one computes

$$\langle n| \, \widehat{\phi}^{2\ell} \, |n\rangle = \langle n| \left(\sqrt{R(\widehat{\mathbf{n}})} \, e^{i\widehat{\theta}} + e^{-i\widehat{\theta}} \sqrt{R(\widehat{\mathbf{n}})} \right)^{2\ell} |n\rangle \tag{2.12}$$

by expanding the binomial then only those terms which are "neutral" will contribute to the diagonal matrix element. At $N = \infty$, the operators $\widehat{\mathbf{x}} = \widehat{\mathbf{n}}/N$ and $\exp(\pm i\widehat{\theta})$

commute and thus we can move all the exponentials to the left in the expansion of the binomial. There will be $(2\ell)!/(\ell!)^2$ such terms thus we conclude

$$\lim_{N\to\infty} \langle n| \,\widehat{\phi}^{2\ell}\, |n\rangle = \binom{2\ell}{\ell} R^\ell \;. \tag{2.13}$$

As demonstrated in [10] it is useful to introduce a function $W(R)$ by

$$\begin{aligned} W(R) &= \frac{1}{4\pi i}\oint \frac{dw}{w}\, V\!\left(\sqrt{R}\,(w+1/w)\right)\;, \tag{2.14}\\[2mm] &= \frac{1}{2\pi}\int_{-1}^{1}\frac{dy}{\sqrt{1-y^2}}\, V\!\left(2\sqrt{R}\cdot y\right) \tag{2.15}\end{aligned}$$

At $N=\infty$ equation (2.8) may be written as

$$x e^{-\mu_B} = W(R)\;, \tag{2.16}$$

where the bare cosmological constant μ_B is defined by $\beta/N = \exp(\mu_B)$. Notice that definition (2.14) is designed to implement the "charge neutrality" condition.

Kazakov's condition [9] for multicriticality requires that the function $W(R)$ have the following behavior

$$W(R) = e^{-\mu_c}[1 - A(\rho_c - R)^m + O((\rho_c - R)^{m+1})] \tag{2.17}$$

The above form for W leads to the function R having a singularity

$$R = \rho_c\left[1 - \frac{1}{\rho_c A^{1/m}}\left(1 - \frac{x}{x_o}\right)^{1/m}\right] + \dots\;, \tag{2.18}$$

where the location of the singularity is $x_o = \exp(\mu_B - \mu_c)$. The critical behavior occurs as x_o approaches one. Following [2] one introduces a scale parameter a to blow up the neighborhood of the singularity

$$1 - x/x_o = a^2 z\;. \tag{2.19}$$

In terms of the scale parameter a, the multicriticality condition (2.17) on W may be stated in an equivalent way by demanding that near criticality an R of the form

$$\sqrt{R(z)} = \sqrt{\rho_c}\left(1 - \frac{1}{2}\,a^{2/m}Q(z)\right) \tag{2.20}$$

leads to

$$W(R) = e^{-\mu_c}\left\{1 - A\rho_c^m a^2 Q^m + O(a^{2+1/m})\right\}\;. \tag{2.21}$$

The conditions for the potential to belong to the m-th universality class are:

$$\frac{1}{2\pi}\int_{-1}^{1}\frac{dy}{\sqrt{1-y^2}}\, y^\ell\, V^{(\ell)}\!\left(2\sqrt{\rho_c}y\right)\begin{cases} \equiv e^{-\mu_c}\,, & \ell = 0 \\[2mm] = 0\,, & 0 < \ell < m \\[2mm] \neq 0\,, & \ell = m \end{cases} \tag{2.22}$$

It is clear from the above that any potential belongs to a well defined universality class. One is actually imposing $(m-1)$ conditions on V since the ϵ^0 term is generic. We will see later that it is convenient to impose a normalization condition $\sqrt{\rho_c} = 1/2$ when studying critical behavior. Thus the minimal potential will be an even polynomial of order $2m$. What is more important is that one can coarsely tune the potential and remain in the same universality class. Grossly adjusting the potential outside the interval $(-1, 1)$ one does not affect (2.22). Additionally, one can finely tune the potential inside $(-1, 1)$ and still remain in the same universality class. Thus for any potential in a given universality class there exist potentials containing any arbitrarily high powers of ϕ which are close to it.

We shall see in the next section that the diffusion equation method leads to conditions for spherical universality which differ slightly from equation (2.22). The variance is due to a different definition of the continuum matrix element.

3   The diffusion equation and the double scaling limit

The main problem in arbitrary genus is how to solve equation (2.8) for R. The difficulty arises because $\hat{x}$ and $\exp(\pm i\hat{\theta})$ do not commute. If we write

$$V(\phi) = \sum_{k=1}^{\infty} v_{2k}\phi^{2k} \tag{3.1}$$

then a typical term in equation (2.8) will entail the computation of $\langle n| \phi^{2k} |n\rangle$. We propose to compute this by studying general matrix elements of the discrete "diffusion" kernel

$$K(n, m|s) = \langle n| \widehat{\phi}^{s} |m\rangle \; . \tag{3.2}$$

It is straightforward to derive the difference equation

$$\begin{aligned} K(n, m|s + 1) &= \langle n| \widehat{\phi}^{s+1} |m\rangle \\ &= \sum_{\ell} \langle n| \widehat{\phi} |\ell\rangle \langle \ell| \widehat{\phi}^{s} |m\rangle \\ &= \sqrt{R_n}\, K(n - 1, m|s) + \sqrt{R_{n+1}}\, K(n + 1, m|s) \end{aligned} \tag{3.3}$$

This is a discrete one dimensional random walk in the space of orthogonal polynomials with hopping amplitude $\sim \sqrt{R}$. We will show that in the continuum limit this equation reduces to a Schroedinger equation in Euclidean time. This diffusion problem will lead to an unambiguous proof of universality. In taking the continuum limit the spatial variable n becomes continuous alongside with the discrete time s.

In higher genus it is convenient to study the m-th universal class by postulating equation (2.20) as an *ansatz* for R. Remembering that $x/x_o = 1 - a^2 z$ leads to the following replacements in equation (3.3):

$$
\begin{aligned}
n &\rightarrow & x &\rightarrow & z \\
n \pm 1 &\rightarrow & x \pm 1/N &\rightarrow & z \mp 1/(Na^2) \\
m &\rightarrow & \cdots &\rightarrow & \zeta
\end{aligned}
$$

These substitutions along with a continuum time defined by $t = \tau s$ lead to the equation

$$
\begin{aligned}
K(z,\zeta|t) &+ \tau\partial_t K(z,\zeta|t) + O(\tau^2) \\
&= \sqrt{\rho_c}\left\{2K(z,\zeta|t) + \left(\frac{1}{Na^2}\right)^2 \partial_z^2 K(z,\zeta|t) - a^{2/m}Q(z)K(z,\zeta|t)\right\} \\
&+ O((Na^2)^{-4}) + O((Na^{2-2/m})^{-1})
\end{aligned}
\tag{3.4}
$$

This equation has a well defined interesting continuum limit if one imposes various conditions. Firstly, in order not to have a large "mass" term as $a \rightarrow 0$ one has to choose $\sqrt{\rho_c} = 1/2$. This is the statement that the total probability of hopping is one to leading order. Only near $\sqrt{\rho_c} = 1/2$ is the random walk critical. Secondly, the diffusion term $\partial_z^2 K$ should be of the same order of magnitude as the potential term QK. This means that as $a \rightarrow 0$ one should define a variable λ by

$$
\frac{1}{Na^2} = \lambda a^{1/m}
\tag{3.5}
$$

with the understanding that λ is held fixed as $a \rightarrow 0$ and $N \rightarrow \infty$. We also see that the diffusion time step is given by $\tau = \sqrt{\rho_c}a^{2/m}$ as expected in any random walk.

The continuum equation which describes the matrix elements of $\widehat{\phi}^s$ is the heat equation

$$
\partial_t K(z,\zeta|t) = \lambda^2 \partial_z^2 K(z,\zeta|t) - Q(z)K(z,\zeta|t)
\tag{3.6}
$$

If we define a Hamiltonian[2] by

$$
\widehat{H} = -\partial_z^2 + \frac{1}{\lambda^2}\, Q(z)
\tag{3.7}
$$

we can rewrite the diffusion kernel as

$$
K(z,\zeta|t) = \langle z|\, e^{-t\lambda^2 \widehat{H}}\, |\zeta\rangle\ .
\tag{3.8}
$$

The diagonal matrix elements of the operator $\widehat{\phi}^{2\ell}$ are given in the continuum limit by

$$
\langle n|\,\widehat{\phi}^s\,|n\rangle \longrightarrow \langle n|\,\widehat{\phi}^s\,|n\rangle_{\text{diff}} = 2\lambda a^{1/m}\langle z|\, e^{-sa^{2/m}\lambda^2 \widehat{H}/2}\, |z\rangle
\tag{3.9}
$$

We have introduced the notation $\langle n|\,\widehat{\phi}^s\,|n\rangle_{\text{diff}}$ for matrix elements defined via the continuum heat kernel. The factor of 2 requires a careful analysis which one can perform in the simplest case $N = \infty$ and $Q = 0$. The matrix element $\langle n|\,\widehat{\phi}^s\,|n\rangle$ is the probability that after s flips of a fair coin one has tossed an equal number of heads and tails

---

[2] Hamiltonians were introduced independently for different purposes by Douglas and Shenker [12], and by Gross and Migdal [3].

and it vanishes if s is odd. Notice however that as one takes the continuum limit, the index n becomes continuous and thus there is no longer any symmetry reason for the odd s matrix element to vanish. Thus our diffusion equation definition of $\langle n | \widehat{\phi}^s | n \rangle_{\text{diff}}$ will not vanish for s odd. In fact it is a smooth function of s for $s > 0$. This means that equations (2.8) and (2.9) should be interpreted in some averaged sense because a point mass distribution is replaced by a density. Equation (3.8) tells us that the correct prescription in the continuum is to replace the operator $\widehat{\phi}$ by $\exp(-a^{2/m}\lambda^2 \widehat{H}/2)$.

We want to point out the following important fact: the commutation relation

$$[\widehat{\mathbf{z}}, e^{\pm i\widehat{\theta}}] = \frac{\mp 1}{Na^2} e^{\pm i\widehat{\theta}} \tag{3.10}$$

leads one to the representation of $\widehat{\theta}$ by

$$\widehat{\theta} = -i \, \frac{1}{Na^2} \, \frac{\partial}{\partial z} \, . \tag{3.11}$$

It is however dangerous to expand naively $e^{\pm i\widehat{\theta}}$ in powers of $\frac{1}{Na^2} \frac{\partial}{\partial z}$. In particular this may violate the "charge" neutrality condition (see equation (2.13)). Our choice of approximation to transform the exact difference equation into a continuum differential equation retains the minimum number of terms which will non-trivially preserve the operator identity $\widehat{\phi}^{s+1} = \widehat{\phi} \cdot \widehat{\phi}^s$ in the continuum limit. This will lead to a consistent calculation of the free energy and of all the correlation functions and we will consequently define the continuum theory by using (3.6) to calculate the matrix elements. Notice that our approach requires the physics to be smooth over a large range of n near N.

In the continuum limit we can see that equation (3.9) is dominated by the short time behavior of the heat kernel which is given by a local expression in the potential. It is well known [13] that the diagonal elements of the heat kernel have an asymptotic expansion

$$\langle z | \, e^{-t(-\partial^2 + u)} \, | z \rangle \sim \frac{1}{\sqrt{4\pi t}} \sum_{k=0}^{\infty} t^k s_k[u] \tag{3.12}$$

where $s_k[u]$ is a polynomial in the potential u and its derivatives. In fact,

$$s_k[u] = \frac{(-1)^k}{k!} \, u^k + \text{derivative terms} \tag{3.13}$$

where the derivative terms are of the form $\partial^{2\delta} u^p$. The non-negative integers δ and p satisfy $2\delta + 2p = 2k$ and the notation $\partial^{2\delta} u^p$ means that the terms have p powers of u and a total of 2δ z-derivatives. The first few s_k's are given by

$$\begin{aligned}
s_0[u] &= 1 \, , \\
s_1[u] &= -u \, , \\
s_2[u] &= \frac{1}{2} \, u^2 - \frac{1}{6} \, \partial_z^2 u \, .
\end{aligned}$$

The short times coefficients $s_k[u]$ are related to the KdV hierarchy [14] in the following way. Introduce a "time" parameter y not to be confused with the short distance expansion time t. Assume that u depends parametrically on y, i.e. $u = u(y, z)$, then the

k-th member of the KdV hierarchy is the equation $\partial_y u = \partial_z s_k[u]$. Up to normalization the KdV equation is $\partial_y u = \partial_z s_2[u]$.

The continuum diagonal matrix element is given by

$$\langle n|\,\widehat{\phi}^{2\ell}\,|n\rangle_{\text{diff}} = \frac{2\lambda a^{1/m}}{\sqrt{4\pi\ell a^{2/m}\lambda^2}} \sum_{k=0}^{\infty} (\ell a^{2/m}\lambda^2)^k s_k[Q/\lambda^2] \tag{3.14}$$

Notice the important cancellation of pre-factors of $\lambda a^{1/m}$ which is necessary to get a non-singular limit as either λ goes to zero or a goes to zero. Thus the $\lambda = 0$ matrix elements are defined by a limiting process because diffusion with diffusion constant $\lambda = 0$ is *not* diffusion. The overall prefactor is $(\pi\ell)^{-1/2}$.

To compare the matrix element evaluation with the diffusion equation evaluation at $\lambda = 0$ we use equation (2.13)

$$\langle n|\,\widehat{\phi}^{2\ell}\,|n\rangle = \frac{1}{2^{2\ell}}\begin{pmatrix} 2\ell \\ \ell \end{pmatrix}\left(1 - \frac{1}{2}\,a^{2/m}Q\right)^{2\ell} \tag{3.15}$$

and the continuum expression

$$\langle n|\,\widehat{\phi}^{2\ell}\,|n\rangle_{\text{diff}} = \frac{1}{\sqrt{\pi\ell}}e^{-\ell a^{2/m}Q}\ . \tag{3.16}$$

To derive the above one uses the observation

$$\lim_{\lambda\to 0}\lambda^{2k} s_k[Q/\lambda^2] = \frac{(-1)^k}{k!}\,Q^k\ . \tag{3.17}$$

The two matrix element expressions agree asymptotically for large values of ℓ because of the asymptotic behavior of the binomial coefficient

$$\begin{pmatrix} 2\ell \\ \ell \end{pmatrix} \sim \frac{1}{\sqrt{\pi\ell}}\left(1 + O(\ell^{-1})\right)\ . \tag{3.18}$$

This asymptotic agreement is understandable because the continuum limit of a random walk is a reasonable approximation if one takes a moderate number of steps. Our diffusion equation is used to define the matrix elements for small number of steps. This is the domain where we expect non-universal behavior to appear.

To discuss universality we analyze equation (2.8) in detail. The $1/\beta$ term on the right hand side of the equation is order $\lambda a^{2+1/m}$ and thus negligible since the left hand side is precisely $(1 - a^2 z)e^{-\mu_c}$. Thus we may write equation (2.8) which determines Q as

$$(1 - a^2 z)e^{-\mu_c} = \sum_{\ell=1}^{\infty}\sum_{k=0}^{\infty}\frac{v_{2\ell}}{\sqrt{4\pi\ell}}(\ell a^{2/m}\lambda^2)^k s_k[Q/\lambda^2] + O(\lambda a^{2+1/m})\ . \tag{3.19}$$

If one defines coefficients $\tilde{v}_k$ by

$$\tilde{v}_k = \sum_{\ell=1}^{\infty}\frac{v_{2\ell}}{\sqrt{4\pi\ell}}\ell^k \tag{3.20}$$

then the equation for Q may be written as

$$(1 - a^2 z)e^{-\mu_c} = \sum_{k=0}^{m} \tilde{v}_k a^{2k/m} \lambda^{2k} s_k[Q/\lambda^2] \tag{3.21}$$

where we have discarded all term of order higher than a^2.

We are now in a position to discuss universality. We ask about the genus zero ($\lambda = 0$) behavior of the above. This is easily computed by using equation (3.17):

$$(1 - a^2 z)e^{-\mu_c} = \sum_{k=0}^{m} \tilde{v}_k a^{2k/m} \frac{(-1)^k}{k!} Q^k \;. \tag{3.22}$$

At genus zero the right hand side of the above is just $W(R)$ thus comparing with equation (2.21) we see that

$$\tilde{v}_k = \begin{cases} e^{-\mu_c} & k = 0 \\[2mm] 0 & 0 < k < m \\[2mm] (-1)^{m+1} e^{-\mu_c} \dfrac{A\, m!}{2^{2m}} & k = m \end{cases} \tag{3.23}$$

Thus the genus zero condition tells us that most of the coefficients $\tilde{v}_k$ vanish and we easily return to equation (3.21) and conclude that the equation for Q is simply:

$$z = A \frac{(-1)^m\, m!}{2^{2m}} \lambda^{2m} s_m[Q/\lambda^2] \;. \tag{3.24}$$

This is the universal equation identified by Gross and Migdal [3, 7].

Conditions (3.23) differ from universality conditions (2.22) due to the difference in definition of the matrix elements. An elementary computation shows that if one uses criterion (2.22) the analog of the right hand side of equation (3.20) is

$$\sum_{\ell=1}^{\infty} \left\{ \frac{1}{2} v_{2\ell} \frac{1}{2^{2\ell}} \begin{pmatrix} 2\ell \\ \ell \end{pmatrix} \ell^k \prod_{j=1}^{k} \left(1 - \frac{j-1}{2\ell} \right) \right\} \tag{3.25}$$

Thus again we see reflections of non-universal behavior even though the universal behavior is identical. The diffusion equation version of equation (2.14) is

$$W_{\text{diff}}(Q) = \frac{1}{4\pi i} \oint \frac{dw}{w} V \left(\frac{1}{2} \left(w + \frac{1}{w} \right) \exp \left(-\frac{1}{2} a^{2/m} Q \right) \right) + O(a^{2+1/m}) \;. \tag{3.26}$$

These considerations can be extended to demonstrate universality for a general potential $U(\phi)$ [8] although there aer some delicate issues concerning the doubling problem discussed by Bachas and Petropoulos (see their contribution to this volume).

References

[1] E. Brezin and V. A. Kazakov. Exactly solvable field theories of closed strings.

Phys. Lett. **B236**, 144 (1990).

[2] M. R. Douglas and S.H. Shenker. Strings in less then one dimension. *Nucl. Phys.* **B335**, 635 (1990).

[3] D. J. Gross and A.A. Migdal. Nonperturbative two–dimensional quantum gravity. *Phys. Rev. Lett.* **64**, 127 (1990).

[4] J.B. Kogut. An introduction to lattice gauge theory and spin systems. *Rev. Mod. Phys.* **51**, 659–714 (1979).

[5] T. Banks, M. R. Douglas, N. Seiberg, and S. H. Shenker. Microscopic and macroscopic loops in non–perturbative two dimensional gravity. *Phys. Lett.* **B238**, 279 (1990).

[6] M.R. Douglas. Strings in less than one dimension and the generalized KdV hierarchies. *Phys. Lett.* **B238**, 176 (1990).

[7] D. J. Gross and A.A. Migdal. A nonperturbative treatment of two dimensional quantum gravity. *Nucl. Phys.* **B340**, 333 (1990).

[8] O. Alvarez and P. Windey. Universality in two dimensional quantum gravity. *Nucl. Phys.* **B348**, 490 (1991).

[9] V. A. Kazakov. The appearance of matter fields from quantum fluctuations of $2D$-gravity. *Mod. Phys. Lett.* **A 4**, 2125–2139 (1989).

[10] D. Bessis, C. Itzykson, and J-B. Zuber. Quantum field theory techniques in graphical enumeration. *Adv. Appl. Math* **1**, 109–157 (1980).

[11] G. Parisi. Unpublished.

[12] M.R. Douglas and S.H. Shenker. Unpublished talk. Soviet-American String Workshop, Princeton, October 30 — November 2, 1989.

[13] P.B. Gilkey. *Invariance Theory, the Heat Equation and the Atiyah-Singer Index Theorem.* Volume 11 of *Mathematics Lecture Series*, Publish or Perish, Inc., 1984.

[14] I.M. Gelfand and L.A. Dikii. Asymptotic behaviour of the resolvent of Sturm-Liouville equations and the algebra of the Korteweg-de Vries equations. *Russian Math. Surveys* **30:5**, 77–113 (1975).

ON TRIANGLES AND SQUARES

C. Bachas

Theory Division, CERN
CH-1211 Geneva 23, Switzerland

One of the amusing features of random discretized surfaces [1] [2][3] is the *geometric* appearance of matter [4]: appropriately choosing the areas assigned to the elementary k-gons that make up the surface, amounts to introducing certain statistical mechanical vertex models, which at criticality can modify the infrared behaviour of the theory. The simplest example is the Lee-Yang edge singularity [5] shown to correspond to the first non-trivial ($k = 3$) multicritical point of the one-matrix model; the known higher order multicritical points are likewise believed to correspond to $(2k - 1, 2)$ non-unitary minimal matter coupled to gravity. The phase structure of the one-matrix model is, however, certainly much richer and still largely unexplored. It can in fact be argued that, if we allow couplings more general than traces of polynomials, the one-matrix model can describe all conceivable string theories: just add any matter explicitly by giving extra (flavour) indices to the matrix that generates triangulations, then integrate out all these matrices but one. This simple argument shows that a full classification of critical behaviour in the one-matrix model may be too ambitious a task.

In fact, even the much more modest problem of classifying critical behaviour for traced-polynomial potentials is still not fully resolved [6]. What I want to discuss here is a little puzzle that has arisen in this context: under certain criticality conditions, automatically satisfied by all (originally studied [2,3]) *even* potentials, one finds in the double scaling limit *two* continuum theories, which do not mix to any finite order in the string coupling constant [7]. As subsequently noticed [8], it is possible to relax the criticality conditions, so that only one of these theories contributes to the singularity of the free energy [1].

---

[1]   We may say that the other theory belongs to the trivial $k = 1$ class and thus has an analytic free energy.

Random Surfaces and Quantum Gravity
Edited by O. Alvarez *et al.*, *Plenum Press, New York, 1991*

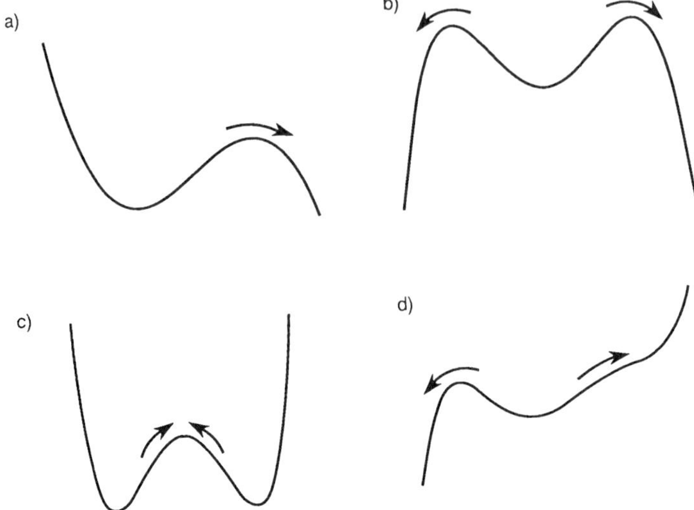

Figure 1 a,b) Critical ϕ^3 and ϕ^4 potentials. c) A critical inverted ϕ^4-potential for which a two-arc eigenvalue distribution merges into one; d) A non-even critical potential giving two decoupled Painlevé equations.

A heuristic explanation of what goes on is shown schematically in figure 1: for a generic potential (fig. 1a) eigenvalues leak out of the central well from one side. For an even potential (fig.1b) on the other hand, they can leak out both to the left and to the right, thus giving two perturbatively independent string theories. This also explains why for the inverted potential of figure 1c, the two "theories" mix [9], since eigenvalues now leak over the barrier to merge a two-arc distribution into one. The above argument must nevertheless be taken with a grain of salt: for instance it is not so obvious why the asymmetric potential of figure 1d, with only one local maximum, also exhibits the doubling phenomenon [10].

Here I want to explain this phenomenon from a somewhat different, more geometrical point of view. I will concentrate on the simplest case exhibiting the doubling: surfaces made out of *squares* , and compare them to surfaces made out of *triangles*. To every surface made out of A squares there clearly correspond 2^A triangulations, since each square can be broken into two triangles in two different ways. The inverse mapping is less trivial: the number of quadrangulations corresponding to a given triangulation $\mathcal{G}_3$, is not only a function of the area but depends on the details of the triangulation. It is equal to the number of different pairings of neighbouring triangles,

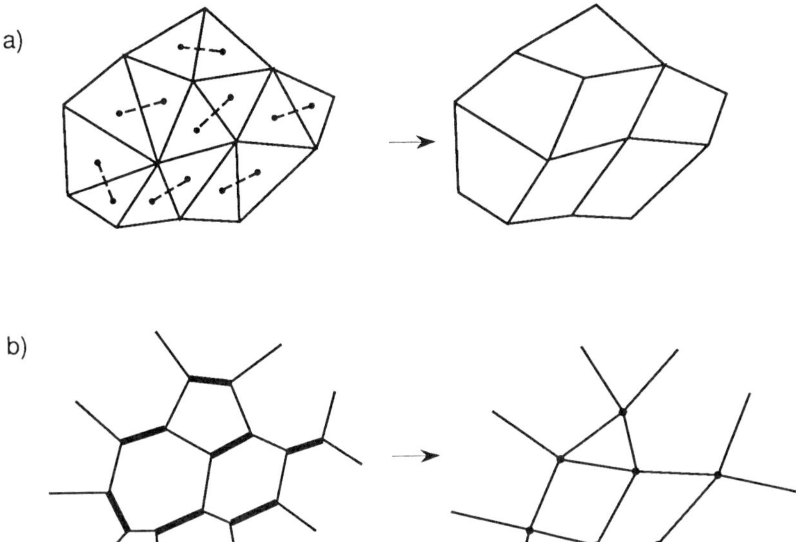

Figure 2 a) A triangulated surface with a perfect matching, and its corresponding quadrangulation. b) The dual ϕ^3 and ϕ^4-diagrams of these surfaces; thick lines stand for propagators collapsing to a point.

or of ways of collapsing lines in the dual ϕ^3-graph so as to form a ϕ^4-graph, as shown in figure 2. In combinatorial optimization this is called the number of *perfect matchings* of the graph [11], and we will denote it by $\mathcal{N}_{perf.m.}(\mathcal{G}_3)$. We can thus relate the measures of quadrangulations and triangulations as follows:

$$\sum_{\mathcal{G}_4} e^{(ln2-\mu)A} = \sum_{\mathcal{G}_3} e^{-\mu A}\mathcal{N}_{perf.m.}(\mathcal{G}_3) \tag{1}$$

Note that we are normalizing the area of a square to 1, and of a triangle to $\frac{1}{2}$ so that the number of triangles is $2A$. The factor $\mathcal{N}_{perf.m.}(\mathcal{G}_3)$ has also a different interpretation: it is the number of ground states of an Ising antiferromagnet on the triangular net [2] . To see why note that the low temperature expansion of a 2d Ising model is a sum over strings in the dual lattice, with energy proportional to length. In the ferromagnetic case these strings are closed and represent domain walls. Frustrated plaquettes, if present, act as string end-points. For an antiferromagnet on a triangular net every (triangular) plaquette is frustrated: ground states are thus in one-to-one correspondence with perfect matchings in the dual graph, as in figure 2.

We can rewrite equation (1) in the following obvious form, in terms of integrals over hermitean $N \times N$ matrices (we drop an irrelevant proportionality factor) :

$$Z_\square = \int [d\phi] e^{-Tr[\frac{1}{2}\varphi^2 - \frac{1}{4N} e^{ln 2 - \mu} \phi^4]} = \int [d\phi][du] e^{-Tr[\frac{1}{2}\varphi^2 + \frac{1}{2}u^2 - \frac{1}{\sqrt{N}} e^{-\frac{1}{2}\mu} \phi^2 u]} \tag{2}$$

Indeed the left-hand side counts quadrangulated surfaces, while the right-hand side counts perfectly matched triangulations, two neighbouring vertices being matched if connected by a u-line. Note that the ϕ^4-vertex has a symmetry factor 4, while the $\phi^2 u$-vertex has no symmetry factor at all. The free energy $ln Z_\square$ exhibits a square-root singularity [14] near the critical cosmological constant $e^{-\mu_\square^{cr}} = \frac{1}{24}$. This model is to be compared to a theory of pure triangulations described by the ϕ^3 model:

$$Z_\triangle = \int [d\phi] e^{-Tr[\frac{1}{2}\varphi^2 - \frac{1}{3\sqrt{N}} e^{-\frac{1}{2}\mu} \phi^3]} \tag{3}$$

The free energy has again a square-root singularity, but at a different cosmological constant $e^{-\mu_\triangle^{cr}} = \frac{1}{12\sqrt{3}}$. The critical cosmological constant $\mu_\triangle^{cr}$ ($\mu_\square^{cr}$) is the entropy of triangulated surfaces (with perfect matchings) per unit area: as μ approaches its critical value, entropy starts to overwhelm energy, and large surfaces dominate the sum.

Let me now digress a bit and do an amusing counting: Equation (1) allows us to write the average number of perfect matchings for given area A as:

$$< \mathcal{N}_{perf.m.}(\mathcal{G}_3) > |_{fixedA} = \frac{2^{-A} n(\mathcal{G}_4)}{n(\mathcal{G}_3)} \tag{4}$$

where $n(\mathcal{G}_4)$ is the number of connected ϕ^4-diagrams of area A, and $n(\mathcal{G}_3)$ the corresponding number of ϕ^3-diagrams. In particular the residual antiferromagnetic entropy at zero temperature on a random surface is simply:

$$\mathcal{S}_{random} = lim_{A \to \infty} \frac{1}{A} log < \mathcal{N}_{perf.m.}(\mathcal{G}_3) > |_A = \mu_\square^{cr} - \mu_\triangle^{cr} \tag{5}$$

Plugging in the critical values gives

$$\mathcal{S}_{random} \simeq 0.1438 \tag{6}$$

We may in fact refine somewhat this calculation by restricting the average to triangulations whose dual graphs have *no tadpoles*. Tadpoles correspond to benign surface-singularities which do not modify the universal critical behaviour of the model, but

---

[2] This should be distinguished from the antiferromagnet with spins placed at the centers of triangles, studied in ref.[12], which need not be fully frustrated.

do renormalize the critical cosmological constant and, as a minute's thought would convince one, greatly reduce the number of perfect matchings. We can subtract tadpoles by adding to the actions in (3) and (2) the linear terms: $\sqrt{N}e^{-\frac{1}{2}\mu}Tr(\phi)$ and $\sqrt{N}e^{-\frac{1}{2}\mu}Tr(u)$ respectively. After a shift and rescaling of the dummy matrix variables one finds that the new critical cosmological constants are given by :

$$\frac{e^{-\widetilde{\mu}_\triangle^{cr}}}{(1+4e^{-\widetilde{\mu}_\triangle^{cr}})^{\frac{3}{2}}} = \frac{1}{12\sqrt{3}} \quad ; \quad \frac{e^{-\widetilde{\mu}_\square^{cr}}}{(1+2e^{-\widetilde{\mu}_\square^{cr}})^2} = \frac{1}{24} \tag{7}$$

and the residual antiferromagnetic entropy is:

$$\widetilde{S}_{random} \simeq 0.3194 \tag{8}$$

This is surprisingly close to the result obtained for a regular triangular lattice [13] [3]

$$S_{regular} \simeq 0.3383 \tag{9}$$

However, if one smooths random surfaces further by subtracting also *self-energy* insertions the entropy decreases to :

$$\widehat{S}_{random} \simeq 0.2613 \tag{10}$$

so that the near equality of (8) and (9) is probably a numerical coincidence. These non-zero entropies demonstrate incidentally that the triangular antiferromagnet has no spin ordering even at zero temperature.

The above calculations illustrate the power of matrix-models for solving counting problems on random graphs. Unfortunately the graphs are dynamical, i.e. the corresponding averages are *annealed*. It would be much more interesting, at least from the point of view of disordered systems and combinatorial optimization , to do the *quenched* averages. This would require introducing replicas, which might be doable provided the matter is not critical, so as not to hit the $c = 1$ barrier. A second comment is in order here: the extensive entropies are not universal but one may ask whether the average number of perfect matchings has any interesting subleading singularities. The answer is I believe no: indeed we know that the leading power-law corrections to both numerator and denominator in equation (4) are equal and cancel out, since both triangulations and quadrangulations lead to the same string susceptibility exponent. Furthermore for planar surfaces the corresponding number of graphs can be computed in closed form [14] and all subleading corrections turn out to be analytic.

---

[3] Our normalization is such that for a regular lattice we are counting entropy per spin.

Let me proceed now to examine the continuum limit of triangulated and quadrangulated surfaces. Using the by now well known orthogonal polynomial techniques
[15] one can express the free energy as

$$lnZ = \sum_{i=1}^{N}(N-i)lnR_i \tag{11}$$

where the R_i satisfy a set of reccursion relations . For the model of pure triangulations
these read :

$$\frac{i}{N} = R_i - e^{-\frac{\mu}{2}}R_i(S_i + S_{i-1})$$
$$0 = S_i - e^{-\frac{\mu}{2}}(S_i^2 + R_i + R_{i+1}) \tag{12}$$

We can use the second equation to eliminate the S_i's, and then expand the first around
$\frac{i}{N} \equiv x = 1 - \delta x$, $\mu = \mu_\triangle^{cr} + \delta\mu$, and $R(x) = R_\triangle^{cr}[1 + \delta R(x)]$, where $\mu_\triangle^{cr} = ln(12\sqrt{3})$
and $R_\triangle^{cr} = \sqrt{3}$. The result in the double-scaling limit [2,3] is the Painlevé equation:

$$\frac{2}{3}(\delta x + \delta\mu) = \delta R_\triangle^2 + \frac{1}{6N^2}\delta R_\triangle'' \tag{13}$$

with primes denoting differentiation with respect to δx. For the model of quadrangulations, eq.(2), the corresponding pair of equations is:

$$\frac{i}{N} = R_i - 2e^{-\mu}R_i(R_{i+1} + R_i + R_{i-1} + S_i^2 + S_iS_{i-1} + S_{i-1}^2)$$
$$0 = S_i - 2e^{-\mu}(S_i^3 + 2S_iR_i + 2S_iR_{i+1} + S_{i+1}R_{i+1} + S_{i-1}R_i) \tag{14}$$

The second equation can now be solved by setting $S_i = 0$. Expanding again the first
as above around the critical values $\mu_\square^{cr} = ln(24)$ and $R_\square^{cr} = 2$ one obtains :

$$(\delta x + \delta\mu) = \delta R_\square^2 + \frac{1}{3N^2}\delta R_\square'' \tag{15}$$

Comparing equations (13) and (15) we note first that they differ in the coefficient
of the left-hand-side: this turns out to be non-universal and can be killed by taking
two more derivatives with respect to x. The difference in the right-hand-side is more
significant: it shows that $\delta R_\square$, proportional to the singular free energy of the model
of quadrangulations, satisfies the same equation as $2\delta R_\triangle$, proportional to *twice* the
singular free energy of triangulated surfaces.

Actually by keeping $S(x)$ formally in the ϕ^4-model, one finds two Painlevé equa-
tions with the normalization of (13) for the two quantities: $\frac{1}{2}(\delta R_\square \pm \frac{S}{\sqrt{2}})$ [7]. Furthermore, a straightforward analysis [16] along the lines of refs. [3,4,7] shows that in
addition to the *even* scaling operators:

$$\mathcal{O}_k^{even}(\phi) = Tr\int_0^1 \frac{du}{u}[1 - \frac{1}{2}u(1-u)\phi^2]^k \tag{16}$$

one can define operators built exclusively out of odd powers of ϕ:

$$\mathcal{O}_k^{odd} = \frac{1}{(k+1)\sqrt{2}} \ Tr(\frac{\partial}{\partial\phi}\mathcal{O}_{k+1}^{even}) = c_k\sum_{j=0}^{k}\frac{j!(-\frac{1}{2})^j}{(2j+1)!(k-j)!}Tr\phi^{2j+1} \tag{17}$$

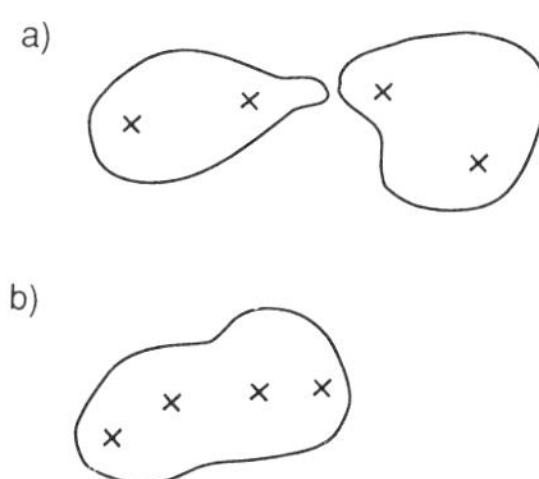

Figure 3. Disconnected (a) and connected (b) contributions to the 4-point function of the puncture operator, which correspond to the two terms on the right-hand-side of the Painleve equation.

which not only have the same dimensions as their even counterparts, but are totally indistinguishable from them, when *taken in pairs* inside correlation functions [4]. As a result the sets of operators $V_k^\pm = \mathcal{O}_k^{even} \pm \mathcal{O}_k^{odd}$ decouple, and correspond to two continuum theories that do not mix to any finite order in the string-loop expansion.

This doubling of continuum degrees of freedom is easy to understand in the language of the $T = 0$ antiferromagnet. Even operators simply drill holes on the surface by removing a number of frustrated triangular plaquettes. The action of an odd operator, on the other hand, is more complex: it is a combined loop and *disorder* operator on the antiferromagnetic lattice. For example $Tr(\phi^3)$ first creates an unfrustrated triangular plaquette, before removing it from the surface. The correlation functions of any odd number of disorder operators vanish, but when taken in pairs they have no effect on the long-distance behaviour of the surface.

Consider now in particular the left-hand side of equation (13) : this has a simple interpretation as the sum of connected and disconnected contributions to the 4-point function of the most relevant puncture operator [3][17] $P = \frac{d}{dx}$, as shown in figure 3 . In a theory with both even and odd punctures, one can easily see that fig. 3b receives twice as many contributions as fig. 3a, since odd punctures must appear in pairs on any connected component of the surface. This explains the factor of two in equation (15). An analogous statement holds for the macroscopic loop equations where connected and disconnected components have in the even-potential case an extra relative factor of two [18].

I conclude with a couple of remarks: First, perturbing the ϕ^4-model with the disorder ϕ^3-operator is equivalent to giving finite fugacity to the dimers of figure 2; at some critical value of the coupling one thus expects to recover the Lee-Yang edge singularity, and corresponding $m = 3$ string equation [5]. Secondly, although the two continuum string theories do not mix in perturbation theory, they could conceivably

---

[4] Note that the odd operators are well defined, and have no constant infinite ambiguity like the even ones.

mix non-perturbatively, if the Z_2-symmetry breaks spontaneously on infinite-genus surfaces. In this case the scaling operators (16) and (17) obtain non-perturbative corrections. It might be possible to use such mixing to give a non-perturbative definition of the theory.

Acknowledgements

I thank A.Bilal, E.Brezin, F.David, D.Gross, V.Kazakov, M.Petropoulos and V.Rivasseau for useful comments, and S.Shenker for asking why triangles differ from squares. I also thank the organizers for a very interesting meeting.

References

[1] J.Ambjorn, B.Durhuus and J.Frohlich, Nucl.Phys.**B257**(1985) 433; F.David, Nucl.Phys.**B257**(1985)45; V.A.Kazakov, Phys.Lett.**150B**(1985)282; V.A.Kazakov, I.K.Kostov and A.A.Migdal, Phys.Lett.**157B**(1985)295.

[2] E.Brezin and V.A.Kazakov, Phys.Lett.**236B**(1990)144; M.R.Douglas and S.H.Shenker, Nucl.Phys.**B335**(1990) 635.

[3] D.J.Gross and A.A.Migdal, P.R.L. **64**(1990)127 and Nucl.Phys.**B340**(1990) 333.

[4] V.A.Kazakov, Mod.Phys.Lett.**A4**(1989)2125 .

[5] M.Staudacher, Nucl.Phys.**B336**(1990)349.

[6] G.M.Cicuta, L.Molinari and E.Montaldi, Mod.Phys.Lett. **A1**(1986)125; J.Phys. **A**: Math.Gen. 23(1990)L421; L.Molinari and E.Montaldi, INFN Milano preprint 1990; M.Douglas, N.Seiberg and S.Shenker, Phys.Lett.**244B**(1990)381; J.Jurkiewicz, Phys.Lett.**245B**(1990)178; G.Bhanot, G.Mandal and O.Narayan, IASSNS-HEP-90/52 preprint, May 1990; K.Demefteri, N.Deo, S.Jain and C.I.Tan, Brown-HET-764 preprint, July 1990;P.M.S.Petropoulos, CPTH-A993.0890 preprint, August 1990; S.Chaudhuri, J.Lykken and T.R.Morris, Fermi-PUB-90/168-T preprint, August 1990; O.Lechtenfeld, R.Ray and A.Ray, IASSNS-HEP-90/69 preprint, September 1990.

[7] C.Bachas and P.M.S.Petropoulos, Phys.Lett.**247B**(1990)363; E.Witten, IASSNS-HEP-90/45 preprint, May 1990.

[8] F.David, Saclay preprint SPhT/90-090, May1990; C.Marzban and R.Raju Viswanathan, Trieste preprint, August 1990.

[9] C.Nappi, IASSNS-HEP-90/61 preprint, August 1990·

[10] P.M.S.Petropoulos, Ecole Polytechnique preprint A.018.1190, November 1990

[11] C.Papadimitriou and K.Steiglitz, Combinatorial Optimization, Algorithms and Complexity, Prentice Hall, New Jersey 1982.

[12] T.R.Morris, Fermilab preprints PUB-90/121-T, June 1990 and PUB-90/136-T, August 1990.

[13] G.H.Wannier, Phys.Rev. **79** (1950)357; R.M.F.Houtappel, Physica **16** (1950) 425.

[14] E.Brezin, C.Itzykson,G.Parisi and J.B.Zuber, C.M.P. **59** (1978) 147.

[15] D.Bessis, C.M.P. **69** (1979) 147;C.Itzykson and J.B.Zuber, J.Math.Phys. **21**

(1980) 109;D.Bessis, C.Itzykson and J.B.Zuber, Adv.Appl.Math. **1** (1980) 109.

[16] C.Bachas, unpublished.

[17] E.Witten, Nucl.Phys.**B340** (1990) 281.

[18] R.Dijkgraaf, H.Verlinde and E.Verlinde, PUPT-1184 and IASSNS-HEP 90/48 preprint, May 1990.

NON-PERTURBATIVE EFFECTS IN 2D GRAVITY AND MATRIX MODELS

François David

Service de Physique Théorique,
CEN Saclay, F91120 Gif sur Yvette Cedex, France

1. Introduction

Two dimensional Euclidean quantum gravity may be formulated as a functional integral over 2–dimensional Riemannian manifolds. This infinite dimensional integral may be discretized in such a way that the topological expansion in terms of the genus of the manifold is mapped onto the $1/N$ expansion of some zero–dimensional matrix model [1]. The $N = \infty$ limit exhibits critical points which can be shown to describe the continuum limit of 2–dimensional gravity on a genus zero manifold, eventually coupled to some matter fields. Recently it was shown that a scaling limit can be constructed [2] . In this limit all the terms of the topological expansion survive and thus one obtains a fully non–perturbative solution for two dimensional gravity. However in the most interesting cases, in particular for pure gravity, the solution is defined as a solution of a non–linear differential equation of the Painlevé type and presents some non–perturbative ambiguities, related to the delicate issue of boundary conditions, which are usually attributed to some "non–perturbative effects" of the theory.

In this talk I shall review some attempts to get a better understanding of these effects. For simplicity and shortness I shall mainly deal with the case of pure gravity, which seems to embody the main problems. The approach that I have followed consists in trying to relate those non–perturbative issues to the non–perturbative effects which are present in the original matrix models.

2. The Scaling Limit

For completeness and in order to have consistent notations, let us recall explicitly how the scaling limit is obtained. We define the partition function for the Hermitian one matrix model as

Random Surfaces and Quantum Gravity
Edited by O. Alvarez *et al.*, *Plenum Press, New York, 1991*

$$Z = \int d\Phi \, e^{-N \, \mathrm{tr}(V(\Phi))}$$

$$\propto \int \prod_{i=1}^{N} d\mu(\lambda_i) \left(\prod_{i<j} (\lambda_i - \lambda_j) \right)^2 \tag{2.1}$$

where $d\mu(\lambda) = d\lambda \, e^{-N \, V(\lambda)}$. Introducing orthonormal polynomials π_n with respects to $d\mu$

$$\langle n|m \rangle = \int d\mu \, \pi_n \, \pi_m = \delta_{nm} \tag{2.2}$$

one obtains by the standard manipulations the expression for the vacuum energy

$$F = \ln Z \simeq \sum_{i=0}^{N-1} (N-i) \, \ln(b_i) \tag{2.3}$$

where the coefficients b_i are related to the matrix elements of the operator Q of multiplication by λ over the π_n's

$$\lambda \, \pi_n = Q_{nm} \, \pi_m = \sqrt{b_{n+1}} \, \pi_{n+1} + a_n \, \pi_n + \sqrt{b_n} \, \pi_{n-1} \tag{2.4}$$

We shall consider the simplest case of the cubic potential, which can be written as

$$V(\lambda) = g\lambda - \frac{\lambda^3}{3} \tag{2.5}$$

From the relation $P + P^t = N \, V'(Q)$, where P is the operator $\frac{\partial}{\partial \lambda}$, one gets the recursion relations

$$
\begin{aligned}
0 &= g - (a_n^2 + b_n + b_{n+1}) \\
\frac{n}{N} &= - b_n \, (a_n + a_{n+1})
\end{aligned}
\tag{2.6}
$$

The large N limit is obtained by taking the continuum limit

$$n/N \; \rightarrow \; x \; ; \; a_n \; \rightarrow \; a(x) \; ; \; b_n \; \rightarrow \; b(x) \tag{2.7}$$

in the recursion relations (2.6). The critical point occurs when $a(x)$ and $b(x)$ becomes singular at $x = 1$. In our case this gives

$$g_c = 3 \, 2^{-2/3} \; , \; a(1) = a_c = -2^{-1/3} \; , \; b(1) = b_c = 2^{-2/3} \tag{2.8}$$

The scaling limit is obtained by rescaling by adequate powers of N

$$
\begin{aligned}
g &= g_c \left(1 + a^2 \, t\right) & n &= N \left(1 - a^2 \, x\right) \\
a_n &= a_c \left(1 - a \, v\right) & b_n &= b_c \left(1 - a^2 \, u\right)
\end{aligned}
\tag{2.9}
$$

and by taking the limit

$$N \rightarrow \infty \; , \; a^{5/2} N = \gamma^{-1}, \; x \text{ and } t \text{ fixed} \tag{2.10}$$

22

γ is here the "string coupling constant" and can be completely absorbed in the normalization. Therefore it will be set to unity. After expanding (2.9) in (2.6) one obtains that $v(x,t) = -u(x,t)$ and that $u(x,t)$ satisfies the "string equation"

$$-\frac{1}{6}\frac{\partial^2 u}{\partial x^2} + u^2 = \frac{2}{3}x + t \qquad (2.11)$$

which is nothing but the Painlevé I equation. From (2.3) the finite part of F in the scaling limit, $F(t)$, is equal to $-\int_0^\infty dx\, x\, u$, and therefore the "susceptibility" $f(t)$ is given by

$$f(t) = F''(t) = (3/2)^2\, u(0,t) \qquad (2.12)$$

and obeys also a Painlevé I equation. Finally the operator Q given by (2.4) becomes the differential operator [3]

$$Q = 2\, b_c^{1/2}\left(N^{2/5} + d^2 - 2u\right) \qquad (2.13)$$

where $d = \partial/\partial x$. Using the free fermion formalism of [4] expectation values of operators in the original matrix model may be expressed as v.e.v. of one body operators for a system on N free fermions with Fock space generated by the one particle states $|n\rangle = \pi_n(\lambda)$. If one starts from the "loop operator" $W(\lambda)$, which is defined as

$$W(\lambda) = N\,\mathrm{Tr}\left(\frac{1}{\lambda - \Phi}\right) \sim \Psi\,\frac{1}{\lambda - Q}\,\Psi^\dagger \qquad (2.14)$$

(where $\Psi^\dagger$ and Ψ are the fermion field operators), the finite part of W in the scaling limit, $w(p)$, is defined by the rescaling

$$w(p) \simeq 2\, b_c^{1/2}\, W(\lambda) \quad ; \quad \lambda = \lambda_c\left(1 + N^{-2/5}p\right) \quad ; \quad \lambda_c = 2\, b_c^{1/2} = 2^{-1/3} \qquad (2.15)$$

The explicit expressions for the one- and two-loop v.e.v. are in the scaling limit (2.9)

$$\langle w(p)\rangle = \int_0^\infty dx\, \langle x|\frac{1}{p - d^2 + 2u}|x\rangle \qquad (2.16)$$

$$\langle w(p)w(q)\rangle = \int_0^\infty dx \int_{-\infty}^0 dy\, \langle x|\frac{1}{p - d^2 + 2u}|y\rangle\langle y|\frac{1}{q - d^2 + 2u}|x\rangle \qquad (2.17)$$

In the large t (or equivalently large x) limit u should fit with the large N solution and therefore should behave as $+t^{1/2}$. The problem is that equation (2.11) admits an infinite family of real solutions with this behavior as $x \to +\infty$. Moreover any such solution must have an infinite number of double poles on the negative real axis (see for instance [5]. The Laurent expansion around each pole x_0 is of the form

$$u(x) = (x - x_0)^{-2} + o((x - x_0)^2) \qquad (2.18)$$

and therefore from (2.12) each pole of f corresponds to a simple zero of the partition function Z. Two solutions of (2.11) have the same large t asymptotic expansion x

$$f(t) = (3/2)^2\, t^{1/2} - \sum_{k=1}^\infty f_k\, t^{(1-5k)/2} \qquad (2.19)$$

but differ by the position of (for instance) their largest pole. Linearizing (2.11) it is easy to see that the difference between two solutions behaves asymptotically as

$$\delta f \ \sim \ t^{-1/8} \ \exp\!\left(-\ \frac{4 \ 3^{3/2}}{5} \ t^{5/4}\right) \tag{2.20}$$

and is therefore exponentially small in the "string coupling constant" $t^{-5/2}$ [6]. This can be related to the fact that the coefficients f_k in (2.19) grow like $(2k)!$ and that the series (2.19) is not Borel summable [7].

Another (but related) problem occurs in the definition of the resolvent $\langle x|(p - d^2 + 2u)^{-1}|y\rangle$. The operator $-d^2 + 2u$ is not defined on the whole real axis because of the poles. A somewhat natural choice, proposed for instance in [4], consists in defining this operator between the largest pole x_0 and $+\infty$. Indeed viewing this operator as the Hamitonian of a particle in the potential u, the potential diverges enough at each pole to prevent tunnelling between the different "sectors". In other word one defines the resolvent by imposing that it vanishes at x_0 and $+\infty$ and plug it into the definition of the correlation functions (2.16),(2.17).

3. Loop Equations

An alternative approach starts from the loop operator $W(\lambda)$ defined by (2.14) or its inverse Laplace transform

$$W(L) \ = \ N \ \mathrm{tr}\!\left(e^{L\Phi}\right) \tag{3.1}$$

which corresponds (moreless) to the operator creating a hole (macroscopic loop) with length L in the two-dimensional worls sheet. The loop equations are the Schwinger–Dyson equations for the matrix model (2.1) and are derived simply by performing the change of variable $\Phi \to \Phi + \epsilon f(\Phi)$ in (2.1) (where $f(z)$ is an analytic function). The Jacobian for this change of variable is

$$J \ = \ 1 + \epsilon \oint \frac{dz}{2i\pi} \ f(z) \ \left(\mathrm{tr}\left(\frac{1}{z - \Phi}\right)\right)^2 + o(\epsilon^2) \tag{3.2}$$

From (3.2) one obtains easily the loop equation [8]

$$N^2 \, V'\!\left(\frac{\partial}{\partial L}\right) \langle W(L)\rangle \ = \ \int_0^L dL' \left\{\langle W(L')\rangle\langle W(L-L')\rangle + \langle W(L')W(L-L')\rangle\right\} \tag{3.3}$$

or by Laplace transform

$$N^2 \left[V'(\lambda)\langle W(\lambda)\rangle\right]_< \ = \ \langle W(\lambda)\rangle^2 + \langle W(\lambda)^2\rangle \tag{3.4}$$

where $[\ \ ]_<$ means the truncation to the powers λ^n with $n < 0$ in the Laurent expansion around $\lambda = \infty$. Including a source term for the loop operators W in the potential V one sees that the loop equation contains implicitely the infinite set of equations of motion for v.e.v. with an arbitrary number of loop operators. Those equations allows to compute recursively the correlation functions at all orders in the topological expansion.

The loop equation (3.4) takes a very simple form in the double scaling limit [9] . Indeed, defining the finite part of $W(\lambda)$ as

$$w(p) = \lambda_c \left(W(\lambda) - \frac{1}{2} V'(\lambda) \right) \tag{3.5}$$

and using (2.8), (2.9) and (2.15), (3.4) becomes

$$\langle w(p) \rangle^2 + \langle w(p)w(p) \rangle = \frac{1}{4} p^3 - \frac{3}{4} t p + \frac{1}{3} \langle P \rangle \tag{3.6}$$

where $\langle P \rangle$ is the v.e.v. of the "puncture operator" $P = -\partial/\partial t$ and depends only on the "renormalized cosmological constant" t. The scaling limit of loop equations involving more loop operators can be obtained in a similar way. For instance we have

$$2\langle w(p) \rangle \langle w(p)w(q) \rangle + \langle w(p)^2 w(q) \rangle + \frac{\partial}{\partial q} \left(\frac{\langle w(q) \rangle - \langle w(p) \rangle}{q - p} \right) = \frac{1}{3} \langle P\, w(q) \rangle \tag{3.7}$$

Those equations can be used to compute recursively (in the topological expansion) correlation functions in the scaling limit (see [10]

The interest of the loop equations is not merely calculational. In [11] and in [12] it was indeed shown that the loop equations can be written as recursion relations which follow from the string equation (2.11) , and from the fact that the partition function is the so-called τ-function of the corresponding KdV hierarchy. Moreover those recursion relations can also be obtained from the formulation of 2-d gravity as a topological fiels theory [13] . Therefore the three approaches (topological gravity, KdV hierarchy and loop equations) are equivalent, at least to all orders of the topological expansion. Let us show for instance explicitly the connection between (3.6) and the results of [11]. One can easily show that $w(p)$ has a large p expansion in powers $p^{-3/2-n}$, with $n \leq 0$, excepted for the one- and two-loops correlators. Defining the "finite part" $\tilde{w}$ of w as its $O(p-3/2))$ part, we get explicitly

$$
\begin{aligned}
\langle w(p) \rangle &= \frac{1}{2} p^{3/2} - \frac{3}{4} t p^{-1/2} + \langle \tilde{w}(p) \rangle \\
\langle w(p)w(p) \rangle &= \frac{1}{16} p^{-2} + \langle \tilde{w}(p)\tilde{w}(p) \rangle
\end{aligned}
\tag{3.8}
$$

From (3.6) we get, if we perform the rescaling $t \to 2/3\, t$

$$\left[\left(p^{3/2} - t p^{-1/2} \right) \langle \tilde{w} \rangle \right]_< + \langle \tilde{w} \rangle^2 + \langle \tilde{w}\tilde{w} \rangle + \frac{1}{16\, p^2} + \frac{t^2}{4\, p} = 0 \tag{3.9}$$

This is[1] Eq. (2.14) of [11] if we identify $p^{3/2} - t p^{-1/2}$ with the derivative of the $m = 2$ singular potential $V'(p)$, and if we shift $\langle \ \rangle \to \frac{1}{2} \langle \ \rangle$ to take into account the "doubling phenomenon" which occurs in matrix models with even potential (see [13] and [14]), which fix the normalization used in [11] for the KdV hierarchy.

One may however expect that the loop equations, which are the equations of mo-

---

[1] up to a factor 2 in the p^{-2} term, for which we have no explanation.

tion for two dimensional gravity, and which have a simple and appealing geometrical interpretation in term of fusion and splitting of loops [8], are valid beyond perturbation theory. This is indeed what occurs in ordinary field theories: non-perturbative effects might change the v.e.v. of some operators but they do not affect the general form of the equations of motion. For pure gravity the equation (3.6) puts very strong constraints on the non-perturbative solutions, and in fact excludes all the real solutions discussed in the previous section. Indeed, if we start from a real solution of (2.11), and if we define the loop correlators by (2.16), (2.17), with the resolvent defined through the operator $Q = d^2 - 2u$ with support between the largest real pole x_0 of u and $+\infty$, the operator Q has a discrete spectrum ($e_0 > e_1 > e_2 > \ldots$), and therefore the resolvent $\langle x|(p - Q)^{-1}|y\rangle$ is a meromorphic function of p with simple poles located on the spectrum of Q. A straightforward calculation shows that the l.h.s. of (3.6) has then double poles with non-vanishing residues. For instance near the first pole we have

$$\langle w(p)\rangle^2 + \langle w(p)^2\rangle \;\simeq\; \frac{1}{(p - e_0)^2} \int_0^\infty dx \, |\psi_0(x)|^2 \tag{3.10}$$

where ψ_0 is the eigenfunction ($Q\psi_0 = e_0\psi_0$). This obviously contradicts (3.6), since the r.h.s. of (3.6) is a polynomial in p and cannot have double poles! In fact the residue of the double pole at $p = e_0$ in (3.10) behaves for large t as $\lambda\exp\left(-\text{cst.}\, t^{5/4}\right)$. Thus the loop equations are violated by non-perturbative terms exactly of the same order as those presents in (2.20). This is of course not a coincidence.

The only way out of this problem is to find a potential u such that the operator Q has a continuous spectrum. A necessary condition is that $u(x)$ is analytic along the whole real axis. As we have seen, this is not possible for any real solution of (2.11). In fact only two *complex conjugate* solutions of (2.11) satisfy this requirement [5]. Those two solutions, which are denoted the "triply truncated solutions", have the following properties. They have have an infinite set of double poles (with Laurent expansion given by (2.18)) in only one fifth of the complex x plane. In the remaining 4/5th, which for one of the solutions is the sector

$$-\frac{6\pi}{5} \;<\; \arg(x) \;<\; \frac{2\pi}{5} \tag{3.11}$$

the function u has at most a finite number of poles and behaves smoothly as $|x| \to \infty$ as $u(x) \sim x^{1/2}$. This analyticity domain contains the whole real axis and one might expect that the loop correlators defined via the resolvent by (2.16) and (2.17), which are of course no more real, satisfy the loop equations. As we shall see in the next section, there are strong evidences that those complex solutions are indeed obtained from the original matrix models, once the problem of the unboundness of the action is properly treated (at the mathematical level...).

4. Non-perturbative Effects in Matrix Models

The main feature of the original potential (2.5) used in the matrix model (2.1) is that it is unbounded from below. This is a general feature for any one matrix model which allows to reach the $m = 2$ critical point (corresponding to pure gravity). This is clear in the original large N solution of the model [15]. This solution relies on the $N = \infty$ eigenvalue density $d\rho(\lambda) = d\lambda\, u(\lambda)$, which must extremize the action

$$F = N^2 \int d\rho(\lambda)\, V(\lambda) - \int d\rho(\lambda) \int d\rho(\mu)\, \ln|\lambda - \mu| \qquad (4.1)$$

From (4.1) the effective potential for *one* eigenvalue is

$$\Gamma(\lambda) = V(\lambda) - \int d\mu\, u(\mu)\, \ln|\lambda - \mu| \qquad (4.2)$$

and the force exerced on one eigenvalue is

$$f(\lambda) = -\Gamma'(\lambda) = -V'(\lambda) + 2\,\mathrm{Re}\big(F(\lambda)\big) \qquad (4.3)$$

where $F(\lambda)$ is nothing but the v.e.v. of the one loop operator

$$F(\lambda) = \int d\mu\, \frac{u(\mu)}{\lambda - \mu} = \lim_{N \to \infty} \frac{1}{N^2}\, \langle W(\lambda) \rangle \qquad (4.4)$$

Extremizing (4.1) leads to the equation

$$f(\lambda) = 0 \qquad \text{if } u(\lambda) \neq 0 \qquad (4.5)$$

which means that the effective potential Γ is constant where eigenvalue density is non zero. The density of eigenvalues $u(\lambda)$ is given simply by the discontinuity of F

$$u(\lambda) = \frac{1}{\pi}\,\mathrm{Im}\big(F(\lambda - i\epsilon)\big) \qquad (4.6)$$

The scaling limit is obtained here by performing the rescaling (2.9) for g and λ and letting $a \to 0$. We obtain for the force

$$f(p) = 2\,\mathrm{Re}\langle w(p) \rangle \quad ; \quad \langle w(p) \rangle^2 = \frac{1}{4}\left(p^3 - 3\,t\,p + \frac{4}{3}\langle P \rangle\right) \qquad (4.7)$$

where $\langle P \rangle$ is some constant. This equation for $\langle w \rangle$ is nothing but the loop equation (3.6) at first order in the topological expansion, where the connected correlator $\langle ww \rangle$ vanishes. The puncture operator $\langle P \rangle$ is fixed by the requirement that $\langle w \rangle$ must have only one cut along $]-\infty, p_0]$. Indeed if this is not the case either u becomes complex, or it has support on two arcs (which is perfectly allowed) but is negative on one of them (which is impossible since u is a density) and moreover the effective potential is not the same on the two arcs. One obtains

$$\langle P \rangle = \frac{3}{2}\,t^{3/2} \;,\; \langle w(p) \rangle = \frac{1}{2}\,(\sqrt{t} - p)\,\sqrt{p + 2\sqrt{t}} \qquad (4.8)$$

Hence the density $u(p)$ and the effective potential $\Gamma(p)$ for one eigenvalue

$$u(p) \;=\; \frac{1}{2\pi}\,\mathrm{Re}\,\left[(\sqrt{t}-p)\sqrt{-p-2\sqrt{t}}\right]$$
$$\Gamma(p) \;=\; \mathrm{Re}\,\left[\frac{2}{5}(3\sqrt{t}-p)(p+2\sqrt{(t)})^{3/2}\right] \tag{4.9}$$

One sees that the effective potential goes to $-\infty$ as $p \to +\infty$ but that the eigenvalues, which are located on $(p < p_0 = -2\sqrt{t})$, are prevented to fall in this well by the "wall" $(-2\sqrt{t} < p < 3\sqrt{t})$ where $\Gamma > 0$, as long as t is positive. At the critical point $t = 0$, this wall disappears and the eigenvalues start to fall, hence the appearence of imaginary parts in the observables.

This classical picture is valid only for $N = \infty$. As long as N is finite, since N^{-1} plays the role of a "Planck constant", eigenvalues may cross the barrier and fall toward $+\infty$. As discussed in [6] this effect is exponentially suppressed at large N, and is therefore non–perturbative. Its amplitude can be estimates very easily by instanton technics[2]. The most probable phenomenon is that *one eigenvalues* crosses the wall while the $N - 1$ others stay at equilibrium. The amplitude for such a process is given by $\exp(-Na^{5/4}\Gamma_{\mathrm{inst}})$, where "inst" corresponds to the configuration where the eigenvalue is at the top of the wall. From (4.9)

$$\Gamma_{\mathrm{inst}} \;=\; \Gamma(\sqrt{t}) \;=\; \frac{4}{5}\,3^{3/2}\,t^{5/4} \tag{4.10}$$

This is exactly the exponential factor in (2.20), which gives the amplitude of the leading non-perturbative effects contained in the string equations.

If one wants to work really at the nonperturbative level with the matrix model (2.1), that is at finite N, the partition function Z can be defined by the method of analytic continuation [16] . For the cubic potential (2.5) we take for the λ_i's in (2.1) a complex integration path going from $-\infty$ to (for instance) $e^{i\pi/3}\infty$, which makes the matrix integral complex but perfectly well defined, for any complex value of g. The large N saddle point described above is not modified by this choice of contour, but now one can show that it is stabilized by this choice of boundary conditions. Indeed, in the scaling limit described above ($n \to \infty$, then $a \to 0$) the contour of integration for the eigenvalues goes now from $-\infty \leftarrow p$ to $p \to e^{2i\pi/5}\infty$. Therefore this choice of boundary condition prevents the fall of the eigenvalues into the well $p \to +\infty$. Indeed, one can find a path which goes from the end point of the support of eigenvalues, $p_0 = -2\sqrt{t}$, to $e^{2i\pi/5}\infty$, and which does not cross a region in the complex p plane where the effective potential Γ is negative.

The existence and the stability of a large N saddle point for *complex t* can easily be studied by complex saddle point methods. One can show that eigenvalues will still be located along the arc given by $\Gamma(p) = 0$, where Γ is given by (4.9) and corresponds now to the real part of the complex effective potential $\int_p 2\langle w \rangle$. There are two natural conditions of stability for this saddle point:

(i) The support of eigenvalues must connect $-\infty$ to the endpoint p_0. One can easily show that this happens only if

---

[2]   as suggested by S. Shenker and J. Zinn-Justin.

$$-\frac{6\pi}{5} \;<\; \mathrm{Arg}(t) \;<\; \frac{6\pi}{5} \tag{4.11}$$

(ii) one can still find a path which goes from p_0 to infinity such that $\Gamma(p) > 0$. With our choice of boundary condition this is possible if

$$-\frac{8\pi}{5} \;<\; \mathrm{Arg}(t) \;<\; \frac{2\pi}{5} \tag{4.12}$$

Thus the large N limit exists only in four-fifth of the complex t plane. One can show that there cannot exist a more complicated limit, such as a two arc phase, in the remaining sector. The instability in the singular sector $2\pi/5 < \mathrm{Arg}(t) < 4\pi/5$ corresponds precisely to instanton effects. Indeed it is on its boundary that the effective action of the instanton considered above vanishes.

The sector where the large N limit exists is *exactly* the same than the sector of analyticity of one of the "triply truncated solution" of (2.11). Since the planar limit is obtained from the scaling limit by letting x and $t \to \infty$, this allows to identify the triply truncated solution with the result of the scaling limit, if one start from the matrix model defined with the complex contour described above. The string susceptibility f and the loop amplitudes $\langle w \rangle$ will of course be complex, but with exponentially small imaginary parts (as $t \to \infty$) proportional to (2.20).

These arguments can be extended to other matrix models and to higher critical points. For instance in [16] the cases of the Painlevé II critical point and the $m = 3$ critical points are discussed in details. The non-perturbative effects in the corresponding string equations can also be attributed to instanton effects in the original matrix models. The same kind of arguments allows to study deformations between (multi)critical models [17] [18] [19] . In all know cases the conclusions of such a saddle point analysis are in perfect agreement with the analysis of non-perturbative effects in the string equations by Borel summation methods [7], and by WKB methods and the study of their monodromy properties [20].

5. Stochastic Quantization and the SUSY 1D String

Let us end by a few simple comments[3] about the proposal by Marinari and Parisi [21] to treat 2D gravity as the ground state of some supersymmetric 1d string model. The idea relies on the fact that in a model of the form (2.1), the average of a observable Q can be written as the ground state expectation value

$$\langle Q \rangle \;=\; \langle 0|Q|0 \rangle \tag{5.1}$$

of the observable Q in a quantum mechanical model with Hamiltonian

$$H_B \;=\; P^2 + V_B(\Phi) \quad ; \quad P = i\partial/\partial\Phi \quad ; \quad V_B = \frac{(V')^2}{4} - \frac{V''}{2} \tag{5.2}$$

---

[3] elaborated while I was writting these notes.

This Hamiltonian H_B is the bosonic part of the supersymmetric quantum mechanical Hamiltonian $\mathbf{H}$ which can be obtained through the stochastic quantization of (2.1) and the associated Fokker-Planck Hamiltonian

$$\mathbf{H} = \begin{pmatrix} H_B & 0 \\ 0 & H_F \end{pmatrix} = \mathbf{Q}^2 \quad ; \quad \mathbf{Q} = \begin{pmatrix} 0 & iP + \frac{V'}{2} \\ -iP + \frac{V'}{2} & 0 \end{pmatrix} \tag{5.3}$$

where $\mathbf{Q}$ is the BRST operator. The l.h.s. of (5.1) make sense only if V is bounded from below. In that case supersymmetry is unbroken, the ground state is bosonic and has zero energy, and (5.1) holds. In the case of interest here, V is unbounded from below but $\mathbf{H}$ is well defined and positive. Supersymmetry is broken and the two degenerate vacua $|0_B\rangle$ and $|0_F\rangle$ have a positive energy. The proposal of [21] (already made in [22]), is to define the v.e.v. of Q by (5.1) (taking of course the bosonic ground state). Some properties of this 1d supersymmetric theory in the scaling limit have been studied in [23], [24]. Of course the equations of motion of the original theory will be violated in the supersymmetric one by terms proportional to the supersymmetry breaking. Indeed the variation of the partition function under a field variation f can be written as

$$\langle -f' + f\, V' \rangle = \langle 0_B|\{\mathbf{Q}, \mathbf{F}\}|0_B\rangle \quad ; \quad \mathbf{F} = \begin{pmatrix} 0 & f \\ f & 0 \end{pmatrix} \tag{5.4}$$

and (5.4) vanishes only if $\mathbf{Q}|0_B\rangle = 0$. However we are dealing with a model of 2d gravity coupled to supersymmetric matter which is perfectly self consistent and which may define a physically interesting theory containing 2d gravity.

Following [21] and [23] we have to consider the ground state of a system of N fermions in the potential $V_B = N(\lambda + (g - \lambda^2)^2/4)$. The $N = \infty$ limit can therefore be studied by the WKB approximation [15]. In the planar scaling limit ($N \to \infty$, then $a \to 0$) the potential V_B becomes

$$v(p) = p^3 - 3\,t\,p \tag{5.5}$$

The particle density $\rho(e, p)$ and the integrated particle density $\rho(p) = \int de\, \rho(e, p)$ are respectively

$$\rho(e, p) = \frac{1}{2\pi\sqrt{e - v(p)}}\, \theta(e - v(p)) \quad ; \quad \rho(p) = \frac{1}{\pi}\sqrt{e_F - v(p)}\, \theta(e_F - v(p)) \tag{5.6}$$

The Fermi energy e_F is fixed by the normalization condition

$$\nu(e_F) = \frac{1}{\pi} \int_{v < e_F} dp\, \sqrt{e_F - v(p)} = 0 \tag{5.7}$$

(where the divergence at $-\infty$ is treated by a finite part prescription).

In the weak coupling region $t > 0$, where SUSY is unbroken, the solution of (5.7) is given by $e_F = -2\,t^{3/2}$, which corresponds to the value of v at the local minimum $p = \sqrt{t}$. Then we recover exactly the large N solution, as expected, since we have $\rho(p) = u(p)$, where $u(p)$ is the eigenvalue density given by (4.9). e_F is identified with $\frac{4}{3}\langle P \rangle$. For $t < 0$ SUSY is spontaneously broken and $v(p)$ has no real local minimum. However (5.7) still has a unique real solution, since $\nu(e)$ is defined on $] -\infty, \infty[$, with

$\nu'(e) > 0$, and since $\nu(e) \sim \pm|e|^{5/6}$ as $e \to \pm\infty$. Since $\nu(0) > 0$, e_F is negative, and scales as $e_F = \mathbf{c}\,(-t)^{3/2}$ with $\mathbf{c}$ some transcendental number.

This has some nasty effects on the physical observables of the theory. Indeed, according to the rule (5.1), the v.e.v. of the loop operator which creates a loop with length ℓ is given (playing with Laplace transform) by

$$\langle w(\ell) \rangle \;=\; \int_{-i\infty}^{i\infty} \frac{dp}{2i\pi}\, \mathrm{e}^{p\ell}\, \sqrt{v(p) - e_F} \tag{5.8}$$

For $\ell > 0$ we wrap the contour around the cut $]-\infty, p_0]$ and we obtain the expected result $\langle w(\ell) \rangle = \int dp\, \rho(p)\, \mathrm{e}^{p\ell}$. For $\ell < 0$, if $t > 0$ we get $\langle w \rangle = 0$, but if $t < 0$ the integrand $\sqrt{v(p) - e_F}$ has a second cut right to the contour of integration and therefore $\langle w(\ell) \rangle$ does not vanish ! Moreover for large negative ℓ it behaves as

$$\langle w(\ell) \rangle \;\sim\; \ell^{-3/2}\, \mathrm{Re}\left(\mathrm{e}^{p_1 \ell}\right) \tag{5.9}$$

where p_1 is one of the two complex conjugate zeros of $(v - e_F)$. The amplitude for a loop with negative length oscillates wildly and can even be negative. The existence of such "unphysical states" is a serious problem if one wants to interpret the 1d SUSY string as a pure 2d gravity theory. In the planar limit they appear only for $t < 0$ but in the scaling limit loops with negative length should have a non-zero, but exponentially small, amplitude for positive t.

6. Conclusion

The various approaches to the scaling limit for two dimensional quantum gravity give different points of view on the non-perturbative effects in the theory. Remarkably those effects can be understood (and to some extend calculated) within the matrix model formulation, and they are deeply connected to the unboundness of the potential. At the present stage my feeling is that pure 2d quantum gravity has a somewhat similar status than QED_4 for negative e^2 [25]. It is a well defined theory in perturbation theory. It is renormalizable and asymptotically free. However the vacuum is unstable under the formation of handles (a process somewhat analogous to $e_+ e_-$ pairs creation for QED) and it seems that no physically acceptable stable vacuum can be reached. The fact that similar issues appear also in critical strings [26] and that $3 + 1$ ordinary gravity is also unstable under conformal modes means that the understanding of this kind of problems is crucial for the elaboration of a quantum theory of gravity.

Acknowledgements

I would like to thank all the participant of the workshop, in particular S. Shenker and J. Zinn-Justin, for their interest, their comments and their questions. I am very grateful to the organizers for this very pleasant **and** exciting conference. Finally I thank Enzo for his patience.

References

[1] J. Ambjørn, B. Durhuus and J. Fröhlich, Nucl. Phys. B257 (1985) 433.
F. Favid, Nucl. Phys. B257 (1985) 45.
V. A. Kazakov, Phys. Lett. 150B (1985) 282; V. A. Kazakov, I. K. Kostov and A. A. Migdal, Phys. Lett. 157B (1985) 295.

[2] E. Brézin and V. A. Kazakov, Phys. Lett. 236B (1990) 2125.
M. R. Douglas and S. H. Shenker, Nucl. Phys. B335 (1990) 635.
D. J. Gross and A. A. Migdal, Phys. rev. Lett. 64 (1990) 27.

[3] M. R. Douglas, Phys. Lett. 238B (1990) 2125.

[4] T. Banks, M. R. Douglas, N. Seiberg and S. H. Shenker, Phys. Lett. B 238 (1990) 279.

[5] E. Hille, "Ordinary Differential Equations in the Complex Domain", Pure and Applied Mathematics, J. Wiley & Sons, 1976.

[6] S. H. Shenker, "The Strength of Nonperturbative Effects in String Theory", these proceedings.

[7] P. Ginsparg and J. Zinn-Justin, these proceedings.

[8] S. R. Wadia, Phys. Rev. D 24 (1981) 970.
A. A. Migdal, Phys. Rep. 102 (1983) 199.

[9] F. David, Mod. Phys. Lett. A5 (1990) 1019.

[10] J. Ambjørn and Y. M. Makeenko, preprint NBI-HE-90-22, May 1990.
J. Ambjørn, J. Jurkiewicz and Y. M. Makeenko, preprint NBI-HE-90-41, August 1990.

[11] R. Dijkgraaf, H. Verlinde and E. Verlinde, "Loop Equations and Virasoro Constraints in Non-Perturbative 2-D Gravity", preprint PUPT-1184 IASSNS-HEP-90/48, May 1990.

[12] M. Fukuma, H. Kaway and R. Nakamaya, "Continuum Schwinger-Dyson Equations and Universal Structures in Two dimensional Quantum Gravity", preprint UT-562 KEK-TH-251, May 1990.

[13] E. Witten, Nucl. Phys. B340 (1990) 281.
R. Dijkgraak and E. Witten, Nucl. Phys. B342 (1990) 486.
E. Verlinde and H. Verlinde, "A Solution of Two Dimensional Topological Gravity", preprint PUPT-1176, 1990.

[14] C. Bachas and P. M. S. Petropoulos, Phys. Lett. 247B (1990) 363.
C. Bachas, "On Triangles and Squares", these proceedings.

[15] E. Brezin, C. Itzykson, G. Parisi and J.-B. Zuber, Commun. Math. Phys. 59 (1978) 35.

[16] F. David, "Phases of the Large N Matrix Model and non-perturbative Effects in 2d Gravity", preprint SPhT/90/090, July 1990.

[17] M.Douglas, N. Seiberg and S. Shenker, Phys. Lett. B244 (1990) 381.

[18] J. Jurkiewicz, Phys. Lett. B 245 (1990) 178.

[19] G. Bhanot, G. Mandal and O. Narayan, "Phase Transitions in 1-Matrix Models", preprint IASSNS-HEP-90-52, May 1990.
G. Mandal, these proceedings.

[20] G. Moore, Commun. Math. Phys. 133 (1990) 261.

[21] E. Marinari and G. Parisi, Phys. Lett. 240B (1990) 375.

[22] J. Greensite and M. Halpern, Nucl. Phys. B242 (1984) 167.

[23] M. Karliner and A. Migdal, "Nonperturbative 2D Quantum Gravity via Super-symmetric String", preprint PUPT-1191, July 1990.

[24] J. Ambjørn, J. Greensite and S. Varsted, "A Non-perturbative Definition of 2D Quantum Gravity by the Fifth Time Action", preprint NBI-HE-90-39, July 1990.

[25] F. J. Dyson, Phys. Rev. 85 (1952) 32.

[26] D. J. Gross and V. Periwal, Phys. Rev. Lett. 60 (1988) 2105.

INTEGRABLE MODELS OF
TWO DIMENSIONAL QUANTUM GRAVITY

P. Di Francesco and D. Kutasov

Joseph Henry Laboratories
Princeton University
Princeton, NJ 08544

1. Introduction

In the last year important progress has been made [1] in non critical string theory
(or equivalently two dimensional Quantum Gravity (QG) coupled to matter). QG
coupled to conformal field theories with a finite number of degrees of freedom was
shown [1], [2] [3] [4] to be closely related to certain integrable systems of differential
equations – the so called KdV hierarchy. Also, a relation to topological gravity was
discovered [5] . Correlation functions for arbitrary topology were obtained, and much
progress has been made in understanding non perturbative phenomena.

This was made possible by a representation of QG coupled to (minimal) matter
in terms of Hermitian matrix integrals [6]. Subsequently, other matrix ensembles
leading to different integrable systems were solved [7] ; many of those still await a
surface interpretation. In any case, while there are many suggestive indications, the
relation of matrix models to Liouville theory is still mysterious (see [8] [9] for recent
discussions).

A different approach, which will be studied here, is to start from the integrable
systems, and identify the basic ingredients which relate solutions of the integrable
differential equations to QG. The main motivations for such an approach are the lack
of an exact solution in many interesting matrix models, and the fact that the matrix
definition is often subtle [10] [11] as well as the need to understand the general features
of the purported solutions to QG, and their classification. A general construction
should be helpful in understanding the results from the point of view of Liouville
theory, which is a very important task, and may help clarify the geometric meaning
of the string equations [12].

The general structure of the solutions to QG is not yet well understood. We will
describe below two examples of the construction, explain their relation to physical

Random Surfaces and Quantum Gravity
Edited by O. Alvarez *et al.*, *Plenum Press, New York, 1991*

systems, describe some predictions for these systems and discuss some of their properties. The two are special cases of the generalized KdV flows [13] [14]. These are two infinite dimensional spaces of couplings with flows defined by certain differential equations, related to the A_n, D_n Dynkin diagrams. The former, when supplemented with a "string equation", were shown [4] to be related to A series modular invariant minimal models coupled to QG; the latter were conjectured [15] to correspond to D modular invariant models coupled to QG.

This note summarizes and extends results obtained in [15]. It is organized as follows: in section 2 we show that the string equation found by Douglas [4] to describe the continuum limit of (A series) multimatrix models is conserved under an infinite set of KdV flows (generalizing [2], [12]). The solutions to the string equation are not unique; they are parametrized by an infinite set of deformations (identified with perturbations of the matrix model/ gravity system by a set of operators), which we find explicitly. These deformations are shown to be given by the KdV flows. In section 3 we solve the above string equations on the sphere at the unitary critical points and calculate some correlation functions. We find that they obey the fusion rules of the underlying CFT. In section 4 we use the experience from sections 2, 3 and, starting from a particular set of D series flows [13], construct a conjectured solution for 2d gravity coupled to D series minimal models. This involves constructing a string equation invariant under the above KdV flows, and showing that deformations of the solutions of this equation are again given by the KdV flows. We show that the solution reproduces known qualitative features of the gravity system. Section 5 is a summary.

2. A series minimal models coupled to gravity

The starting point for our first example is the A series KdV flows in the classification of Drinfeld and Sokolov [13]. Those are obtained in two stages. First we define "time evolution" of a pseudo differential operator

$$L = D + a_0 D^0 + a_1 D^{-1} + a_2 D^{-2} + ... \qquad (2.1)$$

(D^{-1} denotes the formal inverse of $D = \frac{\partial}{\partial x}$, $a_i(x)$ are functions), by:

$$\partial_m L (\equiv \frac{\partial}{\partial \lambda_m} L) = [L_+^m, L] \qquad (2.2)$$

The A series are obtained by further restricting $Q = L^q$ (for some q) to satisfy $Q_- = 0$. One can easily verify that setting $Q_- = 0$ for all λ_m is consistent with the flows (2.2).

The third ingredient needed to obtain a solution to QG is a string equation of the form $f(Q) = 0$, thus further constraining Q, which is essentially determined by the requirement that it should be invariant under the flows (2.2). One can deduce this equation from the structure of (2.2), but since it was first derived from matrix models, and to motivate the more general constructions, we'll present it from that point of view.

In his lecture, M. Douglas has shown how the continuum double scaling limit of chain-interacting multimatrix models with even potentials naturally leads to equations (string equations), governing the perturbative and non perturbative behaviour of the physical system. Those take the form [1]:

$$[P, Q] = 1 \qquad (2.3)$$

where P and Q are differential operators in $D = \partial_x$ (x =renormalized cosmological constant, coupled to the lowest dimension operator in the theory), of respective degrees p and q:

$$P = \sum_{k=0}^{p} v_k D^k \qquad (2.4)$$

$$Q = \sum_{k=0}^{q} u_k D^k \qquad (2.5)$$

and are normalized by $u_q = v_p = 1$; $u_{q-1} = v_{p-1} = 0$. Q and P correspond to insertions of λ (the eigenvalue) and ∂_λ into the matrix integral; (2.3) is the appropriate Heisenberg algebra [4]. In the following we will suppose that p and q are coprimes and that $p > q$ for simplicity. It can be shown from the matrix model that

$$u_{q-2} \propto \partial_x^2 \log Z \qquad (2.6)$$

where Z is the partition function of the model in the continuum double scaling limit (taking $\log Z$ removes the sum over disconnected surfaces).

Our strategy will be the following: by studying the general solutions of (2.3), we will define 'natural' operators in the theory (related to deformations of solutions of (2.3)) and eventually compute their correlation functions on the sphere in some special cases. Their insertions in correlators will be governed [1], [2], [3], [4] by the generalized KdV flows of [13] [14]. We will see that the general features of the correlation functions obtained (for some of the operators) are compatible with correlation functions expected in QG coupled to minimal matter (in particular the critical exponents and fusion rules agree). Some of the operators we will obtain have no continuum interpretation yet (but see [16] [9]).

Let us recall a few definitions and results from pseudo-differential calculus. For any differential operator Q of degree q, there exists a unique pseudo-differential operator of the form (2.1) such that $L^q = Q$. One can also consider other integer powers of L, thus forming fractional powers of Q which commute with Q. We will also denote by the subscript $+$ the differential part of any pseudo-differential operator R, and $R_- = R - R_+$. Finally the coefficient of D^{-1} in R is called the residue of R, denoted $Res(R)$.

Using a theorem of [13], one can show that the most general form P (2.3) can have is:

$$P = \sum_{j=1}^{\infty} \mu_j (Q^{j/n})_+ \qquad (2.7)$$

<hr>

where the μ's are constants (eqn. (2.7) follows from the requirement that deg $[P, Q] \leq n - 2$; the rest of eqns. (2.3) constrain the form of Q). Moving in μ space amounts to considering perturbations of the "critical" string equations $\mu_j = \delta_{j,p}$ to which we first restrict. In that case, (2.3) implicitly contains $q - 1$ integration constants, obtained as follows: using the fact that $Q^{p/q} = L^p$ and $Q = L^q$ commute, rewrite

$$[P, Q] = -[L^p_-, L^q] \tag{2.8}$$

and expand the commutator in powers of D. If $(L^p)_- = c_1(x)D^{-1} + \cdots$, then the leading term in (2.8) $qc_1'D^{q-2}$ has to vanish, giving rise to an integration constant $c_1 = \nu_{q-1}$. Write now $(L^p)_- - \nu_{q-1}L^{-1} = c_2(x)D^{-2} + \cdots$ then by the same argument we get another integration constant $c_2 = \nu_{q-2}$. Repeating this procedure until we reach the D^0 term in the commutator, we get

$$L^p_- = \sum_{\alpha=1}^{q-2} \nu_{q-\alpha}L^{-\alpha} + c_{q-1}(x)D^{-q+1} + \cdots \tag{2.9}$$

and $qc'_{q-1} = 1$ so $c_{q-1} = x/q$, where we have absorbed the last integration constant in x ($\nu_1 = x$).

We will now argue that the dependence of L (and thus Z) on the integration constants $\nu_i \equiv (i/q)\lambda_i$ ($i = 1, .., q - 1$) introduced in (2.9) is governed by the KdV flows (2.2). To establish this we will need a proof of two facts: (a) the string equation (2.3) commutes with the flows (2.2); (b) the flows (2.2) preserve the form (2.9) of L^p_- with $\nu_i = \frac{i}{q}\lambda_i$ (actually it is enough to prove (b) since then (a) follows, but we will prove both for completeness and later use). A proof of (a) and (b) only establishes compatibility of the two sets of deformations (2.9) and (2.2). Given a solution of (2.3) at $\lambda = 0$, (2.2) gives well defined correlation functions while (2.9) has in general ambiguous correlators. Therefore, (2.2) is stronger and one can not use (2.9) as an alternative definition. From our point of view, the KdV flows are an axiom and (a) and (b) are a check of the consistency of the representation (2.9).

Claim a: The string equation (2.3) is compatible with the KdV flows (2.2).

Proof: differentiate (2.3) w.r.t. λ_k using (2.2).

$$\begin{aligned}
\partial_{\lambda_k}[P, Q] &= -\big[[L^k_+, L^p]_-, L^q\big] - \big[[L^p_-, [L^k_+, L^q]]\big] \\
&= -\big[[L^k_+, L^p_-]_-, L^q\big] - \big[[L^p_-, L^k_+], L^q\big] \\
&= \big[[L^k_+, L^p_-]_+, L^q\big] = \big[[L^k, L^p_-]_+, L^q\big]
\end{aligned} \tag{2.10}$$

where we have used the Jacobi identity and (2.3) for the second double commutator, and the fact that $[L^k_+, L^p_+]_- = [L^k_-, L^p_-]_+ = 0$. Using (2.9) for L^p_- we get $[L^k, L^p_-] = k/qD^{k-q} + \cdots$, which has a vanishing differential part for $k \leq q - 1$ implying that (2.10) is zero.

Claim b: The form of L^p_- (2.9) with $\nu_i = \frac{i}{q}\lambda_i$ is preserved by the KdV flows (2.2) (with the same λ_i).

Proof: Introduce the degree zero pseudo-differential operator W, which diagonalizes L, i.e.

$$L = WDW^{-1} \tag{2.11}$$

(such an operator always exists and is uniquely defined up to multiplication by a constant pseudo-differential operator of degree 0 [14]). Then if we consider the most general solution to the string equation: $[R_-, L^q] = 1$, we can solve for R in terms of W (2.11): $R_- = W(S + (x/q)D^{-q+1})W^{-1}$, with $[S, D^q] = 0$. Then $S = \sum_j d_j D^j$, where the d_j are constants. One has $d_{-j} = \nu_{q-j}$, for $j = 1, \cdots, q-1$, so that:

$$R_- = \sum_{\alpha=1}^{q-2} \nu_{q-\alpha} L^{-\alpha} + W(\frac{x}{n} D^{-q+1})W^{-1} + \sum_{\beta \leq -q} d_\beta L^\beta \tag{2.12}$$

Eqn. (2.12) is a refinement of (2.9).

We now express the constants ν_{q-k} as:

$$\nu_{q-k} = Res((L_-^p - \sum_{\alpha=1}^{k-2} \nu_{q-\alpha} L^{-\alpha})L^{k-1}) \tag{2.13}$$

If we pick a solution of (2.3) which is given by (2.9) with $\nu_i = 0$, and then use the KdV flows (2.2) to evolve it in "time", this will generate effectively a set of ν_i which are some functions of λ_i. We want to show that these function are linear $\nu_i \propto \lambda_i$. To do that we consider the variations of the ν's (2.13) w.r.t. the λ's (2.2):

$$\partial_{\lambda_{q-l}} \nu_{q-k} = Res\left\{\left([L_+^{q-l}, L_-^p - \sum_{\alpha=1}^{k-2} \nu_{q-\alpha} L^{-\alpha}]_- - \sum_{\alpha=1}^{k-2} (\partial_{\lambda_{q-l}} \nu_{q-\alpha})L^{-\alpha}\right) L^{k-1}\right\}$$
$$- Res\left\{(L_-^p - \sum_{\alpha=1}^{k-2} \nu_{q-\alpha} L^{-\alpha})[L_-^{q-l}, L^{k-1}]\right\} \tag{2.14}$$

The last term has no residue, as it starts like $D^{-k+1} D^{k-3} +$ lower terms. For the same reason, we can drop the $+$ subscript in the first commutator, and get:

$$\partial_{\lambda_{q-l}} \nu_{q-k} = Res\left\{\left([L^{q-l}, L_-^p - \sum_{\alpha=1}^{k-2} \nu_{q-\alpha} L^{-\alpha}]_- - \sum_{\alpha=1}^{k-2} (\partial_{\lambda_{q-l}} \nu_{q-\alpha})L^{-\alpha}\right) L^{k-1}\right\} \tag{2.15}$$

We saw above (2.12) that we can solve the string equation by $L_-^p - \sum_{\alpha=1}^{k-2} \nu_{q-\alpha} L^{-\alpha} = W(x/q)D^{-q+1}W^{-1} + \sum_{j \leq q} d_j L^j$, hence, using the expression $L^{q-l} = WD^{q-l}W^{-1}$, we get:

$$\partial_{\lambda_{q-l}} \nu_{q-k} = Res((\frac{(q-l)}{q} L^{-l} - \sum_{\alpha=1}^{k-2} \partial_{\lambda_{q-l}} \nu_{q-\alpha} L^{-\alpha})L^{k-1}) \tag{2.16}$$

Let us now proceed by induction. For $k = 1$ the expression above gives a zero answer for $l > 1$ and for $l = 1$, one gets $\partial_{\lambda_{q-1}} \nu_{q-1} = (q-1)/q$ thus establishing that $\nu_{q-1} = \lambda_{q-1}(q-1)/q$. Suppose $\nu_{q-r} = \lambda_{q-r}(q-r)/q$ for all $r < k$. Then for $l < k$, $\partial_{\lambda_{q-l}} \nu_{q-k} = 0$ (2.16) is zero by the induction hypothesis which implies the cancellation of the L^{-l} terms on the r.h.s., and the same holds for $l > k$, by replacing all the ν's in (2.16) in terms of the λ's; the sum on the r.h.s. of (2.16) vanishes then, and we are left with $ResL^{k-l-1} = 0$. Finally for $l = k$, we obtain $\partial_{\lambda_{q-k}} \nu_{q-k} = (q - k)/q$, i.e. $\nu_{q-k} = \lambda_{q-k}(q - k)/q$. This shows that:

$$L_-^p = \sum_{\alpha=1}^{q-2} \frac{q - \alpha}{q} \lambda_{q-\alpha} L^{-\alpha} + \frac{x}{q} D^{-q+1} + \cdots \qquad (2.17)$$

So the variations of L w.r.t. $\lambda_j = (q/j)\nu_j$ are exactly given by the KdV flows (2.2), $j = 1, \cdots, q - 1$ (at least perturbatively).

So far we have restricted ourselves to the subspace of solutions of the string equation (2.3) defined by $\mu_j = \delta_{j,p}$ (see (2.7)). Our next task is to explore the full space of solutions (2.7), i.e. determine the relation of the μ_j in (2.7) to KdV flows (2.2). This can be done by requiring compatibility of (2.3) with *all* the KdV flows (2.2). We will now show that in order for the flows (2.2) with $k > q$ to be compatible with the string equation, we have to set $\mu_j = -\lambda_{q+j}(q + j)/q, j = 1, 2, 3, \cdots$. The proof goes as in the order parameter case. One computes:

$$\partial_{\lambda_k}[P, Q] = [[L^k, \sum_{j=1}^{\infty} \mu_j(L^j)_-]_+ + \sum_{j=1}^{\infty} (\partial_{\lambda_k}\mu_j)L_+^j, L^q] \qquad (2.18)$$

Eq. (2.18) is obtained by a sequence of operations similar to that done in the proof of claim a above (2.10). We can now use (2.12) with $R = \sum \mu_j L^j$ to conclude that:

$$\sum_{j=1}^{\infty} \mu_j(L^j)_- = \sum_{\alpha=1}^{q-2} \frac{q - \alpha}{q} \lambda_{q-\alpha} L^{-\alpha} + W(x/q) D^{-q+1} W^{-1} + \sum_{\beta \leq -q} d_\beta L^\beta \qquad (2.19)$$

Substituting (2.19) into (2.18) we find that the string equation is compatible with the KdV flows iff $\partial_{\lambda_k}\mu_j = -(k/q)\delta_{j,k-q}$ i.e. $\mu_j = -(q + j)/q\lambda_{q+j}$. This completes the proof of the compatibility between the (perturbed) string equation and the (higher) KdV flows.

To summarize, so far we have shown that solutions to (2.3) are not unique but can be deformed in an infinite number of directions, which were identified with generalized KdV flows. From the QG point of view, these perturbations correspond to adding to the action an infinite number of dressed matter operators. Indeed, we are now ready to define an infinite set of operators for our theories, the insertions of which in correlators will be generated by the KdV flows (2.2). They read:

$$\langle \phi_{l_1} \cdots \phi_{l_k} \rangle = \partial_{\lambda_{l_1}} \cdots \partial_{\lambda_{l_k}} \log Z \qquad (2.20)$$

and are related to derivatives of u_{q-2} (2.5) w.r.t. the λ's through (2.6). Technically, these correlators can be computed once a solution L of (2.3) is known, by repeated use of (2.2). For instance, the D^{n-2} coefficients in (2.2) give all the one point functions with one insertion of the "identity" ϕ_1:

$$\langle \phi_1 \phi_k \rangle = (const)Res(Q^{k/q}). \qquad (2.21)$$

In the following we will call order parameters the $q - 1$ first operators $\phi_1, \cdots, \phi_{q-1}$ corresponding to the integration constants in (2.3) in analogy with the lattice model description [17]. In the topological approach [5], these fields are the gravitational primaries, and play a special role.

The picture so far deserves a few comments. An important issue is the $p \to q$ symmetry. Though in the matrix model P and Q appear on a different footing, the (p,q) and (q,p) conformal theories are known to be equivalent. In solving (2.3) one could as well have taken $P = L^p$, $Q = \sum_{j=1}^{\infty} \eta_j (L^j)_+$. This leads of course to the same string equation for Z at the critical point, but the correlation functions are in general different, due to a different definition of the operators . Those coincide in the two pictures only for the $q-1$ order parameter operators of the theory. This leads to some ambiguity in defining the higher operators; in fact, their correlation functions defined by (2.2) do not correspond to correlation functions of dressed primary conformal fields [15]. It is also interesting to notice that the same lack of symmetry is found in the fluctuating lattice integrable model approach of Kostov [17], $p > q$ and $p < q$ referring to dense and dilute versions of the A type integrable lattice model, which are very different

A simple example is the coupling of the Ising model ($p = 4, q = 3$) to gravity [18]. We expect the following operator content: $\phi_1 = Id$, $\phi_2 = \sigma$, $\phi_5 = \epsilon$ (resp. identity, spin and energy operators). The thermal perturbation by ϵ is governed by the interpolating string equation: $[L_+^4 - (5/3)\lambda_5 L_+^2, L^3] = 1$, leading (at least perturbatively [11]) to pure gravity ($p = 3, q = 2$) in the high temperature limit $\lambda_5 \to \infty$. More generally, the unitary conformal series ($p = n + 1, q = n$) have a thermal perturbation by the least relevant operator in the Kac table, $\phi_{1,3}$, which leads along a RG trajectory to the next $(n, n - 1)$ unitary theory. This is realized here by the interpolating string equation:

$$[L_+^{n+1} - \frac{2n - 1}{n}\lambda_{2n-1}L_+^{n-1}, L^n] = 1 \tag{2.22}$$

One can use the flows between the different models to determine the precise relation between u_{q-2} in (2.5) and the free energy. This can be done because the proportionality constant in (2.6) is independent of the infinite number of the flow parameters λ_k and can be determined at a convenient point in coupling space, e.g. one of the one matrix models originally solved in [1]. The result for the unitary series is (see [15] for details): $u_{n-2} = -(n/2)u$ for even potentials in the matrix models (an additional factor of $1/2$ has to be taken for general potentials, [19]). The same consideration applies to the general (p,q) case, where $u_{q-2} = -(q/2)u$.

3. Solution of the unitary (n+1,n) case on the sphere

We want now to derive the sphere correlation functions as expressed through KdV flows (2.2) for the unitary case ($p = n + 1, q = n$). We limit ourselves to correlation functions of order parameters, since as explained above the correlation functions of other operators are not uniquely determined by KdV. The unitary case is simpler than the general (p, q) minimal model from the point of view of Liouville theory since the cosmological constant couples to the true identity operator in the matter sector, and thus the gravitational fusion rules (on the sphere) should be the same as the CFT ones. To calculate the correlation functions, we need to solve (2.3) explicitly on the sphere, and use the value of $L = Q^{1/n}$ to get various correlators. It is easy to see that one only has to retain the leading power of x in all Q and P coefficients, expressed in perturbative "genus" expansions, to account for the sphere. Equivalently these coefficients can be expressed in terms of powers of the string susceptibility $u \propto u_{q-2}$ (2.6) (since the latter also scales as a power of x). The spherical solution is not, in general, determined by (2.3). There is a discrete ambiguity [15], which must be fixed by a physical input. The input suggested in [15] is the requirement that the one point functions of all order parameter operators vanish on the sphere (this is definitely expected to be the case in the continuum treatment). A priori, it is not obvious that there is any solution of (2.3) with this property. The existence and uniqueness of such a solution was established in [15]. We will next describe the solution and outline the evaluation of some correlation functions.

The solution involves introducing a set of pseudo-differential operators :

$$\frac{1}{n}Q_n = \sum_{k=0}^{\infty} \binom{n-k-1}{k-1} \frac{v^k}{k} D^{n-2k} \tag{3.1}$$

with

$$v = -\frac{u}{2}; u = \partial_x^2 \log Z \tag{3.2}$$

It can be shown [15] that on the sphere

$$[Q_n, Q_m] = 0 \tag{3.3}$$

for all n, m. In addition, one has the following crucial property following from the definition (3.1):

$$(Q_n)_- = c_n D^{-n} + O(D^{-n-2}); \quad c_n = -\left(\frac{u}{2}\right)^n \tag{3.4}$$

In [15] it was proved that the unique solution to the string equation with vanishing order parameter one point functions (except that of $\phi_1 = Id$) on the sphere reads: $P = (Q_{n+1})_+$, $Q = (Q_n)_+$. The fact that this choice of P, Q gives a solution follows from (3.3), (3.4). Vanishing of the one point functions on the sphere will be derived below. For the proof of uniqueness we refer the reader to [15]. The sphere string equation is simply $(n + 1)(u/2)^n = x$, giving the critical exponent $\gamma_{str} = -1/n$ in agreement with [20]. Correlators are given, as explained above, by the KdV flows (2.2):

$$\langle \phi_l \phi_1 \rangle = 2Res(L^l)$$
$$\langle \phi_l \phi_m \phi_1 \rangle = 2Res[L_+^l, L_-^m] \tag{3.5}$$
$$\langle \phi_l \phi_m \phi_r \phi_1 \rangle = 2Res\left([[L_-^l, L_+^m], L_-^r] - [L_+^m, [L_+^l, L_-^r]]\right)$$

etc.

The only technical limitation is that we have to approximate $L = Q^{1/n}$ by $M = (Q_n)^{1/n} = Q_1$ on the sphere, in the above expressions. By doing this one makes the following error, computed from (3.4):

$$M^k - L^k = \frac{k}{n} c_n(u) D^{k-2n} + \cdots \tag{3.6}$$

From (3.6) we deduce that for all $k < n$

$$L_-^k = c_k D^{-k} + \ldots \tag{3.7}$$

since (3.7) is satisfied by M_-^k and the difference between the two (3.6) is subleading (for $k < n$). c_k are given by (3.4). Remarkably, (3.7) is the only feature of the explicit solution above, which is needed to calculate all the correlation functions of the order parameters (3.5), with the following results:

$$\langle \phi_l \rangle = 0 \quad \forall l \in \{2, \cdots, 2n - 2\} \tag{3.8a}$$

$$\langle \phi_{2n-1} \rangle = \frac{2n - 1}{2n(n + 1)} x^2 \tag{3.8b}$$

$$\langle \phi_l \phi_m \rangle = 2m \delta_{l,m} u^l \tag{3.8c}$$

$$\langle \phi_l \phi_m \phi_r \rangle = lmr N_{l,m,r} \left(\frac{u}{2}\right)^{\frac{l+m+r-3}{2}} u' \quad \forall l, m, r < n \tag{3.8d}$$

where (3.8b) is the thermal operator $\phi_{1,3}$ expectation value, directly computed from the interpolating sphere string equation: $(n + 1)(u/2)^n - (2n - 1)\lambda_{2n-1}(u/2)^{n-1} = x$. (3.8c) yield the KPZ dimensions [20] for ϕ_l, $\Delta_l = (l-1)/2n$, because 2 point correlators behave like $\langle \phi \phi \rangle \propto x^{2\Delta - \gamma_{str}}$. In (3.8d) the numbers $N_{l,m,r} \in \{0, 1\}$ are the fusion rules of the corresponding conformal field theory (CFT), restricted to the set of order parameters (diagonal of the Kac table).

It is remarkable that the fusion rules of CFT are contained in the KdV correlation functions (3.5). We will derive this fact here; the rest of (3.8) follows (by calculating correlation functions with insertions of ϕ_1 – the dressed identity operator).

One can show that despite appearances, the three point functions in (3.5) are symmetric in m, l, r. Thus without loss of generality we can take $r > l, m$. Then, the first term on the r.h.s. of (3.5) vanishes and we are left with:

$$\langle \phi_l \phi_m \phi_r \phi_1 \rangle = -2 Res\left(\left[L_+^m, [L_+^l, L_-^r] \right] \right) \tag{3.9}$$

For this to be non zero, we must have

$$l + m \geq r + 1 \tag{3.10}$$

Since the three point functions are symmetric in l, m, r we automatically get two circular permutations of (3.10), which must hold for the three point function not to vanish. These are the CFT fusion rules in the $n \to \infty$ limit. At finite n there is another fusion rule : $r + l + m \leq 2n - 1$. Our next task is to derive this from (3.9). To do that, we go back to the general l, m, r (no particular ordering) and assume that

(3.10) (and permutations) hold. We want to show that even then, the three point function (3.5) can vanish. First, since we assume $l + r \geq m + 1$, the first term in (3.5) again vanishes, and the three point function is given (again) by (3.9) (but this time it's symmetric in l, m, r). We can drop the $+$ subscripts from L^m, L^l (again because of (3.10) and permutations). The resulting commutators simplify on the sphere since the derivative D acts only once , so that e.g. $[L^m, L^r_-] = mL^{m-1}[L, L^r_-]$. Using this, one can show that the result (3.9) has the general form $\langle \phi_l \phi_m \phi_r \rangle = mlrF(m + l, r)$. But the l.h.s. of this equation is symmetric under interchange of m, l, r; thus $F(m + l, r)$ must be a function of $m + l + r$ only. We can evaluate this function with any convenient combination of m, l, r. One convenient value to use is to put (say) $r = n$. We know that the n'th flow is trivial, so that all three point functions with $r = n$ vanish. We assumed $l + m \geq r + 1$ (see (3.10)) , so that what this really means is that the three point function vanishes if $l + m + r \geq 2n + 1$. But this is precisely the missing fusion rule we were after. Similarly one can easily calculate the three point functions when they don't vanish by using convenient m, l, r. The result is summarized in (3.8d). One can of course calculate the non vanishing three point functions from the explicit solution (3.1) by plugging in (3.5) (see [15]), but the results involve complicated sums over binomial coefficients. The present derivation avoids evaluating these sums. To calculate correlation functions of ϕ_m $m > n$, it is convenient to use the recursion relations for gravitational descendants in terms of primaries, which originate in topological field theory [5] and therefore hold in the KdV framework as well. New interesting features arise [15]; e.g. the two point functions are not orthogonal; it is very important to understand them in the context of Liouville theory.

Using this technical framework, we have been able to compute the one loop partition function (see [15] for details) as the x^{-2} coefficient in the genus expansion of u:

$$Z_1 = \frac{n - 1}{24} \tag{3.11}$$

One may compare the results (3.8), (3.11), to continuum calculations. The spherical correlation functions are hard to obtain in that approach, but the fusion rules N_{ijk} in (3.8d) must agree. The reason is that in the Liouville approach (2d gravity in conformal gauge), the correlation function factorizes into a Liouville and matter contributions. Fixing the conformal Killing vectors (ghost zero modes) fixes the 3 punctures to fixed positions on the surface (one should not integrate over them) so that the final result is a product of the Liouville and matter correlation functions. If the correlation function in the matter sector alone vanishes, so does the full gravity amplitude. Our results are in agreement with this. The one loop results (3.11) were obtained recently from Liouville theory [22] (see also [23]).

4. D series models coupled to 2d gravity

So far our starting point has been a (multi) matrix model, solvable by orthogonal polynomial techniques, which led to the string equation (2.3). But only chain-interacting matrix models can undergo this treatment. For instance it is not known how to solve cyclic-interaction matrix models in the same way. The simplest example is the three matrix model with potential : $W(M_1, M_2, M_3) = \sum_i V_i(M_i) + c_1 M_1 M_2 + c_2 M_2 M_3 + c_3 M_3 M_1$. For special values of the couplings c this can be mapped onto the 3 states Potts model coupled to gravity, i.e. the $\mathbf{Z}_3$ spins of the model live on the nodes of a fluctuating graph. As a conformal theory the 3 states Potts model corresponds to a $(6, 5)$ theory, but its operator content is different from the usual ('diagonal') unitary $(6, 5)$ theory. This is part of the classification of modular invariant minimal theories [24] with central charges $c(p, q) = 1 - 6(p - q)^2/pq$, including in addition to the diagonal A series (A_{p-1}, A_{q-1}), the D series $(A_{p-1}, D_{\frac{q}{2}+1})$ for q even and the exceptional $(A_{p-1}, E_{6,7,8})$ with $q = 12, 18, 30$. The labels A, D, E refer to an underlying simply laced Lie algebraic structure. In this classification, the 3 states Potts model is (A_4, D_4).

We will now propose an approach to D series coupled to gravity, starting from the generalized KdV flows associated to $SO(2n)$ Lie algebras [13]. Following the logic of section 2, it is natural to look for a string equation which is invariant under the flows [13]. Below we will show that such an equation indeed exists. The data of integrable flows together with a string equation defines then a set of correlators, which we can compute explicitly at least on the sphere (as in section 3). The predictions for critical exponents, fusion rules and one loop partition sum turn out to be in perfect agreement with the KPZ results [20] and some direct Liouville calculations for the D series coupled to gravity. This approach somehow gets rid of the matrix model and is very suggestive that the true symmetry of the model is summarized by a quantized version of a set of integrable flows.

Let us briefly describe the D type KdV flows of [13][2] . Those govern now the evolution of the coefficients of a *pseudo-differential* operator Q, having the form:

$$Q = \bar{Q} D^{-1}, \; deg(Q) = 2n,$$
$$\bar{Q}^* = -Q \tag{4.1}$$
$$(\bar{Q})_- = (-1)^{n+1} u_0 D^{-1} u_0$$

where the involution $*$ leaves functions invariant and satisfies : $D^* = -D$ and $(AB)^* = B^* A^*$, (u_0 is a function). Then one defines two types of flows, "regular flows":

$$\partial_{\lambda_{2k+1}} Q = [(Q^{\frac{2k+1}{2n}})_+, Q] \tag{4.2}$$

and "exceptional flows", the first of which reads:

$$\partial_{\lambda_0} Q = [u_0 D^{-1}, Q] \tag{4.3}$$

We see immediatly that only odd regular flows are allowed in order to preserve the "up to D^{-1}" symmetry of Q of eqn (4.1). Even those seem naively not to make sense,

---

[2] We are going to restrict the discussion to a particular set of flows. In [13] other D type flows were discussed as well. Those don't seem relevant for the particular system we're considering (see [15] for details).

since the r.h.s. of (4.2) doesn't manifestly have the right behaviour under Z_2 (which is (4.1)). Consistency of (4.2) relies on the following property satisfied by any odd power of $L = Q^{\frac{1}{n}}$:

$$(L_+^{2j+1})^* = -D^{-1}L_+^{2j+1}D \tag{4.4}$$

There is an existence theorem for an infinite set of exceptional flows (see [25]) of which (4.3) is the first one. That these flows make sense is a direct consequence of the way they were obtained in [13], resulting from an $SO(2n)$ gauge invariance reduction of a first order differential equation. To check directly consistency of (4.2), (4.3) (namely that they reduce to a finite set of differential equations for the coefficents in Q and nothing new is generated by the flows), one uses a property of odd powers of $L = Q^{1/n}$, restricted to their positive (differential) part (this property follows from (4.4)): if $L_+^{2l+1} = \sum_{k=0}^{2l+1}\beta_k D^k$, then:

$$\sum_{k=0}^{2l+1}(-1)^k\frac{d^k}{dx^k}\beta_k = 0 \tag{4.5}$$

This means in particular that the infinite pseudo-differential tail in Q generated by $(u_0 D^{-1})^2$ has the "regular" variation:

$$\partial_{\lambda_{2k+1}}(u_0 D^{-1})^2 = \{u_0 D^{-1}, \partial_{\lambda_{2k+1}}u_0 D^{-1}\} = \{u_0 D^{-1}, Res[L_+^{2k+1}, u_0 D^{-1}]D^{-1}\} \tag{4.6}$$

which is summarized by one evolution equation:

$$\partial_{\lambda_{2k+1}}u_0 = L_+^{2k+1}(u_0) \tag{4.7}$$

where $R(f)$ denotes the function obtained by action of a differential operator R on a function f. Thus (4.2) reduces to a finite set of compatible evolution equations for the coefficients of Q. The same holds for the exceptional flows, where (4.7) is replaced by:

$$\partial_{\lambda_0}u_0 = (-1)^n Res(Q) \tag{4.8}$$

The next step as in section 2, is to find a string equation invariant under the KdV flows (4.2), (4.3). In analogy to section 2, we will conjecture that this equation has the form:

$$[P, Q] = 1 \tag{4.9}$$

where Q is as in (4.1) and P is a differential operator. We will show later that (4.9) is indeed invariant under the flows. To extract physical content out of this, we again assume that the physical partition function Z is related to $u_{2n-2} \propto \partial_x^2 \log Z$, where u_{2n-2} denotes the first subleading coefficient in Q. The proportionality constant will be fixed later. As opposed to the A case (2.3) where P and Q were differential operators and so was their commutator, leading to a finite number of equations for their coefficients, eqn. (4.9) implies that the commutator of a differential operator P and a pseudo-differential operator Q, which is pseudo-differential itself, has to be equal to 1, i.e. contains an infinite number of differential equations relating the finite number of P and Q coefficients. For (4.9) to make sense, these equations, generated by the presence of a pseudo-differential "tail" in Q, have to be compatible or in other words reduce to a finite set. A necessary condition for (4.9) to make sense is deg $[P, Q] \leq 2n - 2$. This leads [13] to the following form for P:

$$P = (S)_+ = \sum_{j=0}^{\infty} \mu_{2j+1}(Q^{(2j+1)/2n})_+ \tag{4.10}$$

where we have put $\mu_{2j} = 0$ by symmetry. Note that the flows (4.2), (4.3) do not generate any μ_{2j} terms, thus this is consistent. Plugging (4.10) in (4.9) we find that the D^{-n} coefficients ($n \geq 2$) in $[P, Q]$ all vanish if the $n = 2$ one does. To show that, note that the property (4.5) is satisfied by our P operator, thus $[P, (u_0 D^{-1})^2] = \{u_0 D^{-1}, [P, u_0 D^{-1}]\}$ is entirely determined by $[P, u_0 D^{-1}]$, the negative pseudo-differential part of which is simply $P(u_0)D^{-1}$, while its differential part still satisfies (4.5). Thus the coefficient of D^{-2} and lower in (4.9) all vanish if and only if:

$$P(u_0) = 0 \tag{4.11}$$

And (4.9) reduces to a finite set of differential equations relating the coefficients of Q. Note that $P(u_0)$ is a total derivative, due to the property (4.4). Namely one can rewrite $P = -DP^*D^{-1}$, where $Y = -P^*D^{-1}$ is necessarily a differential operator (otherwise $P = DY$ would not be differential), and (4.11) becomes:

$$\frac{d}{dx}Y(u_0) = 0 \tag{4.12}$$

giving rise to an integration constant $\mu_0 = Y(u_0)$.

It is of course of extreme importance to actually check the compatibility between (4.9) and the flows (4.2), (4.3).

For the regular flows, the proof goes exactly as for the A series. Starting from $P = (S)_+$ as in (4.10), one gets n integration constants $\lambda_1 = x, \lambda_3, \cdots, \lambda_{2n-1}$ such that: $(S)_- = \sum_{\alpha=1}^{n-1} \frac{2(n-\alpha)+1}{2n} \lambda_{2(n-\alpha)+1} L^{-2\alpha+1} + (x/2n)D^{-2n+1} + \cdots$. The reason for these jumps of 2 in the powers of L is that the differential equations (4.9) always come in pairs related by the symmetry property of Q (4.1), one being essentially the derivative of the other. We use this expression to rewrite the compatibility equation (2.10) as :

$$\partial_{\lambda_{2k+1}}[P, Q] = \left[[L^{2k+1}, S_-]_+ + \sum_{j=0}^{\infty} \partial_{\lambda_{2k+1}}\mu_{2j+1} L_+^{2j+1}, L^{2n}\right] \tag{4.13}$$

and we find directly zero for $k < n$, whereas for $k > n$ one has to include non zero μ's in (4.10):

$$\mu_{2j+1} = -\frac{2(n+j)+1}{2n}\lambda_{2(n+j)+1} \tag{4.14}$$

which ensure that (4.13) still give zero.

For the the exceptional flow (4.3), the compability check with (4.9) is slightly more subtle. Performing:

$$\begin{aligned}
\partial_{\lambda_0}[P, Q] &= \left[[u_0 D^{-1}, P]_+, Q\right] + \left[P, [u_0 D^{-1}, Q]\right] \\
&= -[Q, P(u_0)D^{-1}]
\end{aligned} \tag{4.15}$$

we get a zero answer, due to the part of the string equation (4.9) generated by the "tail" of Q (4.11). Moreover it is possible to show that the integration constant μ_0 of (4.12) is proportional to the first exceptional KdV flow parameter λ_0 ($\mu_0 = \frac{1}{2}\lambda_0$). This completes the proof of compatibility between the conjectured string equation (4.9) and the D type flows (4.2)-(4.3).

Again, we can consider the most general string equation with P as in (4.10) as some perturbation of the critical ones. Moving in this theory space is a little subtle, due to the lack of symmetry between P and Q. Let us consider the "thermal" perturbation which is expected to interpolate between neighbouring unitary series. We can go from (A_{2n}, D_{n+1}) to (D_{n+1}, A_{2n-2}), by simply considering:

$$[(Q^{\frac{2n+1}{2n}})_+ - \frac{4n-1}{2n}\lambda_{4n-1}(Q^{\frac{2n-1}{2n}})_+, Q] = 1 \qquad (4.16)$$

On the other hand, to go from (D_{n+1}, A_{2n-2}) to (A_{2n-2}, D_n), we first notice that the critical Q operators of both theories denoted resp. $Q(2n)$ and $Q(2n-2)$, share the same $P = Q(2n)_+^{(2n-1)/2n} = Q(2n-2)_+^{(2n-1)/(2n-2)}$. Therefore we can write:

$$[Q(2n) - \alpha_n \lambda_{4n-1} Q(2n-2), P] = 1 \qquad (4.17)$$

where α_n is fixed by the compatibility condition. This shows, like in the A case, that the normalization coefficient between u_{2n-2} in Q and the physical string susceptibility u is determined for all theories in terms of any of them. We just have to identify its value in some simple case. This is done in the following example. We illustrate the D series structure on the simplest example (A_2, D_3), with $deg(P) = 3, deg(Q) = 4$. In the conformal field theory, this is equivalent to the Ising model $(A_2, A_3) = (3, 4)$, so we expect to get the same critical string equations. This was shown to be the case in [15], thus fixing the normalization factor between u_{2n-2} and the string susceptibility u. We get $u_{2n-2} = -2nu$. The order parameter operator content is now: $\phi_1 = Id$, $\phi_3 = \epsilon$ and the exceptional flow leads to $\phi_0 = \sigma$, the $\mathbf{Z}_2$-breaking operator (notice that the $\mathbf{Z}_2$ properties of the model are emphasized here : the "regular" operators are even, and the "exceptional" ones are odd). The D series provides another representation of the critical Ising model, in addition to the A one. The two agree at criticality, but once temperature is taken off criticality they start to differ. In the D case, the interpolating string equation (4.17) yields in the large λ_3 limit the same differential equation as in the pure gravity case, for $u/2$. This factor of 2 is precisely the ratio between the high and low temperature limits of the Ising model free energy [26], suggesting that the A and D spaces represent the high and low temperature phases of the Ising model coupled to gravity.

One can again define a set of operators ϕ_m dual to the couplings λ_m by : $\partial_{\lambda_m}\langle \cdots \rangle = \langle \phi_m \cdots \rangle$. By analogy with the A case, we call order parameters the operators corresponding to the integration constants of (4.9), which include the "odd" operator ϕ_0 associated to the exceptional flow (4.3) and the integration constant λ_0.

To actually solve the string equation of the unitary D series on the sphere, we require again that the order parameter one point functions are zero at criticality. This fixes in particular $\partial_{\lambda_0} u \sim u_0' = 0$, and suppresses the problems due to the tail of Q

(one point functions of $\mathbf{Z}_2$-odd fields must vanish at the critical point to all orders in the genus expansion, thus $u_0 = 0$ to all orders). One can again show that there exists then a unique solution to (4.9) on the sphere, obtained by taking:

$$Q_{sph} = (Q_{2n})_+, P_{sph} = (Q_{2n+1})_+ \tag{4.18}$$

where Q_k's are defined in (3.1) with $u \rightarrow 2u$. Although the solution (4.18) coincides up to this rescaling with the $(2n, 2n + 1)$ A solution of sect.3, it deviates from it once one includes first derivatives of u. The A case has $\mathbf{Z}_2$ covariant operators $P = (P_{sph} - P^*_{sph})/2$ and $Q = (Q_{sph} + Q^*_{sph})/2$, while in the D case this has to be done "up to D^{-1}", so that Q gets a residue : $Q = (Q_{sph}D + DQ^*_{sph})D^{-1}/2$ and $P = [(P_{sph}D - DP^*_{sph})D^{-1}/2]_+$. Nevertheless, as far as the sphere is concerned, all the results of sect.3 apply directly, when properly restricted to the $\mathbf{Z}_2$-even flows λ_{2j+1}. The only new features are the correlations involving the odd field ϕ_0, obtained from (4.8) and $\partial_{\lambda_0} u = u'_0$. These results read for (A_{2n}, D_{n+1}):

$$(2n + 1)u^{2n} = x \quad \rightarrow \quad \gamma_{str} = -\frac{1}{2n} \tag{4.19a}$$

$$\langle \phi_{2l+1}\phi_{2m+1} \rangle = (2l + 1)\delta_{m,l}\, u^{\frac{2l+1}{2n}} \rightarrow \Delta_{2l+1} = \frac{l}{2n} \tag{4.19b}$$

$$\langle \phi_0 \phi_0 \rangle = u^n \quad \rightarrow \quad \Delta_0 = \frac{n-1}{4n} \tag{4.19c}$$

$$\langle \phi_0 \phi_{2l+1}\phi_{2m+1} \rangle = 0 \tag{4.19d}$$

$$\langle \phi_0 \phi_0 \phi_{2l+1} \rangle = (-1)^l(2l + 1)(u^{l+n})' \tag{4.19e}$$

which reproduce all the correct dressed dimensions [20] and the conformal fusion rules when restricted to the diagonal of the Kac table. Similar results are obtained in the (D_{n+1}, A_{2n-2}) case (see [15]). A puzzling aspect of (4.19) is the fact that some three point functions (4.19e) are negative. From Liouville considerations mentioned in sect.3 [8], one would expect positive three point functions. The one loop partition functions can be obtained by going up to second derivatives in u in the above treatment:

$$(A_{2n}, D_{n+1}) \quad Z_1 = \frac{n+1}{24} \tag{4.20a}$$

$$(D_{n+1}, A_{2n-2}) \quad Z_1 = \frac{n^2 - 1}{12(2n - 1)} \tag{4.20b}$$

As for the A series, the results (4.20) were recently verified in Liouville theory [22].

5. Summary

We have presented evidence that the solutions of 2d gravity coupled to minimal models are governed by an ADE classification following from a similar classification of integrable differential equations. It would be very important to understand the results presented above from Liouville theory. There is qualitative agreement; the fusion rules of order parameters on the sphere are precisely those of the underlying CFT; the one loop partition sum agrees with Liouville calculations [22], [23]. It is interesting to note that the structure constants we obtained on the sphere (3.8), (4.19), are much simpler than their CFT counterparts. The correlation functions of higher operators are not well defined in the KdV framework (they are not symmetric in p, q). In both (p, q) and (q, p) pictures, they do not agree with the Liouville predictions [15]. Subtleties arise even in attempting to identify these operators [9], [16].

The topological description of these systems, which is given by twisted $N = 2$ superconformal minimal models coupled to topological gravity [5], [27] gives rise to the *same* D series models as the ones we have encountered here; in particular, the topological correlation functions (at the appropriate multicritical points) are the ones described above.

Since the considerations of this work are algebraic, one can generalize them to supersymmetric systems. A preliminary investigation of those appeared in [28]. A more complete analysis displays additional interesting features and will be described elsewhere.

Acknowledgements

We thank the Aspen Center for Physics for hospitality; one of us (P.D.F.) thanks the Institute for Theoretical Physics, University of California, Santa Barbara. This work was partially supported by NSF grant PHY-85-12793, by a Weizmann fellowship for one of us (D. K.) and by NSF grant PHY-89-04035, supplemented by funds from the National Aeronautics and Space Administration (P.D.F.).

References

[1] D. Gross and A. Migdal, Phys. Rev. Lett. **64** (1990) 127; M. Douglas and S. Shenker, Nucl. Phys. **B335** (1990) 635; E. Brezin and V. Kazakov, Phys. Lett. **B236** (1990) 144 .

[2] T. Banks, M. Douglas, N. Seiberg and S. Shenker, Phys. Lett. **B238** (1990) 279.

[3] D. Gross and A. Migdal, Nucl. Phys. **B340** (1990) 333.

[4] M. Douglas, Phys. Lett. **B238** 176 (1990).

[5] E. Witten, Nucl. Phys. **B340** (1990) 281; J. Distler, Nucl. Phys. **B342** (1990) 523; R. Dijkgraaf and E. Witten, Nucl. Phys. **B342** (1990) 486; E. Verlinde and H. Verlinde, Princeton preprint IASSNS-HEP-90/40, PUPT-1176 (1990).

[6] E. Brezin, C. Itzykson, G. Parisi and J. Zuber, Comm. Math. Phys. **69** (1979) 147; V. Kazakov and A. Migdal, Nucl. Phys. **B311** (1988) 171; I. Kostov, Nucl. Phys. B (Proc. Suppl.) **10A** (1989) 295.

[7] V. Periwal and D. Shevitz, Phys. Rev. Lett. **64** (1990) 1326; R. Myers and V. Perival, ITP preprints 90-50, 90-73 (1990); G. Harris and E. Martinec, Phys. Lett. **B245** (1990) 384; E. Brezin and H. Neuberger, Rutgers preprints RU-90-25, 90-39 (1990).

[8] N. Seiberg, Rutgers preprint RU-90-29 (1990).

[9] J. Polchinsky, Texas preprint UTTG-39-90 (1990).

[10] F. David, Mod. Phys. Lett. **A5** (1990) 1019.

[11] M. Douglas, N. Seiberg and S. Shenker, Phys. Lett. **B244** (1990) 381.

[12] G. Moore, Yale preprint (1990).

[13] V. Drinfel'd and V. Sokolov, Journ. Sov. Math. **30** (1985) 1975.

[14] G. Segal and G. Wilson, Publ. IHES **61** (1985) 1.

[15] P. Di Francesco and D. Kutasov, Nucl. Phys. **B342** (1990) 589.

[16] B. Lian and G. Zuckerman, Yale preprint YCTP-P18-90.

[17] I. Kostov, Nucl. Phys. **B326** (1989) 583.

[18] E. Brezin, M. Douglas, V. Kazakov and S. Shenker, Phys. Lett. **B237** (1990) 43; D. Gross and A. Migdal, Phys. Rev. Lett. **64** (1990) 717; C. Crnkovic, P. Ginsparg and G. Moore, Phys. Lett. **B237** (1990) 196.

[19] C. Bachas and P. Petropoulos, CERN preprint CERN-TH.5714/90; E. Witten, Princeton preprint IASSNS-HEP-90/45.

[20] V. Knizhnik, A. Polyakov and A. Zamolodchikov, Mod. Phys. Lett. **A3** (1988) 819.

[21] F. David, Mod. Phys. Lett. **A3** (1988) 1651; J. Distler and H. Kawai, Nucl. Phys. **B321** (1989) 509.

[22] M. Bershadsky and I. Klebanov, Princeton preprint PUPT-1197 (1990).

[23] A. Gupta, S. Trivedi and M. Wise, Nucl. Phys. **B340** (1990) 475; N. Sakai and Y. Tanii, Tokyo preprint TIT/HEP-160 (1990).

[24] A.Cappelli, C.Itzykson and J.B.Zuber, Comm. Math. Phys **113** (1987)113.

[25] V. Drinfeld and V. Sokolov, Soviet Math. **23** (1981) 461.

[26] V. Kazakov, Phys. Lett. **A119** (1986) 140.

[27] K. Li. , Caltech preprint CALT-68-1662 (1990); R. Dijkgraaf, E. Verlinde and H. Verlinde, Princeton preprint PUPT-1204, IASSNS-HEP-90/71 (1990).

[28] P. Di Francesco, J. Distler and D. Kutasov, Princeton preprint PUPT-1189 (1990).

TOPOLOGICAL STRINGS AND LOOP EQUATIONS

Robbert Dijkgraaf and Herman Verlinde

Joseph Henry Laboratories
Princeton University
Princeton, NJ 08544

1. Introduction

The quantization of generally covariant field theories is a notorious difficult subject. Severe problems arise in the correct treatment of the dynamical metric that is introduced to achieve invariance under general coordinate transformations. Even in two-dimensions, where the issue seems much more accessible and essentially reduces to an understanding of the Liouville dynamics, an exact solution of quantum gravity has until recently remained elusive. However, as is demonstrated by this workshop, the application of an unconventional method, the realization of random surfaces by large N matrix integrals [1]–[3], has led to a remarkable breakthrough [4]–[8]. Soon after the double-scaling limit of the matrix model was solved, an alternative approach to two-dimensional gravity was suggested by Witten along the lines of topological field theory [9].

Topological field theories are exactly solvable systems without local, propagating degrees of freedom [10,11]. Originally, these models seemed to be mainly of mathematical interest, although the possibility was suggested that they are related via some phase transition to the more physical theories. Recently this perhaps rather optimistic point of view has had a surprising confirmation in the context of two dimensional field theory. It has turned out that ordinary quantum gravity and its topological version are related in a very direct way [9,12,13]. With considerable hindsight this might have been expected, since we do not expect propagating modes in two-dimensional gravity. In addition, there is increasing evidence that this correspondence can be extended to gravity coupled to (minimal) matter [14]–[17], and, in fact, it is now believed that ordinary non-critical string theory in $d < 1$ is equivalent to topological string theory in $d < 1$.

In these notes we will try to demonstrate that, due to this relation, the topological formulation provides a very useful starting point for studying the symmetries of string theory in general. In accordance with the general rule that special properties of the world-sheet theory will produce corresponding space-time properties, one expects

Random Surfaces and Quantum Gravity
Edited by O. Alvarez *et al.*, *Plenum Press, New York, 1991*

that the topological symmetry of the two-dimensional theory will be translated into a corresponding target-space symmetry, similar as the 'transmutation' of $N = 2$ supercon-formal invariance into space-time supersymmetry. We will demonstrate that in line with these ideas the topological invariance of the world-sheet theory can be used to derive recursion relations for string amplitudes [13], which can be interpreted as space-time Ward identities or Schwinger-Dyson equations [18,19]. When translated back to the matrix models these relations correspond to loop equations, describing the interactions of macroscopic loops [20,21].

These notes are organized as follows. In section 2 we give a description of the several ingredients of a general topological string theory. In this we follow very closely the usual discussions of the standard string. We then proceed to solve the $d = 0$ model through a set of recursion relations, which relate correlation functions at different genera. In section 4 we exhibit the symmetry structure of the $d = 0$ string and show that the partition function satisfies Virasoro constraints. These constraints are shown to correspond to the loop equations, that naturally arise in matrix models. Steps to generalization of all this to $d < 1$ are given in section 5. Here we also address in more detail the general structure of the matter sector in topological string theory. We show how for the minimal models this can be encoded in a Landau-Ginzburg formulation, and demonstrate the equivalence with the multi-matrix model.

2. Topological String Theory

We will start with a general description of topological string theory, which will be patterned along the familiar lines of ordinary string theory, namely

$$string = matter + Liouville + ghosts. \tag{2.1}$$

Indeed our general philosophy will be that in its formulation and principles topological string theory is completely analogous to the string theories that we are more famil-iar with, in particular the superstring. However, as we will try to demonstrate, it is definitely simpler and more tractable, and in some cases even exactly solvable.

2.1. Topological matter

Following equation (2.1) we will first discuss the matter sector, describing the target space degrees of freedom of the topological string. For this matter theory we will take a two-dimensional *topological conformal field theory*. Such a theory is described by two properties. First, it is conformal invariant, *i.e.*, the trace of the stress-energy tensor vanishes,

$$T_\mu{}^\mu = 0. \tag{2.2}$$

Secondly, it is topological; it has the property that its amplitudes are already generally covariant before coupling to two-dimensional gravity. This topological invariance is due to the presence of a special anti-commuting symmetry generated by a charge Q, satisfying

$$Q^2 = 0. \tag{2.3}$$

This charge plays the role of a BRST-operator, *i.e.*, the spectrum of physical operators is given by the cohomology of Q in the full Hilbert space. The decoupling of the metric from the metric degrees of freedom from the physical amplitudes is achieved because the stress-energy tensor is a Q-exact operator

$$T_{\mu\nu} = \{Q, G_{\mu\nu}\}. \tag{2.4}$$

Note that, although $T_{\mu\nu}$ is not a physical operator, it still makes sense to demand that its trace vanishes.

The simplest example of a theory with the above properties is the topological σ-model describing the motion of the string on a flat target space. On a flat world-sheet, its action is given by

$$S_0 = \int \eta^{ij}(\partial x_i^* \overline{\partial} x_j + \chi_i \overline{\partial} \psi_j + \overline{\chi}_i \partial \overline{\psi}_j). \tag{2.5}$$

Here x_i and x_i^* are world-sheet scalars, and ψ_i and χ_i are their supersymmetric partners of spin zero and one respectively. The (left) Q-transformations are $\delta x_i^* = \epsilon \psi_i$ and $\chi_i = \epsilon \partial x_i$.

In general the conformal invariance allows us to associate chiral currents to the Q-symmetry and the scale invariance. So, we have a spin one current $Q(z)$ and a holomorphic stress-tensor $T(z)$. Since both fields are Q-exact, we can define a $U(1)$ current $J(z)$ and a supercurrent $G(z)$, of the same spin but opposite statistics, by the relations

$$T(z) = \{Q, G(z)\}, \qquad Q(z) = [Q, J(z)]. \tag{2.6}$$

These fields have a unique algebra, which turns out to be isomorphic to a twisted version of the $N = 2$ superconformal algebra, after the identifications $Q = G^+$, $G = G^-$. Here 'twisting' refers to the change

$$T(z) \rightarrow T(z) + \tfrac{1}{2}\partial J(z) \tag{2.7}$$

in the stress-energy tensor [11,15]. One can show that any topological CFT is by twisting related to a $N = 2$ superconformal model and *vice versa*. The central extension of the twisted Virasoro algebra vanishes, but the $U(1)$ current of the twisted theory is coupled to a background charge equal to $d = \hat{c}$, where $\hat{c} = c/3$ is the $N = 2$ central charge. This correspondence with $N = 2$ theory proves to be very useful in the study of topological string theory in general backgrounds.

As we discussed, the primary physical operators in a topological CFT correspond to the cohomology classes of Q. That is, they are defined via the relations

$$Q|\phi_i\rangle = 0, \qquad |\phi_i\rangle \equiv |\phi_i\rangle + Q|\Lambda\rangle, \qquad (2.8)$$

and the condition that $|\phi_i\rangle$ is annihilated by the positive modes of the super-conformal generators. In fact, one can choose representatives that in addition satisfy

$$G_0|\phi_i\rangle = L_0|\phi_i\rangle = 0. \qquad (2.9)$$

In the $N = 2$ terminology, these operators are called chiral primary fields [22]. In general there are only a finite number of such fields, and their $U(1)$ charges, defined by $J_0|\phi_i\rangle = q_i|\phi_i\rangle$, are bounded by

$$0 \leq q_i \leq d. \qquad (2.10)$$

To each chiral primary, we can associate a (chiral $N = 2$) super-field $\Phi(z, \overline{z}, \theta, \overline{\theta})$, where the anti-commuting coordinates θ and $\overline{\theta}$ transform as dz and $d\overline{z}$. So in the expansion

$$\Phi = \phi^{(0)} + \theta\phi^{(1)} + \overline{\theta}\overline{\phi}^{(1)} + \theta\overline{\theta}\phi^{(2)} \qquad (2.11)$$

the component $\phi^{(0)}$ transforms a scalar, whereas $(\phi^{(1)}, \overline{\phi}^{(1)})$ and $\phi^{(2)}$ represent a 1-form and 2-form, respectively. The latter two components can be obtained from $\phi^{(0)}$ by acting with G_{-1} and $\overline{G}_{-1}$.

The local operators have conformal dimensions $(h, \overline{h}) = (0, 0)$ and their operator products are non-singular. Thus we can define the operator algebra, or chiral ring, by

$$\phi_i(z, \overline{z})\, \phi_j(w, \overline{w}) = \sum_k c_{ij}{}^k \phi_k(w, \overline{w}). \qquad (2.12)$$

As we will see in detail in section 5, this chiral ring plays an important role in the solution of the string theories with $d < 1$.

2.2. Coupling to topological gravity

To describe topological strings, we have to couple our matter system to gravity. Let us first discuss this for the case of a flat target space, so that the matter theory is described by (2.5). We introduce a Q-symmetric extension of ordinary gravity, known as topological gravity [23,9,13], consisting of the usual two-dimensional metric $g_{\alpha\beta}$ plus an anti-commuting Q-superpartner $\psi_{\alpha\beta} = \{Q, g_{\alpha\beta}\}$. With the help of these fields, we can write down the covariantized form of the sigma-model action S_0

$$S = \int \sqrt{g}\, g^{\alpha\beta}\eta^{ij}(\partial_\alpha x_i^* \partial_\beta x_j + \chi_{i,\alpha}\partial_\beta\psi_j + \hat{\psi}_{\alpha\gamma}\chi_i^\gamma\partial_\beta x_j^*), \qquad (2.13)$$

where $\hat{\psi}_{\alpha\beta} = \psi_{\alpha\beta} - \frac{1}{2}g_{\alpha\beta}\psi_\gamma^\gamma$ is the traceless part of the spin 2 gravitino $\psi_{\alpha\beta}$. The above action is invariant under diffeomorphisms and local Weyl rescalings, as well as under the Q-superpartners of these transformations.

Generalizing Polyakov's prescription in ordinary string theory, we may now define scattering amplitudes in topological string theory as the functional integral over the 'matter fields' (x_i, χ_i, ψ_i) and over the topological metric $(g_{\alpha\beta}, \psi_{\alpha\beta})$

$$\int [dg_{\alpha\beta}d\psi_{ab}][dx^i d\psi^i]e^{-S[g_{\alpha\beta},\psi_{\alpha\beta},x^i,\psi^i]}\prod_i V_i \tag{2.14}$$

Here the V_i are the physical, *i.e.* gauge-invariant, vertex operators in the model. In order to evaluate the string amplitude (2.14) we fix the invariance under the group of super-diffeomorphism by imposing the conformal gauge

$$g_{\alpha\beta} = e^\varphi \delta_{\alpha\beta},$$
$$\psi_{\alpha\beta} = \psi e^\varphi \delta_{\alpha\beta}. \tag{2.15}$$

In addition, because Q-symmetry ensures that the path-integral measure does not have a conformal anomaly, we can also gauge fix the super-Weyl transformations, $g_{\alpha\beta} \to \lambda g_{\alpha\beta}$ and $\psi_{\alpha\beta} \to \psi_{\alpha\beta} + \hat{\lambda}g_{\alpha\beta}$, by imposing the condition that the curvature of $g_{\alpha\beta}$ is equal to some *given* background curvature $\hat{R}$

$$\bar{\partial}\partial\varphi = \hat{R},$$
$$\bar{\partial}\partial\psi = 0. \tag{2.16}$$

Here the second gauge condition is the Q-variation of the first. The resulting gauged fixed action of topological string theory consists of three pieces

$$S = S_{matter} + S_{Liouville} + S_{ghost}, \tag{2.17}$$

where S_{matter} is given in (2.5), and

$$S_{Liouville} = \int(\bar{\partial}\pi\partial\varphi + \hat{R}\pi + \bar{\partial}\chi\partial\psi), \tag{2.18}$$

$$S_{ghost} = \int(b\bar{\partial}c + \beta\bar{\partial}\gamma) + \text{c.c.} \tag{2.19}$$

Here π and χ are Lagrange-multipliers imposing the gauge condition (2.16) and have spin 0. The ghost-fields b and c are the usual anti-commuting ghosts of associated with the conformal gauge, and β and γ are commuting Q-partners of the same spin.

2.3. Physical vertex operators

To obtain the spectrum of physical operators of the topological string, we have to consider the BRST cohomology. The BRST-generator Q_{brst} is given by the sum of the global Q-supersymmetry charges of the separate models plus the usual BRST-charge constructed out of the ghosts and super-conformal generators T and G

$$\begin{aligned}
Q_{brst} &= Q_S + Q_V, \\
Q_S &= Q_m + \oint (b\gamma + \psi\partial\pi), \\
Q_V &= \oint \left[c(T_{m+L} + \tfrac{1}{2}T_{gh}) + \gamma(G_{m+L} + \tfrac{1}{2}G_{gh}) \right].
\end{aligned} \qquad (2.20)$$

Q_{brst} is nilpotent regardless of the dimension d of the topological matter system. So, unlike other string theories, topological string theory does not have a critical dimension, at least in the usual sense.

The physical vertex operators in topological string theory can be written in the general form

$$\sigma_n(\phi_i) = \sigma_n \cdot \phi_i, \qquad (2.21)$$

where both σ_n and ϕ_i are superfields of the form (2.11). The operators ϕ_i are the topological 'primary' fields, or chiral primaries of the matter sector. The other fields σ_n exist for all $n \geq 0$ and are constructed out of the Liouville and ghost fields, and are described below. One may call the operators $\sigma_n(\phi_i)$ with $n \geq 1$ the 'descendants' of the primaries ϕ_i. These gravitational physical operators σ_n are given by

$$\sigma_n = \mathcal{R}^n. \qquad (2.22)$$

Here $\mathcal{R}$ is the superfield version of the curvature form on the Riemann surface

$$\mathcal{R} = \gamma_0 + \tfrac{1}{2}(\theta\partial\psi - \overline{\theta}\overline{\partial}\psi) + \theta\overline{\theta}\partial\overline{\partial}\varphi, \qquad (2.23)$$

where γ_0 is the commuting local Lorentz ghost, which can be expressed in terms of the ghost and Liouville fields [13]; we do not need its explicit form here. Note that the superfield curvature $\mathcal{R}$ is a 0-form, and its top-component is indeed the usual curvature 2-form $R^{(2)} = \partial\overline{\partial}\varphi$. Geometrically, $\mathcal{R}(z,\theta)$ represents a 2-form on the moduli space of the punctured Riemann surface on which we consider the correlation function [9].

3. Virasoro Recursion Relations in Topological Gravity

We will now describe the solution of the simplest topological string theory, namely pure topological gravity. So our task is to calculate all correlation functions

$$\langle \sigma_{n_1} \ldots \sigma_{n_s} \rangle_g \qquad (3.1)$$

on a genus g surface Σ. As in ordinary string theory, these amplitudes are given by the integral over the moduli of Σ as well as over the positions of the operators, or in other words, over the moduli space $\mathcal{M}_{g,s}$ of genus g surfaces with s punctures. Since each operator σ_{n_i} represents a $2n_i$-form on this space and $\dim \mathcal{M}_{g,s} = 6g - 6 + 2s$, the integrand is a volume form only if

$$\sum_i (n_i - 1) = 3g - 3. \tag{3.2}$$

Note that the partition function in topological gravity vanishes, except at genus one.

3.1. The puncture- and dilaton equation

The two simplest physical operators in topological gravity are the puncture operator $\mathcal{P} = \sigma_0$ and the dilaton σ_1

$$\mathcal{P} = 1,$$
$$\sigma_1 = \mathcal{R}.$$

Both these operators have very simple properties, and can be eliminated from correlation functions via two useful recursion relations.

Naively, one would expect that the puncture operator $\mathcal{P}$ cannot have any non-vanishing amplitudes, since these involve an integral over the anti-commuting coordinates and $\int d^2\theta \, 1 = 0$. However, a closer examination reveals that when the puncture $\mathcal{P}$ approaches one of the other operators in the correlator there is a contact term contribution. Specifically, one finds that [14,13]

$$\int_{D_\epsilon} \mathcal{P} \, |\sigma_n\rangle = |\sigma_{n-1}\rangle, \tag{3.3}$$

where D_ϵ denotes the infinitesimal neighbourhood of the operator insertion σ_n. This contact term leads to the so-called 'puncture equation' [14]

$$\langle \mathcal{P} \prod_{i=1}^{s} \sigma_{n_i} \rangle_g = \sum_{j=1}^{s} \langle \sigma_{n_j - 1} \prod_{i \neq j} \sigma_{n_i} \rangle_g \tag{3.4}$$

A second useful equation is the 'dilaton equation,' which simply states that the integral of the dilaton field $\sigma_1 = \mathcal{R}$ yields the Euler number of the punctured surface

$$\langle \sigma_1 \prod_{i=1}^{s} \sigma_{n_i} \rangle_g = (2g - 2 + s) \langle \prod_{i=1}^{s} \sigma_{n_i} \rangle_g \tag{3.5}$$

In this equation the contribution $2g - 2$ comes from the curvature of the genus g surface, and the additional term s comes from a contact term interaction, similar as that of the puncture

$$\int_{D_\epsilon} \sigma_1 |\sigma_n\rangle = |\sigma_n\rangle. \tag{3.6}$$

Note that the dilaton-equation can be interpreted as the Ward identity of translations $\pi \to \pi + \xi$ in the π-field.

3.2. The contact term algebra

We would now like to generalize these observation to derive similar recursion relations for the other physical fields. The idea is to arrange things in such a way that essentially all contributions to the integral over the position of the operators is contained in the contact interactions. To this end we will make use of the fact that in topological gravity we have a freedom of choosing the distribution of curvature on the Riemann surface in any way we want, since Weyl-rescalings are gauge transformations. One particular choice we can make is to locate all curvature at the position of the operators. More precisely, in view of the ghost-number selection rule (3.2), one finds that the only consistent way to do this to put $\frac{2}{3}(n-1)$ units of curvature at each operator σ_n. This prescription amounts to the replacement

$$\sigma_n \to \tilde{\sigma}_n = e^{\frac{2}{3}(n-1)\pi}\sigma_n. \tag{3.7}$$

Although this redefinition looks rather innocent, it has rather drastic consequences. Namely, the new operators $\tilde{\sigma}_n$ now have a mixed function, since, when integrated, they *measure* curvature, as seen from (2.22), whereas they also *create* curvature. As was shown in [13], this has the consequence that their mutual contact terms become *asymmetric*. More precisely, one finds that

$$\int_{D_\epsilon} \tilde{\sigma}_n|\tilde{\sigma}_m\rangle - \int_{D_\epsilon} \tilde{\sigma}_m|\tilde{\sigma}_n\rangle = \tfrac{2}{3}(n-m)|\tilde{\sigma}_{n+m-1}\rangle. \tag{3.8}$$

(A heuristic 'derivation' of this and the following results will be given in the next subsection.) This shows that the evaluation of contact terms is non-commutative, and that, in fact, their algebra is isomorphic to the Virasoro algebra. The separate contact terms are found to be

$$\int_{D_\epsilon} \tilde{\sigma}_n|\tilde{\sigma}_m\rangle = \tfrac{2}{3}(n+\tfrac{1}{2})|\tilde{\sigma}_{n+m-1}\rangle. \tag{3.9}$$

It turns out that this non-commutative contact term algebra can be used to derive new recursion relations, which determine all correlation functions in topological gravity. These recursion relations are generalizations of the puncture and the dilaton equation, and read as follows

$$\langle \tilde{\sigma}_n \prod_{m\in S} \tilde{\sigma}_m \rangle_g = \sum_{j\in S} \tfrac{2}{3}(j+\tfrac{1}{2})\langle \tilde{\sigma}_{j+n-1} \prod_{m\neq j} \tilde{\sigma}_m \rangle_g + \tfrac{2}{9}\sum_{j=1}^{n-1} \langle \tilde{\sigma}_{j-1}\tilde{\sigma}_{n-j-1} \prod_{m\in S} \tilde{\sigma}_m \rangle_{g-1}$$

$$+\tfrac{2}{9}\sum_{j=1}^{n-1} \sum_{\substack{X\cup Y=S \\ g_1+g_2=g}} \langle \tilde{\sigma}_{j-1} \prod_{m\in X} \tilde{\sigma}_m \rangle_{g_1} \langle \tilde{\sigma}_{n-j} \prod_{m\in Y} \tilde{\sigma}_m \rangle_{g_2} \tag{3.10}$$

This relation expresses that, when one integrates the operator $\tilde{\sigma}_n$ over the two-dimensional surface, the only non-vanishing contributions come from contact interactions. The first term on the right-hand side are the contributions at the other operators, the second and third terms are the so-called factorization terms, which account for the contact interactions of σ_n with the possible nodes in the surface. In the connected term this node is formed by pinching one of the handles of the surface and the disconnected term is the contribution from a node dividing the surface in two. The form of these terms is determined by ghost number conservation and consistency with the contact term algebra.

3.3. An interpretation of the Virasoro algebra

Clearly, this solution of topological gravity raises some immediate questions. In particular, where does this mysterious Virasoro algebra come from? In the following we will make a speculative attempt to provide an answer to this question, by giving a very simple and explicit realization of this Virasoro-contact term algebra in terms of the topological Liouville fields. In this description, the physical operators will be different than described previously, but we believe that the two representations are equivalent and can be related via an appropriate picture changing procedure, although we have not proved this. Therefore, the following should be read as a possible *effective* description of topological gravity, which will have the virtue of being simple and providing an interpretation of the Virasoro symmetry.

Consider the topological Liouville theory is described by the action (2.18), where we have put the background curvature $\hat{R} = 0$. The action is invariant under the Q-supersymmetry transformations $\delta\varphi = \epsilon\,\psi$ and $\delta\chi = \epsilon\,\pi$. Let us now define the following alternative set of physical operators for $n \neq 1$

$$\sigma_n = \tfrac{2}{3}(n-1)e^{\frac{2}{3}(n-1)\pi}(\partial\pi\overline{\partial}\varphi + \partial\chi\overline{\partial}\psi). \tag{3.11}$$

while the dilaton $\sigma_1 = \partial\overline{\partial}\varphi$ remains the same as before. It is straightforward to check that the integral of the above operators is indeed BRST-invariant, so that they indeed correspond to physical degrees of freedom of the theory. In fact, the operators (3.11) are equal to the variation of the action S under the infinitesimal transformation

$$\pi \;\rightarrow\; \pi + \xi_n e^{\frac{2}{3}(n-1)\pi},$$

$$\partial\chi \;\rightarrow\; \partial\chi + \xi_n \tfrac{2}{3}(n-1)e^{\frac{2}{3}(n-1)\pi}\partial\chi. \tag{3.12}$$

The σ_n can therefore be interpreted as the generators of the group of reparametrizations $\pi \rightarrow \tilde{\pi}(\pi)$ under which $\partial\chi$ transforms as a 1-form. In the above description it is immediately clear that there are only contact interactions: the σ_n are all proportional to the equation of motion, and therefore vanish except at the positions of other operators. The contact terms at the other operators are in fact equal to the variation of these operators under the 'reparametrization' (3.12)

$$\int_{D_\epsilon} \xi_n \sigma_n |\mathcal{O}\rangle = \delta_{\xi_n}|\mathcal{O}\rangle. \tag{3.13}$$

This exactly reproduces the result (3.9), if one takes into account that, due to the fact that π is coupled to a background charge $Q = 2$, in the operator-state correspondence one has[*]

$$e^{q\pi} \leftrightarrow e^{(q+1)\pi_0}|0\rangle. \tag{3.14}$$

In a similar way, a careful consideration of the interaction with the nodes of the 2-dimensional surface can also explain the factorization terms in the recursion formula (3.10).

One possible objection against this simple naive picture of topological gravity could be that there is no apparent reason for restricting the set of physical operators (3.11) to $n \geq 0$. However, one has to remember that the operators $e^{q\pi}$ create $-q$ units of curvature in the surface and in a local point on the surface one can concentrate at most 1 unit of positive curvature. This implies that the operators with $q \leq -1$ do not correspond to local operators, and explains why the Virasoro algebra is truncated from below.[†]

In complete analogy with ordinary strings, one could try to interpret the topological Liouville fields π and φ as target space coordinates of the topological string, thus providing it with a two-dimensional target-space. This interpretation might be a possible starting point for constructing a second quantized string field theory, in the spirit of [25], of which the recursion relations (3.10) are the Ward identities or Schwinger-Dyson equations. We will return to this line of thought in the next section[‡].

4. Multi-Critical Points and Loop Equations

4.1. Topological strings in non-trivial backgrounds

We have seen that the spectrum of topological gravity contains an infinite set of weight $(1,1)$ vertex operators $\sigma_n(z, \overline{z})$. In principle, these operators can be added to the world-sheet Lagrangian with coupling constants t_n

$$S \rightarrow S - \sum_n t_n \int \sigma_n. \tag{4.1}$$

As in ordinary string theory we will have an interpretation of a topological string moving in a non-trivial background parametrized by the couplings t_n. Here the space of backgrounds is infinite dimensional. We can define, at least formally, the string free energy $F[t]$ through its perturbative expansion

$$F[t] = \sum_g \lambda^{2g-2} F_g[t], \tag{4.2}$$

---

[*]This immediately follows from the definition of the operator-state mapping using the path-integral on the disk. The shift $+1$ then comes from the Euler number of the disk [24].

[†]This above argument is the in fact the same as (one of several) arguments showing that in ordinary Liouville theory the vertex operators $e^{q\phi}$ only exist for $q \leq -Q/2$, [24].

[‡]See also Tom Banks' contribution to this proceedings [26].

with λ the string coupling constant and with genus g contributing

$$F_g[t] = \left\langle \exp\left(\sum_n t_n \int \sigma_n\right)\right\rangle_g \qquad (4.3)$$

The full string partition function is $Z[t] = \exp F[t]$. Connected string correlation functions in a specific background $t = \tilde{t}$ are defined as

$$\langle \sigma_{n_1} \ldots \sigma_{n_s}\rangle = \left.\frac{\partial^s \log Z}{\partial t_{n_1} \cdots \partial t_{n_s}}\right|_{t=\tilde{t}} \qquad (4.4)$$

where contributions of fixed genus are read off in the expansion in λ.

Of particular interest are the couplings t_0 to the puncture operator — the cosmological constant, that we will also denote as x in the subsequent — and the coupling t_1 to the dilaton operator σ_1. The value of this latter coupling is rather arbitrary, since we can absorb it in a redefinition of λ and the normalization of the operators. It will turn out to be very convenient to make a shift $t_1 \rightarrow t_1 + 1$, so that the value $t_1 = -1$ corresponds to the trivial background that we considered up to now. With this choice the origin in coupling space will represent a singular point, since the dilaton equation (3.5) tells us that correlators scale with some inverse power of t_1 and we cannot set $t_1 = 0$. We note that the topological model allows us to treat the cosmological constant perturbatively, in contrast with ordinary non-critical string theory.

One might worry about the convergence properties of the all these formal expressions, even in the limited context of fixed genus. In this respect it is enlightening to consider the special case of genus zero. Here it can be shown that for *any* background we have $\langle \sigma_n \mathcal{P}\rangle_0 = u^{n+1}/(n+1)$ with $u \equiv \partial_x^2 F_0 = \langle \mathcal{P}\mathcal{P}\rangle_0$ [14]. Together with the puncture equation (3.4) this gives us a topological derivation of the so-called genus zero string equation [7,8]

$$\sum_{n=0}^{\infty} t_n u^n = 0. \qquad (4.5)$$

This algebraic relation certainly tells us for which couplings the genus zero partition function is well-defined. Similar analysis can be given for higher genus, and one finds no new obstructions. (We cannot say much from the perspective of topological string theory about the correct *non*-perturbative definition of the partition function.)

The free energy $F[t]$ satisfies two interesting differential equations as a result of ghost charge conservation (3.2) and the dilaton equation (3.5). In terms of the couplings t_n these relations read respectively

$$\sum_n (n-1)t_n \frac{\partial F_g}{\partial t_n} = (3g-3),$$

$$\sum_n t_n \frac{\partial F_g}{\partial t_n} = (2g-2). \qquad (4.6)$$

By taking a suitable linear combination of these two equations, such that the coefficient in front of t_k vanishes, we immediately observe that the axes $t_n = \delta_{n,k}t$ in coupling space are special. Here we meet critical behavior, *i.e.*, correlation functions scale with respect to the cosmological constant x as

$$\langle \sigma_{n_1} \cdots \sigma_{n_s} \rangle_g \sim x^{\sum \delta_n - s - (2-\gamma)(g-1)}, \qquad \gamma = -\frac{1}{k}, \; \delta_n = \frac{n}{k}. \tag{4.7}$$

In these multi-critical models the operator σ_k plays the role of the dilaton field; it has scaling dimension one and is a 'marginal' operator in the sense that adding it to the action does not influence the scaling behaviour. The results for these scaling dimensions are in agreement with the one matrix model and the KPZ result [27] for $(2, 2k-1)$ minimal CFT's on random surfaces. We will see in a moment that we can identify the one matrix model *exactly* with the critical points of the topological string. So, in particular we have a very simple correspondence with topological gravity (at $k = 1$) and standard quantum gravity (at $k = 2$). One obtains the latter by simply adding $\sigma_2 - \sigma_1$ to the action! In fact, it is possible to continuously interpolate between the two. Another interesting point is the Lee-Yang edge singularity at $k = 3$.

4.2. Schwinger-Dyson equations

To make further contact with the matrix model solutions, we recall that in the last section we showed that $d = 0$ topological string theory is complete solvable. At the 'topological point' $k = 1$ all correlation functions can be determined through the recursion relations (3.10). Quite surprisingly, these relations can be expressed in terms of *linear* quadratic differential equations for the partition function [18]

$$L_n Z[t] = 0, \qquad n \geq -1, \tag{4.8}$$

where the generators L_n are defined by

$$
\begin{aligned}
L_{-1} &= \sum_{m=1}^{\infty} (m + \tfrac{1}{2}) t_m \frac{\partial}{\partial t_{m-1}} + \tfrac{1}{8}\lambda^{-2} t_0^2, \\
L_0 &= \sum_{m=0}^{\infty} (m + \tfrac{1}{2}) t_m \frac{\partial}{\partial t_m} + \tfrac{1}{16}, \\
L_n &= \sum_{m=0}^{\infty} (m + \tfrac{1}{2}) t_m \frac{\partial}{\partial t_{m+n}} + \tfrac{1}{2}\lambda^2 \sum_{m=1}^{n} \frac{\partial^2}{\partial t_{m-1} \partial t_{n-m}}.
\end{aligned}
\tag{4.9}
$$

The generators satisfy (part of) the Virasoro algebra

$$[L_n, L_m] = (n - m) L_{n+m}. \tag{4.10}$$

We already met this algebra in (3.8). Of course, we cannot have constraints $L_n Z = 0$ with $n < -1$, since there would be a central extension (with $c = 1$).

Before we discuss the equations (4.8) in more detail, let us first remark that an identical set of constraints can be derived from the matrix model solutions [18,19].

This proves the full equivalence of $d = 0$ topological string theory and the one-matrix model. In the matrix model these equations actually correspond to the Schwinger-Dyson equations, so we should interpret our solution of topological string theory as the *string equations of motion*. This is of course very gratifying. Since the recursion relations (3.10) and similarly the equations (4.8) relate contributions of different genus, we are more or less forced to consider all genera at once. So we are really dealing with a string theory, not a two-dimensional field theory.

As a further application we can expand these Virasoro constraint around one of the critical points in coupling space, $t_n = \delta_{n,k} t$, instead of the (singular) origin of coupling space, as we did in (4.8). In this way we get again a set of recursion relations for the physical operators σ_n. For one-point functions these relations look like

$$\langle \sigma_{n+k} \rangle = \frac{L_n Z}{Z}, \qquad n \geq -1. \tag{4.11}$$

and similar expressions for multi-point functions. For $k = 1$ this allows us to solve for the correlation functions of all operators, as we have seen in the previous section. However, if $k > 1$, the n-point functions of the fields $\sigma_0, \ldots, \sigma_{k-1}$ cannot be found in this way: they do not appear on the left-hand side of our equation. These operators correspond exactly with the primary fields in the $(2, 2k-1)$ CFT, which the multi-critical points are supposed to describe. In this spirit we can call the operators σ_n with $n \geq k$ *redundant*, since they can be eliminated with the aid of the string equations of motion.

The Virasoro constraints (4.8) become more recognizable if we introduce a two-dimensional chiral scalar field $\Phi(z)$, with a $\mathbf{Z}_2$-twist $\Phi(e^{2\pi i} z) = -\Phi(z)$. If we regard t_n and $\partial/\partial t_n$ as the coherent state realizations of the creation and annihilation operators in the mode expansion of the current $\partial_z \Phi(z)$,

$$\partial \Phi(z) = \sum_{n=0}^{\infty} (n + \tfrac{1}{2}) z^{n-\frac{1}{2}} \lambda^{-1} t_n + \sum_{n=0}^{\infty} z^{-n-\frac{1}{2}} \lambda \frac{\partial}{\partial t_n}, \tag{4.12}$$

the above generators become the standard $c = 1$ Virasoro generators, defined as the mode expansion of the stress-energy tensor

$$T(z) = -\tfrac{1}{2} : \partial \Phi \partial \Phi : (z) + \tfrac{1}{16} z^{-2}. \tag{4.13}$$

When expanded around $t_n = \tilde{t}_n$, the Virasoro constraints (4.8) can now be interpreted to determine a state $|\Omega_{\tilde{t}}\rangle$ in the Fock space of the twisted boson [18], see also [28]. Since this space contains no translation invariant states, it is clear that this interpretation breaks down at $\tilde{t}_n = 0$.

It is of course tempting to identify the complex coordinate z with the exponential $e^{\frac{2}{3}\pi}$ of the dual Liouville field, and the chiral boson $\Phi(z)$ with the string field $\Phi(\pi)$. Following the line of reasoning of section 3.3, one would arrive at a 'space-time action' of the form

$$S = \int d\pi d\varphi (\partial_\pi \Phi \partial_\varphi \Phi + \Lambda \partial_\varphi \Phi + V(\Phi) + J\Phi). \tag{4.14}$$

Here the dynamical fields are Φ and Λ, $V(\Phi)$ some interaction potential, and J the string source. In the standard construction of string field theory, where the quadratic piece of the action is given by $\langle\Psi|Q_{brst}|\Psi\rangle$, the first term in the above action comes from $\langle\Psi|Q_V|\Psi\rangle$ and the second one from $\langle\Psi|Q_S|\Psi\rangle$. This second term imposes the condition that the physical fields should not depend on the (zero modes of) the Liouville coordinate φ. The potential $V(\Phi)$ is in principle completely determined by the Virasoro symmetry. The dilaton equation implies that $V(\Phi)$ is homogeneous of degree 2, and this property, together with the fact that all terms in the action except $J\Phi$ are translationally invariant, is equivalent to the ghost selection rule.

4.3. Loop equations

The above considerations lead us to consider the formal expressions

$$w(z) = \sum_{n=0}^{\infty} \sigma_n z^{-n-\frac{3}{2}}, \tag{4.15}$$

together with their sources

$$J(z) = \sum_{n=0}^{\infty} t_n z^{n+\frac{1}{2}}. \tag{4.16}$$

In terms of these fields we have $\partial_z \Phi(z) = w(z) + \partial_z J(z)$. The Schwinger-Dyson equations now read

$$\left[J'(z)\langle w(z)\rangle\right]_< + \tfrac{1}{2}\lambda^2\langle w^2(z)\rangle + \tfrac{1}{2}\langle w(z)\rangle^2 + \frac{\lambda^2}{16z^2} + \frac{t_0^2}{8z} = 0, \tag{4.17}$$

where the subscript $<$ indicates a truncation to negative powers of z. This mysterious equation becomes much more familiar if we write $w(z)$ as the Laplace transforms of the *loop operator* $w(\ell)$

$$w(\ell) = \sum_{n=0}^{\infty} \frac{\ell^{n+\frac{1}{2}}}{\Gamma(n+\frac{3}{2})}\sigma_n. \tag{4.18}$$

Then the above equation translates into a *loop equation* [20] for the operator $w(\ell)$. This becomes particularly transparent if we expand around the multicritical point $J(z) = z^{k+\frac{1}{2}}$. We obtain [18]

$$\frac{1}{2}\left(\frac{\partial}{\partial\ell}\right)^{k-\frac{1}{2}}\langle w(\ell)\rangle = \frac{\lambda^2\ell}{16} + \frac{t_0^2}{8} + \int_0^\infty d\ell'\,\ell'J(\ell')\langle w(\ell+\ell')\rangle$$
$$\frac{1}{2}\int_0^\ell d\ell'\left(\lambda^2\langle w(\ell')w(\ell-\ell')\rangle + \langle w(\ell')\rangle\langle w(\ell-\ell')\rangle\right). \tag{4.19}$$

This equation has the usual interpretation of a loop equation [20]: a variation of the boundary (the left-hand side) either causes the loop to join another loop, or makes the loop split into two parts (the right-hand side). The two constant contributions are a result of extra symmetries at genus zero and one. A similar loop equation has been derived in [21] from the matrix point of view.

66

The correct incorporation of loops in topological gravity is still much of a mystery. It is clear from considerations of the matrix models that their asymptotic expansion should be of the above kind [7]. Even at $k = 1$ loop operators receive contributions at each genus, being an expansion in all σ_n's, and so are naturally defined in a string theory where one is forced to sum over all genera. For instance, the recursion relations tell us that in topological gravity

$$
\left\langle \mathcal{P}\,\mathcal{P}\,w(\ell) \right\rangle = \sqrt{\frac{\ell}{\pi}}\lambda^{-2} \exp\left(\frac{1}{24}\lambda^2 \ell^3 + \ell x \right),
$$

$$
\left\langle \mathcal{P}\,w(\ell_1)w(\ell_2) \right\rangle = \sqrt{\frac{\ell_1 \ell_2}{\pi^2}}\lambda^{-2} \exp\left(\frac{1}{24}\lambda^2(\ell_1^3 + \ell_2^3) + x(\ell_1 \ell_2) \right) \tag{4.20}
$$

$$
\times \sum_{n=0}^{\infty} \frac{n!\,2^{-n}}{(2n+1)!}\lambda^{2n}\left[\ell_1 \ell_2(\ell_1 + \ell_2) \right]^n.
$$

We further note that, since the operators σ_n correspond to the Chern classes c_1^n on the moduli space of the punctured surface [9], we formally have $w(\ell) = \sqrt{\frac{\ell}{\pi}}(1 - c_1 \ell)^{-1}$.

5. Topological Strings in $d < 1$

Up to now we have been mainly be concerned with pure topological gravity, or if one wishes $d = 0$ topological string theory. The next logical step is to consider the coupling to a nontrivial matter sector. As we have indicated in section 2, any twisted $N = 2$ model, with central charge c, can be used for this purpose. It will give rise to a topological string theory in $d = \frac{c}{3}$ 'dimensions.' In case the matter represents a twisted $N = 2$ non-linear sigma model, d corresponds to the complex dimension of the target space. Particularly interesting is the case $d < 1$. It has been observed [14] that the multi-matrix models allow for an interpretation as topological string theories with some unidentified matter sector. Recently the issue of the 'missing matter' was elegantly addressed by Li, who proposed to take twisted $N = 2$ minimal models [16]. Subsequently, substantial evidence for this interpretation of the matrix chain models was presented in [17], where the KdV structure has been unearthed. This gives additional evidence of the close relation of topological and ordinary string theory, at least in low dimension.

In general the spectrum of physical states will consist of matter operators ϕ_i, with $U(1)$ charges $0 \leq q_i \leq d$, together with their gravitational descendants $\sigma_n(\phi_i)$. The pure matter operators, or 'primary' fields, are the chiral primary fields of the $N = 2$ model. We would like to calculate arbitrary correlation functions

$$
\langle \sigma_{n_1}(\phi_{i_1}) \cdots \sigma_{n_s}(\phi_{i_s}) \rangle_g \tag{5.1}
$$

on a genus g surface. Because of the $U(1)$ charge selection rule these correlation functions will only be non-zero if

$$
\sum (n + q_i - 1) = (d - 3)(1 - g). \tag{5.2}
$$

Before we proceed, let us remark that this charge conservation rule together with the dilaton equation (3.5) can again be used to look for multi-critical behaviour when we add the operators $\sigma_n(\phi_i)$ with coefficients $t_{n,i}$ to the action. If the new dilaton is given by $\sigma_m(\phi_j)$ with $m + q_j = \delta$, we find the equation

$$x\frac{\partial F_g}{\partial x} = \sum_{n,i}\left(\frac{n + q_i}{\delta} - 1\right) t_{n,i}\frac{\partial F_g}{\partial t_{n,i}} + \left(\frac{d - 1}{\delta} - 2\right)(g - 1). \qquad (5.3)$$

From this equation we read off that the fields $\sigma_n(\phi_i)$ have scaling dimensions $(n + q_i)/\delta$, and that $\gamma_{string} = (d - 1)/\delta$. We have seen that putting the coupling constant $t_{2,0}$ to the field σ_2 to a non-zero value 'transmuted' topological gravity to standard two-dimensional quantum gravity. So this critical point, with $\delta = 2$, might quite generally be of special interest. Indeed, we will see that in the case of $d < 1$ minimal models this critical point corresponds to the coupling of *unitary* minimal CFT's to standard gravity. At these points we have $\gamma_{string} = \frac{1}{2}(d - 1)$.

5.1. Correlation functions of primary fields

An interesting subset of correlators are the n-point function of the primary fields ϕ_i. The interest in these quantities is two-fold. First, as was shown by Witten [9], the knowledge of these primary correlation functions is sufficient to determine the tree-level and one-loop correlators of general type (5.1), due to a certain set of recursion relations. Secondly, and this is a particular relevant point for $d < 1$ where the primary correlators are only non-vanishing at genus zero, these correlation functions tell us what terms can appear in the generalized factorization algebra, *i.e.*, give us information about the analogue of the Virasoro algebra we discovered in $d = 0$.

Let us stress that the primary correlation functions can be computed entirely within the matter sector. The only reference to the coupling to gravity is the selection rule (5.2). This rule can be simply understood as the prescription that three operators should be represented as zero-forms, whereas all other fields should be taken as two-forms and consequently be integrated over. This makes the correlation function before integration a volume form on the moduli space $\mathcal{M}_{0,s}$ of the s-punctured sphere. The invariance under $SL(2,\mathbf{C})$ transformations and their fermionic Q-partners can be used to show that this prescription is well-defined; it does not matter which three operators we represent as zero-forms.

It will be very convenient to consider the generating functional of all these correlators

$$c_{ijk}(t) = \left\langle \phi_i\phi_j\phi_k \exp\left(\sum_n t_n \int \phi_n\right)\right\rangle. \qquad (5.4)$$

Here we introduced coupling coefficient t_n with respect to the two-form operators $\int \phi_n$. Higher s-point functions can be obtained by taking the relevant partial derivatives. Since even with these perturbations we are still left with a topological theory, albeit no longer a conformal invariant one, general principles tell us that the $c_{ij}{}^l(t)$ are still the structure coefficients of a commutative, associative algebra — the perturbed chiral ring — if we raise the index with the metric $\eta_{ij} = c_{0ij}$. Symmetry of the four-point function gives us the integrability equation

$$\frac{\partial c_{ijk}}{\partial t_l} = \frac{\partial c_{ijl}}{\partial t_k},\tag{5.5}$$

which proves the existence of a function $F(t)$ — the free energy of the string — that satisfies

$$c_{ijk}(t) = \frac{\partial^3 F}{\partial t_i \partial t_j \partial t_k}.\tag{5.6}$$

Equation (5.5) similarly shows that the metric $\eta_{ij}(t)$ is in fact independent of the couplings t_i. The two-form version of the identity vanishes, and consequently so do all derivatives with respect to t_0:

$$\frac{\partial \eta_{ij}}{\partial t_k} = \frac{\partial c_{ijk}}{\partial t_0} = 0.\tag{5.7}$$

5.2. Landau-Ginzburg and KdV

There are many strategies that can be used to determine the generating function $F(t)$, and consequently all primary correlation functions, in specific cases. A particularly interesting class of $N = 2$ models are the ones that allow a Landau-Ginzburg description [29,30]. That is, before twisting we have an action of type

$$S = \int d^2z\, d^4\theta \, K(X_i, \overline{X}_i) + \int d^2z\, d^2\theta\, W(X_i) + c.c.\tag{5.8}$$

with X_i the LG superfields and $W(X_i)$ the so-called superpotential. The superconformal model corresponds to the renormalization group fixed point of the LG theory.* For these models the (unperturbed) chiral ring can be neatly summarized as the quotient ring $\mathbb{C}[x_i]/\nabla W$. The perturbed chiral ring can now be uniquely determined if we assume that it will be described by a perturbed superpotential.

This is of particular relevance for the minimal models with $d < 1$. Recall that the members of the $N = 2$ discrete series have $d = \frac{k}{k+2}$ and are labeled by a simply-laced Lie group of type ADE with Coxeter number $h = k+2$. The spectrum of chiral primary fields consists of operators ϕ_i with $U(1)$ charges

$$q_i = \overline{q}_i = \frac{i}{k+2}\tag{5.9}$$

where $i + 1$ is an exponent of the Lie group G. (We will write $i + 1 \in Exp(G)$.) All these models have LG descriptions, and the superpotentials $W(x_i)$ read in these cases

$$A_n \quad : \quad W = x^{n+1},$$

---

*It has recently been shown by C. Vafa that the above action can also be used after twisting [31].

$$
\begin{aligned}
D_n &: \quad W = x^{n-1} + xy^2, \\
E_6 &: \quad W = x^3 + y^4, \\
E_7 &: \quad W = x^3 + xy^3, \\
E_8 &: \quad W = x^3 + y^5.
\end{aligned}
\tag{5.10}
$$

We will now demonstrate for the specific example of the models in the A-series how one determines the perturbed chiral ring. We define the perturbed superpotential as

$$
W(x,t) = \frac{x^{k+2}}{k+2} - \sum_{i=0}^{k} g_i(t) x_i.
\tag{5.11}
$$

Here the functions $g_i(t)$ are *a priori* arbitrary functions of the couplings t_i that we will determine later. The fields ϕ_i, which before the perturbation corresponded to the monomials x^i, are now defined by

$$
\phi_i(x,t) = -\frac{\partial W}{\partial t_i}.
\tag{5.12}
$$

In general these are polynomials in x and the couplings t_i, with leading behavior $\phi_i = x^i + O(t)$. For a given superpotential their algebra can be calculated without any difficulty.

Our strategy to determine these ϕ_i's is the following. We know by equation (5.7) that after the perturbation the metric, *i.e.* the two-point function, remains unchanged, and we want to impose this fact as a constraint on the polynomials. To this end we first have to define a metric for arbitrary W. It has to satisfy

$$
\langle \phi_i \phi_j \rangle = c_{ij}{}^l(t) \langle \phi_l \rangle = c_{ij}{}^k(t),
\tag{5.13}
$$

since by $U(1)$ charge conservation only the field with maximal charge $q = d$, and so $l = k$, can have a one-point function on the sphere, which can be set to one. For a *given* superpotential there is a *unique* inner product that is well-defined on the ring $C[x]/W'(x)$ with this property. It reads

$$
\langle \phi_i \phi_j \rangle = \int \frac{dx}{2\pi i} \frac{\phi_i(x)\phi_j(x)}{W'(x)} \equiv \mathrm{res}\left(\frac{\phi_i \phi_j}{W'} \right).
\tag{5.14}
$$

If we now impose the condition $\langle \phi_i \phi_j \rangle = \delta_{i+j,k}$, the 'orthogonal polynomials' can be expressed as

$$
\phi_i(x,t) = \frac{1}{i+1} \frac{\partial [L^{i+1}]_+}{\partial x},
\tag{5.15}
$$

where L is the fractional root of the superpotential, $L^{k+2} = (k+2)W$, and the subscript $+$ indicates a truncation to positive powers of x. Up to now we still have left undetermined the functions $g_i(t)$. However, we can now demand consistency between (5.12) and (5.15), and obtain the 'KdV equation'

$$\frac{\partial W}{\partial t_i} = -\frac{1}{i+1}\frac{\partial[L^{i+1}]_+}{\partial x}. \tag{5.16}$$

This fixes the relation between g_i and t_i. In fact one has $t_i = -\mathrm{res}(L^{k-i+1})/(k-i+1)$ [17]. We have now determined all unknown variables and can straightforwardly calculate the perturbed chiral ring by

$$c_{ijl}(t) = \mathrm{res}\left(\frac{\phi_i \phi_j \phi_l}{W'}\right). \tag{5.17}$$

After some algebra [17], this can be partly integrated, and the partition function can be seen to satisfy

$$\frac{\partial F}{\partial t_i} = \frac{\mathrm{res}(L^{k+i+3})}{(i+1)(k+i+3)}, \tag{5.18}$$

which agrees with the answer of the matrix model. The other minimal models can be treated similarly [17].

5.3. W-constraints and loop equations

Armed with the knowledge of the genus zero primary correlation functions, we can now try to attack the problem of solving the general correlation functions (5.1). We would expect that, at least for $d < 1$, much of our analysis of the pure topological gravity still holds with the appropriate modifications. The major difference is that in $d = 0$ the only non-zero primary correlator was the three point function of the puncture operator, while in general we have many more. Indeed, the $k = 0$ version of equation (5.18) reads

$$\frac{\partial F}{\partial t_0} = \tfrac{1}{2}t_0^2, \tag{5.19}$$

which is the appropriate reduction of the puncture equation $L_{-1}Z = 0$ with all couplings to descendants set to zero. As we have seen, the coupling to gravity gave rise to a full set of *bilinear* relations $L_n Z = 0$. The fact that these equation are bilinear was a consequence of the occurrence of contact and factorization terms which always included a sphere with three punctures.

In $d < 1$ there are many more non-vanishing contributions of primary fields. One property of (5.18), that generalizes to all minimal $d < 1$ models, is that is can be written as

$$\frac{\partial F}{\partial t_i} = F_i(t), \tag{5.20}$$

where F_i is a polynomial of degree $i + 2$ in the couplings [17]. In fact, if we include the dilaton coupling, $F_i(t)$ will be even a homogeneous polynomial. One can now propose

[18,19] that (5.20) is the truncation of the equations

$$W_n^{(i+2)} Z = 0, \qquad n \geq -i - 1, \tag{5.21}$$

where the generators $W^{(i+2)}$ span a so-called W-algebra. Note that the spins of the generators correspond exactly to the orders of the Casimirs of the Lie group G that labeled the minimal model. So the relevant W-algebra is the one based on G. The validity of these relations has been put on firmer grounds in the analysis of the contact algebra in [16] and the fact that W-constraints can be proved rigorously in the multi-matrix model [32].

The interpretation of these equations can be patterned exactly to our discussion of the $d = 0$ case. We introduce the Laplace transforms of the loop operators and their currents

$$w_i(z) = \sum_n \sigma_n(\phi_i) z^{-n-1-\epsilon_i}, \qquad J_i(z) = \sum_n \sigma_n(\phi_i) z^{n+\epsilon_i}, \tag{5.22}$$

with $\epsilon_i = \frac{i+1}{k+2}$. They correspond to the negative resp. positive modes of currents $\partial_z \Phi_i(z)$, twisted by element $e^{2\pi i \epsilon_i} \in \mathbf{Z}_{k+2}$. The W-algebra generators are single-valued in z and of spin j, where j is the order of a Casimir of G,

$$W^{(j)}(z) = \sum_i c_{i_1 \cdots i_j} : \partial \Phi_{i_1} \cdots \partial \Phi_{i_j} : (z) \tag{5.23}$$

This includes in particular a Virasoro algebra of central charge $k + 1$, generated by the elements $W^{(2)}$. If we rewrite the relation (5.21) in terms of the loop operators $w_i(\ell)$, that are now 'colored' with an index i, we retrieve loop equations that are of the usual form for the neutral loop $w_0(\ell)$, but that describe multiple splitting and joining for the colored versions, as dictated by the genus zero primary correlation functions [32].

These loop equations can again be used to determine all correlation functions at the topological point. To discuss there role at the multi-critical points, it is convenient to relabel the operators as $\sigma_n(\phi_i) = \Psi_{np+i+1}$ with $p = k + 2$. If we now consider a critical point where the operator Ψ_{p+q} is marginal, with (p,q) relative prime, the scaling dimensions of Ψ_n will be given by $(n-1)/(p+q-1)$ and $\gamma_{string} = -2/(p+q-1)$ according to (5.3). In complete analogy with the $d = 0$ case we define the physical operators as those operators whose correlation functions cannot be reduced to more elementary ones by use of the W-constraints. We find for the set of fields

$$\Psi_{-rp+sq}, \qquad s \in Exp(G), \ 1 \leq r \leq [sq/p]. \tag{5.24}$$

This is *exactly* the spectrum of (p,q) minimal model labeled by the modular invariant (G, A). The scaling dimensions obey KPZ [27]. This gives in our opinion another convincing argument that the minimal ADE topological string theories adequately describe all minimal models coupled to gravity.

5.4. The limit $k \to \infty$

Let us briefly address the limit $k \to \infty$ of the minimal models, which would describe a theory with $d = 1$, and can be regarded as the classical limit. To this end it is useful to state a recursion relation that the polynomials $\phi_i(x, t)$ obey [17]

$$x\phi_i = \phi_{i+1} + \sum_{j=0}^{i-1} t_{j-i+k+1}\phi_j. \tag{5.25}$$

This relation is very helpful to determine the form of the polynomials inductively in the order of the couplings. One concludes thus that the correlation function $\langle \phi_{i_1} \ldots \phi_{i_s} \rangle$ is a polynomial in the labels i_n of order $s - 3$. If we take the limit $i_n, k \to \infty$ with, say, i_n/k fixed, this correlation function seems to diverge. However, as discussed in [17], we have the freedom to rescale the dilaton coupling and multiply every s-point function with a power k^{3-s}, which makes the correlators clearly finite. It is an amusing exercise to calculate these quantities explicitly. For instance for the four-point function we found in the $k \to \infty$ limit

$$\langle \phi_{q_1} \phi_{q_2} \phi_{q_3} \phi_{q_4} \rangle = \min\{q_i, 1 - q_i\}. \tag{5.26}$$

Here we labeled the fields with their $U(1)$-charges, which satisfy for a general s-point function

$$\sum_i q_i = s - 2. \tag{5.27}$$

Higher point functions become rather complicated, because one has to study many different channels. It is an interesting open problem what the correct interpretation of this limit is, in particular whether it is related to one-dimensional strings [6].

Acknowledgements

We would like to thank E. Verlinde and E. Witten, in collaboration with whom most of the above results were derived, for stimulating discussions. It is furthermore a pleasure to acknowledge fruitful discussions with T. Banks, Ph. Di Francesco, J. Distler, S. Giddings, D. Gross, I. Klebanov, D. Kutasov, K. Li, P. Nelson, A. Polyakov, N. Seiberg, S. Shenker, A. Strominger, C. Vafa, and N. Warner. We finally thank the organizers of this remarkable inspiring workshop for the opportunity to present our results. The research of R.D. is supported by DOE Grant DE-AC02-76WRO3072, and that of H.V. by NSF Grant PHY80-19754.

References

[1] V. Kazakov, Phys. Lett. **159B** (1985) 303; F. David, Nucl. Phys. **B257** (1985) 45; V. Kazakov, I. Kostov, and A. Migdal, Phys. Lett. **157B** (1985) 295; J. Fröhlich, "The Statistical Mechanics Of Surfaces," in *Applications Of Field Theory To Statistical Mechanics*, ed L. Garrido (Springer, 1985).

[2] F. David, Phys. Lett. **159B** (1985) 303; D. Boulatov, V. Kazakov, and A. Migdal, Nucl. Phys. **B275 [FS117]** (1986) 543; A. Billoire and F. David, Phys. Lett. **186B** (1986) 279; J. Jurkievic, A. Krzywicki, and B. Peterson, Phys. Lett. **186B** (1986) 273; J. Ambjorn, B. Durhuus, J. Fröhlich, and P. Orland, Nucl. Phys. **B270 [FS16]** (1986) 457.

[3] V. Kazakov, Mod. Phys. Lett. **A4** (1989) 2125.

[4] E. Brézin and V. Kazakov, Phys. Lett. **B236** (1990) 144; M. Douglas and S. Shenker, Nucl. Phys. **B335** (1990) 635; D.J. Gross and A. Migdal, Phys. Rev. Lett. **64** (1990) 127.

[5] E. Brézin, M. Douglas, V. Kazakov, and S. Shenker, Phys. Lett. **237B** (1990) 43; D.J. Gross and A. Migdal, Phys. Rev. Lett. **64** (1990) 717; C. Crnković, P. Ginsparg, and G. Moore, Phys. Lett. **237B** (1990) 196.

[6] E. Brezin, V. Kazakov, and Al.B. Zamolodchikov, Nucl. Phys. **B338** (1990); D.J. Gross and N. Miljković, Phys. Lett **238B** 91990) 217; P. Ginsparg and J. Zinn-Justin, Phys. Lett **240B** (1990) 333; D.J. Gross and I.R. Klebanov, Nucl. Phys. **B344** (1990) 475.

[7] D.J. Gross and A. Migdal, Nucl. Phys. **B340** (1990) 333; T. Banks, M. Douglas, N. Seiberg, and S. Shenker, Phys. Lett. **238B** (1990) 279.

[8] M. Douglas, Phys. Lett. **238B** (1990) 176; P. Di Francesco and D. Kutasov, Nucl. Phys. **B342** (1990) 589.

[9] E. Witten, Nucl. Phys. **B340** (1990) 281.

[10] E. Witten, Commun. Math. Phys. **117** (1988) 353.

[11] E. Witten, Comm. Math. Phys. **118** (1988) 411.

[12] J. Distler, Nucl. Phys. **B342** (1990) 523.

[13] E. Verlinde and H. Verlinde, 'A Solution Of Two Dimensional Topological Quantum Gravity,' preprint IASSNS-HEP-90/40, PUPT-1176 (1990).

[14] R. Dijkgraaf and E. Witten, Nucl. Phys. **B342** (1990) 486.

[15] T. Eguchi and S.-K. Yang, 'N=2 Superconformal Models as Topological Field Theories,' Tokyo-preprint UT-564.

[16] K. Li, 'Topological Gravity with Minimal Matter,' Caltech-preprint CALT-68-1662; 'Recursion Relations in Topological Gravity with Minimal Matter,' Caltech-preprint CALT-68-1670.

[17] R. Dijkgraaf, E. Verlinde, and H. Verlinde, 'Topological Strings in $d < 1$,' Princeton preprint PUPT-1204, Nucl. Phys. **B** to be published.

[18] R. Dijkgraaf, E. Verlinde, and H. Verlinde, 'Loop Equations and Virasoro Constraints in Non-Perturbative 2D Quantum Gravity,' preprint PUPT-1184, Nucl. Phys. **B** to be published.

[19] M. Fukuma, H. Kawai, and R. Nakayama, 'Continuum Schwinger-Dyson Equations and Universal Structures in Two-Dimensional Quantum Gravity,' Tokyo preprint UT-562 (1990).

[20] A.A. Migdal, Phys. Rep. **102** (1983) 199.

[21] F. David, 'Loop Equations and Non-Perturbative Effects in Two Dimensional Quantum Gravity,' Saclay preprint SPhT/90-043 (1990).

[22] W. Lerche, C. Vafa, and N.P. Warner, Nucl. Phys **B324** (1989) 427.

[23] J. Labastida, M. Pernici, and E. Witten, Nucl. Phys. **B310** (1988) 611; D. Montano and J. Sonnenschein, Nucl. Phys. **B313** (1989) 258.

[24] N. Seiberg, 'Notes on Quantum Liouville Theory and Quantum Gravity,' Rutgers preprint RU-90-29 (1990), this proceedings; J. Polchinsky, 'Remarks on the Liouville Field Theory,' Texas preprint UTTG-19-90, to be published in the proceedings of Strings 90.

[25] J. Polchinsky, 'Critical Behavior of Random Surfaces in One Dimension,' University of Texas preprint UTTG-15-90 (1990); S.R. Das and A. Jevicki, 'String Field Theory and Physical Interpretation of $D = 1$ Strings,' Brown University preprint BROWN-HET-750 (1990); A. Sengupta and S. Wadia, 'Excitations and Interactions in $d = 1$ String Theory,' Tata preprint TIFR/TH/90-13 (1990); D.J. Gross and I. Klebanov, 'Fermionic String Field Theory of $c = 1$ Two-Dimensional Quantum Gravity,' Princeton preprint PUPT-1198 (1990); A. Sen, 'Matrix Models and Gauge Invariant Field Theory of Subcritical Strings,' Tata preprint TIFR/TH/90-35 (1990).

[26] T. Banks, 'Matrix Models, String Field Theory and Topology,' Rutgers preprint RU-90-52, in this proceedings.

[27] V. Knizhnik, A. Polyakov, A. Zamolodchikov, Mod. Phys. Lett. **A 3** (1988) 819; F. David, Mod. Phys. Lett. **A 3** (198) 1651; J. Distler and H. Kawai, Nucl. Phys. **B321** (1989) 509.

[28] G. Moore, 'Geometry of the String Equations,' Yale-preprint YCTP-PA-90; 'Matrix Models of 2D Gravity and Isomonodromic Deformation,' Rutgers preprint RU-90-53, in this proceedings.

[29] E. Martinec, Phys. Lett. **217B** (1989) 431; 'Criticality, Catastrophe and Compactifications,' V.G. Knizhnik memorial volume, 1989.

[30] C. Vafa and N. Warner, Phys. Lett. **218B** (1989) 51.

[31] C. Vafa, to appear.

[32] J. Goeree, 'W Constraints in 2-D Quantum Gravity,' Utrecht preprint THU-19 (1990).

THE TWO-MATRIX MODEL

Michael R. Douglas

Laboratoire de Physique Theorique
Ecole Normale Superieure
Paris, France 75005,
and
Department of Physics and Astronomy
Rutgers University
Piscataway, NJ 08855-0849

Abstract: We study the two-matrix model, show that it contains multicritical points
of type (p, q) for general p and q, describe the generalized Toda flows which interpolate
between them, and discuss the continuum limit.

Recent work [1] has provided solutions of many models of matter coupled to
two-dimensional lattice gravity. In this talk we will study the two-matrix model
in somewhat more detail than in [2]. The general formalism of that work will be
seen to be applicable, with a continuum limit described by two linear differential
operators satisfying a Heisenberg algebra, and correlations given by KP flows. These
are equivalent to generalized KdV flows for relevant operators.

We find that the two-matrix model contains multicritical points of type (p, q) with
general p and q, in particular with $q > 3$. We will also describe the generalized Toda
flows which interpolate between these models at finite N. This raises the question
of whether there is a consistent description of such flows in the scaling limit. Such a
description would involve flows generated by irrelevant operators which were different
from generalized KdV in that they changed the order of Q. Although it seems unlikely
that any such structure could be universal, the possibility remains open.

The model is

$$Z = \int d^{N^2} A \, d^{N^2} B \, e^{-\mathrm{Tr}\ V_1(A) - \mathrm{Tr}\ V_2(B) + c \mathrm{Tr}\ A\ B} \tag{1}$$

Random Surfaces and Quantum Gravity
Edited by O. Alvarez *et al., Plenum Press, New York, 1991*

and we will rescale the matrices to set $c = 1$. Further couplings Tr $A^k B^l$ and more general orderings of A and B might have some interest, either to make contact with the Schwinger-Dyson equations or to get more general critical points such as the six-vertex model, [3] and might be tractable using the theorem of [4], but will not be considered here. So, [5]

$$Z = \int \prod d\alpha_i d\beta_i \Delta(\alpha)\Delta(\beta) e^{-\sum_i [V_1(\alpha_i) - V_2(\beta_i) + \alpha_i \, \beta_i]}. \tag{2}$$

In principle this model can be solved at genus zero by saddle point; this is tricky because (rather counter-intuitively) if we order the α's so that $\alpha_i < \alpha_j$ for $i < j$, the saddle point does not satisfy $\beta_i < \beta_j$ for $i < j$, and we cannot describe it by two spectral densities $u_1(\alpha)$ and $u_2(\beta)$. Rather [6] we must go to $U(N)$ invariants $\sum_i \alpha_i^k \beta_i^l$. To my knowledge this has not been worked out explicitly, and understanding this technique better would be helpful in attacking more complicated models with tree-like couplings.

Since we want an exact solution we follow Mehta and collaborators [7] and use orthonormal polynomials:

$$\int d\alpha d\beta f_m(\alpha) e^{-V_1(\alpha) - V_2(\beta) + \alpha\beta} \tag{3}$$

These can be constructed from the moments of the integrand, and so given V_1 and V_2 they are uniquely defined, up to a possible transformation

$$\begin{aligned} f_m(\alpha) &\to c_m f_m(\alpha), \\ g_m(\beta) &\to \frac{1}{c_m} g_m(\beta). \end{aligned} \tag{4}$$

(As usual we need the integral to make sense to make this statement. For present purposes we will ensure this by analytic continuation in the couplings if necessary.) For $V_1 \neq V_2$, we will have $f_m \neq g_m$.

Define $Q(1)$ and $Q(2)$ to be the operators α and β inserted in eqn. ((3)), and $P(1)$ and $P(2)$ to be $d/d\alpha$ and $d/d\beta$ inserted before and after the exponential respectively. Then

$$\begin{aligned} P(1) &= -V_1'(Q(1)) + Q(2); \\ P(2) &= V_2'(Q(2)) - Q(1) \end{aligned} \tag{5}$$

and the commutator $[P(1), Q(1)] = 1$ is equivalent to $[Q(2), Q(1)] = 1$. Unlike the one-matrix model, these operators have no simple adjointness properties, unless we take the special case $V_1 = V_2$, in which case $Q(1)^+ = Q(2)$ (resp. P).

With this choice of basis and polynomial potentials, we will have the usual statement that these matrices have finitely many non-zero diagonals. We will refer to a matrix with non-zero entries up to m diagonals below and n above the main diagonal as having order (m, n). The notation "$(m, *)$" will mean we have some finite but unspecified bound on n. The result of multiplying or commuting two matrices has the sum of their orders. One sees easily that $Q(1)$ is $(*, 1)$, $P(1)$ is $(*, -1)$, $Q(2)$

is $(1, *)$ and $P(2)$ is $(-1, *)$. The equations of motion imply that $Q(2)$ is actually $(1, \mathrm{degree} V_1 - 1)$ and so on.

In the continuum limit, these operators will become finite order differential operators. Given a matrix with k non-zero diagonals with adjustable coefficients, we might expect a model to exist in which the limit is a differential operator of order k. In particular by using high order potentials it would seem that we could find $P(1)$ and $Q(1)$ each of arbitrarily high order, realizing the arbitrary model (p, q) of [2]. For example, Tada and Yamaguchi [8] have analyzed the general model with a sixth order potential and found the point $(4, 5)$. It is not hard to see that with $V_1 = V_2$ we can realize all $(q, q + 1)$. Write $Q(1) = S + A$ with S the leading self-adjoint part in the scaling limit and A the leading skew-adjoint part; then $Q(2) = S - A$ and the string equation is $2[S, A] = 1$. The order of these operators is determined by the matrix elements of Q at leading order, i.e. in the approximation $Q_{n,n+i}$ independent of n. In this order we can just write

$$Q(Z) = \sum_i Q(2)_{n,n+i} Z^i \tag{6}$$

with a "shift operator" Z; Q is determined by the condition that $P(2)$ be $(-1, *)$, i.e. that

$$P(Z) = V'(Q(Z)) - Q(1/Z) \tag{7}$$

be analytic at $Z = 0$.

This equation is non-linear in Q but linear inhomogeneous in the couplings of the potential $V = \sum c_k M^k$. Furthermore it is upper triangular if written as a matrix equation on the vector c_k. Therefore, given an arbitrary $Q(Z)$, for example one chosen to cancel all up to a specified number q of derivatives in $Q(e^{d/dx})$, there exists a potential which will produce it. Using such a potential will therefore give a scaling limit of type (q, p) with $p > q$. Generically we have $p = q + 1$, while further choices in our prescribed Q should allow fixing p. Since P and Q are S and A (not necessarily in that order), it is clear for the Z_2-symmetric case that $p + q$ will be odd.

There remains an important point which we have not checked. We implicitly assumed that the coefficient functions in our operator Q would be continuous in the scaling limit, to reach a (p, q) model. For general potentials, there are many possible alternate scaling limits, in which Q goes to a matrix differential operator in the scaling limit. The correct limit will be the one which minimizes the free energy, and we should check that a choice of Q exists for which our simple scaling ansatz is correct. Unfortunately this is rather difficult with the existing formalisms. It would be much easier if we had a criterion depending only on quantities defined in the continuum limit, unlike the absolute free energy.

Returning to the finite N solution, we would like to describe correlation functions by some sort of operator flow as in [9]. The argument for the one-matrix model goes through with minor changes. We can describe correlation functions if we can describe the variation of the matrix elements of Q under arbitrary perturbations. This requires expressing the orthonormal polynomial bases in the perturbed model in terms of the bases in the original model. Write

$$\delta f_m(\alpha) = \sum_n \delta A_{m,n} f_n(\alpha),$$
$$\delta g_m(\beta) = \sum_n \delta B_{m,n} g_n(\beta); \tag{8}$$

then the new bases $f + \delta f$ and $g + \delta g$ will be orthonormal if

$$\delta A + \delta B = 0. \tag{9}$$

So, just as in the one-matrix model, all perturbations act on operators as

$$\delta O = [\delta A, O]. \tag{10}$$

On the other hand we no longer have $\delta A^+ = -\delta A$.

Given the perturbation

$$\delta(V_1(\alpha) + V_2(\beta)) = \delta c_k \beta^k, \tag{11}$$

the new basis satisfies

$$(g + \delta g)_m = g_m(1 - \delta c_k \beta^k) + \text{lower order polynomials}. \tag{12}$$

Therefore
$$\delta A = \delta c_k(Q(2)^k + U_k) \tag{13}$$

where U_k is some $(0, *)$ (upper triangular) matrix. It must preserve the fact that $Q(1)$ is $(*, 1)$ and $Q(2)$ is $(1, *)$. Since $[Q(2), Q(2)^k] = 0$, the second requirement will be true for any U_k. The first requirement can be satisfied only if δA is $(*, 0)$, so U_k is exactly $-Q(2)^k$ above the diagonal. It is not determined on the main diagonal unless we fix the ambiguity mentioned earlier in the normalizations of f_m and g_m. One could set $Q(1)_{n,n+1} = Q(2)_{n+1,n}$, for example; in the continuum limit this ambiguity corresponds to redefinitions $\delta Q = [Q, f]$ which allow defining away one scaling function.

We have found that δA is the lower triangular part of $\delta V_2(Q(2))$. Similarly perturbations of V_1 act by commutators with minus the upper triangular part of $\delta V_1(Q(1))$. We see that if we stay within the subspace $V_1 = V_2$, we will get commutators with skew-adjoint operators as in the one-matrix model, but moving outside this subspace will give general operators. There is no constraint on the total order of Q.

In the appendix we show that these flows commute. Since they are canonical transformations, they are compatible with the string equation. We therefore have a complete solution of the model at finite N, and it remains to understand the continuum limit.

This will be taken in the standard way, [10] which necessarily takes our matrices to finite order differential operators. The result will be a system described by two differential operators, the limits of $Q(1)$ and $Q(2)$, and commuting flows given by commutators with differential operators whose coefficient functions are differential polynomials in the coefficients both of $Q(1)$ and $Q(2)$. This is equivalent to allowing the coefficients to be differential polynomials in the coefficients of (say) $Q(1)$ and $P(1)$. The string equation will then set the coefficients of P to be differential polynomial in the coefficients of Q, so the result is that flows in the couplings are an infinite commuting set of flows on scalar differential operators by commutators. One can show that any such set must be some subset of the KP flows – in other words, there exists a pseudo-differential operator L of order 1 such that the flows are commutators with L_+^k.

Further progress is more difficult than in the one-matrix model, for two reasons. First, in the one-matrix model one knows that all flows preserve the order $(1, 1)$ nature of the matrix Q and therefore the second order nature of its continuum limit. This constraint is exactly the one which reduces the KP flows to the KdV flows. There is no equally simple constraint in the two-matrix case, but to define correlators of arbitrary operators in a universal way in multi-matrix models, we need to find some substitute for it.

Second, it is clear that there is no general definition of scaling operators in the two-matrix model which will work for all critical points with all p and q. The most we could hope for would be a general definition for all p and a fixed q. This makes it hard to analyze flows changing q.

If we use a potential of the minimal possible degree to construct the model $(q, q + 1)$, then since the scaling operators of order less than q can be made with lower order perturbations of the potential, they can be chosen to preserve the order of Q in the continuum limit. This means that $L_+^k = Q_+^{k/q}$ for $k < q$. This is enough information to specify the coefficient functions in L of dimension less than q, but it could be that L differed from $Q^{1/q}$ in coefficient functions of higher dimension. This would mean that flows of dimension greater than q, or in other words insertions of irrelevant operators, could change the order of Q.

In general one expects to find difficulties in defining the continuum limit of irrelevant operators, whose couplings must be scaled by negative powers of the lattice spacing a. The result would seem to necessarily depend on terms in the continuum limit of Q and P which would go to zero as $a \rightarrow 0$, and are non-universal. This ambiguity is represented in the KP formalism as the choice of higher order terms in L, which are not constrained by consistency of the formalism.

Compatibility with the string equation is not a constraint. Because the flows are canonical transformations, any choice would be compatible with the string equation. If we think in terms of determining the correlation functions by solving the differential equations coming from perturbing the string equation, the point is that we need to make an assumption about whether irrelevant couplings couple to new scaling functions which are zero right at the critical point to get a definite answer.

For a general L we will lose the simple expression we had for P in the one-matrix model, a sum linear in the couplings. Now P is still determined at finite N by an equation of motion linear in the couplings, and this may be an important constraint if we can take it over to the continuum limit. This is not yet clear.

The reduction of the KP structure we have derived to the generalized KdV flows with $L = Q^{1/q}$ is the simplest and most natural possibility from many points of view. It is not completely clear that it is the most natural definition of irrelevant operators in the matrix model. On the other hand, even if we find another flow structure which interpolates between the (p, q) models, it is difficult to imagine that the same structure would result from the chains with more matrices, which surely also contain all scaling limits (p, q).

Clearly this is not a proof that a universal flow structure including all p and q does not exist in the continuum and it would be quite interesting if it did. The physics of such a structure is rather unclear. One of the mysterious and fascinating results from lattice 2D gravity has been that "pure" gravity theories defined as sums over specially weighted lattices can in fact be equivalent to gravity coupled to matter in the continuum limit, [11] and the results here provide more examples. We still do not understand this very deeply even for the one-matrix model, and such understanding will probably have to wait for a more precise connection between the lattice definition and the continuum integral over metrics.

I would like to thank E. Brezin and V. Kazakov for helpful discussions.

After the completion of this work I learned from E. Martinec that he has studied the two-matrix model in [9] with very similar conclusions.

Appendix A

We show here that the operator flows corresponding to varying the couplings commute. Let the variations be δ_1 and δ_2, corresponding to commutators with R_1 and R_2. Then

$$
\begin{aligned}
[\delta_2, \delta_1]O &= [[O, R_1], R_2] + [O, \delta_1 R_2] \\
&\quad - [[O, R_2], R_1] - [O, \delta_2 R_1] \\
&= [O, [R_1, R_2]] + [O, \delta_1 R_2 - \delta_2 R_1]
\end{aligned} \tag{A.1}
$$

which should be zero for any O, so we need

$$
[R_1, R_2] = \delta_2 R_1 - \delta_1 R_2. \tag{A.2}
$$

Write O_+ for the upper diagonal part of a matrix O and O_- for the lower diagonal part. Clearly $(A_+ B_+)_- = 0$. An identity we will use below follows from $[Q^k, Q^l] = 0$:

$$
[Q^k_+, Q^l_+] + [Q^k_-, Q^l_-] = -[Q^k_+, Q^l_-] - [Q^k_-, Q^l_+]. \tag{A.3}
$$

By linearity, it suffices to consider two cases: where the variations are both in V_1, and where one is in V_1 and one in V_2. We start with the first, and let $R_1 = Q(1)^k_+$ and $R_2 = Q(1)^l_+$ for some k and l, then we want (let $Q(1) = Q$)

$$
[Q^k_+, Q^l_+] = [Q^k, Q^l_+]_+ - [Q^l, Q^k_+]_+. \tag{A.4}
$$

Writing $Q^k = Q^k_+ + Q^k_-$ (resp. for Q^l) and using the identity above on $([Q^k_-, Q^l_+] + [Q^k_+, Q^l_-])_+$ establishes the result.

The second case is

$$
[Q(1)^k_+, Q(2)^l_-] = [Q(1)^k, Q(2)^l_-]_+ - [Q(2)^l, Q(1)^k_+]_-. \tag{A.5}
$$

which after expanding in $+$ and $-$ parts is evident.

References

[1] For complete references, see the upcoming review in Physics Reports by V. Kazakov and I. Kostov, or the other talks in this volume.

[2] M. R. Douglas, Phys.Lett. 238B (1990) 176.

[3] G. Moore, private communication.

[4] J. J. Duistermaat and G. J. Heckman, Invent.Math. 69 (1982) 259.

[5] C. Itzykson and J.B. Zuber, J.Math.Phys. 21 (1980) 411.

[6] A. Jevicki and B. Sakita, Nucl. Phys. B185 (1981) 89.

[7] S. Chadha, G. Mahoux and M. L. Mehta, J. Phys. A: Math. Gen. 14 (1981) 579; M. L. Mehta, Comm. Math. Phys. 79 (1981) 327.

[8] T. Tada and M. Yamaguchi, Tokyo preprint UT-Komaba 90-17, June 1990.

[9] L. Alvarez-Gaumé, C. Gomez and J. Lacki, CERN preprint CERN-TH.5875/90, September 1990; M. R. Douglas, to be published in the proceedings of the Texas A&M March 1990 String theory workshop; A. Gerasimov, A. Marshakov, A. Mironov, A. Morozov and A. Orlov, Lebedev institute preprint, July 1990; E. Martinec, Chicago preprint EFI-90-67, September 1990; E. Witten, IAS preprint IASSNS-HEP-90/45, May 1990.

[10] E. Brézin and V. Kazakov, Phys.Lett. 236B (1990) 144; M. Douglas and S. Shenker, Nucl.Phys. B335 (1990) 635; D. Gross and A. Migdal, Phys.Rev.Lett. 64 (1990) 127.

[11] V. A. Kazakov, Mod.Phys.Lett. A4 (1989) 2125.

ACTION PRINCIPLE AND LARGE ORDER BEHAVIOR
OF NON-PERTURBATIVE GRAVITY

P. Ginsparg[1] and J. Zinn-Justin[2]

[1]MS-B285
Los Alamos National Laboratory
Los Alamos, NM, USA 87545

[2]Service de Physique Théorique de Saclay
F-91191 Gif-sur-Yvette Cedex, France

0. Introduction

In these lectures we will consider some features of the large order behavior of
the perturbation series of recently exactly solved models [1–3] of 2d gravity coupled
to $d < 1$ matter. Our motivation is to learn what we can about non-perturbative
features of quantum gravity and string theory using standard mathematical techniques
for studying the asymptotic behavior of perturbation series. In particular we wish
ultimately to probe whether we indeed have the proper definition of the physical
theory of interest (i.e. whether we are studying non-perturbative gravity as opposed
to non-perturbative matrix models), and to infer which general properties of the nonperturbative ground state may carry over to strings embedded in dimensions $d \gtrsim 1$.
In addition, we show how the formalism of ref. [4] for 2d gravity coupled to minimal
(p, q) conformal matter results from an action principle. Viewed as a more intrinsic
formulation of 2d quantum gravity, abstracted from a matrix model underpinning,
this may result in clues to a proper formulation of string field theory.

In section 1, we review the basic formulation of the models of interest, and in
section 2 we discuss the action formulation. The results of these two sections are
adapted from [5]. There is some overlap in our methods and results with the classic
results (e.g. [6,7]) on the generalized KdV hierarchy, and the formulation in terms of
an action principle has also been treated recently in [8]. In section 3, we review some
basic results concerning large order behavior and Borel summability of perturbation

Random Surfaces and Quantum Gravity
Edited by O. Alvarez *et al., Plenum Press, New York, 1991*

series, and then combine with the results of section 2 to analyze in particular the properties of 2d gravity perturbation series[9]. In section 4, we recall some comments from [10] on analogous properties that appear for 2d gravity coupled to $d = 1$ matter.

1. Known results

We shall show in these first two sections that the coupled differential equations for the partition function summed over topologies in the recently proposed non-perturbative formulation of 2d quantum gravity [1–3] coupled to (p, q) minimal conformal matter [4] follow from an action principle. The basic action for a (p, q) model will be seen to take the general form $S_{(p,q)} = \int \mathrm{Res}\big[Q^{p/q+1} + \sum_{k=0}^{q-2} \sum_{\alpha=0}^{q-2} t_{(k),\alpha} Q^{k+(\alpha+1)/q}\big]$, where Q is a qth order differential operator and the $t_{(k),\alpha}$'s are sources for operator insertions. The action $S_{(p,q)}$ will also be seen to embody the essential features of the problem (including the relation to generalized KdV hierarchies) in a compact form.

1.1. $(q-1)$-matrix models

The free energy of a particular $(q-1)$-matrix model is written [11]

$$
\begin{aligned}
Z &= \ln \int \prod_{i=1}^{q-1} \mathrm{d}M_i \; \mathrm{e}^{-\mathrm{tr}\left(\sum_{i=1}^{q-1} V_i(M_i) - \sum_{i=1}^{q-2} c_i\, M_i M_{i+1}\right)} \\
&= \ln \int \prod_{\substack{i=1,q-1 \\ \alpha=1,N}} \mathrm{d}\lambda_i^{(\alpha)} \; \Delta(\lambda_1)\mathrm{e}^{-\sum_{i,\alpha} V_i\left(\lambda_i^{(\alpha)}\right) + \sum_{i,\alpha} c_i \lambda_i^{(\alpha)} \lambda_{i+1}^{(\alpha)}} \Delta(\lambda_{q-1}) \,,
\end{aligned}
\tag{1.1}
$$

where the M_i $(i = 1, q-1)$ are $N \times N$ hermitian matrices, the $\lambda_i^{(\alpha)}$ $(\alpha = 1, \ldots, N)$ their eigenvalues, and $\Delta(\lambda_i) = \prod_{\alpha<\beta}(\lambda_i^{(\alpha)} - \lambda_i^{(\beta)})$ is the Vandermonde determinant.

Following [11], we introduce operators Q_i and P_i that represent the insertions of λ_i and $\mathrm{d}/\mathrm{d}\lambda_i$ respectively in the integral (1.1). These operators necessarily satisfy $[P_i, Q_i] = 1$. In the $N \to \infty$ limit, it was argued in [4] that P and Q become differential operators of finite order, say p, q respectively (where we assume $p > q$), and these continue to satisfy

$$
[P, Q] = 1 \,.
\tag{1.2}
$$

In the continuum limit of the matrix problem (i.e. the "double" scaling limit, with couplings in (1.1) tuned to certain critical values), Q becomes a differential operator of the form

$$
Q = \mathrm{d}^q + \big\{v_{q-2}(x), \mathrm{d}^{q-2}\big\} + \cdots + 2v_0(x) \,,
\tag{1.3}
$$

where $\mathrm{d} = \mathrm{d}/\mathrm{d}x$. The continuum scaling limit of the multi-matrix models is thus abstracted to the mathematics problem of finding solutions of (1.2).

In [4], the function v_{q-2} was identified (up to normalization) in the continuum scaling limit with the "specific heat", i.e. the second derivative of the gravity partition function Z with respect to x. Equivalently we can write this as $v_{q-2} \propto \langle \mathcal{P}\mathcal{P} \rangle$, in terms of the 2-point function of a puncture operator $\mathcal{P}$ [3,12]. In [13], it was argued that the normalization is in general given by $Z'' = (4/q)v_{q-2}$.[1]

The differential equations (1.2) may be constructed as follows. For p, q relatively prime, a pth order differential operator that can satisfy (1.2) is constructed as a fractional power of the operator Q of (1.3). Formally, a qth root may be represented within an algebra of formal pseudo-differential operators (see, e.g. [15]) as

$$Q^{1/q} = \mathrm{d} + \sum_{i=1}^{\infty} \left\{ e_i, \mathrm{d}^{-i} \right\} \, , \tag{1.4}$$

where d^{-1} is defined to satisfy $\mathrm{d}^{-1} f = \sum_{j=0}^{\infty} (-1)^j f^{(j)} \, \mathrm{d}^{-j-1}$. The differential equations describing the (p,q) minimal model are given by

$$[Q_+^{p/q}, \, Q] = 1 \, , \tag{1.5}$$

where $Q_+^{p/q}$ indicates the part of $Q^{p/q}$ with only non-negative powers of d.

1.2. One-matrix models

To illustrate the procedure we reproduce now the results for the one-matrix models, which can be used to generate (p,q) of the form $(2l-1, 2)$. These models are obtained by taking Q to be the hermitian operator

$$Q = K \equiv \mathrm{d}^2 - u(x) \, . \tag{1.6}$$

The formal expansion of $Q^{l-1/2} = K^{l-1/2}$ (an anti-hermitian operator) in powers of d is given by

$$K^{l-1/2} = \mathrm{d}^{2l-1} - \frac{2l-1}{4} \left\{ u, \mathrm{d}^{2l-3} \right\} + \dots \, . \tag{1.7}$$

We decompose $K^{l-1/2} = K_+^{l-1/2} + K_-^{l-1/2}$, where $K_+^{l-1/2} = \mathrm{d}^{2l-1} + \dots$ contains only

---

[1] Actually, this is the normalization for even potentials. As discussed in [14], the normalization for general odd potentials would come out to be $Z'' = (2/q)v_{q-2}$.

non-negative powers of d, and the remainder $K_-^{l-1/2}$ has the expansion

$$K_-^{l-1/2} = \sum_{i=1}^{\infty} \left\{ e_{2i-1}, \mathrm{d}^{-(2i-1)} \right\} = \left\{ R_l, \mathrm{d}^{-1} \right\} + O(\mathrm{d}^{-3}) + \ldots \ . \tag{1.8}$$

Here we have identified $R_l \equiv e_1$ as the first term in the expansion of $K_-^{l-1/2}$. For $K^{1/2}$, for example, we find $K_+^{1/2} = \mathrm{d}$ and $R_1 = -u/4$.

The prescription (1.5) with $p = 2l-1$ corresponds to calculating the commutator

$$\left[K_+^{l-1/2}, K \right] = \left[K, K_-^{l-1/2} \right] = \text{leading piece of } \left[K, 2R_l\, \mathrm{d}^{-1} \right] = 4R_l' \ . \tag{1.9}$$

After integration, the equation $\left[K_+^{l-1/2}, K \right] = 1$ thus takes the simple form

$$(l + \tfrac{1}{2}) R_l[u] = x \ , \tag{1.10}$$

where we have rescaled x and u for later convenience. (This rescaling is enabled by the property that all terms in R_l have fixed grade, namely $2l$.)

The quantities R_l in (1.8) are easily seen to satisfy a simple recursion relation. From $K^{l+1/2} = KK^{l-1/2} = K^{l-1/2}K$, we find

$$K_+^{l+1/2} = \frac{1}{2} \left(K_+^{l-1/2}K + KK_+^{l-1/2} \right) + \left\{ R_l, \mathrm{d} \right\} \ .$$

Commuting both sides with K and using (1.9), simple algebra gives [6]

$$R_{l+1}' = \frac{1}{4} R_l''' - uR_l' - \frac{1}{2} u'R_l \ . \tag{1.11}$$

While this recursion formula only determines R_l', by demanding that the R_l ($l \neq 0$) vanish at $u = 0$, we obtain

$$R_0 = \frac{1}{2} \ , \qquad R_1 = -\frac{1}{4}u \ , \qquad R_2 = \frac{3}{16}u^2 - \frac{1}{16}u'' \ ,$$

$$R_3 = -\frac{5}{32}u^3 + \frac{5}{32}\left(uu'' + \tfrac{1}{2}u'^2 \right) - \frac{1}{64}u^{(4)} \ ,$$

$$R_4 = \frac{35}{256}u^4 - \frac{35}{128}\left(uu'^2 + u^2u'' \right) + \frac{7}{256}\left(2uu^{(4)} + 4u'u''' + 3u''^2 \right) - \frac{1}{256}u^{(6)} \ . \tag{1.12}$$

The R_l's satisfy as well a functional relation that allows us to write eq. (1.10) as the variation of an action. The "residue" and "trace" [16] of a formal pseudo-differential operator $A = \sum_{i=-\infty}^{k} a_i(x)\, \mathrm{d}^i$ are defined as

$$\mathrm{Res}\, A \equiv a_{-1} \ ,$$

$$\mathrm{Tr}\, A \equiv \int \mathrm{d}x \ \mathrm{Res}\, A = \int \mathrm{d}x \, a_{-1} \ . \tag{1.13}$$

This trace can be interpreted as the "logarithmic divergence" of $\mathrm{tr}\, A = \int \mathrm{d}x \, \langle x|A|x \rangle$, from which follows the cyclicity property $\mathrm{Tr}AB = \mathrm{Tr}BA$ for any two pseudo-differential operators A, B (also easily verified directly by considering the trace of basis elements $a(x)\mathrm{d}^m$ and $b(x)\mathrm{d}^n$).

Since $R_{l+1} = \frac{1}{2} \operatorname{Res} K^{l+1/2}$, we see that

$$\frac{\delta}{\delta u} \int dx\, R_{l+1}[u] = \frac{\delta}{\delta u} \frac{1}{2} \operatorname{Tr} K^{l+1/2} = -(l + \tfrac{1}{2})\frac{1}{2} \operatorname{Res} K^{l-1/2}$$
$$= -(l + \tfrac{1}{2})R_l[u] \, . \tag{1.14}$$

The differential equation (1.10) therefore results as the variational derivative with respect to u of the action

$$S = \int dx \left(R_{l+1} + xu \right) \, .$$

(We treat the above integral formally here and ignore throughout that physically relevant boundary conditions on u typically preclude existence of such integrals.)

For the general massive model interpolating between multicritical points (corresponding to taking $P = \sum t_{(k)} K_+^{k-1/2}$), the string equation is [1,3,17]

$$x = \sum_{k=1} \left(k + \tfrac{1}{2}\right) t_{(k)}\, R_k[u] \, . \tag{1.15}$$

Using (1.14), eq. (1.15) can be seen to follow from the action

$$S = 2 \int dx \left(\sum_{k=1} t_{(k)} R_{k+1}[u] + t_0\, R_1 \right)$$
$$= \operatorname{Tr}\left(\sum_{k=1} t_{(k)}\, Q^{k+1/2} + t_0\, Q^{1/2} \right) \, . \tag{1.16}$$

where $t_0 \equiv -4x$, $Q = K$. We shall see shortly that the form of the action (1.16) generalizes to (p,q) models.

As described in [3,17], the dependence of the specific heat u on the parameters $t_{(k)}$ is given by the higher KdV flows

$$\frac{\partial}{\partial t_{(k)}} u = \frac{\partial}{\partial x} R_{k+1}[u] = \frac{1}{4}\left[K_+^{k+1/2}, K \right] \, . \tag{1.17}$$

Using the commutativity of the higher KdV flows, it is straightforward to verify consistency of (1.17) with (1.15). Since $u = \langle \mathcal{P}\mathcal{P} \rangle$, we have $(\partial/\partial t_{(k)})\langle \mathcal{P}\mathcal{P}\rangle = \langle \mathcal{P}\mathcal{P}O_{(k)}\rangle = (\partial/\partial x)\langle \mathcal{P}O_{(k)}\rangle$, where $O_{(k)}$ is the appropriate scaling operator [3,17] that couples to $t_{(k)}$. (1.17) therefore identifies $R_{k+1}[u] = \langle \mathcal{P}O_{(k)}\rangle$ as the 2-point function of the puncture operator $\mathcal{P}$ with $O_{(k)}$, and we can rewrite the string equation (1.15) and the action (1.16) in terms of these 2-point functions.

2. (p,q) string actions [5]

We now turn to consider the first few cases of higher matrix models. The Ising model has a natural realization as a two-matrix model [18] in which the two matrices represent the $+/-$ states of an Ising spin. Adding a third matrix to represent vacancies gives a realization of the tricritical Ising model [19].

2.1. The critical Ising model (4,3)

For the simplest two-matrix model [20,4] (see also appendix C of [5]), we take

$$Q = \mathrm{d}^3 - \frac{3}{4}\{u,\mathrm{d}\} + \frac{3}{2}w = K_+^{3/2} + \frac{3}{2}w \ ,$$
$$P = Q_+^{4/3} = K^2 + \{w,\mathrm{d}\} + v \ ,$$

(2.1)

where w is a $\mathbb{Z}_2$ breaking field, ultimately resulting in coupling to a magnetic field. For generality, we can also perturb P by a term $-tK$, which represents a deviation from criticality of the Ising model, i.e. in the direction of pure gravity (as described by the Painlevé equation $R_2[u] \sim x$ — recall that $\left[K_+^{3/2}, K\right] = 4R_2'$). Calculating the commutator $1 = [P,Q]$ of (1.5), and integrating appropriate linear combinations of the resulting equations, we find

$$0 = 4R_2 + \frac{3}{2}v \ ,$$

(2.2a)

$$-h = w'' - 3uw - 3tw \ ,$$

(2.2b)

$$x = -8R_3 + \frac{3}{2}uv - \frac{1}{4}v'' + \frac{3}{2}w^2 + 4t\,R_2 \ ,$$

(2.2c)

where h is a constant of integration corresponding to an external magnetic field.

From eqns. (2.2a–c), we read off the scaling properties $u \sim [x]^{1/3}$, $v \sim [x]^{2/3}$, $w \sim [x]^{1/2}$, so that $h \sim [x]^{5/6}$, $t \sim [x]^{1/3}$. Correlation functions of the spin field and energy operator (which have dressed gravitational weights [21] equal to $\frac{5}{6}$ and $\frac{1}{3}$) can be calculated by taking derivatives of the solution u with respect to h and t.

We observe that (2.2a,b,c) can be derived as derivatives respectively with respect to v, w, u of the action

$$S_{\mathrm{Ising}}(u,v,w) = \int \mathrm{d}x \left[\frac{16}{7}R_4 + 4R_2 v + \frac{3}{2}R_0 v^2 - \frac{8}{5}tR_3 \right.$$
$$\left. + \frac{1}{2}w'^2 + \frac{3}{2}(u+t)w^2 - hw - xu \right] .$$

(2.3)

2.2. The tricritical Ising model (5,4)

For the simplest case that requires a three-matrix model, we parametrize the two operators as

$$Q = K^2 + \{w,\mathrm{d}\} + v \ ,$$

(2.4a)

$$P = Q_+^{5/4} = K_+^{5/4} + \frac{5}{4}\{w,\mathrm{d}^2\} + \frac{5}{8}\{v,\mathrm{d}\} - \frac{5}{4}uw \ ,$$

(2.4b)

where w is again a $\mathbb{Z}_2$ breaking field that results in coupling to a magnetic field. Again after integrating suitable linear combinations of the equations derived by setting $1 = [P,Q]$, we find

$$T = 4R_3 + \frac{5}{8}v'' - \frac{5}{4}uv + \frac{5}{4}w^2 \ ,$$

(2.5a)

$$h = \frac{1}{2}w^{(4)} - \frac{5}{4}(uw)'' - \frac{5}{4}uw'' - \frac{5}{2}vw + \frac{5}{4}u^2w \ ,$$

(2.5b)

$$x = 8R_4 + \frac{1}{16}v^{(4)} + \frac{5}{8}v^2 + \frac{15}{8}u^2v - \frac{5}{8}(uv'' + u'v' + vu'')$$
$$- \frac{5}{4}ww'' + \frac{5}{4}w^2u + Tu \ ,$$

(2.5c)

where T and h are constants of integration.

We recognize $(2.5a, b, c)$ as functional derivatives with respect to v, w, u of the action

$$
\begin{aligned}
S_{\text{t.c. Ising}}(u, v, w) = -\frac{9}{8} \int \mathrm{d}x \Big(& -\frac{16}{9} R_5 - 4v R_3 - \frac{5}{2} R_1 v^2 + \frac{5}{16} v'^2 \\
& + \frac{1}{4}(w'')^2 - \frac{5}{4} v w^2 + \frac{5}{8} w^2 u^2 - \frac{5}{4} u w w'' + T(\tfrac{1}{2}u^2 + v) - hw - xu \Big) .
\end{aligned}
\tag{2.6}
$$

To perturb the tricritical Ising model in the direction of the next lower model, i.e. the Ising model, we use instead of $(2.4b)$

$$
\widetilde{P} = P - V Q_+^{3/4} = P - V \left(K_+^{3/4} + \frac{3}{2} w \right) ,
\tag{2.7}
$$

where Q remains as in $(2.4a)$. The equations that follow from $\left[\widetilde{P}, Q \right] = 1$ are the tricritical Ising equations $(2.5a, b, c)$ plus V times respectively the Ising equations $(2.2a, b, c)$. (For V "large" in some sense, the equations cross over to the Ising equations.) The action for the full perturbed tricritical Ising system is given in terms of the actions (2.3) and (2.6) of the independent systems by the combination

$$
\widetilde{S}(u, v, w) = S_{\text{t.c. Ising}} + V S_{\text{Ising}} .
\tag{2.8}
$$

The genus zero equations for the tricritical Ising model are given by ignoring the derivatives in $(2.5a, b, c)$. From these equations, we read off the scaling properties $u \sim [x]^{1/4}$, $v \sim [x]^{1/2}$, $w \sim [x]^{3/8}$, so that $h \sim [x]^{7/8}$, $T \sim [x]^{3/4}$, $V \sim [x]^{1/4}$. $7/8$ is the gravitationally dressed weight [21] of the spin field (the $(2,2)$ operator with undressed conformal weight $3/80$) in the tricritical Ising model, $3/4$ is the dressed weight of the energy operator (the $(3,3)$ operator with undressed conformal weight $1/10$), and $1/4$ is the dressed weight of the vacancy operator (the $(3,2)$ operator with undressed conformal weight $3/5$). Derivatives of the free energy with respect to the parameters h, T, V generate correlation functions of these operators.

2.3. Ising and tricritical Ising actions reconsidered

To illustrate how the actions (2.3) and (2.6) generalize, we show that the action (2.8) (which leads to eqs. $(2.5a, b, c)$) can equivalently be written

$$
S_{\text{t.c. Ising}} = \operatorname{Tr} \left(Q^{9/4} + t_{(1),2} \, Q^{7/4} + t_2 \, Q^{3/4} + t_1 \, Q^{1/2} + t_0 \, Q^{1/4} \right) ,
\tag{2.9}
$$

where we take $Q = K^2 + \{w, \mathrm{d}\} + v$ as in $(2.4a)$. We first note that the constants t_0, t_1, t_2 in (2.9) couple respectively to

$$
\begin{aligned}
\operatorname{Res} Q^{1/4} &= -u/2 \\
\operatorname{Res} Q^{1/2} &= w \\
\operatorname{Res} Q^{3/4} &= \frac{3}{4}(v + \frac{8}{3} R_2) = \frac{3}{4}(v + \frac{1}{2} u^2 - \frac{1}{6} u'') ,
\end{aligned}
\tag{2.10}
$$

so can be identified as $t_0 \sim x$, $t_1 \sim h$, $t_2 \sim T$ (i.e. they are proportional to the integration constants in (2.6)). $\operatorname{Tr} Q^{7/4}$ will turn out to be proportional to S_{Ising} of (2.3), so that $t_{(1),2} \sim V$.

So we need to show that $\operatorname{Tr} Q^{9/4}$ gives identically the portion of $S_{\text{t.c. Ising}}$ of (2.6) independent of the couplings to T, h, and x. In terms of the coefficients e_i in the expansion

$$Q_-^{5/4} = \{e_1, \mathrm{d}^{-1}\} + \{e_2, \mathrm{d}^{-2}\} + \{e_3, \mathrm{d}^{-3}\} + \dots ,$$

the equations derived from (1.5) are equivalent to

$$
\begin{aligned}
1 = [P, Q] = [Q_+^{5/4},\, Q] &= [Q,\, Q_-^{5/4}] \\
&= \left[\mathrm{d}^4 - \{\mathrm{d}^2, u\} + \{w, \mathrm{d}\},\ \{\mathrm{d}^{-1}, e_1\} + \{\mathrm{d}^{-2}, e_2\} + \{\mathrm{d}^{-3}, e_3\} + \dots\right] \\
&= 4\{e_1', \mathrm{d}^2\} + 4\{e_2', \mathrm{d}\} + 2\left(4e_3' - 2u'e_1 - 4ue_1' + e_1'''\right) .
\end{aligned}
$$

$$(2.11)$$

But taking the variational derivatives of the first term in (2.9), we find

$$
\begin{aligned}
\frac{\delta}{\delta v(x)} \operatorname{Tr} Q^{9/4} &= \frac{9}{4} \operatorname{Res} Q^{5/4} = \frac{9}{2} e_1 , \\
\frac{\delta}{\delta w(x)} \operatorname{Tr} Q^{9/4} &= \frac{9}{4} \operatorname{Res}\{\mathrm{d}, Q^{5/4}\} = 9 e_2 , \\
\frac{\delta}{\delta u(x)} \operatorname{Tr} Q^{9/4} &= \frac{9}{4} \operatorname{Res}\left(-\{\mathrm{d}^2, Q^{5/4}\} + 2u\, Q^{5/4}\right) = -9\left(e_3 + \tfrac{1}{2} e_1'' - u e_1\right) .
\end{aligned}
$$

$$(2.12)$$

The relations in (2.11) are easily expressed as linear combinations of the derivatives of the variations (2.12). Moreover by combining (2.10) and (2.12), we see that setting to zero the variation of the action (2.9) yields the equations $e_1 \sim T$, $e_2 \sim h$, $e_3 + uT \sim x$. These are equivalent to the integrated form of the equations generated by the commutator (2.11), i.e. to eqs. (2.5a, b, c). ($\operatorname{Tr} Q^{9/4}$ may also be calculated directly to establish the normalization chosen in (2.6).)

The same argument just used to show that $\operatorname{Tr} Q^{9/4}$ can be used to generate the tricritical Ising equations can equally be used to verify that $\operatorname{Tr} Q^{7/4}$ is proportional to the Ising action (2.3), thus identifying $t_{(1),2} \sim V$. (In general $\operatorname{Tr} Q^{p/q+1}$ will generate the portion of the action for a (p, q) model independent of integration constants. In [5], we established a proportionality between the actions $\operatorname{Tr} Q^{p/q+1}$ and $\operatorname{Tr} \widetilde{Q}^{q/p+1}$ generated by operators $Q, \widetilde{Q}$ of order q, p respectively. In particular, this shows why $\operatorname{Tr} Q^{7/4}$ can serve instead of the action $\operatorname{Tr} Q^{7/3}$ constructed from the 3rd order operator Q of (2.1).)

The form of (2.9) suggests adding more general terms of the form $t_{(k),\alpha}\, Q^{k+(\alpha+1)/4}$. For terms such as $t_{(1),1}\, Q^{3/2}$, $t_{(1),0}\, Q^{5/4}$, for example, $t_{(1),1}$ and $t_{(1),0}$ would have grades 3 and 4, hence couple respectively to operators of dressed weight $d = \frac{3}{8}, \frac{1}{2}$. We could also consider adding higher terms such as $t_{(2),2}\, Q^{11/4}$, for which $t_{(2),2}$ (of grade -2) would couple to an operator of weight $-\frac{1}{4}$, coincident with the dressed weight of the supersymmetry generator of the tricritical Ising model (the $(3,1)$ operator with undressed conformal weight $3/2$).

2.4. Action for $(q-1)$-matrix models and generalized KdV flows

The action generalizing (2.9) is

$$S = \mathrm{Tr}\left(Q^{p/q+1} + \sum_{k=1}^{q-2}\sum_{\alpha=0}^{q-2} t_{(k),\alpha}\, Q^{k+(\alpha+1)/q} + \sum_{\alpha=0}^{q-2} t_\alpha\, Q^{(\alpha+1)/q}\right), \qquad (2.13)$$

with Q as in (1.3). The middle terms can be interpreted as perturbing P by the operator $t_{(k),\alpha}\, Q_+^{k-1+(\alpha+1)/q}$ (similar to the t perturbation added to the Ising model in (2.2), perturbing to pure gravity, and the V perturbation to the tricritical Ising model in (2.7), perturbing to ordinary Ising). The remaining t_α's correspond to the constants of integration generated by solving (1.5). They couple to the functions $u_\alpha \equiv \mathrm{Res}\, Q^{(\alpha+1)/q}$, which are simple functions of the v_α's of (1.3) (and their derivatives), and replace them as the basic variables in the theory. (The u_α's also have the natural interpretation as 2-point functions, $u_\alpha = \langle \mathcal{P}O_\alpha\rangle$, of certain scaling operators O_α with the puncture operator $\mathcal{P}$, as we shall confirm.) Finally, when $(k-1)q + \alpha + 1$ and q are relatively prime, $t_{(k),\alpha}$ also corresponds to a perturbation of the (p,q) model in the direction of the $(p',q) = \big((k-1)q+\alpha+1,q\big)$ model. This is because $\mathrm{Tr}\, Q^{k+(\alpha+1)/q}$ in this case emerges equally as the action $S_{((k-1)q+\alpha+1,q)}$ for a $\big((k-1)q+\alpha+1,q\big)$ model.

The proof that the action (2.13) gives the same equations as are generated by $[P,Q]=1$ with $P = Q_+^{p/q}+\dots$ is a simple generalization of the argument that showed that the vanishing of the variational equations in (2.12) is equivalent to (2.11). The details may be found in [5]. As pointed out in [8], the action S is also proportional to partition function Z for these models.

Since the multi-matrix model action (2.13) generalizes the action (1.16) for the one-matrix models, we mention briefly how the "generalized" KdV flows associated to the multi-matrix case (as suggested in [4]) extend the KdV flows (1.17) of the one-matrix models. The "$\big((k),\alpha\big)^{\mathrm{th}}$ flow" of the qth KdV hierarchy is generated by

$$\frac{\partial}{\partial t_{(k),\alpha}}Q = \left[Q_+^{k+(\alpha+1)/q}, Q\right], \qquad (2.14)$$

(for convenience, we now employ a normalization in which $t_0 = x$) which can be interpreted equivalently as an infinite set of compatible differential equations for the functions v_α characterizing the differential operator Q of (1.3). It is easily seen that

$$\langle \mathcal{P}\mathcal{P}O_{(k),\gamma}\rangle \propto \frac{\partial}{\partial t_{(k),\gamma}}v_{q-2} \propto \frac{\mathrm{d}}{\mathrm{d}x}\frac{\partial}{\partial v_0}\mathrm{Tr}\, Q^{k+1+(\gamma+1)/q} \propto \frac{\mathrm{d}}{\mathrm{d}x}\, \mathrm{Res}\, Q^{k+(\gamma+1)/q},$$

where we have identified $\partial/\partial t_{(k),\alpha}$ with the insertion of the scaling operator $O_{(k),\alpha}$. (Our notation $(k),\alpha$ is chosen to conform to the descendant structure proposed in [12,22]: $O_{(k),\alpha}$ has a natural interpretation as the gravitational descendant $\sigma_k(O_\alpha)$ of the operator O_α.) Up to normalization, it follows that

$$\mathrm{Res}\, Q^{k+(\alpha+1)/q} = \langle \mathcal{P}O_{(k),\alpha}\rangle \qquad (2.15)$$

(generalizing the result $\mathrm{Res}\, K^{k+1/2} = \langle \mathcal{P}O_{(k)}\rangle$ discussed after (1.17)). For $k=0$, this identifies the variables $u_\alpha = \mathrm{Res}\, Q^{(\alpha+1)/q}$ as two-point functions, $u_\alpha = \langle \mathcal{P}O_\alpha\rangle$. Phrased in terms of the u_α's, the commutativity [7] of the generalized KdV flows insures the consistency of the operator interpretation: infinitesimally these flows correspond to the insertion of scaling operators. Here this consistency follows from the relation $(\partial/\partial t_{(k),\alpha})u_\beta \propto \langle O_{(k),\alpha}O_\beta\mathcal{P}\rangle$.

2.5. Other generalizations

The models considered here are interpreted as the so-called A-series of conformal $d < 1$ matter coupled to 2d gravity. In [13], there is proposed an analogous formulation for the D-series coupled to gravity. For this formulation, an analogous action principle $S = \int \mathrm{Res} Q^{p/q+1}$ can be written down and shown to generate the equations of motion[23].

The matrix model formulation (1.1) that leads to the fundamental continuum equation (1.2) can also be considered before taking the double scaling limit. In that case, (1.2) remains a matrix equation which can be shown as well to emanate from a "discretized" action whose continuum limit is equivalent to the action (2.13). Some properties of this action in discrete form are discussed in [24].

3. Large order behavior and Borel summability

In this section we show how one can determine the large order behavior of the topological expansion of the $d < 1$ models by a straightforward analysis of the differential equation satisfied by their partition functions. After recalling some standard facts about divergent series, we discuss in detail the simplest case of pure gravity, and sketch the generalization to other cases.

3.1. Divergent series and Borel transforms

We start by recalling some standard features of divergent series, Borel summability, and summation methods (see, for example, pp. 840–842 of [25] for a recent treatment with physical applications). Consider a function $f(w)$, analytic in some sector S (say $|\mathrm{Arg}\, w| \leq \alpha/2$, $|w| \leq |w_0|$) in which it has an asymptotic expansion

$$f(w) \approx \sum_0^\infty f_k\, w^k \; . \tag{3.1}$$

This means that the series diverges for all non-vanishing w, but in S there is a bound of the form

$$\left| f(w) - \sum_{k=0}^N f_k\, w^k \right| \leq C_{N+1}\, |w|^{N+1} \qquad \text{for all } N \; , \tag{3.2}$$

and for definiteness we assume that $C_N = M\, A^{-N}\, (\beta N)!$. Though the series diverges, it can be used to estimate $f(w)$ for $|w|$ small by taking the truncation of the series (3.1) at a value of N that minimizes the bound (3.2). This gives the best possible estimate of $f(w)$, with a finite error $\varepsilon(w) = \min_{\{N\}} C_N\, |w^N| \sim \exp -(A/|w|)^{1/\beta}$. An asymptotic series does not in general define a unique function since we can always add to it any function analytic and smaller than $\varepsilon(w)$ in the sector S.

When the angle α defining the sector S above satisfies $\alpha > \pi\beta$, however, a classical theorem of analytic functions applies to show that a function analytic in S and bounded there by $\varepsilon(w)$ must vanish identically. This is the only case in which the asymptotic series defines a unique function $f(w)$, and for which there exist methods to reconstruct the function from the series. One such method is based on the Borel transform $B_f(w)$ of $f(w)$, defined from the expansion (3.1) by

$$B_f(w) = \sum_0^\infty b_k \, w^k \equiv \sum_0^\infty \frac{f_k}{(\beta k)!} w^k \; . \tag{3.3}$$

According to the assumptions following (3.1), we have $|f_k/(\beta k)!| < M \, A^{-k}$, so $B_f(w)$ is analytic at least in a circle of radius A and uniquely defined by the series. Then the integral representation

$$f(w) = \int_0^\infty dt \; e^{-t} \, B_f(w \, t^\beta) \tag{3.4}$$

converges in the sector $|\mathrm{Arg}\, w| \le \alpha/2$ for $|w|$ small enough, and yields the unique function which has the asymptotic expansion (3.1) in the domain S.

In general, however, the function $B_f(w)$ may have poles and cuts running from its singularities off to infinity on the complex plane. If there is a singularity on the positive real axis, we say that the original series for $f(w)$ is not Borel summable since we cannot run the contour in (3.4) along the real axis. Cuts in $B_f(w)$ are indicative of possible "non-perturbative" effects (i.e. exponential in an inverse string coupling, $\kappa^{-1} \sim x^{(2l+1)/2l}$ in the notation of (1.10)), and generally the choice of contour in (3.4) from the origin to ∞ in the $\mathrm{Re}\, t > 0$ half-plane reflects possible "non-perturbative" ambiguities.

We now recall how the large order behavior of $f(w)$ may be extracted from such non-perturbative behavior. The coefficients of the series defining $B_f(w)$ satisfy

$$b_k = \frac{1}{2i\pi} \oint_C \frac{ds}{s^{k+1}} B_f(s) \underset{k \to \infty}{\propto} \frac{1}{2i\pi} \int_r^\infty ds \, \frac{1}{s^{k+1}} \, \mathrm{disc}\, B_f(s) \; , \tag{3.5}$$

where the contour C encloses the origin. For k large, the behavior of b_k is related to values of $B_f(s)$ near the point on the contour where $|s|$ is minimal, so by deforming the contour to run along the cuts of $B_f(s)$ to infinity, we see that the integral above is dominated by the discontinuity of $B_f(s)$ along the cut corresponding to the singularity r closest to the origin. After Borel transformation it follows that the large order behavior of the original series is given by

$$f_k = (\beta k)! \, b_k = \int_0^\infty dt \, e^{-t} \, t^{\beta k} \, b_k = \frac{1}{2i\pi} \int_0^\infty dt \, e^{-t} \oint \frac{ds}{s^{k+1}} B_f(st^\beta)$$

$$\underset{k \to \infty}{\propto} \int_0^\infty dt \, e^{-t} \int_{r/t}^\infty \frac{ds}{s^{k+1}} \, \mathrm{disc}\, B_f(st^\beta) \propto \int_0^\infty \frac{ds}{s^{k+1}} \int_r^\infty dt \, e^{-t} \, \mathrm{disc}\, B_f(st^\beta)$$

$$= \int_0^\infty \frac{ds}{s^{k+1}} \left(f_+(s) - f_-(s) \right) \; ,$$

$$\tag{3.6}$$

where $f_\pm(s)$ are the Borel transforms corresponding to integrations in (3.4) on opposite sides of the cut.

In the cases of interest to follow here, both $f_\pm$ will satisfy the same differential equation, and the (exponentially small) difference

$$\epsilon(w) = f_+(w) - f_-(w) = \int_r^\infty dt \, e^{-t} \operatorname{disc} B_f(w \, t^\beta) \tag{3.7}$$

will be determined by the corresponding linearized equation. Knowledge of the leading behavior of $\epsilon(w)$ can be used to infer a great deal about $f(w)$ and $B_f(s)$. For example, when $\epsilon(w)$ has leading behavior

$$\epsilon(w) \sim w^{-b/\beta} \, e^{-(A/w)^{1/\beta}} \,, \tag{3.8}$$

we find from (3.6) that $f_k \sim \Gamma(\beta k + b)$ for k large. Moreover we see from (3.7) that the above behavior for $\epsilon(w)$ results from the singular behavior $B_f(s) \sim (1 - s/A)^{-b}$ near $s = A$, where A is the singularity nearest the origin in the Borel plane. For large k this means that $b_k \sim A^{-k} k^{b-1}$ and hence we have the refined estimate $f_k \sim (\beta k)! \, A^{-k} k^{b-1}$. This is typical of large order behavior of perturbation theory in quantum mechanics and field theory, where A is a classical instanton action.[2] In what follows, we can now bypass the intermediate steps and use eqs. (3.6)–(3.8) directly to determine the asymptotics of f_k and the locations of singularities of $B_f(s)$.

3.2. Pure gravity

For pure gravity, the differential equation satisfied by the second derivative of the partition function is ((1.10) with $l = 2$, after suitable rescaling)[3]

$$u^2(z) - \frac{1}{3} u''(z) = z \,. \tag{3.9}$$

If $u(z)$ has an asymptotic expansion for z large, it satisfies $u(z) = \pm\sqrt{z} + O\left(z^{-2}\right)$. The solution that corresponds to pure gravity has a z large expansion of the form

$$u(z) = z^{1/2}\left(1 - \sum_{k=1}^\infty u_k \, z^{-5k/2}\right), \tag{3.10}$$

where the u_k are all positive.[4]

---

[2] As stressed by Shenker[26], however, the value of β provides an important distinction between string theory and field theory. In field theory, $\beta = 1$ and the non-perturbative effects in the exponential in (3.8) go as the inverse loop coupling, $w^{-1} = 1/g^2$. In string theory, on the other hand, we shall see that $\beta = 2$, leading to much larger non-perturbative effects in the exponential that go as the inverse square root of the loop coupling, $w^{-1/2} = 1/\kappa$.

[3] We exchange x for z in what follows since we continue to the complex plane. Our normalization in (3.9) corresponds to a matrix model with an even potential; for an odd potential the second term is instead $-\frac{1}{6}u''$.

[4] The first term, i.e. the contribution from the sphere, is dominated by a regular part which has opposite sign. This is removed by taking an additional derivative of u, giving a series all of whose terms have the same sign — negative in the conventions of (3.9). The other solution, with leading term $-z^{1/2}$, has an expansion with alternating sign which is presumably Borel summable, but not physically relevant.

To determine the large order behavior of the expansion we argue as in (3.3)–(3.6). We consider the Borel transform of the expansion, defined by

$$B(s) = \sum_{k=1}^{\infty} \frac{u_k}{(\beta k)!} s^k \, , \tag{3.11}$$

in which β is chosen so that the series (3.11) is convergent in a circle of finite radius. Then a solution (in general complex) to equation (3.9) is obtained from the integral

$$\tilde{u}(z) = z^{1/2}\left(1 - \int_0^{\,} dt\ e^{-t} B\!\left(t^\beta\, z^{-5/2}\right)\right) \, , \tag{3.12}$$

provided there exists a suitable contour of integration from the origin to infinity in the $\mathrm{Re}\,t > 0$ half-plane on which the integral converges.

The functions $u_\pm(z)$ defined respectively by integration in (3.12) above and below the cut both satisfy equation (3.9). For z large, their difference $\epsilon \equiv u_+ - u_-$ is exponentially small compared to their average $u_0 \equiv (u_+ + u_-)/2$. $\epsilon(z)$ is therefore a solution of the equation obtained by linearizing (3.9). Taking the differences of the equations (3.9) satisfied by u_+ and u_-, we find that ϵ satisfies

$$\epsilon''(z) - 6u_0(z)\,\epsilon(z) = 0 \, , \tag{3.13a}$$

where u_0 is determined by

$$u_0^2(z) - \frac{1}{3}u_0''(z) + \epsilon^2(z) = z \tag{3.13b}$$

(and $\epsilon^2(z)$ in (3.13b) can be ignored to leading order in large z). To leading order, the function ϵ is also proportional to the difference between any Borel sum of the series and the exact non-perturbative solution of the differential equation (up to even smaller exponential corrections corresponding to multi-instanton like effects).

Eq. (3.13a) can easily be solved by the WKB method for z large. Substituting the ansatz $\epsilon'/\epsilon = r u_0^{1/2} + b\, u_0'/u_0$, we find $r^2 = 6$ and $b = -1/4$. Dividing by the leading term $u_0 \sim z^{1/2}$ to remove the overall factor $z^{1/2}$ in (3.10) gives[5]

$$\frac{\epsilon(z)}{z^{1/2}} \propto z^{-5/8}\, e^{-\frac{4\sqrt{6}}{5} z^{5/4}} \left(1 + \ldots\right) \, . \tag{3.14}$$

---

[5] In [27], it is confirmed that the coefficient in the above exponential coincides with the action for a single eigenvalue climbing to the top of the barrier in the matrix model potential. This allows us to interpret the exponential piece of the solution to (3.9) as an instanton effect. This is ultimately because the large order behavior is related to the instanton effects which appear for a quartic potential unbounded below, i.e. having quartic term $(1/g)\mathrm{tr}M^4$ with $g < 0$. A peculiarity of the usual large N limit is that because the instanton effects are of order $e^{-NK(g)}$, they disappear for any value of g in the interval $g_c < g < 0$ (the number of planar diagrams grows only geometrically with the order). As $g \to g_c$, however, $K(g)$ vanishes as $(g_c - g)^{5/4}$, so in the double scaling limit $NK(g)$ remains finite and turns out equal to the exponent in (3.14).

In terms of the expansion parameter (string loop coupling) $\kappa^2 = z^{-5/2}$, the ratio (3.14) reads

$$\sigma\big(z(\kappa)\big) \equiv \frac{\epsilon(z)}{z^{1/2}} \; \propto \; \kappa^{1/2}\, e^{-\frac{4}{5}(\sqrt{6}/\kappa)} \; . \tag{3.15}$$

The above solution is valid for z large, i.e. κ small, so we may apply (3.6) to find that the large order behavior in (3.10) is given by

$$u_k \underset{k\to\infty}{\propto} \int_0 \frac{d\kappa}{\kappa^{2k+1}}\, \sigma(\kappa) \; \propto \; \left(\frac{5}{4\sqrt{6}}\right)^{2k} \Gamma(2k - \tfrac{1}{2}) \; . \tag{3.16}$$

(The constant of proportionality in the above cannot be determined by this method.) The asymptotic $\Gamma(2k - \tfrac{1}{2})$ behavior is a slight refinement of the $(2k)!$ behavior determined in [1,3,28].

From the discussion following (3.8), we can see directly from (3.15) that $\beta = 2$, and the reality of r^2 has implied a singularity on the real axis in the Borel plane. This obstruction to Borel summability is consistent with the large order behavior in (3.16), in which all terms have the same sign.

Remark. Note that other singularities of the Borel transform are related to higher order corrections in the expansion of ϵ. Because the coupled equations $(3.13a,b)$ for u_0 and ϵ have a well-defined parity in ϵ, the exponential behavior of successive non-perturbative corrections to ϵ will be of the form $\exp(-n\, C\, z^{5/4})$, where $C = (4/5)\sqrt{6}$ and n is an odd integer. This follows from iterating $(3.13a,b)$: $\exp(-n\, C\, z^{5/4})$ terms in ϵ in $(3.13b)$, with n odd (including the leading $\exp(-C\, z^{5/4})$ piece), result only in $\exp(-m\, C\, z^{5/4})$ terms in u_0 with m even, and vice versa, in $(3.13a)$.

3.3. Ising / Yang–Lee

We now consider the fourth order differential equations [20,4] for the Yang–Lee edge singularity and the critical Ising model. After suitable rescaling (different for the two cases, (1.10) with $l = 3$ and $(2.2a\text{--}c)$ with $h = t = 0$), the equations are written

$$u^3 - uu'' - \tfrac{1}{2}u'^2 + au^{(4)} = z \; , \tag{3.17}$$

where $a = \tfrac{1}{10}, \tfrac{2}{27}$ respectively for Yang–Lee and critical Ising.

The asymptotic expansion takes the form

$$u(z) = z^{1/3}\left(1 + \sum_{k=1} u_k\, z^{-7k/3}\right) \; .$$

Substituting $u = u_0 + \epsilon$ (where now $u_0 \sim z^{1/3}$), we find that the linearized equation for the discontinuity $\epsilon(z)$, at leading order for z large, reads

$$a\epsilon^{(4)} + 3z^{2/3}\epsilon - z^{1/3}\epsilon'' - \frac{1}{3}z^{-2/3}\epsilon' = 0 \; .$$

The ansatz $\epsilon'/\epsilon = ru_0^{1/2} + bu_0'/u_0$ now gives $b = -3/4$ and yields a solution with the asymptotic form

$$\sigma \equiv \frac{\epsilon(z)}{z^{1/3}} \; \propto \; z^{-7/12}\, e^{-\frac{6}{7}rz^{7/6}} \; , \tag{3.18}$$

where r satisfies

$$ar^4 - r^2 + 3 = 0 \ . \tag{3.19}$$

The solutions are $r^2 = 9/2, 9$ in the case of the Ising model and $r^2 = 5 \pm i\sqrt{5}$ in the Yang–Lee edge case.

In terms of the expansion parameter $\kappa = z^{-7/6}$, we have $\sigma(\kappa) \propto \kappa^{1/2} \exp\left(-\frac{6}{7}(r/\kappa)\right)$, and the large order behavior according to (3.6) is

$$u_k \underset{k \to \infty}{\propto} \int_0 \frac{\mathrm{d}\kappa}{\kappa^{2k+1}} \, \sigma(\kappa) \ \propto \ \int_0 \frac{\mathrm{d}\kappa}{\kappa^{2k+1/2}} \, \mathrm{e}^{-\frac{6}{7}(r/\kappa)} \ \propto \ \left(\frac{7}{6r}\right)^{2k} \Gamma(2k - \tfrac{1}{2}) \ . \tag{3.20}$$

Remarkably enough, the large order behavior takes the same form, $\Gamma(2k - \frac{1}{2})$, as in (3.16).

From (3.19), we see that r^2 is complex when $a > 1/12$, and from (3.20) we see that this is related to the non-unitarity of the Yang–Lee edge case, since it leads to asymptotic coefficients u_k that are not positive definite. This is also related to the Borel summability in the Yang–Lee edge case, since from the discussion following (3.8) we see that the poles nearest the origin are a finite distance off the real axis in the Borel plane. There thus exists a real and physically acceptable Borel sum, presumably equal to the solution of [29]. In the (unitary) Ising case, on the other hand (with r^2 real), the terms of the series are all positive, there is a singularity on the real axis in the Borel plane, and the series is not Borel summable.

We shall shortly generalize the conclusion concerning Borel summability in the ($l = 3$) Yang–Lee case to all the l odd one-matrix models. We can also show that it is not affected by higher order exponential corrections. If we look for such corrections by expanding in ϵ, we find that the coefficient r in the exponential is replaced by $n_+ r + n_- r^*$, but because the equations have a well-defined parity in ϵ, $n_+ + n_-$ must necessarily be odd.[6] All such allowed terms will not occur (some of them can correspond to singularities of the Borel transform B in other sheets of the complex plane), but we see in any event that no singularity will appear on the positive real axis and an integral like (3.12) may be performed to define a unique real function. This function solves the differential equation of interest and is therefore the natural candidate for the partition function of the original matrix problem.

3.4. $l = 3$ Yang–Lee perturbed by $l = 2$ pure gravity[7]

Starting from some matrix model potential $V(M)$, we define polynomials $P_i(\lambda)$ orthogonal with respect to the measure $\mathrm{d}\lambda \exp -\left(NV(\lambda)/g\right)$. These polynomials satisfy recursion relations whose coefficients R_i are determined in the large N limit

---

[6] The solution ϵ_1 of the linearized equation is a linear combination of the two exponentials involving r and r^*, each multiplied by power series. The leading order corrections to u_0 involve ϵ_1^2 and thus correspond to exponentials involving $2r$, $2r^*$, and $r + r^*$. The argument then proceeds by iteration, just as remarked at the end of subsection *3.2* concerning the analogous properties of solutions to eqs. (3.13a,b).

[7] We thank in particular S. Shenker for discussions regarding this subsection.

$(i/N \to x,\ R_i \to NR(x))$ by an equation of the form $gx = W(R)$ [28]. To describe a perturbation of an $l = 3$ model in the direction of an $l = 2$ model we write, after suitable rescalings,

$$W(R) = -(1 - R)^3 - \xi_2(1 - R)^2 \ . \tag{3.21}$$

Recalling [28] that a potential $V(M) = \sum_p g_p M^{2p}$ in general leads to $W(R) = 2\sum_p \frac{(2p-1)!}{((p-1)!)^2} g_p R^p$, up to a constant term the matrix model potential associated to (3.21) is

$$2V(\lambda) = 3\lambda^2 - \frac{1}{2}\lambda^4 + \frac{1}{30}\lambda^6 + \xi_2\left(2\lambda^2 - \frac{1}{6}\lambda^4\right) \ . \tag{3.22}$$

The double scaling limit in this case [20] involves taking $R = 1 + N^{-2/7}u$, and $\xi_2 = N^{-2/7}T$. The resulting (all-genus) string equation that describes the flow of the $l = 3$ model (3.17) to the $l = 2$ model (3.9) is

$$u^3 - uu'' - \frac{1}{2}u'^2 + \frac{1}{10}u^{(4)} + T\left(u^2 - \frac{1}{3}u''\right) = z \ . \tag{3.23}$$

The scaling limit corresponds to take $z \sim (g - g_c)N^{6/7}$, $T \sim u \sim z^{1/3} \sim N^{2/7}$. Thus, at leading order (genus zero), z and T are large with z/T^3 fixed in such a way that the leading order equation becomes

$$u^3 + Tu^2 = z \ . \tag{3.24}$$

To treat this (mixed) case more easily, we change variables to $\tilde{z} \equiv z/T^3$, $\tilde{u}(\tilde{z}) \equiv u/T$, and $\kappa^2 \equiv T^{-7}$. Eq. (3.23) reads with these notations

$$\tilde{u}^3 + \tilde{u}^2 - \kappa^2\left(\tilde{u}\tilde{u}'' + \frac{1}{2}\tilde{u}'^2 + \frac{1}{3}\tilde{u}''\right) + \frac{1}{10}\kappa^4\tilde{u}^{(4)} = \tilde{z} \tag{3.23$'$}$$

(where perturbation theory corresponds to an expansion in small κ) and eq. (3.24) becomes

$$\tilde{u}^3 + \tilde{u}^2 = \tilde{z} \ . \tag{3.24$'$}$$

By the same linearization procedure ($\tilde{u} \mapsto \tilde{u} + \epsilon$) as used before, we find at leading order

$$(3\tilde{u}^2 + 2\tilde{u})\epsilon - \kappa^2\left(\tilde{u} + \frac{1}{3}\right)\epsilon'' + \frac{1}{10}\kappa^4\epsilon^{(4)} = 0 \ .$$

Substituting the WKB ansatz $\epsilon'/\epsilon = \kappa^{-1}\tilde{u}^{1/2}r$ gives an algebraic equation for r,

$$0 = 3 - r^2 + \frac{1}{10}r^4 + \frac{1}{\tilde{u}}\left(2 - \frac{1}{3}r^2\right) \ . \tag{3.25}$$

r^2 depends on $\tilde{z} = z/T^3$ implicitly through $\tilde{u} = u/T$. In the limits $\tilde{u}$ large and small (and hence via (3.24)$'$ $\tilde{z}$ large and small), we recover the cases $l = 3$ and $l = 2$ respectively. For some value of $\tilde{z}$ and hence of $\tilde{u} = u/T$ classical, the solution of eq. (3.25) goes from complex to real. Thus for T large enough, the equation for r^2 has real roots and the Borel summability is lost.

Now we examine when the potential (3.22) begins to admit real instantons. Taking the derivative with respect to λ^2, we find the equation for minima of (3.22) away from the origin

$$0 = 3 - \lambda^2 + \frac{1}{10}\lambda^4 + \xi_2\left(2 - \frac{1}{3}\lambda^2\right) . \tag{3.26}$$

Above some value of ξ_2, which corresponds to the critical value of $\tilde{u}^{-1} = T/u$ in (3.25) for which the roots in r^2 become real, we see that the potential (3.22) allows real instantons, giving a physical interpretation for the loss of Borel summability.

This argument relating the loss of Borel summability to the appearance of real instantons is straightforwardly generalized to the case of perturbations of any l-odd by any l-even model.

3.5. The tricritical Ising model

In the case of the tricritical Ising model we perform an analogous computation by substituting $u \to u + \epsilon_u$, $v \to b + \epsilon_v$ into eqs. (2.5a–c) with $T = h = 0$. We introduce at leading order the ansatz $\epsilon_u'/\epsilon_u = ru^{1/2}$, $\epsilon_v'/\epsilon_v = ru^{1/2}$, and recall that at leading order $v \sim -u^2/2$. This gives a system of two linear equations for ϵ_u and ϵ_v/u,

$$\begin{aligned}
\left(-\tfrac{1}{2}r^4 + 5r^2 - 10\right)\epsilon_u + \left(5r^2 - 10\right)\frac{\epsilon_v}{u} &= 0 , \\
\left(r^4 - 10r^2 + 20\right)\left(-\tfrac{1}{2}r^2 + 2\right)\epsilon_u + \left(r^4 - 10r^2 + 20\right)\frac{\epsilon_v}{u} &= 0 .
\end{aligned} \tag{3.27}$$

Imposing the vanishing of the determinant of the 2×2 matrix of coefficients yields an equation for r^2. Two roots come from the overall factor of $r^4 - 10r^2 + 20$ in the second equation above, and the other two roots are then determined by the equation $r^4 - 5r^2 + 5 = 0$. The four solutions for r^2 are

$$r^2 = 5 \pm \sqrt{5}, \; \tfrac{1}{2}(5 \pm \sqrt{5}) .$$

Since these give real values of r, the theory is not Borel summable. This is again as expected for a unitary theory, in which the coefficients in the asymptotic expansion have fixed sign. The large order behavior is, up to the value of r^2, the same as found in the earlier cases (eqs. (3.16) and (3.20))

$$u_k \propto \left(\frac{9}{8r}\right)^{2k} \Gamma\left(2k - \tfrac{1}{2}\right) . \tag{3.28}$$

3.6. The general one-matrix problem

We consider the string equation (1.10), $R_l[u] \sim z$. To examine the leading large order behavior of perturbation theory, it is only necessary to know the terms in $R_l[u]$ of the form

$$R_l[u] = \frac{1}{l} A_{ll} u^l + \sum_{j=1}^{l-1} A_{lj} u^{j-1} u^{(2l-2j)} + \ldots \tag{3.29}$$

(i.e. that contain at most one derivative of u factor. The next leading contribution is given by terms such as $u^{j-2}u^{(2l-2j-1)}u'$, i.e. with a single factor of u' as well). From the recursion relation (1.11), we have for example that the coefficient of the highest derivative term $u^{(2l-2)}$ in (3.29) is given by $A_{l1} = -4^{-l}$ and the coefficient of u^l is $A_{ll}/l = (-1)^l (2l-1)!!/(2^{l+1} l!)$.

Denoting the discontinuity of $u(z)$ by $\epsilon(z)$ and substituting $u = u_0 + \epsilon$ as before (now $u_0 \sim x^{1/l}$), to leading order for z large we find that ϵ satisfies

$$0 = \sum_{j=1}^{l} A_{lj} \, u_0^{j-1} \, \epsilon^{(2l-2j)} \ . \tag{3.30}$$

Substituting the WKB ansatz $\epsilon'/\epsilon = r u_0^{1/2}$ then gives

$$0 = \sum_{j=1}^{l} A_{lj} \, r^{2l-2j} \ , \tag{3.31}$$

an $(l-1)^{\text{st}}$ order equation for r^2 with real coefficients. For l even, this gives an odd–order equation that will have at least one real positive solution for r^2 (positive since A_{l1} and A_{ll} have opposite sign for l even).

Since $A_{l1} = -4^{-l}$, we change variables to $\rho \equiv 4/r^2$. Then the r.h.s. of (3.31) takes the form $-A_l(\rho)/(4\rho^{l-1})$, where $A_l(\rho) = 1 + \dots$ is an $(l-1)^{\text{st}}$ order polynomial in ρ, and (3.31) is equivalently expressed as

$$0 = A_l(\rho) \ . \tag{3.31$'$}$$

The relation
$$\frac{\partial R_l}{\partial u} = -(l - \tfrac{1}{2}) R_{l-1} \ . \tag{3.32}$$

allows us to determine $A_l(\rho)$ explicitly. (In contrast to the functional variation $\delta/\delta u$ in (1.14), $\partial/\partial u$ in (3.32) denotes the derivative of a functional of u only with respect to *constant* variations of u. This derivative satisfies $\partial K^\lambda/\partial u = -\lambda K^{\lambda-1}$, and hence $\partial K_\pm^{l-1/2}/\partial u = -(l - \tfrac{1}{2}) K_\pm^{l-3/2}$, implying (3.32).) From (3.29) and (3.32) we find that

$$A_l'(\rho) = -(l - \tfrac{1}{2}) A_{l-1}(\rho) \ , \tag{3.33}$$

so together with the normalization condition $A_l(0) = 1$, we have

$$A_l(\rho) = (1 - \rho)^{l-1/2}\big|_{l-1} \ , \tag{3.34}$$

where the subscript means that the small–ρ series expansion of $(1-\rho)^{l-1/2}$ is truncated at order $l - 1$.[8]

By induction, it is straightforward to prove from (3.34) that for $\rho \le 1$,

$$A_{2l}(\rho) < (1-\rho)^{2l-1/2} \ , \qquad A_{2l+1}(\rho) > (1-\rho)^{2l+1/2} \ ,$$

---

[8] (3.34) is also proportional to $\xi^{-1} C_{2l-1}^{1-l}(\xi)$ where $\xi^2 = 1/\rho$ and C is a Gegenbauer polynomial defined by analytic continuation in l. [3]

and furthermore that

(i) $A_{2l}(\rho)$ has only a single real zero, which falls in the interval $[0,1]$; and

(ii) $A_{2l+1}(\rho)$ has only a single minimum, which falls in the interval $[0,1]$ where the function is positive. We see that $A_{2l+1}(\rho)$ cannot vanish for ρ real.

For l even, the series therefore cannot be Borel summable because the equation (3.31) for r^2 always has a real positive solution. Since it has only one such solution, we expect the ambiguity of the corresponding solution to the differential equation to be characterized by one parameter.

For l odd, on the other hand, the equation (3.31) for r^2 has no real solutions and therefore we expect the solution of the differential equation to be determined by the perturbative expansion.

Now let us recall the explicit expressions for the integrand $\mathrm{e}^{-NV(M)}$ of the critical matrix models,

$$V(M) = \mathrm{tr} \int_0^1 \frac{\mathrm{d}s}{s} W\big(s(1-s)M^2\big) \, ,$$

with[9]

$$W(R) = g_c - \alpha(r_c - r)^l \, .$$

It was pointed out in [29] that according to whether l is odd or even, the original matrix integral is well-defined or not because $V(M)$ is always positive in the first case while in the latter case it becomes negative for M large. We thus find a direct correspondence between the property of Borel summability and the existence of the original integral.[10] We expect as well that the results here, including the explicit values of the exponentials, can be recovered from instanton calculations (as mentioned for the $l = 2$ case in the footnote before eqn. (3.14)). When the potential is unbounded from below the instanton action is real and the series therefore non-Borel summable, in accord with our comments above.

Although there is no general proof, there is good evidence for another difference between the odd and even cases: Before taking the large N continuum limit, the operators P_i and Q_i (described before eq. (1.2)) have a continuous spectrum and form a regular representation of the canonical commutation relations. This property still holds for the the operator $\mathrm{d}^2 - u$ when l is odd, while for l even the operator has a discrete spectrum. A careful analysis of the scaling limit in the $l = 2$ case [32] has shown that the spectrum cannot be discrete for physically interesting cases.

---

[9] Note that these formulæ reproduce the relation between $V(M)$ and $W(R)$ given after (3.21). There we chose normalizations such that $\alpha = R_c = 1$. The sign of the term linear in R in $W(R)$ is always determined by the sign of the quadratic term in $V(M)$ that gives a convergent gaussian integral.

[10] In [30], it was further shown that it is impossible to flow by means of the perturbation of eq. (3.21) from the real solution of the $l = 3$ model found in [29] to a solution of the $l = 2$ pure gravity model. This result generalizes to show that flows from real solutions of arbitrary l-odd models to solutions of l-even are impossible [31] (see also [27]), further distinguishing the l-odd and l-even cases.

The subleading terms in $R_l[u]$ mentioned after (3.29) are immediately deduced from the leading terms by noting that since $R_l[u]$ is derived from an action (see eq. (1.14)), the operator acting on ϵ is hermitian. Therefore the operator $u^{j-1}d^{2l-2j}$ should be replaced by the symmetrized form $\frac{1}{2}\{u^{j-1}, d^{2l-2j}\}$, correcting (3.30) to

$$0 = \sum_{j=1}^{l} A_{lj}\left(u^{j-1}\epsilon^{(2l-2j)} + \tfrac{1}{2}(2l-2j)(j-1)u^{j-2}\,u'\,\epsilon^{(2l-2j-1)}\right) . \tag{3.35}$$

To characterize more precisely the large order behavior, to next order we set

$$\frac{\epsilon'}{\epsilon} = ru^{1/2} + b\frac{u'}{u} ,$$

from which it follows, to the same order, that

$$\frac{\epsilon^{(k)}}{\epsilon} = r^k u^{k/2} + r^{k-1}u^{(k-3)/2}\,u'\,k\left(b + \frac{1}{4}(k-1)\right) . \tag{3.36}$$

Substituting into (3.35), we find

$$0 = \sum_{j=1}^{l} A_{lj}\left(r^{2l-2j} + u^{-3/2}\,r^{2l-2j-1}\,u'(l-j)\big[j-1+2\big(b+\tfrac{1}{4}(2l-2j-1)\big)\big]\right) . \tag{3.37}$$

We see that r remains a solution to (3.31) and $b = (3 - 2l)/4$, independent of r. Dividing ϵ by the leading term $u \propto z^{1/l}$ results in

$$\sigma \equiv \frac{\epsilon(z)}{z^{1/l}} \propto z^{-(2l+1)/4l}\,e^{-\frac{2l}{2l+1}rz^{(2l+1)/2l}} , \tag{3.38}$$

generalizing (3.14) and (3.18). In terms of the expansion parameter $\kappa = z^{-(2l+1)/2l}$, we find

$$\sigma(\kappa) \propto \kappa^{1/2}\,e^{-\frac{2l+1}{2l}(r/\kappa)} ,$$
$$u_k \underset{k\to\infty}{\propto} \int_0^{} \frac{d\kappa}{\kappa^{2k+1/2}}\,e^{-\frac{2l+1}{2l}(r/\kappa)} \propto \left(\frac{2l+1}{2lr}\right)^{2k}\Gamma\left(2k-\tfrac{1}{2}\right) . \tag{3.39}$$

We see that the $\Gamma(2k - \frac{1}{2})$ factor in (3.16), (3.20), and (3.28) is general, owing in the case of one-matrix models to the special form of the subleading term in the equation (3.35) satisfied by ϵ (due to the fact that the original equations descended from an action principle). Again we note that the method employed here also shows that some models (for which r has no real solutions) are not unitary since terms in the perturbation series at large order do not have a fixed sign.

3.7. General (p,q) model

In the case of the general (p, q) model (eqs. (1.2) and (2.13)) there results a system of coupled linear differential equations for the variations ϵ_u, $\epsilon_{v_i}(x)$ associated with the functions $u(x)$, $v_i(x)$. As in the case of the tricritical Ising model considered in *3.5*, at leading order we set $\epsilon'_u/\epsilon_u = ru^{1/2} = \epsilon'_{v_i}/\epsilon_{v_i}$. We obtain, taking into account the leading relations between u and the v_i, a linear system for ϵ_u and each of the ϵ_{v_i}'s multiplied by a power of u determined by the grading (generalizing eq. (3.27)). Imposing again the vanishing of the determinant of the linear system gives an equation for the coefficient r (and to leading order all functions ϵ_{v_i} are thus proportional to ϵ_u multiplied by the power of u determined by the grading).

To determine more precisely the behavior of ϵ_u we have to consider subleading terms. As in the one-matrix case they can be determined by a hermiticity argument. Since the equations for u, v_i derive from an action (2.13), the linear equations for ϵ_u, ϵ_{v_i} define a hermitian operator. Eliminating for example all the ϵ_{v_i} yields an equation to next leading order for ϵ_u which can be expressed as a hermitian operator acting on ϵ_u (as was the case leading to (3.35)). The coefficient of the subleading term that led to the $\Gamma(k - \frac{1}{2})$ behavior found in (3.39) for the one-matrix models, since it only depended on the hermiticity of the operator acting on ϵ, is thus universal for all the (p, q) models.

For any given q, we expect the general (p, q) model to be Borel summable or not according to the $\mathrm{mod}\,2$ parity of p. (The unitary $(m + 1, m)$ models will never be Borel summable for reasons already explained.) The same $\mathrm{mod}\,2$ grading will as well determine which models may flow to one another.

3.8. Possible moveable singularities

All the differential equations for the one-matrix models (ordinary KdV, generated by the R_l's in subsection *1.2*) have as moveable singularities double poles. For all the equations investigated so far, the residues of these double poles are even integers. Although it was shown in [32] that solutions to these equations with poles on the real axis are not physically relevant (i.e. do not correspond to the gravity model of interest beyond perturbation theory), complex poles do exist even for the relevant solutions. For completeness we briefly recall here some previous results on possible poles of these solutions and add results for the Ising and tricritical Ising models.

For the one-matrix models, we set

$$u(x) \sim a/x^2 \ , \qquad R_l[u] \sim \rho_l(a)/x^{2l} \ ,$$

and using the recursion relation (1.11) gives

$$\rho_{l+1}(a) = \frac{2l + 1}{2(l + 1)} \big[a - l(l + 1)\big]\rho_l(a) \ .$$

The solutions of the equation $R_l[u] = x$ have double poles with residues belonging to the set $2, ..., j(j + 1), ..., l(l - 1)$ (all of which are even integers).

For the critical and tricritical Ising models, a similar analysis shows that the possible residues are $2, 10$ and $2, 6, 8, 14, 30$ respectively, again even integers.

These results are calculated using a normalization for the differential equation generated by an even matrix model potential, and the poles above all correspond to double zeroes of the partition function. For generic models with non-even potentials, on the other hand, the residues are all divided by two and the partition function has only simple zeroes.

4. Large order behavior and Borel summability for $d = 1$

We conclude here with some comments about non-perturbative 2d quantum gravity coupled to a single bosonic field ($d = 1$ matter) [10,33]. As for $d < 1$ models, the continuum theory is formulated from a matrix realization of the discretized model by taking a double scaling limit in which the 2d cosmological constant g tends towards a critical value g_c (at which the perturbation series diverges) and the string coupling $1/N \to 0$. For $d < 1$ [1–3], the combination $\xi = (g - g_c)N^{2m/(2m+1)}$ was held fixed to give a smooth scaling limit (for a non-unitary model with string susceptibility $\gamma = -1/m$). Due to the $g \to g_c$ enhancement of the high genus contribution, this scaling limit is interpreted as the continuum limit for the theory summed over all topologies. In the case of $d = 1$, we shall see that a scaling limit may be defined by holding fixed instead a scaling parameter $\alpha \propto \ln(g - g_c)/((g - g_c)N)$ [10,33].

In [34], it was pointed out that the vacuum energy E_0 of a matrix model solved by large N techniques in [35] could be interpreted as the partition function Z for 2d gravity coupled to $d = 1$ matter. This vacuum energy is calculated as the groundstate energy of a matrix hamiltonian, which, if we ignore "angular" excitations, becomes a sum $H = \sum_{i=1}^{N} H(\lambda_i)$ of independent hamiltonians for the eigenvalues, where

$$H(\lambda) = -\frac{1}{2N}\frac{\mathrm{d}^2}{\mathrm{d}\lambda^2} + \frac{N}{g}V\left(\lambda\sqrt{g}\right) , \tag{4.1}$$

and V is the original matrix model potential. The groundstate energy is given by the sum $E_0 = \sum_{i=1}^{N} e_k$, where the e_k are the first N eigenvalues of the one-body hamiltonian $H(\lambda)$. (This is the energy of an ideal Fermi gas of N particles at zero temperature, with the Fermi level given by the highest energy e_N.) In the large N limit, it is convenient to rescale to eigen-energies $\varepsilon(t) \equiv e_k g/N$, where $t \equiv gk/N$. Then the (rescaled) ground state energy, which gives the partition function Z for the gravity theory, may be written $E = E_0/N^2 = g^{-2}\int_0^g \mathrm{d}t\, \varepsilon(t)$, from which we see that the leading singular behavior of the specific heat $C \propto \partial^2 Z/\partial g^2$ is proportional to $\varepsilon'(g)$. To obtain a non-trivial scaling limit to the continuum, we look for a singularity in the function $\varepsilon(g)$ at some $g = g_c$ for which the perturbation series diverges. This emphasizes graphs with an infinite number of vertices, which, after rescaling of area, can be used to define the continuum limit theory of interest.

To leading order in small $\delta \equiv 2\pi(g_c - g)$, the leading singular behavior is found to be given by

$$\varepsilon \sim \frac{\delta}{\ln\delta}\left(1 + \frac{2}{\ln\delta}f(\alpha)\right) , \tag{4.2}$$

where the expansion parameter $\alpha = -g_c\ln\delta/(2N\delta)$. By explicit treatment of the Schrödinger equation defined by the Hamiltonian (4.1) (for a generic potential with a quadratic deviation from its critical point), the function $f(\alpha)$ was determined [10,33] to be given by the expansion coefficients

$$f_{2k} = (-1)^k(2^{2k-1} - 1)\frac{B_{2k}}{2k(2k-1)} , \tag{4.3}$$

where the B_k's are Bernoulli numbers. In this formulation logarithmic corrections already present at tree level persist to all higher genus, suggesting a significant difference from $d < 1$ matter.

In [10], we have shown that the large order behavior of the expansion in powers of α can also be obtained by considering barrier penetration effects (for a recent review of semiclassical methods in this context, see [25]). These effects are typically of the form $k!/A^k$, where the A is an instanton action given by the integral $\int dx \sqrt{V(x) - E}$ between the turning points. Here, after rescaling, the energy of the Fermi sea is -1 and the potential is $-x^2$. The instanton action is thus $A = 2 \int_{-1}^{+1} dx \sqrt{1 - x^2} = \pi$, integrated between the turning points $x = \pm 1$.

In the case at hand, we find that the expansion coefficients of the function $f(\alpha) = \sum^\infty f_{2k} \alpha^{2k}$ have the asymptotic behavior

$$f_{2k} \underset{k \to \infty}{\sim} \frac{1}{\pi^{2k}} \Gamma(2k - 1) \, , \tag{4.4}$$

giving naively the contribution to the partition function from surfaces of genus k. This is in agreement with the asymptotic form of the Bernoulli numbers in (4.3) although the present point of view allows us to understand the $(2k)!$ behavior as a result of barrier penetration effects.

The $(2k)!$ large order behavior is also the generic behavior for the $c < 1$ models (more precisely we found $f_{2k} \sim \Gamma(2k - \frac{1}{2})$ in the previous section). The perturbative expansion by itself then does not fully determine the partition function and instead misses some essential non-perturbative feature of the problem. In our $d = 1$ formulation, we expect that the barrier penetration effects, induced by the property that the critical potential necessarily has a maximum, will play an important role.

It is also possible to consider "multicritical" models, determined by the hamiltonian (4.1) with a potential $V(\lambda)$ that has its first $s - 1$ derivatives vanishing at the critical point. These models have string susceptibility $\gamma = -(s - 2)/(s + 2)$, and their physical interpretation is in general unclear. They do however provide a useful arena for studying how perturbations from one model to another can destroy Borel summability. We expect again a mod 2 grading that determines which models are Borel summable, and as well which models may flow to one another.

Acknowledgements

During most of this work P.G. was supported by NSF contract PHY-82-15249; P.G. is also supported by DOE contract W-7405-ENG-36, DOE OJI grant FG-84ER40171, and by the A. P. Sloan foundation. We thank S. Shenker, T. Banks, and I. Klebanov for discussions, and J. Cohn and P. Mende for comments on the manuscript while it was being completed at the Aspen Center for Physics.

References

[1] M. Douglas and S. Shenker, Nucl. Phys. B335 (1990) 635.

[2] E. Brézin and V. Kazakov, Phys. Lett. B236 (1990) 144.

[3] D. Gross and A. A. Migdal, Phys. Rev. Lett. 64 (1990) 127; Nucl. Phys. B340 (1990) 333.

[4] M. R. Douglas, Phys. Lett. B238 (1990) 176.

[5] P. Ginsparg, M. Goulian, M. R. Plesser, and J. Zinn-Justin, Nucl. Phys. B342 (1990) 539.

[6] I. M. Gel'fand and L. A. Dikii, Russian Math. Surveys 30:5 (1975) 77.

[7] I. M. Gel'fand and L. A. Dikii, Funct. Anal. Appl. 10 (1976) 259.

[8] A. Jevicki and T. Yoneya, Mod. Phys. Lett. A5 (1990) 1615.

[9] P. Ginsparg and J. Zinn-Justin, "Large order behavior of non-perturbative gravity", LANL/Saclay preprint (1990), to appear in Phys. Lett. B.

[10] P. Ginsparg and J. Zinn-Justin, Phys. Lett. B240 (1990) 333.

[11] M. L. Mehta, Comm. Math. Phys. 79 (1981) 327;
S. Chadha, G. Mahoux and M. L. Mehta, J. Phys. A14 (1981) 579.

[12] E. Witten, Nucl. Phys. B340 (1990) 281.

[13] P. Di Francesco and D. Kutasov, Nucl. Phys. B342 (1990) 589.

[14] C. Bachas and P.M.S. Petropoulos, "Doubling of equations and universality in matrix models of random surfaces," CPTH-A-964-0490.

[15] V. G. Drinfel'd and V. V. Sokolov, Jour. Sov. Math. (1985) 1975;
G. Segal and G. Wilson, Pub. Math. I.H.E.S. 61 (1985), 5.

[16] M. Adler, Inv. Math. 50 (1979) 219.

[17] T. Banks, M. Douglas, N. Seiberg, and S. Shenker, Phys. Lett. B238 (1990) 279.

[18] V. Kazakov, Phys. Lett. 119A (1986) 140;
D. Boulatov and V. Kazakov, Phys. Lett. 186B (1987) 379.

[19] M. Kreuzer and R. Schimmrigk, "Tricritical Ising model, generalized KdVs, and nonperturbative 2d-quantum gravity", ITP Santa Barbara preprint NSF-ITP-90-30 (1990);
H. Kunitomo and S. Odake, "Nonperturbative analysis of three matrix model", Univ. Tokyo preprint UT-558 (1990);
K. Fukazawa, K. Hamada, and H. Sato, "Phase structures of 3-Matrix chain model", UT-Komaba 90-11 / HUPD-9006 (1990).

[20] E. Brézin, M. Douglas, V. Kazakov, and S. Shenker, Phys. Lett. B237 (1990) 43;
D. Gross and A. A. Migdal, Phys. Rev. Lett. 64 (1990) 717;
C. Crnković, P. Ginsparg, and G. Moore, Phys. Lett. B237 (1990) 196.

[21] V. G. Knizhnik, A. M. Polyakov, and A. B. Zamolodchikov, Mod. Phys. Lett. A3 (1988) 819;
F. David, Mod. Phys. Lett. A3 (1988) 1651;
J. Distler and H. Kawai, Nucl. Phys. B321 (1989) 509.

[22] R. Dijkgraaf and E. Witten, Nucl. Phys. B342 (1990) 486.

[23] P. di Francesco and P. Ginsparg, unfinished discussion.

[24] P. Ginsparg and J. Zinn-Justin, to appear.

[25] J. Zinn-Justin, *Quantum Field Theory and Critical Phenomena*, Oxford Univ. Press (1989).

[26] S. Shenker, presentation at Cargèse workshop, these proceedings.

[27] F. David, "Phases of the large N matrix model and non-perturbative effects in 2d gravity," Saclay preprint SPhT/90-090 (1990).

[28] D. Bessis, C. Itzykson, and J.-B. Zuber, Adv. Appl. Math. 1 (1980) 109.

[29] E. Brézin, E. Marinari, and G. Parisi, Phys. Lett. B242 (1990) 35.

[30] M. Douglas, N. Seiberg, and S. Shenker, Phys. Lett. B244 (1990) 381.

[31] G. Moore, "Geometry of the string equations," Yale preprint YCTP-P4-90 (1990).

[32] F. David, Mod. Phys. Lett. A5 (1990) 1019.

[33] E. Brézin, V. A. Kazakov, and Al. B. Zamolodchikov, Nucl. Phys. B338 (1990) 673;
G. Parisi, Phys. Lett. B238 (1990) 209, 213; Europhys. Lett. 11 (1990) 595;
D. J. Gross and N. Miljkovic, Phys. Lett. B238 (1990) 217.

[34] V. A. Kazakov and A. A. Migdal, Nucl. Phys. B311 (1988) 171.

[35] E. Brézin, C. Itzykson, G. Parisi and J.-B. Zuber, Comm. Math. Phys. 59 (1978) 35.

W- GEOMETRY

C. Itzykson

Service de Physique Théorique de Saclay
Laboratoire de la Direction des Sciences de la Matière
du Commissariat à l'Energie Atomique
F-91191 Gif-sur-Yvette Cedex France

1. During the meeting I presented work done with J.-B. Zuber on matrix combinatorics applied to triangulated surfaces. Since other contributors to this volume cover this subject, rather than paraphrasing our paper, I thought it would be more appropriate to discuss here some investigations carried in collaboration with P. Di Francesco and J.-B. Zuber on the relation between geometry, differential operators and classical *W*- algebras. The latter were introduced by Zamolodchikov as a generalization of the Virasoro algebra in the context of conformal quantum field theory. Their status remains however still dubious (at least for this author) and it is difficult to pinpoint a precise geometric definition. Some still very preliminary and incomplete indications at the end of this paper were worked out in collaboration with M. Bauer.

Except for a few remarks, I will concentrate on *W*-algebra pertaining to the linear group, although following the work of Drinfeld and Sokolov -our basic reference- it is possible to generalize the notion to other simple Lie groups. Consequently the underlying geometry will be the projective or affine one.

Such a classical subject as linear differential operators in one variable is a very rich one. It embraces uniformization of algebraic curves, differential Galois theory, Kortweg de Vries integrable systems and their generalizations, including applications to the Schottky problem (of characterizing Jacobian varieties) or recent findings in two-dimensional quantum gravity, Painlevé equations... In a nutshell it may be interpreted as a linearization of many interesting non linear problems. One could also mention the background of quantum mechanics with its use of differential algebras as a natural realization of operators.

We will focus here on one main theme: to investigate those relations which have a covariant character. In other words we wish to emphasize properties which are independent of a specific choice of coordinate, thus treating differential operators as intrinsic geometric objects.

At first we need not be specific about the basic field, either the real or complex numbers, most relations involving in fact rational coefficients. We deal with generic linear differential operators with regular coefficients, i.e. admitting as many derivatives as necessary. According to the context those are defined in a fixed neighborhood either in the real or complex domain. In some case we need a non trivial integration cycle C. This could extend over the entire real axis (assuming coefficients vanishing fast enough at infinity) or over an interval for periodic coefficients or even on a closed curve encircling some singularity in the complex case. One could also treat the whole subject by purely algebraic means and formal series.

Random Surfaces and Quantum Gravity
Edited by O. Alvarez *et al., Plenum Press, New York, 1991*

2. Linear differential operators give local maps of functions on functions (this will be qualified later). They are intrinsically determined (up to multiplication by a fixed function) by their kernel, a finite dimensional vector space. By quotienting this space through dilatations one can achieve two goals. On the one hand one can normalize the operator of order n to the form

$$Q_n = d^n + \sum_{p=2}^{n} a_p d^{n-p}$$

where d stands for the derivative $\frac{d}{dx}$ if x is the variable, by multiplying any function in the source space by an appropriate (x-dependent) factor. On the other hand, we retain only projective properties of the kernel (up to a finite cyclic extension which accounts for the difference between SL_n and PSL_n which does not affect infinitesimal properties). Hence above a fixed neighborhood U we can see the operator as defining a finite vector subspace E_n in the infinite dimensional vector space of regular functions, then its projectivization PE_{n-1}.

The normalizing procedure can also be understood as a means of selecting definite transformation properties of the source functions under diffeormorphisms (of U) and therefore of the coefficients of the operators as follows. Assume at first the source functions to have conformal weight λ, i.e. for a map $x_1 \longleftrightarrow x_2$ we have $f_1 \longleftrightarrow f_2$ such that $f_1(x_1) dx_1^\lambda = f_2(x_2) dx_2^\lambda$, it is natural to assume that Q_n increases the weight by n. Call $\mathcal{F}_\lambda$ the set of regular λ-differentials in U. Then

$$Q_n : \mathcal{F}_\lambda \longrightarrow \mathcal{F}_{\lambda+n}$$

Let us insist however to have a normalized operator in any coordinate. This requires the logarithmic derivative of the Wronskian of n linearly independent solutions (which is minus the coefficient of the term in d^{n-1}) to vanish in any coordinate. The Wronskian gives a map

$$\overset{n}{\Lambda} \mathcal{F}_\lambda \longrightarrow \mathcal{F}_{n\lambda + \frac{n(n-1)}{2}}$$

and its constancy can only be achieved if the target weight is zero, i.e. if

$$\lambda = \frac{1-n}{2}$$

If n is even $\frac{1 \mp n}{2}$ are half integral and a square root of the non vanishing jacobian has to be chosen in a coordinate change, this however does not affect projective properties based on ratios.

For p functions $f_0, ..., f_{p-1}$ write the Wronskian as

$$W_p(f_0, ..., f_{p-1}) = \det f_k^{(\ell)}, \qquad 0 \le k, \ell \le p-1$$

where $f^{(\ell)} = d^\ell f$. If $f_0, ..., f_{n-1}$ stand for a basis of $E_n = \ker Q_n$ we have

$$f \longrightarrow Q_n f = \sum_0^n a_p d^{n-p} f = \frac{W_{n+1}(f, f_0, ..., f_{n-1})}{W_n(f_0, ..., f_{n-1})}$$

with $a_0 = 1$, $a_1 = -W_n'(f_0, ..., f_{n-1})/W_n(f_0, ..., f_{n-1})$ and

$$f \in \mathcal{F}_{\frac{1-n}{2}} \qquad Q_n f \in \mathcal{F}_{\frac{1+n}{2}}$$

Since W_n is a scalar $W_n'/W_n \in \mathcal{F}_1$. Its vanishing is therefore a coordinate free property. For simplicity we assume in the sequel W_n equal to unity. Said otherwise if we start with arbitrary a_1, $f \in \mathcal{F}_\lambda$ then $f \exp - \frac{1}{n} \int^x a_1 \in \mathcal{F}_{\frac{1-n}{2}}$ and after this change of function Q_n is normalized. This discussion assumes therefore $W_n \neq 0$. It could happen that for a specific choice of basis $f_0, ..., f_{n-1}, W_n$ vanishes at certain points, these would have to be deleted as they correspond to singularities in the coefficients of Q_n.

From the above formula for the coefficients a_k one can derive their transformation properties under diffeormorphisms, in particular if $x \longleftrightarrow \tilde{x}$, $a_2 \longleftrightarrow \tilde{a}_2(\tilde{x})$ with

$$\tilde{a}_2(\tilde{x}) = a_2(x) \left(\frac{dx}{d\tilde{x}}\right)^2 + \frac{n(n^2-1)}{12} \{x, \tilde{x}\}$$

where $\{x, \tilde{x}\}$ stands here for the Schwarzian derivative

$$\{x, \tilde{x}\} = \frac{x'''}{x'} - \frac{3}{2}\left(\frac{x''}{x'}\right)^2 \qquad x \equiv x(\tilde{x})$$

More generally in infinitesimal form, $x = \tilde{x} + \epsilon(\tilde{x})$, we have (with $a_0 - 1 = a_1 = 0$)

$$\delta a_k = \epsilon a_k' + k \epsilon' a_k + \sum_{\ell=0}^{k-1} \left[\frac{n-1}{2}\binom{n-\ell}{k-\ell} - \binom{n-\ell}{k-\ell+1}\right]\epsilon^{(k+1-\ell)} a_\ell$$

3. Rather than identifying the data coded in Q_n with the vector space $E_n = \ker(Q_n)$, we can equally well think of a map $U \longrightarrow V_n$ where V_n is a fixed vector space equipped with a basis $\underline{e}_0, ..., \underline{e}_n$ by treating n independent solutions $f_0, ..., f_{n-1}$, as the coordinates of a vector

$$x \in U \xrightarrow{Q_n} \underline{f}(x) = \sum_{p=0}^{n-1} f_p(x)\underline{e}_p \in V_n$$

Thus we have (a segment of) a curve in V_n and after projectivization in PV_{n-1}. Allowing invertible linear transformations among the f_p's or $\underline{e}_p'$s (with constant coefficients) means that, in U, Q_n yields the neighborhood of a point on a curve in projective space of dimension $n-1$ (the lift in V_n being obtained by requiring $W_n = 1$) *up to an arbitrary projective transformation.* In other words Q_n codes the invariant (local and differential) aspects of this map. Note that one can draw a parallel with polynomial equations.

In the case of an algebraic curve we may remark that it not customary to embed it in projective space through $\frac{1-n}{2}$ (meromorphic) differentials. One uses rather 1 (or 3-) holomorphic differentials (the canonical embeddings) but in principle nothing prevents one from using other choices. For instance take the canonical embedding of the (genus 3) Klein curve giving the smooth quartic $uw^3 + vu^3 + wv^3 = 0$ in CP_2. For $n = 3$ using meromorphic vector fields it becomes a quintic $x_1^2 x_2^3 + x_2^2 x_3^3 + x_3^2 x_1^3 = 0$ with 3 cusps at the base points.

The case $n = 2$ where PV_1 is one dimensional deserves special mention as it is related to the classical uniformization problem of algebraic curves studied in great depth by Poincaré and the late XIXth century mathematicians. In its normalized form Q_2 is a Schrödinger operator

$$Q_2 = d^2 + a_2(x)$$

acting on $-1/2$ differentials. Thanks to its "anomalous" transformation property, the "potential" a_2 can be made to vanish in an appropriate coordinate u so that $a_2(x) = \frac{1}{2}\{u, x\}$. In the uniformization coordinate u the general element of the kernel is a first degree polynomial, hence u is in general the ratio of two solutions defined up to a PSL_2 transformation and we have

$$a_2 = \frac{1}{2}\left\{\frac{f_1}{f_0}, x\right\}$$

which agrees with the above formula as a ratio of determinants for $W_2 = 1$. To describe an algebraic (complex) curve one has to look at the non trivial global homotopy group $\bar{\mathbf{C}} - \{\text{sing } a_2\}$ acting as a monodromy group on $u = f_1/f_0$, giving a discrete subgroup of PSL_2. In the best of all worlds this is a Fuchsian group with a fixed circle. The standard example corresponds to the hypergeometric equation with three (regular) singularities at 0, 1, ∞ as described in textbooks.

4.   The transformation properties of the coefficients in Q_n are quite complicated except for a_2, which behaves like the energy momentum in conformal field theory. (This can be traced to the fact that we deal with projective representations of diffeormorphisms). This justifies to look for an invertible transformation from $a_2, a_3, ..., a_n$ to $w_2 \equiv a_2, w_3, ..., w_n$ where the w's are differential polynomials in the a's and vice versa (by differential polynomial we mean a polynomial both in the functions and finitely many of their derivatives) in such a way that w_k, $k \geq 3$ behaves like a k-differential ($w_k(x)\mathrm{d}x^k = \tilde{w}_k(\tilde{x})\,\mathrm{d}\tilde{x}^k$), the only anomaly appearing in $w_2 = a_2$ in the form of the only non trivial 2-cocycle, the Schwarzian derivative. As a result the w's will generate the same differential algebra as the a's.

The w's are not uniquely determined by the above properties even if we require that to leading order $w_k = a_k + ...$, for it is clear that if $k = k_1 + k_2$ with $k_1, k_2 \geq 3$ we can add to w_k a term proportional to $w_{k_1} w_{k_2}$ of if $k = k_1 + k_2 + 1$, $k_1 \neq k_2$, $k_1, k_2 \geq 3$ we can add to w_k a term in $k_1 w_{k_1} w'_{k_2} - k_2 w'_{k_1} w_{k_2}$ etc... . It turns out that $w_2 = a_2$, w_3, w_4, w_5 are in fact unique.

We can however make a canonical choice which excludes the former ambiguity by requiring that w_k be globally linear in the a'_ℓs and their derivatives ($\ell \geq 3$) with coefficients which are differential polynomials in a_2 up to an inhomogeneous term, itself a differential polynomial in a_2. We obtain this result as follows. We first pick a coordinate u in which a_2 vanishes. This requires solving a second order differential equation

$$\left(d^2 + \frac{6a_2}{n\,(n^2 - 1)} \right) f = 0 \qquad u = \frac{f_1}{f_0}$$

We then write the differential operator as

$$Q_n = \Delta_2^{(n)}(0) + \sum_{k=3}^{n} \Delta_k^{(n)}(w_k, 0)$$

where

$$\Delta_2^{(n)}(0) = d_u^n$$
$$\Delta_k^{(n)}(w_k, 0) = \sum_{\ell=0}^{n-k} \alpha_{k\ell} w_k^{(\ell)}(u) d_u^{n-k-\ell} \qquad k \geq 3$$

The coefficients

$$\alpha_{k,\ell} = \frac{\dbinom{k+\ell-1}{\ell}\dbinom{n-k}{\ell}}{\dbinom{2k+\ell-1}{\ell}}$$

are chosen in such a way that by returning to a general x-coordinate

$$Q_n(u) \longrightarrow Q_n(x) = \varphi^{\frac{n+1}{2}} Q_n(u)\varphi^{\frac{n-1}{2}} \qquad\qquad \varphi(x) = \frac{\mathrm{d}u}{\mathrm{d}x}(x)$$
$$Q_n(x) = \Delta_2^{(n)}(a_2) + \sum_{k=3}^{n} \Delta_k^{(n)}(w_k, a_2)$$

where the Δ's depend only on the Schwarzian derivative of the map, i.e. on $a_2 = \frac{1}{12}n\left(n^2 - 1\right)$ $\{u, x\}$, and its derivatives (as well as on the w's and their derivatives). Setting $b = \frac{1}{\varphi}\frac{d\varphi}{dx}$, the Schwarzian derivative is

$$\{u, x\} = \frac{db}{dx} - \frac{1}{2}b^2$$

so that with $j = \frac{n-1}{2}$

$$\Delta_2^{(n)}(x) = (d_x - jb)\,(d_x - (j-1)b)\cdots(d_x + jb)$$

and under a variation δb which leaves $\{u, x\}$ fixed i.e. such that $\delta b' - b\delta b = 0$ or

$$[d_x - (k+1)]\,\delta b = \delta b\,[d_x - kb]$$

we have

$$\delta\Delta_2^{(n)}(x) \;= -\sum_{-j}^{j}(d_x - jb)\cdots(d_x - (k+1)b)\,\delta b\,(d_x - (k-1)b)\cdots(d_x + jb)$$

$$= -\left(\sum_{-j}^{+j}k\right)\delta b\,(d_x - (j-1)b)\cdots(d_x + jb) = 0$$

One uses the same reasoning for

$$\Delta_k^{(n)}(x) \;= \varphi^{\frac{n+1}{2}}\left(\sum_{\ell=0}^{n-k}\alpha_{k\ell}w_k^{(\ell)}(u)d_u^{n-k-\ell}\right)\varphi^{\frac{n-1}{2}}$$

$$= \varphi^{\frac{n+1}{2}}\sum_{\ell=0}^{n-k}\alpha_{k\ell}\left[\left(\varphi^{-1}d_x\right)^{\ell}\varphi^{-k}w_k(x)\right]\left(\varphi^{-1}d_x\right)^{n-k-\ell}\varphi^{\frac{n-1}{2}}$$

$$= \sum_{\ell=0}^{n-k}\alpha_{k\ell}\left[\mathcal{D}^{\ell}w_k\right]\mathcal{D}^{n-k-\ell}$$

where $\mathcal{D}$ is a covariant derivative which acts from p differentials to $(p+1)$ ones as $d_x - pb$, i.e.

$$\left[\mathcal{D}^{\ell}w_k\right] = (d_x - (k+\ell-1)b)\cdots(d_x - kb)\,w_k$$

and

$$\mathcal{D}^{n-k-\ell} - (d_x - (n-k-\ell-j-1)b)\cdots(d_x + jb)$$

Requiring $\alpha_{k,0} = 1$ and $\delta\Delta_k^{(n)}(x) = 0$ under a variation δb with fixed a_2 leads to the recursion relation

$$\ell(\ell + 2k - 1)\alpha_{k,\ell} = (k+\ell-1)(n+1-k-\ell)\alpha_{k,\ell-1}$$

with the solution given above. To relate w's and a's one identifies w_3 as the coefficient of d^{n-3} in $Q_n - \Delta_2^{(n)}(a_2)$, w_4 as the coefficient of d^{n-4} in $Q_n - \Delta_2^{(n)}(a_2) - \Delta_3^{(n)}(w_3, a_2)$ and so on. This yields

$$w_2 \;= a_2$$

$$w_k \;= \sum_{\ell=2}^{k}B_{k,\ell}a_\ell^{(k-\ell)} + \sum_{\substack{0 \le p_1 \le p_2 \le \ldots \le p_r \\ \Sigma p_i + 2r = k}}C_{p_1,\ldots,p_r}a_2^{(p_1)}\cdots a_2^{(p_r)}$$

$$+ \sum_{\substack{0 \le p_1 \le p_2 \le \ldots \le p_r \\ 3 \le \ell \le k - \Sigma p_i - 2r}}D_{p_1,\ldots,p_r}a_2^{(p_1)}\cdots a_2^{(p_r)}a_\ell^{(k-\ell-\Sigma p_i - 2r)}$$

The coefficients $B_{k,\ell}$ (the only ones necessary in the u-coordinate) are given by

$$B_{k,\ell} = (-1)^{k-\ell} \frac{\dbinom{k-1}{k-\ell}\dbinom{n-\ell}{k-\ell}}{\dbinom{2k-2}{k-\ell}}$$

while conversely

$$a_k \;=\; \sum_{\ell=2}^{k} A_{k\ell} w_\ell^{(k-\ell)} + \text{ nonlinear terms}$$

$$A_{k,\ell} \;=\; \alpha_{\ell,k-\ell} = \frac{\dbinom{k-1}{k-\ell}\dbinom{n-\ell}{k-\ell}}{\dbinom{k+\ell-1}{k-1}}$$

Both $A_{k,\ell}$ and $B_{k,\ell}$ vanish when $\ell > k$ while one checks that

$$\sum_\ell B_{k\ell} A_{\ell m} = \delta_{k,m}$$

Explicitly

$$
\begin{aligned}
w_2 &= a_2 \\
w_3 &= a_3 - \frac{n-2}{2} a_2' \\
w_4 &= a_4 - \frac{n-3}{2} a_3' + \frac{(n-2)(n-3)}{10} a_2'' - \frac{(n-2)(n-3)(5n+7)}{10n(n^2-1)} a_2^2 \\
w_5 &= a_5 - \frac{n-4}{2} a_4' + \frac{3(n-3)(n-4)}{28} a_3'' - \frac{(n-2)(n-3)(n-4)}{84} a_3''' \\
&\quad + \frac{(n-3)(n-4)(7n+13)}{14n(n^2-1)} \left[(n-2)a_2 a_2' - 2a_3 a_2\right] \\
\ldots
\end{aligned}
$$

and it is a tedious but straightforward matter to extend the list. Of course for a given n we need only $w_2, ..., w_n$.

Because for generic n we made a definite, but arbitrary, choice of w's, it is at first glance frivolous to try to find a direct geometric interpretation of the w's. However a statement such as $w_k = 0$, $k \geq 3$ has an intrinsic meaning and implies that the curve is rational in U. For small n the w's are unique and we shall try an interpretation in the last section.

Moreover it is an interesting matter to count the number of linearly independent k-differentials ($k \geq 3$) as differentials polynomials in w's (or a's). Such a formula has been obtained by N. Sochen and J.-B. Zuber.

5. Differential operators admit a natural involution

$$Q_n \longmapsto Q_n^* = \sum a_p^* d^{n-p} = \sum (-d)^{n-p} a_p \;,$$

which act as antiautomorphism: $(Q_1 Q_2)^* = Q_2^* Q_1^*$. Note that we *do not* conjugate the coefficients would they happen to be complex. The above decomposition $Q_n = \sum_{k=2}^{n} \Delta_k^{(n)}$ enjoys the property $\Delta_k^{(n)*} = (-1)^{n-k} \Delta_k^{(n)}$.

116

The relation between the kernels of Q_n and Q_n^* (if Q_n is normalized to $a_0 - 1 = a_1 = 0$ so is $(-1)^n Q_n^*$) follows from a factorization of Q_n. Let $\{f_0\}$, $\{f_1, f_0\}$, ..., $\{f_{n-1}, ..., f_0\}$ be an increasing sequence of subspaces of dimensions $1, 2, ..., n$ in $E_n = \ker Q_n$ (a flag). Set

$$W_0 = 1 \qquad W_p \equiv W_p(f_{p-1}, ..., f_0) \qquad W_n = \text{cst}$$

then

$$Q_n = (d - b_n)(d - b_{n-1}) .... (d - b_1)$$

where

$$b_p = \left(\ell n \, \frac{W_p}{W_{p-1}} \right)' \quad ,$$

so that $\sum_1^n b_p = \frac{W_n'}{W_n} = 0$. We observe that W_p is $\frac{p(p-n)}{2}$ differential but of course not a differential polynomial in the a's since it depends on the choice of a flag. This makes factorizations not unique as opposed to the polynomial case. (The map $b \longrightarrow a$ is sometimes called a Miura transformation, it might as well be called a Ricatti transformation). The stabilizer of a flag in GL_n is a Borel sub-group of upper triangular matrices B_n so that the manifold of complete factorizations in GL_n/B_n is of dimension $n^2 - \frac{1}{2}n(n+1) = \frac{1}{2}n(n-1)$.

From

$$(-1)^n Q_n^* = (d + b_1) ... (d + b_n)$$

it follows that $\ker Q_n^*$ is spanned by the Wronskians W_{n-1} (since $W_n = \text{cst}$) for sets of $(n-1)$ linearly independent elements in $\ker Q_n$ i.e. we have the duality formula

$$\ker Q_n^* = \overset{n-1}{\Lambda} \ker Q_n$$

For $n = 3$ when we interpret $\ker Q_3$ as a curve in P_2 (two-dimensional projective space) we recognize that $\ker Q_3^*$ is associated to the dual curve. In general this suggests the consideration of the family of adjoint curves, describing tangents, osculating 2-planes, 3-planes etc... in $P_{\binom{n}{k}-1}$ and raises the question of finding the corresponding differential operators (or sets of operators). Since the corresponding Grassmannians are intersections of quadrics, we deal in the simplest case with a pair of differential equations which can be reduced to a pseudo-differential operator!

As a first instructive and non trivial example take $n = 4$ and $k = 2$, i.e. the set of tangent lines to a curve in P_3. These are points on a quadric in P_5. If

$$Q_4 = d^4 + a_2 d^2 + a_3 d + a_4$$

corresponds to the original curve and if for two independent elements f and g in $\ker Q_4$, we use the Plucker coordinates

$$u = 2(f'g'' - f''g') \qquad v = (fg' - f'g)$$

one finds the coupled differential system

$$\begin{aligned} (d^3 + a_2 d + a_3)\, v &= du \\ (d^3 + da_2 - a_3)\, u &= 2[a_4, d]_+\, v \end{aligned}$$

or equivalently the pseudo differential operator (obtained by eliminating u)

$$\begin{aligned} Q_4 \longrightarrow \hat{Q} \; &= d^5 + [a_2, d^3]_+ + \left[a_3' - a_2'' - 2a_4 + \frac{1}{2} a_2^2, d \right]_+ \\ &\quad - (a_3 - a_2')\, d^{-1}\, (a_3 - a_2') \end{aligned}$$

As in the six-dimensional vector space a quadratic form was left invariant and as SL_4 acting on $V_4 \wedge V_4$ is isomorphic to SO_6, we have explicitly exhibited here the relation between the differential operator Q_4 associated to the Lie algebra A_3 and the pseudo differential operator $\hat{Q}$ ($\hat{Q}^* = -\hat{Q}$) pertaining to $D_3 \approx A_3$ (Drinfeld and Sokolov). In agreement with this interpretation we see that the above system admits a quadratic differential invariant

$$d\left\{ uv'' - u'v' + u''v + a_2 vu - 2a_4 v^2 - \frac{u^2}{2} \right\} = 0$$

It is an interesting excercise to show that the bracket does transform as a scalar (v has weigth -2 but u is a scalar only mod v').

6. If one starts with a curve in projective space P_{n-1} in the neighborhood of a point and introduces homogeneous coordinates, one has a natural ring of functions but it is not invariant pointwise with respect to projective transformations in P_{n-1}. On the other hand if the curve is parametrized and the homogeneous coordinates are promoted to $\frac{1-n}{2}$ differentials, with constant Wronskian, we have the set of differential polynomials in the coefficients of the corresponding Q_n as projective invariants. We have seen that it requires some work to find the appropriate combinations which transform covariantly under diffeomorphisms. It is interesting to organize the differential polynomials using a natural basis suggested by the analogy between the expression of the coefficients a_k and the characters of the linear group although I must admit that this line of thought did not lead to any further understanding of the covariance properties. Let therefore a sequence of integers $0 \leq \ell_0 \leq \ell_1 \leq \ldots \leq \ell_{n-1}$ correspond to a Young tableau $Y\{\ell\}$ and let $\{f_{n-1}, \ldots, f_0\}$ be a basis of ker Q_n. The following "differential" characters

$$a_Y = \frac{1}{W_n\left(f_{n-1}, \ldots, f_0\right)} \det \begin{vmatrix} d^{n-1+\ell_{n-1}} f_{n-1} & d^{n-1+\ell_{n-1}} f_0 \\ \vdots & \vdots \\ d^{0+\ell_0} f_{n-1} & d^{0+\ell_0} f_0 \end{vmatrix}$$

are independent of the choice of basis, and are such that

$$a_\phi = 1 \qquad a_{\boxminus \}p \text{ boxes}} = (-1)^p a_p$$

Clearly the a_Y's are differential polynomials in $a_0 = 1$, $a_1 = 0$, $a_2, \ldots, a_n$, and reduce to ordinary characters when $f_k = \exp q_k x$, in which case the $a_k's$ become up to a sign the (constant) fundamental symmetric functions in the $q's$. The latter are nothing but the n (generally distinct) solutions of the characteristic equation $\sum_0^n a_k q^{n-k} = 0$ since the Wronskian is up to a factor $\exp \sum q_k$ the discriminant of this equation.

More generally on the n-th Cartesian product $\overset{n}{\times} U$ we can define the antisymetric Slater determinant (or fermionic wave function, or Plucker coordinate)

$$\psi_n\left(z_{n-1}, \ldots, z_0\right) = \det \begin{vmatrix} f_{n-1}\left(z_{n-1}\right) & \cdots & f_0\left(z_{n-1}\right) \\ \vdots & & \vdots \\ f_{n-1}\left(z_0\right) & \cdots & f_0\left(z_0\right) \end{vmatrix}$$

Changing $\{f_{n-1}, \ldots, f_0\}$ to linearly independent combinations (with constant coefficients) only multiplies ψ by a constant (the determinant of the transformation). This ψ_n is a generating function for the differential characters as follows from the Taylor expansion around a coinciding point. Set $z_p = x + y_p$, then

$$\psi_n\left(x + y_{n-1}, \ldots, x + y_0\right) = W_n(x) \prod_{0 \leq i < j \leq n-1} \left(y_j - y_i\right)$$

$$\times \sum_{Y\{\ell\}} \frac{1}{\prod_0^{n-1}(p+\ell_p)!} \mathrm{ch}_Y(y) a_Y(x)$$

118

where $\mathrm{ch}_Y(y)$ are the polynomial characters of the linear group, i.e. with $r_p = p + \ell_p$ an increasing sequence of non negative integers

$$\mathrm{ch}_Y = \frac{\det \begin{vmatrix} y_{n-1}^{r_{n-1}} & \cdots & y_0^{r_{n-1}} \\ \vdots & & \vdots \\ y_{n-1}^{r_0} & \cdots & y_0^{r_0} \end{vmatrix}}{\det \begin{vmatrix} y_{n-1}^{n-1} & \cdots & y_0^{n-1} \\ \vdots & & \vdots \\ 1 & & 1 \end{vmatrix}}$$

In particular $da_Y = \Sigma a_{Y'}$ where Y' correspond to those irreducible representations of the linear group which occur in the tensor product of the one pertaining to Y by the fundamental (n-dimensional) representation. Symbolically

$$da_Y = \Sigma a_{Y'} \longleftrightarrow Y \otimes \square = \oplus Y'$$

If under a diffeormorphism $x = g(t)$, $f(x)\mathrm{d}x^{\frac{1-n}{2}} = \tilde{f}(t)\mathrm{d}t^{\frac{1-n}{2}}$ we write for the polar part

$$\frac{g'(u)^{\frac{n+1}{2}}}{[g(u) - g(t)]^{r+1}} = \sum_{k=0}^{r} \frac{A_{r,k}(t)}{(u-t)^{k+1}} + \mathrm{reg.}$$

we can find the transformation properties of the differential characters in closed form as

$$\frac{a_Y(x)}{\prod_0^{n-1} r_p!} = \sum_{Y' \leq Y} A_{Y,Y'}(t) \frac{\tilde{a}_{Y'}(t)}{\prod_0^{r-1} k_p!}$$

with the partial ordering meaning that if $Y \longrightarrow \{0 \leq r_0 < r_1 < \ldots < r_{n-1}\}$ and $Y' \longrightarrow \{0 \leq k_0 < k_1 < \ldots < k_{n-1}\}$ then $Y' \leq Y$ means $k_0 \leq r_0$, $k_1 \leq r_1, \ldots, k_{n-1} \leq r_{n-1}$. Finally

$$A_{YY'}(t) = \det \left(A_{r_p,k_s} \right)$$

In infinitesimal form $x = t + \epsilon(t)$, to leading order

$$A_{r,r} = 1 + \left(\frac{n-1}{2} - r \right) \epsilon'(t)$$

$$A_{r,k} = \frac{1}{(r-k+1)!} \left[(r-k+1)\frac{n+1}{2} - r - 1 \right] \epsilon^{(r-k+1)}(t) \quad 0 \leq k < r$$

The rule $A_{YY'} \neq 0$ only for $Y' \leq Y$ is of course in agreement with the transformation proprieties of the quantities $a_2, a_3, \ldots, a_n$. Note also that $A_{YY} = g'(t)^{-\Sigma \ell_p} = \left(\frac{\mathrm{d}t}{\mathrm{d}x} \right)^{\Sigma \ell_p}$. Unfortunately we were not able, using this formalism, to find a simple algorithm to compute the set of k differential forms.

7. Except for the discussion of the behaviour of differential operators under diffeomorphisms, we have hardly justified up to now the title of this contribution. We want now to study general W-deformations of these operators. The symbol W seems quite appropriate given the prominent role of Wronskian. For this purpose we first recast the previous analysis as follows. In infinitesimal form the maps of $\mathcal{F}_{\frac{1 \mp n}{2}}$ on themselves

$$f \longrightarrow F = (1 + X_1)f \qquad g \longrightarrow G = (1 + Y_1)g$$

with

$$X_1 = \epsilon d + \frac{1-n}{2}\epsilon' \qquad Y_1 = \epsilon d + \frac{1+n}{2} = -X_1^*$$

transform the map $f \longrightarrow g = Q_n f$ into $F \longrightarrow G = (Q_n + \delta Q_n) F$ where $Q_n + \delta Q_n$ is again an n-th order normalized operator with

$$\delta Q_n = Y_1 Q_n - Q_n X_1$$

We are therefore led to look for W-generalizations as pairs of differential operators (we drop an infinitesimal scale factor) such that the variation of a *generic* Q_n

$$\delta Q_n = Y Q_n - Q_n X \tag{1}$$

be a differential operator of order less than or equal to $n - 2$, with X and Y defined up to the addition of a common (irrelevant) constant term. Indeed Q_n commutes with the multiplication by a constant.

It is good to remember that isospectral flows are just of this form with the requirement $X = Y$. Since we drop this constraint X and Y will now depend on some arbitrary functions. In a very loose sense this is a "gauged" form of the generalized KdV flows (pertaining to the linear group).

For the sequel it is convenient to use freely the algebraic concept of pseudo-differential operator. We rely on the exposition of Drinfeld and Sokolov where the reader can find some of the original references (some older ones seem to be to Pincherle and Schur). We already introduced the symbol d^{-1} such that

$$d^{-k}f = \sum_{i=0}^{\infty}(-1)^i \binom{k+i-1}{i} f^{(i)}d^{-i-k} \qquad \left(d^{-1}\right)^* = -d^{-1}$$

Let R_+ stand for the differential part of a pseudo differential operator R and $R_- = R - R_+$. The coefficient in d^{-1} (written to the right or to the left, they are equal) is called the residue of R and enjoys the fundamental property

$$\mathrm{Res}\,[R_1, R_2] = \text{total derivative}$$

which leads to the definition of a commutative trace on pseudo differential operators

$$\mathrm{tr}\,R = \int_C \mathrm{Res}\,(R)$$

Any non zero pseudo differential operator has a finite formal inverse R^{-1} and the $*$ involution extends to pseudo differential operators. Since Y is a differential operator, from (1) written as

$$Y = Q_n X \, Q_n^{-1} + \delta Q_n Q_n^{-1}$$

it also follows that

$$Y = \left(Q_n X \, Q_n^{-1}\right)_+$$

showing that Y is entirely determined by X and of the same order. Moreover since the order of $\delta Q_n \leq n - 2$, the residue of $\delta Q_n Q_n^{-1}$ vanishes, hence X is constrained by the unique condition

$$\mathrm{Res}\,Q_n X \, Q_n^{-1} = 0$$

If we write $X = \tilde{X} + c$ with c a function the above reads

$$\text{Res } Q_n \tilde{X} \, Q_n^{-1} + nc'(x) = 0$$

showing that we can pick a fiducial $\tilde{X}$, requiring for instance that it admits constants in its kernel: $\tilde{X} \cdot 1 = 0$ in terms of which

$$X = \tilde{X} - \frac{1}{n} \int_c^x \text{Res } \left(Q_n \tilde{X} \, Q_n^{-1} \right)$$

Finally we see that if we add to X any term of the form $P \, Q_n$ (P differential operator) then Y increases by $Q_n P$ while δQ_n is unaffected. By the Euclidean division algorithm applied to differential operators we can reduce X (and Y) to differential operators of order at most $n - 1$. Given the above linear constraint on X this shows that we can break the most general W-deformation into a linear combination of $n - 1$ independent ones, hence that it depends on $n - 1$ arbitrary functions (of definite weight as we shall see). This number $n - 1$ (as the number of w's) is, not surprisingly, the rank of SL_n.

To illustrate the above considerations on the first of these deformations, we have X_1, Y_1 of order 1 with

$$\tilde{X}_1 = \epsilon d \qquad X_1 = \epsilon d - \frac{1}{n} \int^x \text{Res } \left(Q_n(\epsilon d) Q_n^{-1} \right)$$

Hence

$$X_1 = \epsilon d - \frac{1}{n} \int^n \frac{n(n-1)}{2} \epsilon'' = \epsilon d - \frac{n-1}{2} \epsilon'$$

We note that the constraint $\text{Res } Q_n X Q_n^{-1} = 0$ relates the coefficients in X to those of Q_n. As a result if δ_X stands for the deformation of Q_n (or rather its coefficients) the commutator of two such operations reads

$$[\delta_X, \delta_{X'}] = \delta_{[X',X]} + \delta_X X' - \delta_X X$$

Where consistently by $\delta_X X'$ we mean the δ_X deformation of its dependence on Q_n.

We now want to find a basis of these deformations $\delta_k(\eta) \equiv \delta_{X_k(\eta)}$ each one depending on a single function of weight $-k$, i.e. such that,

$$[\delta_k(\eta), \delta_1(c)] = \delta_k \left(c\eta' \quad k\eta c' \right)$$

which diagonalizes the effect of diffeomorphisms. This is achieved most efficiently by using an appropriate Poisson bracket (or Hamiltonian) structure on the manifold of Q_n's in such a way (as it will turn out) that

$$\delta_k(\eta) \cdot Q_n(x) = \sum_{p=2}^{n} \int_c \mathrm{d}y \; \eta(y) \left\{ w_{k+1}(y), \, a_p(x) \right\} d_x^{n-p}$$

Of course one could also directly try to solve the above condition.

8. The goal is now to describe a Hamiltonian structure on (integrals of) differential polynomials in the coefficients of the generic n-th order differential operator. We shall briefly reproduce the construction given by Drinfeld and Sokolov. Assuming $a_1 = 0$, it is sufficient to deal with linear forms

$$\ell_U(Q_n) = \int \mathrm{d}x \sum_{i=2}^{n} a_i(x) u_i(x)$$

where the collection of cofficients can be associated to a pseudo differential operator

$$U = d^{1-n}u_2 + d^{2-n}u_3 + \cdots + d^{-1}u_n$$

in such a way that

$$\ell_U(Q_n) = \operatorname{Tr} Q_n U$$

Since a_1 vanishes one can freely add to U an arbitrary term $d^{-n}u_1$. This freedom will be used in the sequel.

The Poisson bracket structure to be defined is derived from a first order matrix formalism, substituting the differential operator Q_n, which then allows a neat generalization to other Lie groups. Suppose we are given a $n \times n$ matrix valued function on the cycle $C : x \longrightarrow A(x)$ and consider an (action) functional

$$S(A) = \int_C L(A)$$

where $L(A)$ (the Lagrange function) is a differential polynomial in the coefficients of A. For two such maps let

$$(B,C) = (C,B) = \int_C \operatorname{tr} BC$$

where tr is the ordinary matrix trace. With dots denoting derivatives $\dot{B} \equiv [d, B]$ we have obviously

$$\left(\dot{B},C\right) + \left(B,\dot{C}\right) = 0 \qquad ([A,B],C) + (B,[A,C]) = 0$$

Thus

$$([d + A, B], C) + (B, [d + A, C]) = 0$$

Now define the standard functional derivative $\frac{\delta S}{\delta A}(x) \equiv \operatorname{grad}_A S$

$$\left.\frac{\mathrm{d}}{\mathrm{d}t}S(A + tH)\right|_{t=0} = \int_C \operatorname{tr}\left(\operatorname{grad}_A S, H\right)$$

for $H(x)$ an arbitrary matrix function on C. This means

$$(\operatorname{grad}_A S)_{ij} = \sum_{p \geq 0}(-d)^p \frac{\partial L}{\partial A_{ji}(x)}$$

where we observe the transposition of indices implied by the trace. Define the Poisson bracket of two such actions as

$$\{S_1, S_2\} = (\operatorname{grad}_A S_1, \ [d + A, \ \operatorname{grad}_A S_2])$$

which is a natural generalization of the standard one dimensional prescription. One readily verifies that it is antisymmetric and satisfies the Jacobi identity.

To the differential operator (we need not assume for the time being that $a_1(Q_n)$ vanishes)

$$Q_n = d^n + a_1 d^{n-1} + a_2 d^{n-2} + \ldots$$

one associates a first order matrix differential operator

$$d + A \equiv d + a - J = \begin{pmatrix} d + a_1 & a_2 & a_3 & \cdots & a_n \\ -1 & d & 0 & \cdots & 0 \\ & -1 & d & \cdots & 0 \\ & & \ddots & \ddots & \vdots \\ & & & -1 & d \end{pmatrix}$$

where d multiplies the identity matrix, J has 1 in the first diagonal below the main one and a has $a_1, ..., a_n$ in the first row. Clearly to $f \in \ker Q_n$ there corresponds a vector $\underline{f} = \left(f^{(n-1)}, f^{(n-2)}, ..., f \right)^T$ such that

$$(d + A)\underline{f} = (Q_n f, 0, ..., 0)^T = 0$$

and conversely. Hence the kernels are in one to one correspondence.

The above formalism endows the actions (integrals of differential polynomials in the a'_ks) of a Poisson bracket structure which, as we shall verify, in the case of linear functionals turns out to be

$$\begin{aligned} \{\ell_U, \ell_V\} &= \mathrm{Tr}\left((Q_n U)_+ Q_n V - V Q_n (U Q_n)_+ \right) \\ &= \ell_V \left((Q_n U)_+ Q_n - Q_n (U Q_n)_+ \right) \end{aligned}$$

the so-called second Gelfand Diki Hamiltonian structure. The analogy of this formula with the deformation equation is striking and suggests a mean to construct pairs (X, Y) of differential operators corresponding to the W-algebra. Before setting this point it is however interesting to present the interpretation given by Drinfeld and Sokolov of the matrix formulation to enable one to generalize it to other geometries (or other Lie groups).

The combination $d + A$ leads to the interpretation of $x \longrightarrow A(x)$ as a matrix valued gauge potential with values in the Lie algebra $g\ell_n$ ($s\ell_n$ when a_1 vanishes). Now $A = a - J$, so one needs to discuss both $a(x)$ and J. Diagonal elements are generating a maximal abelian (i.e. Cartan) subalgebra (for the time being we ignore the distinction between $g\ell_n$ and $s\ell_n$). Each matrix with a unique 1 just below the diagonal may be interpreted as a generator corresponding to a simple (positive) root while matrices with non vanishing elements above the diagonal correspond to the nihilpotent Lie algebra generated by elements corresponding to the negative roots. If we include the diagonal ones we get a Borel subalgebra $\mathcal{B}$. Let us assume in general that $a \in \mathcal{B}$ and that J is the sum of Chevalley generators indexed by simple positive roots. Finally let $\mathcal{N}$ denote the group generated by the above nihilpotent algebra. In the above presentation they are matrices with zero below the diagonal and one along it. If $N \in \mathcal{N}$ then

$$N^{-1} J N = J + b \qquad b \in \mathcal{B}$$

From this it follows that under a gauge transformation under the subgroup $\mathcal{N}$, i.e. with $x \longrightarrow N(x) \in \mathcal{N}$

$$d + A \longrightarrow d + \tilde{A} = N^{-1}(d + A)N$$

we have

$$\begin{aligned} \tilde{A} &= N^{-1} A N + N^{-1}(dN) = \tilde{a} - J \\ \tilde{a} &= N^{-1} a N + N^{-1}(dN) - \left(N^{-1} J N - J \right) \end{aligned}$$

with each term in the r.h.s. belonging to $\mathcal{B}$ if a does. Therefore the condition $a \in \mathcal{B}$ is stable under $\mathcal{N}$-gauge transformation. It allows therefore to consider actions $S(a) \equiv S(A)$ with $a \in \mathcal{B}$ invariant under the above gauge transformations. We call them for short gauge invariant and write $S(a)$ or $S(A)$ since only a is allowed to vary.

The definition of $\mathrm{grad}_A S$ is now ambiguous since one is only allowed to consider variations $S(A + tH)$, $H \in \mathcal{B}$, so using the invariant quadratic form in the general case as a substitute for the trace in the definition of (A, B)

$$\frac{\mathrm{d}}{\mathrm{d}t} S(A + tH)\big|_{t=0} = (H, \mathrm{grad}_A S)$$

and taking into account that $(H, \mathrm{Lie}\,\mathcal{N}) = 0$ for $H \in \mathcal{B}$ we see that $\mathrm{grad}_A S$ is only defined up to an element in Lie $\mathcal{N}$ (here strictly upper triangular matrices). Nevertheless this indetermination does not affect the Poisson bracket between two such functionals. Indeed for

$N = 1 + \eta M$, $M \in$ Lie $\mathcal{N}$, η infinitesimal constant, we have to first order

$$d + A_\eta = d + A + \eta[d + A, M]$$

$$\frac{d}{d\eta} S\left(A_\eta\right)|_{\eta=0} = \frac{d}{d\eta} S(A)|_{\eta=0} = 0 = (\text{grad}_A S, [d + A, M])$$

which insures that the definition of the bracket

$$\{S_1, S_2\} = (\text{grad}_A S_1, [d + A, \text{grad}_A S_2])$$

is insensitive to changing $\text{grad}_A S_i \longrightarrow \text{grad}_A S_i + M_i$. Furthermore one also wants to show that the bracket is itself ($\mathcal{N}$-) gauge invariant. Let $x \longrightarrow N(x) \in \mathcal{N}$ induce such a transformation, $A \longrightarrow \tilde{A}$. For B also in $\mathcal{B}$ we have (η constant)

$$S_i(A + \eta B) = S_i\left(\widetilde{A + \eta B}\right) = S_i\left(\tilde{A} + \eta N^{-1}BN\right)$$

Hence

$$\frac{d}{d\eta} S_i(A + \eta B)|_{\eta=0} = (\text{grad}_A S_i, B) = \left(\text{grad}_{\tilde{A}} S_i, N^{-1}BN\right)$$

$$= \left(N \, \text{grad}_{\tilde{A}} S_i N^{-1}, B\right)$$

and

$$\text{grad}_A S_i = N \, \text{grad}_{\tilde{A}} S_i N^{-1} \qquad \text{mod Lie } \mathcal{N}$$

and since Lie $\mathcal{N}$ is invariant under the endjoint action of $\mathcal{N}$ we have

$$\text{grad}_{\tilde{A}} S_i = N^{-1} \text{grad}_A S_i N \qquad \text{mod Lie } \mathcal{N}$$

Therefore

$$\{S_1, S_2\}\left(\tilde{A}\right) = \left(N^{-1}\text{grad}_A S_1 N, \left[N^{-1}(d + A)N, N^{-1}\text{grad}_A S_2 N\right]\right)$$

$$= \{S_1, S_2\}(A)$$

Finally using $\mathcal{N}-$ gauge invariance one can bring A, hence $a \in \mathcal{B}$, into a unique canonical form. For instance in the linear case a recursive reasoning shows that one can take either

$$a_{\text{can}} = \begin{pmatrix} a_1 a_2 ... a_n \\ \\ 0 \end{pmatrix} \quad \text{or} \quad a^{\text{can}} = \begin{pmatrix} & & a_n^* \\ 0 & & -a_{n-1}^* \\ & & \vdots \\ & & (-1)^{n-1}a_1^* \end{pmatrix}$$

Note that switching from Q_n to $(-1)^n Q_n^*$ amounts to change $(-1)^{n-p}a_p^* \longrightarrow (-1)^p a_p^*$ in a^{can}.

In the case of an arbitrary simple Lie algebra $\mathcal{L}$ one can grade it as follows. Write the Chevalley-Serre basis indexed by simple roots as h_i, x_i, y_i with i ranging over the nodes of the Dynkin diagramm

$$[h_i, h_j] = 0 \qquad\qquad [x_i, y_j] = \delta_{ij} h_i$$
$$[h_i, x_j] = n_{ij} x_j \qquad\qquad [h_i, y_j] = -n_{ij} y_j$$

$$(\text{ad } x_i)^{1-n_{ij}} x_j = (\text{ad } y_i)^{1-n_{ij}} y_j = 0 \quad i \neq j$$

with integers n_{ij} given in terms of the diagramm.

Then

$$\mathcal{L} = \bigoplus_{-k \le j \le k} \mathcal{L}^{(j)} \qquad x_i \in \mathcal{L}^{(1)} \qquad y_i \in \mathcal{L}^{(-1)} \qquad h_i \in \mathcal{L}^{(0)}$$

and the grading is compatible with commutation. The integer $h = k + 1$ is the Coxeter number and the (positive) integer j is a Coxeter exponent if and only if the injective map $(J \equiv \sum_i x_i)$

$$\text{ad } J \quad \mathcal{L}^{(-j-1)} \longrightarrow \mathcal{L}^{(-j)}$$

is *not* surjective with $\text{mult}(j) = \dim\mathcal{L}^{(-j)} - \dim\mathcal{L}^{(-j-1)}$ (hence $\sum \text{mult}(j) = \dim\mathcal{L}^{(0)} = \text{rank}\mathcal{L}$).

From the above, a canonical form for a can be chosen such that it has non vanishing elements in a complementary of the images of ad J in each $\mathcal{L}^{(-j)}$, thus a number of elements equal to the rank, each one with a grade corresponding to a Coxeter exponent.

In the linear case let us briefly discuss the relation with the differential operator $Q_n\,(a_{\text{can}})$. To find the application $d + A \longrightarrow Q_n$, one considers that $d + A$ is a matrix in the differential ring generated by smooth functions and d, loosely speaking that matrix elements are quantum mechanic operators and one writes

$$d + A = \begin{pmatrix} \underline{\alpha} & \beta \\ \Delta & \underline{\gamma} \end{pmatrix}$$

with $\underline{\alpha}$ and $(\underline{\gamma})$ a row (column) vector with $n-1$ entries, β a 1×1 matrix, Δ a $(n-1) \times (n-1)$ matrix with -1 along the diagonal and zero below, hence invertible. One then sets

$$d + A \longrightarrow Q_n \equiv \beta - \underline{\alpha}\Delta^{-1}\underline{\gamma}$$

It follows that Q_n is an n-th order differential operator. This arises by comparing the kernels when writing a column vector on which $d + A$ acts as $(\underline{g}, f)^T$ where $\underline{g}$ has $n - 1$ entries, then from the system

$$\underline{\alpha}.\underline{g} + \beta f = 0$$
$$\Delta \underline{g} + \underline{\gamma} f = 0$$

one obtains by elimination

$$\left(\beta - \underline{\alpha}\Delta^{-1}\underline{\gamma}\right) f = 0$$

One shows that Q_n is independent from ($\mathcal{N}$-) gauge transformation and that when $a = a_{\text{can}}$ it reduces to the required form.

The final (and painful) exercise is to express the Hamiltonian structure in terms of the coefficients a_k. We present the calculation again in the linear case starting with a linear functional $\ell_U\,(Q_n)$ corresponding to a unique gauge invariant functional $S_U(A)$. For $H(x) \in \mathcal{B}$

$$S_U(A + tH) = \int \text{Res}\,(Q_n(t)U)$$

with

$$d + A + tH = \begin{pmatrix} \underline{\alpha}(t) & \beta(t) \\ \Delta(t) & \underline{\gamma}(t) \end{pmatrix}$$

If $H = \begin{pmatrix} \underline{\xi} & \zeta \\ \delta & \underline{\eta} \end{pmatrix}$

$$\underline{\alpha}(t) = \underline{\alpha} + \underline{\xi}t, \qquad \beta(t) = \beta + \zeta t, \qquad \underline{\gamma}(t) = \underline{\gamma} + \underline{\eta}t,$$

$$\Delta(t) = \Delta + \delta t, \qquad \Delta^{-1}(t) = \Delta^{-1} - t\Delta^{-1}\delta\Delta^{-1} + \dots$$

implying

$$(\mathrm{grad}_A S_U, H) = \int \mathrm{Res}\frac{\mathrm{d}}{\mathrm{d}t}Q_n(t)|_{t=0}\, U$$

where

$$\frac{\mathrm{d}Q_n(t)}{\mathrm{d}t}\Big|_{t=0} = \zeta - \underline{\xi}\Delta^{-1}\underline{\gamma} - \underline{\alpha}\Delta^{-1}\underline{\eta} + \underline{\alpha}\Delta^{-1}\delta\Delta^{-1}\underline{\gamma}$$

i.e.

$$\mathrm{grad}_A S_U = \mathrm{Res}\begin{pmatrix} -\Delta^{-1}\underline{\gamma}U & \Delta^{-1}\underline{\gamma} \otimes U\underline{\alpha}\Delta^{-1} \\ U & -U\underline{\alpha}\Delta^{-1} \end{pmatrix}$$

where we note the transposition of the sizes of the blocks. Since when P is a pseudo differential operator

$$d\,\mathrm{Res}P = \mathrm{Res}[d, P]$$

we find for $S_V(A)$ corresponding to another linear functional

$$[d + A, \mathrm{grad}_A S_V] = \mathrm{Res}\left[\begin{pmatrix} \underline{\alpha} & \beta \\ \Delta & \underline{\gamma} \end{pmatrix}, \begin{pmatrix} -\Delta^{-1}\underline{\gamma}V & \Delta^{-1}\underline{\gamma} \otimes V\underline{\alpha}\Delta^{-1} \\ V & -V\underline{\alpha}\Delta^{-1} \end{pmatrix}\right]$$

Remembering that $Q_n = \beta - \underline{\alpha}\Delta^{-1}\underline{\gamma}$ one sees that

$$[d + A, \mathrm{grad}_A S_V] = \mathrm{Res}\left\{\begin{pmatrix} Q_n V & -Q_n V\underline{\alpha}\Delta^{-1} \\ 0 & 0 \end{pmatrix} - \begin{pmatrix} 0 & -\Delta^{-1}\underline{\gamma}V Q_n \\ 0 & V Q_n \end{pmatrix}\right\}$$

Explicitly with $a \equiv a_{\mathrm{can}}$

$$\underline{\alpha} = (d + a_1, a_2, ..., a_{n-1}) \qquad \beta = a_n$$

$$\Delta^{-1} = -\begin{pmatrix} 1 & d & d^2 & ... & d^{n-2} \\ & 1 & d & ... & d^{n-3} \\ 0 & & & \ddots & 1 \end{pmatrix}$$

hence

$$\underline{\alpha}\Delta^{-1} = -\left(d + a_1, d^2 + a_1 d + a_2, ..., d^{n-1} + a_1 d^{n-2} + ... + a_{n-1}\right)$$

$$= -\left(\left(Q_n d^{1-n}\right)_+, ..., \left(Q_n d^{-1}\right)_+\right)$$

$$\Delta^{-1}\underline{\gamma} = -\begin{pmatrix} d^{n-1} \\ \vdots \\ d \end{pmatrix}$$

Therefore

$$\{\ell_U, \ell_V\}(A) = \{S_U, S_V\}(A) = \int \mathrm{tr}\,(\mathrm{grad}_A S_U, [d + A, \mathrm{grad}_A S_V])$$

$$= \int \mathrm{tr}\,\mathrm{Res}\Lambda_1\,\mathrm{Res}\Lambda_2$$

with

$$
\Lambda_1 = \left(
\begin{array}{cc}
\begin{pmatrix} d^{n-1}U \\ \vdots \\ dU \\ U \end{pmatrix} &
\begin{array}{c}
\left[\begin{array}{c} d^{n-1} \\ \vdots \\ d \end{array}\right] \otimes U \left[\left(Q_n d^{1-n}\right)_+ , ..., \left(Q_n d^{-1}\right)_+\right] \\[2ex]
U \left[\left(Q_n d^{1-n}\right)_+ , ..., \left(Q_n d^{-1}\right)_+\right]
\end{array}
\end{array}
\right)
$$

$$
\Lambda_2 = \left(
\begin{array}{cc}
\left[Q_n V, Q_n V \left(Q_n d^{1-n}\right)_+ , ..., Q_n V \left(Q_n d^{-2}\right)_+\right] & Q_n V \left(Q_n d^{-1}\right)_+ - d^{n-1} V Q_n \\[2ex]
0 & - \left[\begin{array}{c} d^{n-2} V Q_n \\ \vdots \\ dV Q_n \\ V Q_n \end{array}\right]
\end{array}
\right)
$$

This means that (with $\operatorname{Res}\left(Q_n d^{-1}\right)_+ = 1$)

$$
\{\ell_U, \ell_V\}(A) = \int \sum_{k=1}^{n} \left[\operatorname{Res}\left(d^{k-1}U\right) \operatorname{Res} Q_n V \left(Q_n d^{-k}\right)_+ \right.
$$
$$
\left. - \operatorname{Res}\left(U \left(Q_n d^{-k}\right)_+ \operatorname{Res}\left(d^{k-1} V Q_n\right)\right)\right]
$$

Since for α a function $\alpha \operatorname{Res} P = \operatorname{Res} \alpha P = \operatorname{Res} P \alpha$ the first sum under the integral reads

$$
\operatorname{Res}\left(Q_n V \sum_{k=1}^{n} \left(Q_n d^{-k}\right)_+ \operatorname{Res}\left(d^{k-1}U\right)\right)
$$
$$
= \operatorname{Res}\left(Q_n V \sum_{k=1}^{n} \left(Q_n d^{-k}\right)_+ u_{n+1-k}\right)
$$
$$
= \operatorname{Res}\left(Q_n V \left(Q_n \sum_{k=1}^{n} d^{-k} u_{n+1-k}\right)_+\right) = \operatorname{Res}\left(Q_n V \left(Q_n U\right)_+\right)
$$

where we have written $U = \sum_{k=1}^{n} d^{-k} u_{n+1-k}$. Similarly the second sum reads

$$
\operatorname{Res}\left(U \left(Q_n \left(V Q_n\right)_-\right)_+\right)
$$

Thus one form of the bracket is

$$
\{\ell_U, \ell_V\} = \operatorname{Tr}\left[Q_n V \left(Q_n U\right)_+ - U \left(Q_n \left(V Q_n\right)_-\right)_+\right]
$$

But

$$
\operatorname{Tr}\left[U \left(Q_n \left(V Q_n\right)_-\right)_+\right] = \operatorname{Tr}\left[U Q_n \left(V Q_n\right)_-\right] - \operatorname{Tr}\left[U \left(Q_n \left(V Q_n\right)_-\right)_-\right]
$$

and the second term is readily seen to vanish. In the first we can replace similarly $U Q_n$ by $\left(U Q_n\right)_+$ then $\left(V Q_n\right)_-$ by $V Q_n$ and use $\operatorname{Tr} AB = \operatorname{Tr} BA$ to rewrite finally

$$
\{\ell_U, \ell_V\} = \operatorname{Tr}\left[\left(Q_n U\right)_+ Q_n V - V Q_n \left(U Q_n\right)_+\right]
$$

as claimed above.

9. We have obtained the Poisson structure corresponding to $g\ell_n$ rather than $S\ell_n$ which corresponds to $a_1\left(Q_n\right) = 0$. In order to apply the above expressions to find a pair, tentatively $X = \left(U Q_n\right)_+$, $Y = \left(Q_n U\right)_+ = \left(Q_n X Q_n^{-1}\right)_+$ for a deformation of Q_n, we want the condition $a_1 = 0$ to apply also to the deformed operator. Indeed such a pair X, Y would in general lead to $\delta a_1 \neq 0$ or equivalently $\operatorname{Res}\left(Q_n X Q_n^{-1}\right) \neq 0$.

But recall that a term $d^{-n}u_1$ in U does not affect the functional $\ell_U(Q_n)$ when a_1 vanishes but would affect the above brackets. One is therefore free to adjust u_1 to

$$u_1 = \frac{1}{n} \int^x \text{Res}\,[U, Q_n]$$

to remove the variation δa_1 in $\delta Q_n = (Q_n U)_+ Q_n - Q_n (U Q_n)_+$. So that we use the above brackets with the understanding that v_1 vanishes and u_1 is given by the above formula.

We are now ready to express the W-deformations using Poisson brackets as follows. Let $\ell_V(Q_n)$ stand for an arbitrary linear functional of Q_n

$$\ell_V(Q_n) = \int \sum_2^n a_i v_i = \text{Tr}\,Q_n V$$

$$V = d^{1-n} v_2 + ... + d^{-1} V_n$$

Choose the non-linear function $(\epsilon(x)$ of weight $-k)$

$$H_k = \int \epsilon w_{k+1}$$

Compute its gradient and associate to it a pseudo differential operator

$$U_k = \sum_{i=1}^{n-1} d^{-i} \frac{\delta H_k}{\delta a_{n+1-i}} + \frac{1}{n} d^{-n} \int^x \text{Res}\left[\sum_{i=1}^{n-1} d^{-i} \frac{\delta H_k}{\delta a_{n+1-i}}, Q_n \right]$$

where the functional derivative (the gradient) $\frac{\delta H}{\delta a_p}$ is a short hand for

$$\sum_r (-1)^r \left(d^r \frac{\delta H}{\delta a_p^{(r)}} \right)$$

we have then the k-th "W-flow" (depending on ϵ)

$$\delta_k \ell_V(Q_n) = \{H_k, \ell_V(Q_n)\} = \ell_V(\delta_k Q_n)$$

$$\delta_k Q_n = (Q_n U_k)_+ Q_n - Q_n (U_k Q_n)_+$$

Since we have the expression of w_{k+1} we derive the pairs X_k, Y_k of differential operators explicitly as

$$X_k = \tilde{X}_k - \frac{1}{n} \int^x \text{Res}\left(Q_n \tilde{X}_k Q_n^{-1} \right)$$

$$Y_k = \tilde{Y}_k + \frac{1}{n} \int^x \text{Res}\left(Q_n^{-1} \tilde{Y}_k Q_n \right)$$

$$\tilde{X}_k = \left(\left(\sum_{i=1}^{n-1} d^{-i} \frac{\delta H_k}{da_{n+1-i}} \right) \left(d^n + \sum_{j=2}^n a_j d^{n-j} \right) \right)_{++}$$

$$\tilde{Y}_k = \left(\left(d^n + \sum_{j=2}^n a_j d^{n-j} \right) \left(\sum_{i=1}^{n-1} d^{-i} \frac{\delta H_k}{\delta a_{n+1-i}} \right) \right)_{++}$$

where the symbol $++$ means that we keep the differential part omitting the constant term $(\tilde{X}_k \cdot 1 = \tilde{Y}_k \cdot 1 = 0)$.

Using the notations of section 4 the terms in $\tilde{X}_k$, $\tilde{Y}_k$ that are independent of a's or w's take a simple form

$$\tilde{X}_k(0) \quad = \sum_{s=1}^{k} B_{k+1,s+1}\epsilon^{(k-s)}d^s$$

$$\tilde{Y}_k(0) \quad = (-1)^k \tilde{X}_k^*(0) = \sum_{s=1}^{k} C_{k+1,s+1}\epsilon^{(k-s)}d^s$$

where

$$C_{k+1,s+1} \quad = (-1)^k \sum_{\ell=0}^{j-1}(-1)^{\ell+s}\binom{\ell+s}{\ell} B_{k+1,\ell+s+1}$$

$$= B_{k+1,s+1}(n \longrightarrow -n)$$

where use has been made of the identity

$$B_{k+1,s+1} = (-1)^{k-s}\sum_{\ell=0}^{k-s}\binom{n-s-1}{\ell} B_{k+1,\ell+s+1}$$

The first few pairs X_k, Y_k are displayed in the following table.

$$X_1 \;\; = \epsilon d - \frac{n-1}{2}\epsilon'$$

$$Y_1 \;\; = \epsilon d + \frac{n+1}{2}\epsilon'$$

$$X_2 \;\; = \epsilon d^2 - \frac{n-2}{2}\epsilon'd + \left\{\frac{2}{n}\epsilon a_2 + \frac{1}{12}(n-1)(n-2)\epsilon''\right\}$$

$$Y_2 \;\; = \epsilon d^2 + \frac{n+2}{2}\epsilon'd + \left\{\frac{2}{n}\epsilon a_2 + \frac{1}{12}(n+1)(n+2)\epsilon''\right\}$$

$$X_3 \;\; = \epsilon d^3 - \frac{n-3}{2}\epsilon'd^2 + \left\{\frac{(n-2)(n-3)}{10}\epsilon'' + \frac{6}{5}\frac{3n^2-7}{5n\left(n^2-1\right)}a_2\epsilon\right\}d$$

$$+ \left\{\frac{3}{n}w_3\epsilon - \frac{3(n+2)(n-7)}{10n(n+1)}a_2'\epsilon - \frac{(n-3)(4n-7)}{5n(n+1)}a_2\epsilon' - \frac{(n-1)(n-2)(n-3)}{5!}\epsilon'''\right\}$$

$$Y_3 \;\; = \epsilon d^3 + \frac{n+3}{2}\epsilon'd^2 + \left\{\frac{(n+2)(n+3)}{10}\epsilon'' + \frac{6}{5}\frac{3n^2-7}{5n\left(n^2-1\right)}a_2\epsilon\right\}d$$

$$+ \left\{\frac{3}{n}w_3\epsilon + \frac{3(n-2)(n+7)}{10n(n-1)}a_2'\epsilon + \frac{(n+3)(4n-7)}{5n(n-1)}a_2\epsilon' + \frac{(n+1)(n+2)(n+3)}{5!}\epsilon'''\right\}$$

With the explicit expressions of the w's, X's and Y's we can now from the Poisson brackets of w's among themselves obtaining thus the classical W-algebra (for SL_n). For small n it already appears in the literature. Of course it always contains the Virasoro algebra (as a Lie subalgebra) together with the brackets expressing that w_k has weight k. Thus using distributions

$$\{w_2(y), w_2(x)\} \quad = \left(w_2'(x) + 2w_2(x)d_x + \frac{n\left(n^2-1\right)}{12}d_x^3\right)\delta(x,y)$$

$$\{w_2(y), w_2(x)\} \quad = (w_k' + kw_k(x)d_x)\,\delta(x,y)$$

The other brackets are of course easier to compute in the coordinate u where $a_2 \equiv w_2$ vanishes. If

$$\{w_k(v), w_\ell(u)\}_{w_2=0} = \Delta\left(w_j(u), d_u\right)\delta(u,v)$$

with Δ an ordered polynomial in d_u, $w_j(u)$ and their derivatives, then in the generic coordinate

$$\{w_k(y), w_\ell(x)\} = \varphi^k \Delta \left(\varphi^{-j} w_k(x), \varphi^{-1} d_x\right) \varphi^{\ell-1} \delta(x, y)$$

with $\varphi = \left(\frac{du}{dx}\right)$ and the δ-distribution has contributed an extra factor φ^{-1}. The differential operator Δ has of course to satisfy specific constraints in order that the final expression depend only on the Schwarzian derivative of the map hence on w_2. As an illustration for generic n in the u-coordinate we have

$$\{w_3(v), w_3(u)\}|_{w_2=0} = \left(2\left[w_4, d\right]_+ - \frac{(n-2)(n-1)n(n+1)(n+2)}{6!} d^5\right) \delta(u, v)$$

$$\{w_3(v), w_4(u)\}|_{w_2=0} = \left(5 w_5 d + 2 w_5' - \frac{(n-3)(n+3)}{70}\left(14 w_3 d^3 + 14 w_3' d^2 + 6 w_3'' d + w_3'''\right)\right) \delta(u, v)$$

$$\{w_4(v), w_4(u)\}|_{w_2=0} = \left(3\left[w_6, d\right]_+ - \frac{n^2 - 19}{30}\left(3\left[w_4, d^3\right]_+ - 2\left[w_4'', d\right]_+\right)\right.$$
$$\left. -3\frac{n-3}{n} w_3 d w_3 + \frac{(n-3)(n-2)(n-1)n(n+1)(n+2)(n+3)}{20.7!} d^7\right) \delta(u, v)$$

where on the r.h.s. all w's are evaluated at u and $d \equiv d_u$. By restoring the w_2 dependence and truncating to n (i.e. $w_\ell = 0$ for $\ell > n$) one obtains the w-algebra.

In particular one can show that the "central term" i.e. the term independent of w's is diagonal in the indices k, ℓ and of the form

$$(-)^k \frac{\binom{n+k-1}{n-k}}{\binom{2k-2}{k-1}} \delta_{k\ell} d_x^{k+\ell-1} \delta(x, y)$$

10. We will not extend the algebraic machinery beyond the previous indications, referring to our paper for the discussion of the orthogonal case. Rather in this last and still very speculative part I would like to look more closely at the simplest case of W_3 (for SL_3) and ask what is really this kind of structure.

Apart from diffeormorphisms generated by $w_2 = a_2$ corresponding to the pair

$$X_1 = \epsilon d - \epsilon' \qquad\qquad Y_1 = \epsilon d + 2\epsilon'$$

we have in this case transformations generated by

$$w_3 = a_3 - \frac{1}{2} a_2'$$

with

$$X_2 = \epsilon d^2 - \frac{1}{2}\epsilon' d + \frac{2}{3}\epsilon a_2 + \frac{1}{6}\epsilon''$$
$$Y_2 = \epsilon d^2 + \frac{5}{2}\epsilon' d + \frac{2}{3}\epsilon a_2 + \frac{5}{3}\epsilon''$$

while

$$\{w_2(y), w_2(x)\} = \left(w_2'(x) + 2 w_2(x) d_x + 2 d_x^3\right) \delta(x, y)$$

$$\{w_2(y), w_3(x)\} = \left(w_3'(x) + 3 w_3(x) d_x\right) \delta(x, y)$$

$$\{w_3(y), w_3(x)\} = -\frac{1}{6}\left(d_x^5 + \frac{5}{2}\left[w_2(x), d_x^3\right]_+ - \frac{3}{2}\left[w_2''(x) d_x\right]_+ + 2\left[w_2^2, d\right]_+\right) \delta(x, y)$$

While this is *not* the general situation, let us assume nevertheless that $\ker Q_3$ describes a smooth projective curve in CP_2 encoded by the vanishing of an irreducible homogeneous polynomial $P(u_1, u_2, u_3)$ of degree n in three variables. Hence the curve has genus $(n-1)(n-2)/2$ (we assume no double points, no cusps,..., an example would be the Fermat curve $u^n + v^n + w^n = 0$).

Thus w_3 is a meromorphic 3-differential characterized by its zeroes and poles. Once the curve is parametrized (by patches), the poles only occur where the Wronskian vanishes i.e. at the $3n(n-2)$ inflexion points and it is readily seen that they all occur as triple poles (this arises from the necessity of absorbing a_1 into a multiplicative factor on the coordinates). The poles are given by the vanishing of the Hessian

$$H_1(u_1, u_2, u_3) = \det \frac{\partial^2 P}{\partial u_i \partial u_j}$$

a homogeneous polynomial of degree $3(n-2)$ which will appear cubed in the denominator of w_3. On the other hand we have to understand what the zeroes stand for. For this we first get rid of a_1, then go to a coordinate where a_2 vanishes (call it t), so that locally the operator is $Q_3 = d_t^3 + w_3$. If we assume that w_3 has a single zero at $t = 0$ in the vicinity of the origin, i.e. $w_3 \sim t$ it is readily seen that we can take three solutions of $Q_3 f = 0$ of the form

$$\begin{aligned} f_0 &\sim 1 + 0\left(t^4\right) \\ f_1 &\sim t + 0\left(t^5\right) \\ f_2 &\sim t^2 + 0\left(t^6\right) \end{aligned}$$

Hence

$$f_2/f_0 \sim \left(\frac{f_1}{f_0}\right)^2 + 0\left(\frac{f_1}{f_0}\right)^6$$

which means that the osculating *non degenerate* conic (the would be conic with 5 points of contact at the origin; recall that a conic is defined by requiring that it passes through 5 points) has in fact six points of contact with the curve at the origin. These are analogous to Weirstrass points (the inflexion points). Let H_2 a homogeneous polynomials vanishing only on these points. Then

$$w_3 = \frac{H_2}{H_1^3}$$

The homogeneity degree of w_3 is then $\deg H_2 - 3 \deg H_1$. Now it is known that for a smooth curve of degree $n > 3$ the coordinates can be taken of weight $\frac{1}{n-3}$. This is one for a quartic (genus 3, corresponding to the canonical embedding) and it can be verified in general by finding three holomorphic differentials with zeroes of order $n - 3$. For instance for a Fermat curve $u_1^n + u_2^n + u_3^n = 0$ seen as a n-fold branch cover over the $x = \frac{u_2}{u_1}$ plane with $y = \frac{u_3}{u_1}$ (or vice versa) take

$$\omega_1 = \frac{\mathrm{d}x}{y^{n-1}} \qquad n - 3 \text{ fold zeroes over the } n - \text{ points at infinity}$$

$$\omega_2 = -\frac{\mathrm{d}y}{x^2} \qquad n - 3 \text{ fold zeroes at } n - \text{th roots of } y = -1, \; (x = 0)$$

$$\omega_3 = \frac{\mathrm{d}x}{y^2} \qquad n - 3 \text{ fold zeroes at } n - \text{th roots of } x = -1, \; (y = 0)$$

then $\omega_1 : \omega_2 : \omega_3 = u_1^{n-3} : u_2^{n-3} : u_3^{n-3}$ (of course $n(n-3) = 2g - 2$).

In order that ω_3 be indeed a 3-differential it is thus required that

$$3(n-3) = \deg H_2 - 3 \deg H_1$$

or

$$\deg H_2 = 3[n - 3 + 3(n - 2)] = 3[4n - 9]$$

giving a total of $3n[4n - 9]$ points where a conic touches a smooth curve with a contact of order six. I dont known whether this is a classical result (it surely is) but at any rate for $n = 3$ (27) and $n = 4$ (84) it agrees with values which are quoted in textbooks. Thus at least when Q_3 describes a smooth curve in $\mathbf{C}P_2$ we have a global view of the meromorphic 3-differential w_3 and with more elaborate geometry one could find analogs in higher dimension.

Finally let us inquire about the deformations

$$\left\{ \int \eta w_3, \; Q_3 \right\}$$

with $Q_3 = d^3 + \frac{1}{2}[w_2, d]_+ + w_3$. If we ignore covariance and take $\eta = \mathrm{cst}$, this is the isospectral KdV flow

$$\begin{aligned}
\dot{w}_2 &= 2w_3' \\
\dot{w}_3 &= -\frac{1}{6}[w_2''' + 4w_2 w_2']
\end{aligned}$$

where the dots stand for derivatives with respect to the parameter along the flow. Taking an extra derivative we obtain for w_2 the Boussinesq equation

$$\ddot{w}_2 = -\frac{1}{3}\left[w_2'' + 2w_2^2\right]''$$

On the other hand if η varies (as it should, having weight -2) we get

$$\begin{aligned}
\delta_\eta w_2 &= 3\eta' w_3 + 2\eta w_3' \\
\delta_\eta w_3 &= -\frac{1}{6}\left[\eta^{(5)} + 5\eta'' w_2 + \frac{15}{2}\eta'' w_2' + \frac{9}{2}\eta' w_2'' + \eta w_2''' + 4\eta' w_2^2 + 4\eta w_2 w_2'\right]
\end{aligned}$$

Even it we have chosen a coordinate where w_2 vanishes, but not w_3, then a general w_3-flow will have $\delta_\eta w_2 \neq 0$. But this can be compensated by changing the new coordinate, so let us look at the case $w_2 = 0$, $\delta_\eta w_2 = 0$. It therefore follows that $\eta^3 w_3^2 = \mathrm{cste}$, hence η can be taken as one of the cube roots of w_3^{-2} (this has the correct weight at least away from the zeroes or poles of w_3). Choosing one of these cube roots we get a W_3-flow

$$\frac{\mathrm{d}w_3}{\mathrm{d}t} = \left(\frac{\mathrm{d}}{\mathrm{d}x}\right)^5 \frac{1}{w_3^{2/3}}$$

with coordinates (of weight -1) changing as

$$\frac{\mathrm{d}}{\mathrm{d}t} f = -\left[4\eta'' - 6\eta'\delta + 6\eta d^2\right] f, \qquad \eta = \frac{1}{w_3^{2/3}}$$

Up to this 3-fold ambiguity this flow should have a geometric interpretation in terms of the embedded curve in $\mathbf{C}P_2$ involving osculating conics. I find it very suggestive that if we look at two such neighboring conics, three of their common points have limits distinct from the point on the curve and by drawing lines to the osculating point define three deformation directions. This construction becomes singular at inflexion points or when there is a sixfold contact!

References

V.G. Drinfeld and V.V. Sokolov
  "Lie algebras and equations of Kortweg-de Vries type"
  Journ. Sov. Math. **30** (1985) 1975-2036.

W-algebras were introduced by

A.B. Zamolodchikov
  "Infinite additional symmetries in two-dimensional conformal quantum field theory"
  Theor. Math. Phys. **65** (1985) 1205-1213.

Our geometric inspiration came from

G. Sotkov and M. Stanishkov
  "Affine geometry and W_n-gravities"
  Preprint ISAS-70/90/EP (June 1990).

Further references can be found in

P. Di Francesco, C. Itzykson, J.-B. Zuber
  "Classical W-algebras"
  Preprint SPhT/90-149 and PUPT-1211 (Oct. 1990) submitted to Communications in Mathematical Physics.

The paper quoted in the introduction is

C. Itzykson, J.-B. Zuber
  "Matrix integration and combinatorics of modular groups"
  to appear in Communications in Mathematical Physics.

STRINGS EMBEDDED IN DYNKIN DIAGRAMS

I.K. Kostov

Service de Physique Théorique,[†]
Direction des Sciences de la Matière
CEN-Saclay, F-91191 Gif-sur-Yvette Cedex, France

ABSTRACT

We present a closed set of loop equations for strings having as a target space a Dynkin diagram of ADE type. The critical exponents for the order parameters are extracted from the exact expressions for the one-loop correlations in the scaling limit.

1. INTRODUCTION

The latest progress of the theory of non-critical strings has been prepared by the conjecture, advanced some years ago [1], that the functional integral of the $2d$ gravity can be discretized as a sum over planar graphs. Such a discretization allowed to apply powerful methods of calculation borrowed from the theory of random matrices [2]. Some of the corresponding matrix models can be solved exactly after being reduced to the problem of non-interacting fermions in an external potential in one space dimension. This fermionic representation proved to be very helpful in practical calculations but its connection with the original model seems quite formal. Thus it is not clear at all whether the non-perturbative phenomenae in the problem of fermions have any string interpretation.

It is therefore desirable to have a formalism in which the string excitations appear explicitly. Such a formalism can be developed starting with the Dyson-Schinger equations for the loop amplitudes.

In the simplest case of zero embedding dimensions (pure gravity) a closed string state is characterized only by the length of the loop in the sense of the intrinsic geometry of the world sheet. The equations of motion for the loop amplitudes [3] are equivalent to the Dyson-Schwinger equations for the correlation functions in the corresponding random matrix model and can be solved order by order in the topological expansion. Their renormalized version in the scaling limit was recently presented by F. David [4]. It was also pointed out that the loop

---

[†] On leave from the Institute for Nuclear Research and Nuclear Energy, 72 Bld Lenin, 1784 Sofia, Bulgaria

Random Surfaces and Quantum Gravity
Edited by O. Alvarez *et al., Plenum Press, New York, 1991*

equations can be put in the form of Virasoro constraints on the partition function [5,6]. It could be very important to understand the exact meaning of this Virasoro algebra.

The case of non-trivial embedding space seems much more difficult to investigate because of the complexity of the space of string excitations. Any closed loop in the embedding space represents a possible string state. Miraculously enough, it is nevertheless possible to find a closed system of equations acting in the subspace of loops occupying a single point in the embedding space. This allows to carry out an analysis similar to that of pure gravity. The miracle happens if we take as an embedding space a Dynkin diagram of simply laced Lie algebra or the corresponding extended Dynkin diagram. This choice is sufficient to reproduce the whole spectrum of string theories immersed in less then one dimension [7]. It includes as well the case of strings compactified on circles of different radii and orbifolds. Our approach is in a sense complementary to that of M. Douglas [8] using chains of random matrices. Its main advantage is the smooth transition to the one-dimensional string which correspond to an infinitely long Dynkin diagram.

The aim of this lecture is to present the construction of strings embedded in Dynkin diagrams which was originally proposed in ref.[9], with as less as possible indigestible details. In this presentation we shall use exclusively the language of the loop equations. To make the reader familiar with it, we begin with the analysis of the simplest possible model of random surfaces which corresponds to the Gaussian ensemble of random matrices. This model is the first in the series we are going to introduce later, corresponding to surfaces without embedding. It however does not describe pure gravity since the intrinsic area of the surface is zero. Nevertheless the continuum limit for the loop amplitudes exists; it corresponds to taking infinitely large loops. The absence of area can be explained by a cancellation of the fluctuations of the gravitational field (that is, the intrinsic metric of the world sheet) and a ghost field with central charge $C = -2$. Indeed, the fact that the bosonic string embedded in -2 dimensions is related to the Gaussian matrix model is well known [10]. Recently the gaussian model of random surfaces was reconsidered in the context of topological field theories [11]. Its interpretation as a two-dimensional field theory was elaborated in ref.[12].

Before deriving the loop equations for the ADE strings (we call them so because they have as embedding space a Dynkin diagram of a classical Lie group belonging to the series A,D or E) we shall describe the space of loop excitations. These are labelled by the eigenvalues of the connectivity matrix of the Dynkin diagram which correspond in turn to the so-called Coxeter exponents. Roughly speaking, the Coxeter exponents form the momentum space dual to the "coordinate space" represented by the Dynkin diagram.

The eigenvectors of the connectivity matrix become the primary scaling operators in the continuum limit. They represent the order parameters of the matter field and their scaling dimensions correspond to the points along the diagonal of the conformal grid [7]. In the limit of an infinite number of points the order parameters become the usual plane waves spanning the Hilbert space of states of the one-dimensional string. An important point is that no other excitations are present in the partition function. Thus the cosmological constant in the non-unitary versions of our models remains coupled to the identity operator and not to the operator of minimal dimension as in the version of M. Douglas [8]. Therefore we obtain a different value for the string susceptibility exponent.

We extract the spectrum of scaling dimensions from the exact solution of the loop equation for the one loop amplitude in the planar limit. It indeed satisfies the KPZ scaling [7]. Finally, we give the complete set of loop equations for in presence of loop sources and find their scaling limit.

2. THE LOOP AMPLITUDES IN THE GAUSSIAN MODEL

The aim of this section is to introduce the formalism of the loop equations and to fix our notations. It contains the solution of the gaussian model of random surfaces by means of the loop equations, without any reference to the random matrices. The reader familiar with the loop equations can go directly to section 3.

The gaussian model describes random surfaces with zero area. The partition function for closed surfaces is of course zero. The loop amplitudes are nevertheless non-trivial and give the number of the pairwise identifications of the points of the boundary with certain topology. A planar surface with a single boundary looks therefore as a branched tree (Fig.1).

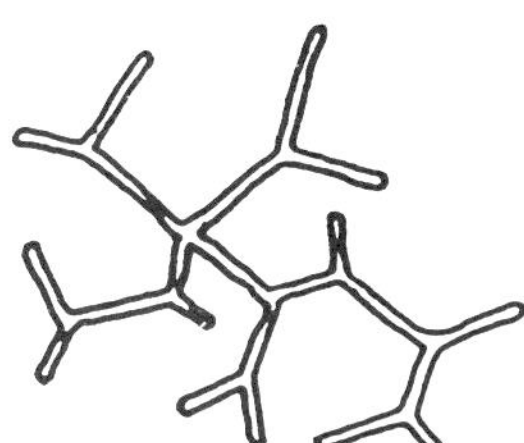

Figure 1. a surface of zero area forms a branched free.

Let us first consider the simplest loop amplitude $W(\ell)$ representing the partition function of the surfaces with the topology of a disc having a boundary of length ℓ. We assume that the boundary is supplied with a reference point. The general form of the equation of motion for $W(\ell)$ is fixed by the conditions of locality and planarity [13]. An infinitesimal variation of boundary results as an infinitesimal change of ℓ. Taking into accound the contribution of the degenerated surfaces such that the reference point coincides with another point of the boundary (in fact, all surfaces in our model are degenerated!) we write

$$\frac{\partial}{\partial \ell} W(\ell) = \int_0^\ell \mathrm{d}\ell' W(\ell') W(\ell - \ell') \tag{1}$$

This equation becomes algebraic for the Laplace transform

$$\hat{W}(P) = \int_0^\infty \mathrm{d}\ell W(\ell) \mathrm{e}^{-P\ell} \tag{2}$$

of the loop amplitude. Namely

$$P\hat{W}(P) - 1 = \hat{W}^2(P) \tag{3}$$

The solution

$$\hat{W}(P) = \frac{1}{2}\left(P - \sqrt{P^2 - 4}\right) \tag{4}$$

is chosen so that at $P \longrightarrow \infty$, $\hat{W}(P) \sim 1/P$. This condition corresponds to the normalization $W(\ell = 0) = 1$ [13].

Similarly we can define the multiloop amplitudes and write the corresponding equations of motion. The n-point amplitude can be expressed symbolically as

$$\hat{W}(P_1, ..., P_n)_H = \frac{\partial}{\partial P_1} ... \frac{\partial}{\partial P_n} \left\langle \hat{\phi}(P_1) ... \hat{\phi}(P_n) \right\rangle_H \tag{5}$$

where $\hat{\phi}(P)$ is the operator creating a loop with "momentum" P and $\langle\ \ \rangle_H$ means the sum over connected surfaces with H handles. The derivatives in P_i, $i = 1,...,n$, reflect the fact that each of the n loops has reference point on it.

The partition function with a loop source $\ell J(\ell)$ and a string coupling constant g is defined as

$$F[g; J] = \sum_{H=0}^{\infty} g^{2H-2} \left\langle \exp g \int_0^{\infty} d\ell\ \ell J(\ell)\phi(\ell) \right\rangle_H \tag{6}$$

The string interaction constant g is thus coupled to the Euler characteristic $\chi = 2-2H-n$ of the world sheet.

Define the one-loop amplitude in presence of a loop source as

$$W(\ell) = g\frac{\delta F}{\delta J(\ell)} \tag{7}$$

It satisfies the following loop equation which is a direct generalization of (1)

$$\frac{\partial}{\partial \ell}W(\ell) = \int_0^{\ell} d\ell' W(\ell')W(\ell - \ell')+$$
$$g \int_0^{\ell} d\ell' \delta W(\ell - \ell')/\delta J(\ell') + g \int_0^{\infty} d\ell'\ell' J(\ell')W(\ell + \ell') \tag{8}$$

All three contact terms on the r.h.s. have a clear geometrical interpretation in terms of splitting and joining of loops (Fig.2).

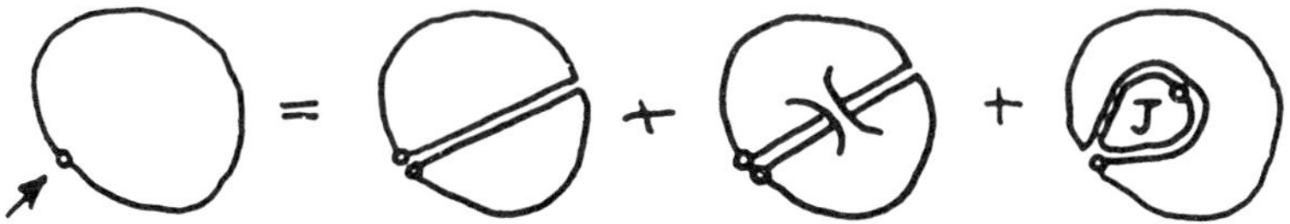

Figure 2. graphical representation of the loop equation (8).

The factor of ℓ' in the last term on the r.h.s. counts for the possibility of joining the second loop at any point along its length.

The Laplace image of (7)

$$\hat{W}(P) = \frac{\partial}{\partial P} \left\langle \hat{\phi}(P) \right\rangle = g\frac{\delta F}{\delta J(P)}; \quad \hat{J}(P) = \int_0^{\infty} d\ell\ e^{P\ell} J(\ell) \tag{9}$$

can be found as the solution of the equation

$$P\hat{W}(P) = 1 + g\delta\hat{W}(P)/\delta\hat{J}(P) + g \int_C \frac{dQ\hat{W}(Q)}{2\pi i(P - Q)} \frac{\partial\hat{J}(Q)}{\partial Q} \tag{10}$$

The contour integral results to the truncation of $\hat{W}(P)\hat{J}'(P)$ to the Laurent powers with $n \leq -1$. The contour C encloses the singularities of $\hat{W}(P)$ ant not these of $\hat{J}(P)$. In any case, all quantities are well defined if $\hat{J}(P)$ is an entire function of P.

Solving Eq.(10) order by order in $\hat{J}$ and g we find

$$W(P,Q)_{H=0} = \frac{1}{4}\frac{(P + Q)^2 \left(\sqrt{P^2 - 4} + \sqrt{Q^2 - 4}\right)^{-2} - 1}{\sqrt{P^2 - 4}\sqrt{Q^2 - 4}} \tag{11}$$

$$W(P,Q,R)_{H=0} = \frac{PQ + PR + QR + 4}{(P^2 - 4)^{3/2}(Q^2 - 4)^{3/2}(R^2 - 4)^{3/2}} \tag{12}$$

$$W(P)_{H=1} = \frac{1}{4P^2}\left(P^2 - 4\right)^{-5/2} \tag{13}$$

138

etc... . These expressions however carry some redundant information. In fact we are interested only in the singular piece of these amplitudes in the vicinity of the critical "momentum" $P = 2$ which determines the large ℓ behaviour of the $W's$.

In order to define the scaling limit we, as usual, introduce a cut-off a which can be consider as the elementary length. The renormalized length of the boundaries L and its conjugated "momentum" z are defined as

$$L = a\ell, \quad P = 2 + az; \quad a \longrightarrow 0 \tag{14}$$

In the general case $\hat{W}(P)$ has a single cut along the real axis and the vicinity of the right endpoint of the cut is relevant for the continuum limit. The change of variables (14) blows up this vicinity to the whole complex plane. The left endpoint of the cut goes to $-\infty$ and thus a fractional power is added to the Laurent expansion of the loop amplitudes.

The cut-off a disappears from the loop equation if we define the renormalized loop correlator $\hat{W}(z, j)$ and the renormalized string interaction constant G as

$$\hat{W}(P) = \frac{1}{2}P + a^{1/2}\hat{w}(z) \tag{15}$$

$$G = a^{-3/2}g \tag{16}$$

Then equation (10) takes the form

$$z = \hat{w}^2(z) + G\delta\hat{w}(z)/\delta j(z) + G \oint_C \frac{dt}{2\pi i(z-t)}\hat{w}(t)\frac{\partial j(t)}{\partial t} \tag{18}$$

where we have put $j(z) = J(P)$. The source does not renormalize because its dimension is taken by the string interaction constant in the exponential on the r.h.s. of (6). The contour C encloses the cut of $\hat{W}(z)$ going along the negative real axis.

The gaussian model does not contain any parameters like cosmologican constant, etc... and this makes possible to calculate all multiloop correlators (5). They are given by totally symmetric functions of the arguments $z_1, ..., z_n$, homogeneous of degree $3 - 3H - 5/2n$. We remind that H is the number of handles on the world sheet. The explicit formula for the correlators of the loop field $\hat{\phi}(z)$ in the continuum limit is

$$\left\langle \hat{\phi}(z_1)\hat{\phi}(z_2)...\hat{\phi}(z_n) \right\rangle_H - (\text{cst}) \left[\left(\frac{\partial}{\partial z}\right)^{3H-3+n} \prod_{i=1}^{n}(z_i + z)^{-1/2} \right]_{z=0} \tag{19}$$

if $3H - 3 + n \geq 0$. For the cases $n = 1$ and $n = 2$, $H = 0$ we have from Eqs.(6) and (11)

$$\left\langle \hat{\phi}(z) \right\rangle_{H=0} = \frac{2}{3}z^{3/2} \tag{20}$$

$$\left\langle \hat{\phi}(z_1)\hat{\phi}(z_2) \right\rangle_{H=0} = \ln\left(\sqrt{z_1} + \sqrt{z_2}\right) \tag{21}$$

The formula (19) is an evident generalization of a formula for the planar loop amplitudes in pure gravity presented in [14].

To conclude let us remind that the string susceptibility exponent γ_{str} is defined by the dimension of G

$$G = g\, a^{-\nu(1-1/2\gamma_{\text{str}})} \tag{22}$$

where ν is the fractal dimension of the world sheet of the string (not necessarily equal to two). The meaning of Eq.(22) is that the scaling dimension of the operator $\hat{\phi}(z)$ creating a loop is $\frac{1}{2}\gamma_{str}$. Comparing Eqs.(22) and (16) we conclude that in the gaussian model

$$\gamma_{str} = -1, \quad \nu = 1 \tag{23}$$

The central charge of the matter field according to the KPZ formula [7]

$$C = 1 - 6\gamma_{str}^2 / (1 - \gamma_{str}) \tag{24}$$

is $C = -2$. The fractal dimension of the world sheet is one because the surface consists only of boundary.

3. KINEMATICS OF STRINGS EMBEDDED IN DYNKIN DIAGRAMS

In this section we give a prescription for introducing matter fields in the gaussian model. We regret to restrict ourselves to the case of central change $C \leq 1$.

It is well known [13] that the rational $2d$ conformal theories with central charge $C \leq 1$ are classified by the simply laced Lie algebras (i.e., these of the classical series A, D, E). For each such theory V. Pasquier constructed a lattice statistical model whose local degrees of freedom are labelled by the points of the Dynkin diagram of the corresponding Lie algebra [16]. In fact, the construction of V. Pasquier can be applied to any graph with the structure of one-dimensional simplicial complex. Such a graph is defined by its connectivity matrix C with matrix elements corresponding to all pairs of points (σ, σ') :

$$C_{\sigma\sigma'} = \begin{pmatrix} \text{the number of the lines of the graph having} \\ \text{as extremities the points } \sigma \text{ and } \sigma' \end{pmatrix} \tag{25}$$

Curiously enough, it turns out that long range correlations in the corresponding statistical model appear only if all eigenvalues of C are less or equal to two. The only connectivity matrices satisfying this condition are the Dynkin diagrams of simply laced Lie algebras and the corresponding extended Dynkin diagrams.

In ref.[9] we generalized the models of V. Pasquier to lattices with random geometry and constructed the equation of motion for the planar loop amplitudes. We have to admit that reading ref.[9] may be a pleasure only for a physicist with special test for complicated geometrical constructions. That is why we present here a simpler approach, based entirely on the loop equations. To the readers who will find it not sufficiently rigorous we recommend ref.[9].

Let $\mathcal{D}$ be a simply laced Dynkin diagram and $C_{\sigma\sigma'}$ be its connectivity matrix. Up to a sign the matrix C coincides with the non-diagonal part of the Cartan matrix. The graph $\mathcal{D}$ will be the target space of our string theory. This means that each string configuration represents a map of the world sheet of the string into the graph $\mathcal{D}$. We assume that this map is continuous in the sense the coordinate $\sigma \in \mathcal{D}$ of a point on the world sheet can jump only to a neighbour point $\sigma' \in \mathcal{D}$ (such that $C_{\sigma\sigma'} \neq 0$) when we move the point slowly.

Let us now concentrate on the possible string excitations that appear in a time slice of the world sheet. These correspond to closed or open paths in the graph $\mathcal{D}$.

In order to identify the ground state and the excited states we have to solve the problem of a random motion on $\mathcal{D}$. The propagation kernel for a random walk consisting of T steps on the graph $\mathcal{D}$ is directly from its connectivity matrix

$$K_{\sigma\sigma'}(T) = \begin{pmatrix} \text{the number of walks of } T \text{ steps} \\ \text{starting at } \sigma \text{ and ending at } \sigma' \end{pmatrix}$$

$$= \sum_{\sigma_1, \sigma_2, \ldots, \sigma_{T-1}} C_{\sigma\sigma_1} C_{\sigma_1\sigma_2} \ldots C_{\sigma_{T-1}\sigma'}$$

$$= \left(C^T\right)_{\sigma\sigma'} \tag{26}$$

Introducing the eigenvectors $S^\sigma_{(m)}$ of the connectivity matrix

$$\sum_{\sigma'} C_{\sigma\sigma'} S^{\sigma'}_{(m)} = \beta_{(m)} S^{\sigma'}_{(m)} \tag{27}$$

we can write the kernel (26) as a sum of projectors onto the eigenstates

$$K_{\sigma\sigma'}(T) = \sum_m \left[\beta_{(m)}\right]^T S^\sigma_{(m)} S^{\sigma'}_{(m)} \tag{28}$$

The eigenvalues of C have the form

$$\beta_{(m)} = 2\cos(\pi\, m/p), \quad m \in \mathcal{D}^*, \tag{29}$$

where p is the Coxeter number. The set $\mathcal{D}^*$ of the so-called Coxeter exponents is listed in Fig.3 for all simply laced Dynkin diagrams $\mathcal{D}$. They make sense of discrete momenta while $S^\sigma_{(m)}$ are the corresponding plane waves.

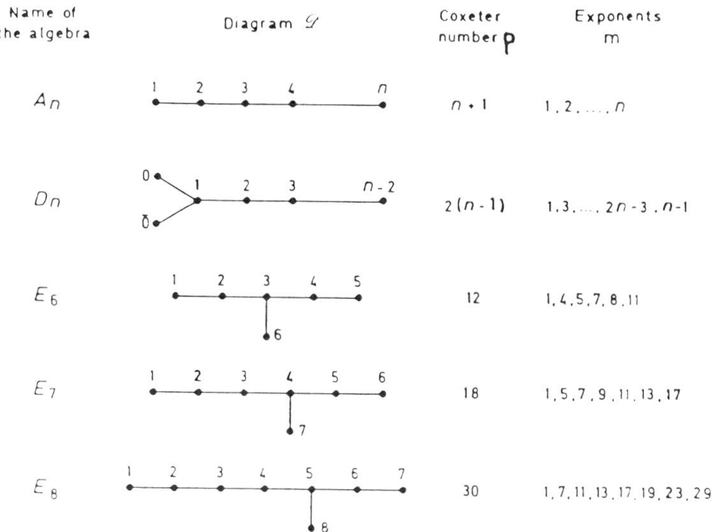

Figure 3. Dynkin diagrams and their Coxeter exponents.

For example, for the diagram of A_{p-1}

$$S^\sigma_{(m)} = \sqrt{\frac{2}{p}}\sin(\pi\, m\sigma/p), \quad m = 1, 2, ..., p - 1 \tag{30}$$

The wave functions of the states with "momenta" m and $p - m$ are related by

$$S^\sigma_{(m)} + (-1)^\sigma S^\sigma_{(p-m)} = 0 \tag{31}$$

The ground state will be the only one to survive in the r.h.s. of (28) in the limit $T \longrightarrow \infty$. It corresponds to the largest eigenvalue

$$\beta \equiv \beta_{(1)} = 2\cos\pi/p \tag{32}$$

Note that for the extended Dynkin diagrams the largest eigenvalue is always $\beta = 2$. This is the upper bound of $\beta_{(1)}$ which is reached for $p \longrightarrow \infty$.

The eigenstate $S^\sigma_{(1)}$ gives the probability amplitude to find the random particle at the point σ after infinite number of steps. It will be the vacuum state in the unitary version of the ADE strings. The order parameters are then the wave functions of the excited states normalized by this of the vacuum state

$$\chi^\sigma_{(m)} = \frac{S^\sigma_{(m)}}{S^\sigma_{(1)}} \tag{33}$$

The eigenvectors $S^\sigma_{(m)}$ form a complete set of orthonormalized states in the Hilbert space

$$\sum_{\sigma \in \mathcal{D}} S^\sigma_{(m)} S^\sigma_{(m')} = \delta_{mm'}, \qquad \sum_{m \in \mathcal{D}^*} S^\sigma_{(m)} S^{\sigma'}_{(m)} = \delta^{\sigma\sigma'} \tag{34}$$

It is possible to project the theory onto a subspace spanned by $S_{(m)}$ with $m = k(p-q)$, for some $q \leq p-1$. The role of the vacuum state in this subspace is played by the excited state $S_{(p-q)}$. This subspace will be used in the construction of the non-unitary models. The order parameters in a general (p,q) model are given by

$$\chi^\sigma_{(k)} = \frac{S^\sigma_{(k(p-q))}}{S^\sigma_{(p-q)}} \tag{35}$$

They satisfy a closed algebra

$$\chi^\sigma_{(k)} \chi^\sigma_{(k')} = \sum_{k''} C_{kk'k''} \chi^\sigma_{(k'')} \tag{36}$$

We can formally define the partition function as the sum over all embeddings of the world sheet in the graph $\mathcal{D}$. Each embedding is described by an ensemble of non-intersecting self-avoiding loops dividing the world sheet into domains of constant σ (Fig.4). The values σ' and σ'' on the two sides of a domain wall should be such that $C_{\sigma'\sigma''} = 1$.

Figure 4. a possible configuration of domain walls for a
world sheet with the topology of a disk and constant value
of σ along its boundary.

We assume that the energy of a domain wall is proportional to its length. Therefore each string configuration will contribute a factor $\exp[-2P_0 \ell_{tot}]$ where ℓ_{tot} is the total length of the domain walls and P_0 is a coupling constant which turns out to be the cosmological constant of our string.

4. DYNAMICS OF STRINGS EMBEDDED IN DYNKIN DIAGRAMS. LOOP EQUATIONS AND CONTINUUM LIMIT

The loop amplitudes will depend in general on arbitrary closed loops immersed in the Dynkin diagram $\mathcal{D}$. It is however possible to write closed equations for a very special kind of loops. These are the loops occupying a single point σ of $\mathcal{D}$.

Let us denote by $\hat{\phi}_\sigma(\ell)$ the operator creating a loop of length ℓ situated at the point $\sigma \in \mathcal{D}$. The n-loop amplitude will be defined in the same way as in the gaussian model (Eq.(5)). It will depend on the "momenta" $P_1, ..., P_n$ along the boundaries as well as on their coordinates $\sigma_1, ..., \sigma_n$ in $\mathcal{D}$. The generating function for the multiloop amplitudes is the partition function with $p - 1$ loop sources $J_\sigma(\ell)$, $\sigma = 1, ..., p - 1$:

$$F[g; J] = \sum_{H=0}^{\infty} g^{2H-2} \left\langle \exp\left(g \sum_{\sigma=1}^{p} \int_0^\infty d\ell\, \ell J_\sigma(\ell)\phi(\ell) \right) \right\rangle_H \tag{37}$$

The loop amplitude

$$W_\sigma(\ell) = g\frac{\delta F}{\delta J_\sigma(\ell)} \tag{38}$$

satisfies the following equation of motion

$$\frac{\partial}{\partial \ell} W_\sigma(\ell) = \int_0^\ell d\ell' W_\sigma(\ell') W_\sigma(\ell - \ell') +$$

$$+ g \int_0^\ell d\ell' \delta W_\sigma(\ell - \ell')/\delta J_\sigma(\ell') + g \int_0^\infty d\ell' \ell' J_\sigma(\ell') W_\sigma(\ell + \ell')$$

$$+ \sum_{\sigma'} C_{\sigma\sigma'} \int_0^\infty d\ell'\, e^{-2P_0 \ell'} [W_\sigma(\ell + \ell') W_{\sigma'}(\ell') + g\delta W_\sigma(\ell + \ell')/\delta J_{\sigma'}(\ell')] \tag{39}$$

The last term on the r.h.s. counts the possibility of a domain wall touching the boundary at the reference point. If we cut the world sheet along this domain wall, a piece of length ℓ' will be added to the original loop, and a new loop of length ℓ' will be created (Fig.5). We shall obtain either two disconnected surfaces or a new boundary. The new loop is situated at a point σ' such that $C_{\sigma\sigma'} = 1$.

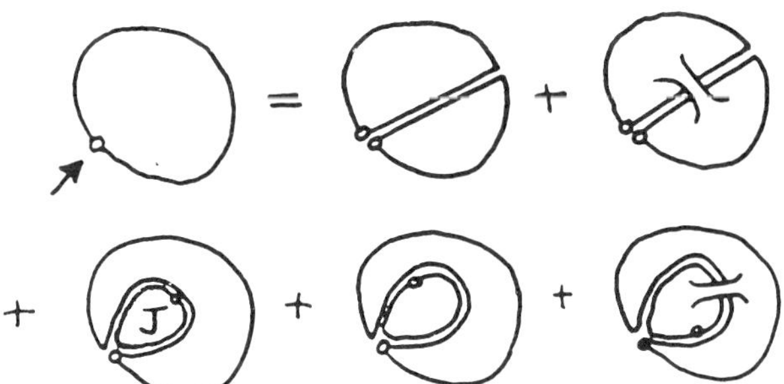

Figure 5. graphical representation of the loop equation.

It is convenient to introduce the Laplace images

$$\hat{W}_\sigma(P) = \int_0^\infty d\ell\, W_\sigma(\ell) e^{-P\ell} = g\frac{\delta F}{\delta \hat{J}_\sigma(P)} \tag{40}$$

$$\hat{J}_\sigma(P) = \int_0^\infty \mathrm{d}\ell \, J_\sigma(\ell) \mathrm{e}^{P\ell} \tag{41}$$

Then Eq.(39) takes the form

$$P\hat{W}_\sigma(P) - W_\sigma(0) = \left[\hat{W}_\sigma(P)\right]^2 + g\delta\hat{W}_\sigma(P)/\delta J_\sigma(P) + \oint_C \frac{\mathrm{d}Q}{2\pi i(P-Q)}$$
$$\left[g\frac{\partial\hat{J}_\sigma(Q)}{\partial Q}\hat{W}_\sigma(Q) + \sum_{\sigma'} C_{\sigma\sigma'}\left(\hat{W}_{\sigma'}(2P_0 - Q)\hat{W}_\sigma(Q) + g\delta\hat{W}_\sigma(Q)/\delta J_{\sigma'}(2P_0 - Q)\right)\right] \tag{42}$$

where the contour C encloses the singularities of $\hat{W}_\sigma(P)$. We are looking for a solution having a single cut along the real axis by the following reason. When $P_0 \longrightarrow +\infty$ the domain walls are suppressed and we have p copies of the gaussian model. The solution for $W_\sigma(P)$ than has a cut along the interval $[-2, 2]$. When P_0 decreases, up to some critical value P_* we are still in the phase of the gaussian model and the only thing to change is the position of the cut.

The Laplace-transformed loop equation (42) depends explicitly on the initial conditions $W_\sigma(\ell = 0)$, $\sigma \in \mathcal{D}$. Fixing the loop amplitude for a loop of length zero is equivalent to choosing a ground state for our string. This is the probability amplitude for a random particle moving in the target space. Therefore for the (p, q) model we set

$$W_\sigma(\ell = 0) = S^\sigma_{(p-q)} \tag{43}$$

The choice of the true ground state, $p - q = 1$, corresponds to a unitary matter field. Choosing an excited state leads to a non-unitary model. In this case only the first part of the orthonormality relations (34) can be used since the Hilbert space of states is spanned only by the vectors $S_{(m)}$ with $m = k(p - q)$. It is therefore convenient to work directly in the space of Coxeter exponents $\mathcal{D}^*$. With the help of Eqs.(35) and (36) we transform Eq.(42) into a system of equations for the Fourier images

$$\hat{W}_{(k)}(P) = \sum_{\sigma\in\mathcal{D}} S^\sigma_{(k(p-q))}\hat{W}_\sigma(P) \tag{44}$$

labelled by the set of integers

$$\mathcal{D}^*_{(p,q)} = \{k \,|\, k(p-q) \in \mathcal{D}^*\} \tag{45}$$

The vectors $S^\sigma_{(m)}$ with $m = k(p - q)$, $k \in \mathcal{D}^*_{(p,q)}$ span the Hilbert space of states for the (p, q) models. Each of them defines an order parameter $\chi^\sigma_{(k)}$ (Eq.(35)).

The new loop amplitudes (44) satisfy the set of equations

$$P\hat{W}_{(k)}(P) = \delta_{k,1} + \sum_{k',k''} C_{k'k''k}\left\{\hat{W}_{(k')}(P)\hat{W}_{(k'')}(P) + g\delta\hat{W}_{(k')}(P)/\delta J_{(k'')}(P)\right.$$
$$+ \oint_C \frac{\mathrm{d}Q}{2\pi i(P-Q)}\left[g\frac{\partial\hat{J}_{(k'')}(Q)}{\partial Q}\hat{W}_{(k')}(Q) + 2\cos\left(\pi k'(p-q)/p\right)\hat{W}_{(k')}(2P_0 - P)\right.$$
$$\left.\left.\cdot\hat{W}_{(k'')}(P) + 2\cos\left(\pi k'(p-q)/p\right)g\delta\hat{W}_{(k')}(2P_0 - P)/\delta J_{(k'')}(P)\right]\right\} \tag{46}$$

The next thing to do is to find the critical value P_* of the parameter P_0 and the scaling behaviour of the loop amplitudes in the vicinity of this point.

The geometrical interpretation of the critical singularity is the following. In the limit $P_0 \longrightarrow \infty$ all domain walls are suppressed and the string is trapped at a single point of the target space. The critical behaviour of the loop amplitudes is that of the gaussian model considered in section 2. When P_0 decreases, domain walls appear and the string starts to fluctuate along the Dynkin diagram. The size of the fluctuations is measured by the characteristic length of a domain wall. At the critical value P_* of P_0 this length becomes infinite and a condensation of domain walls occurs. The scaling limit at the critical point is reached if we stretch the boundaries of the world sheet to infinity, in the same time tuning P_0 so that the length of the boundary and the characteristic length of the domain walls remain of the same order.

In order to fix the critical point P_*, let us consider the simplest case $g = 0$, $J \equiv 0$. In this case the system of equations (46) reduces to a single equation for the planar loop amplitude $\hat{W}_{(1)}(P)$ corresponding to the identity operator ($k = 1$) while all other amplitudes are zero.

Setting

$$\hat{W}(P) \equiv \hat{W}_{(1)}(P), \quad \beta \equiv \beta_{(p-q)} \tag{47}$$

we have

$$P\hat{W}(P) = 1 + \hat{W}^2(P) + \beta \oint_C \frac{\mathrm{d}Q}{2\pi i(P - Q)} \hat{W}(Q)\hat{W}(2P_0 - Q) \tag{48}$$

Let P_L, P_R be the endpoints of the cut along the real P-axis where $\hat{W}(P)$ has a non-vanishing imaginary part. Then the Cauchy integral in Eq.(48) makes sense only if $P_R < P_0$; otherwise the cuts of $\hat{W}(Q)$ and $\hat{W}(2P_0 - Q)$ would overlap. The contour C encloses the cut (P_L, P_R) leaving outside its mirror image $(2P_0 - P_R, 2P_0 - P_L)$.

We can get rid of the Cauchy integral in (48) by applying the following trick. Write Eq.(48) again with P replaced by $2P_0 - P$. The last term can be given the same form as in (48) but the contour C will be replaced by its mirror image $\bar{C}$ enclosing the cut $[2P_0 - P_R, 2P_0 - P_L]$. Summing up the two equations and applying the Cauchy theorem to the integral along the contour $C + \bar{C}$ we obtain the following functional equation for $\hat{W}(P)$

$$P\hat{W}(P) + (2P_0 - P)\hat{W}(2P_0 - P) = 2 + W^2(P) + \hat{W}^2(2P_0 - P) + \beta\hat{W}(P)\hat{W}(2P_0 - P) \tag{49}$$

Eq.(49) allows to evaluate immediately the value of $W(P)$ at the point P_0

$$W(P_0) = \left(P_0 - \sqrt{P_0^2 - 2(2 + \beta)}\right)/(2 + \beta) \tag{50}$$

We chose the minus sign in order to have the right asymptotic $W \sim 1/P$ at infinity. The singularity of $W(P_0)$ appears at the point

$$P_* = \sqrt{2(2 + \beta)} \tag{51}$$

where the two cuts coalesce, i.e., $P_R = P_0 = 2P_0 - P_R = P_*$.

The solution of Eq.(46) at the critical point $P_0 = P_*$ has been found in ref.[17]. For $P \sim P_*$ it behaves as $(P - P_*)^{1-\theta}$ where the exponent θ is related to β by $\beta = 2\cos\pi\theta$. Below we present the exact solution of (48), resp. (49), in the scaling limit.

Let us rescale the quantities $P, P_0, \hat{W}$ by introducing a regulator a with dimension of length (as before $L = \ell a$ is the renormalized length)

$$P = P_* + az \tag{52}$$

$$P_0^2 = P_*^2 + a^{2-2\theta}\left(2\hat{W}(P_*)\right)^{-1}\Lambda \tag{53}$$

$$\hat{W}(P) = \hat{W}(P_*) + a^{1-\theta}\hat{w}(z) \tag{54}$$

Taking the limit $a \longrightarrow 0$ and keeping the renormalized cosmological constant Λ finite we expand the infinitesimal vicinity of the critical point $P = P_*$ to the whole complex plane. The cut in the z-plane is going from some point z_R on the negative real axis to $-\infty$ and the scaling loop amplitude $\hat{w}(z)$ will exhibit a singularity at infinity. Therefore the Laurent expansion of $\hat{w}(z)$ will contain fractional powers of z.

Inserting (52-54) in Eq.(49) we find that the terms of order 1 and $a^{1-\theta}$ vanish and the term of order $a^{2-2\theta}$ gives

$$\hat{w}(z)^2 + \hat{w}(-z)^2 + \beta \hat{w}(z)\hat{w}(-z) = \Lambda \tag{55}$$

The next power is $a^{2-\theta}$ and can be neglected provided θ is strictly positive.

Eq.(55) is the loop equation in the scaling limit for $g = 0$ and $J = 0$.

If we parametrize

$$\beta = 2\cos\pi\theta \tag{56}$$

$$z = z_R \mathrm{ch}\, u, \quad z_R = \Lambda^{\frac{1}{2-2\theta}} \tag{57}$$

then the solution of Eq.(55) is

$$\hat{w}(z) = \frac{\sqrt{\Lambda}}{\sin\pi\theta} \mathrm{ch}(1-\theta)u \tag{58}$$

The variable z sweeps the whole complex plane if we restrict u to the half-strip $\mathrm{Re}\ u > 0$, $0 < \mathrm{Im}\ u < 2\pi$. Looking at the dimensions of z and $\hat{w}$ we see that θ in Eq.(56-58) and θ in Eq.(53-54) are identical.

Taking $\theta = (p-q)/p$ for the (p,q) model we arrive at the following explicit expression for $\hat{w}(z)$

$$\hat{w}(z) = \frac{1}{\sin\pi q/p}\left[\left(\frac{z + \sqrt{z^2 - \Lambda^{p/q}}}{2}\right)^{q/p} + \left(\frac{z - \sqrt{z^2 - \Lambda^{p/q}}}{2}\right)^{q/p}\right] \tag{59}$$

The loop amplitude (59) has a cut from $z_L = -\infty$ to $z_R = -\Lambda^{p/2q}$. The position of the cut determines the large L behaviour of the inverse Laplace transform $w(L)$

$$w(L) \sim L^{\frac{q-p}{p}} e^{z_R L} \tag{60}$$

Thus the characteristic length of the loop is $|z_R|^{-1}$ while the characteristic volume of the world sheet is $\Lambda^{-1} = |z_R|^{-\nu}$ with $\nu = 2q/p$.

The loop amplitude (58) is related to the scaling operator $\hat{\phi}(z)$ creating a loop by

$$\hat{w}(z) = \frac{\partial}{\partial z} \left\langle \hat{\phi}(z) \right\rangle_{H=0} \tag{61}$$

The r.h.s. scales as $z_R^{-1+\nu(1-1/2\gamma_{\mathrm{str}})}$ which together with the scaling of $\hat{w}$ gives

$$\gamma_{\mathrm{str}} = \frac{q-p}{q}, \quad \nu = \frac{2q}{p} \tag{62}$$

Thus we can add to (52-54) the scaling law for the string interaction constant

$$g = G a^{\nu(1-1/2\gamma_{\mathrm{str}})} = G^{1+q/p} \tag{63}$$

The central charge of the matter field is, according to Eq.(24)

$$C = 1 - 6\frac{(p-q)^2}{pq} \tag{64}$$

The scaling dimensions Δ_k of the order parameters $\chi_{(k)}^\sigma$, $k \in \mathcal{D}_{(p,q)}^*$ (Eq.(33)) can be extracted from the scaling behaviour of the loop amplitude

$$W_{(k)}(P,*) = \frac{\partial}{\partial P}\left\langle \hat{\phi}(P)\chi_{(k)} \right\rangle \tag{65}$$

This amplitude satisfies a loop equation which has been studied in ref.[9]. Its solution in the scaling limit reads

$$\hat{w}_{(k)}(z,*) = \frac{\left(z + \sqrt{z^2 - \Lambda^{p/q}}\right)^{\frac{k(p-q)}{p}} - \left(z - \sqrt{z^2 - \Lambda^{p/q}}\right)^{\frac{k(p-q)}{p}}}{\sqrt{z^2 - \Lambda^{p/q}}}$$
$$= \Lambda^{\frac{k(p-q)}{2q} - \frac{p}{2q}}\frac{\mathrm{sh}\left(\frac{k(p-q)}{p}u\right)}{\mathrm{sh}\, u} \tag{66}$$

In particular, for $k = 1$ the amplitude (65) involves surfaces with the topology of a disc and a puncture on the world sheet. Therefore $\hat{w}_{(1)}(z,*) = \partial\hat{w}(z)/\partial\Lambda$.

The power of Λ in Eq.(66) should be $1 - \frac{1}{2}\gamma_{\mathbf{str}} - (1 - \Delta_k) - 1/\nu$. Therefore

$$\Delta_k = \frac{(k-1)(p-q)}{2q}, \quad k \in \mathcal{D}_{(p,q)}^* \tag{67}$$

These values correspond to the diagonal ($k = r = s$) of the table of scaling dimensions forming the KPZ spectrum [7]

$$\Delta_{rs} = \frac{|ps - qr| - (p-q)}{2q} \tag{68}$$

The non-diagonal operators are not present in the partition function (37) by construction. In order to include more operators we have to extend the space of loops and consider as well loops that occupy more than one point of the target space. Such loops are sources of domain walls. If we are going along the boundary, each time we jump to another point of the target space a domain wall is created. The number of created domain walls should be even if the boundary is closed.

The correlation function of two sources of $2r$ domain walls can be evaluated. The corresponding scaling dimension is [18]

$$\Delta_r = \frac{r}{2} - \frac{p-q}{2q} \tag{69}$$

It corresponds to taking $s = 0$ in (68).

Now let ut return to the loop equation in presence of loop source and for $g > 0$. Inserting Eqs.(52-54) and (63) in (42) we find the renormalized loop equation for the amplitude $\hat{w}_\sigma(z, j(z))$

$$[\hat{w}_\sigma(z)]^2 + G\delta\hat{w}_\sigma(z)/\delta j_\sigma(z) + \oint_C \frac{dt}{2\pi i(z-t)}\left[G\frac{\partial j_\sigma(t)}{\partial t}\hat{w}_\sigma(t) + \sum_{\sigma'} C_{\sigma\sigma'}\right.$$
$$\left.(\hat{w}_{\sigma'}(-z)j_\sigma(z) + G\delta\hat{w}_{\sigma'}(-z)/\delta j_\sigma(z)\right] = \Lambda S_{(p-q)}^\sigma \tag{70}$$

In conclusion, we would like to make the following two remarks.

1) The value $1 - p/q$ of the string susceptibility γ_{str} (Eq.(62)) is different from $\gamma_{str}^M = -2/(p + q - 1)$ obtained in the matrix models [8]. In fact, there is no contradiction since the cosmological constants Λ and Λ^M of these two realisations of the $C < 1$ strings are coupled to different operators. In the formalism using a chain of random matrices or, briefly, M-formalism, all scaling operators are present in the effective action. Therefore in the scaling limit the cosmological constant Λ^M will couple to the operator of smallest dimension Δ_{min}

$$\Delta_{min} = \min_{r,s} \frac{|pr - qs| - (p - q)}{2q} = \frac{1 - p + q}{2q} \tag{71}$$

Therefore Λ^M is measured in units of $(\text{volume})^{-1+\Delta_{min}} \sim \Lambda^{1-\Delta_{min}}$ which implies

$$\left(2 - \gamma_{str}^M\right)\left(1 - \Delta_{min}\right) = 2 - \gamma_{str} \tag{72}$$

2) The loop equation (55) for the scaling part of the amplitude was obtained provided $\theta = (p-q)/p$ is strictly positive number. In the limit $q = p-1 \longrightarrow \infty$ which corresponds to the string immersed in a one-dimensional space, the next term of order $a^{2-\theta}$ becomes important and Eq.(55) should be replaced by

$$\hat{w}^2(z) + \hat{w}^2(-z) + 2\cos\pi\theta\,\hat{w}(z)\hat{w}(-z) - a^\theta\,|z_R|^\theta\,z\left(\hat{w}(z) - \hat{w}(-z)\right) = \Lambda \tag{73}$$

The solution of this equation is a linear combination of the functions $\text{ch}(1 - \theta), \text{ch}(1 + \theta)u$ and $\text{ch}\,u$ where u is related to z by Eq.(57). In the limit $\theta \longrightarrow 0$ it becomes

$$\hat{w}(z) = \pi^{-2}\sqrt{\Lambda}\left[u^2\text{ch}\,u + u\,\text{sh}\,u\,\ln(a/\sqrt{\Lambda})\right]$$

$$= \frac{1}{\pi^2}\left[z\left(\ln\frac{z + \sqrt{z^2 - \Lambda}}{\sqrt{\Lambda}}\right)^2 + \sqrt{z^2 - \Lambda}\ln\frac{a}{\sqrt{\Lambda}}\ln\frac{z + \sqrt{z^2 - \Lambda}}{\sqrt{\Lambda}}\right] \tag{74}$$

The explicit dependence of the physical quantities on the cut-off a (scaling violation) has been already noticed in the context of another realisation of the $D = 1$ string [19].

REFERENCES

[1] V. Kazakov, Phys. Lett. **159B** (1985) 303;
F. David, Nucl. Phys. **B257** (1985) 45;
V. Kazakov, I. Kostov and A. Migdal, Phys. Lett. **157B** (1985) 295;
J. Fröhlich, "The Statistical Mechanics of Surfaces", in Applications of Field Theory To Statistical Mechanics, ed. L. Garrido (Springer, 1985).

[2] For a list of references and a good presentation of some latest results see:
D. Gross and A.A. Migdal, Nucl. Phys. **B340** (1990) 333.

[3] A.A. Migdal, Phys. Rep. **102** (1983) 199.

[4] F. David, Mod. Phys. Lett. A, vol.5, n°13 (1990) 1019.

[5] R. Dijkgraaf, H. Verlinde and E. Verlinde, "Loop Equations and Virasoro constraints in Non-perturbative $2 - D$ Quantum gravity", Princeton University preprint PUPT-1184, May 1990.

[6] M. Fukuma, H. Kawai and R. Nakayama, "Continuum Schwinger - Dyson equation and Universal Structures in Two-Dimensional Quantum Gravity", Tokyo preprint UT-562 (1990).

[7] V. Knizhnik, A. Polyakov and A.B. Zamolodchikov, Mod. Phys. Lett. **A3** (1988) 819;
F. David, Mod. Phys. Lett. **A3** (1988) 1651;
J. Distler and H. Kawai, Nucl. Phys. **B321** (1989) 509.

[8] M. Douglas, Phys. Lett. **B238** (1990) 176.

[9] I. Kostov, Nucl. Phys. **B326** (1989) 583.

[10] F. David, Phys. Lett. **B159** (1985) 303;
I. Kostov and M. Mehta, Phys. Lett. **B189** (1987) 118.

[11] E. Witten, Nucl. Phys. **B340** (1990) 281.
R. Dijkgraaf and E. Witten, Nucl. Phys. **B342** (1990) 486.

[12] J. Distler, Nucl. Phys. **B342** (1990) 523.

[13] V. Kazakov, Mod. Phys. Lett. **A4** (1989) 2125.

[14] J. Ambjorn, J. Jurkiewicz and Ju.M. Makeenko, NBI preprint, August 1990.

[15] A. Cappelli, C. Itzykson and J.B. Zuber, Comm. Math. Phys. **113** (1987) 113.

[16] V. Pasquier, Nucl. Phys. **B285** [FS89] (1987) 162; J. Phys. **35** (1987) 5707.

[17] M. Gaudin and I. Kostov, Phys. Lett. **B220** (1989) 200.

[18] B. Duplantier and I. Kostov, Phys. Rev. Lett. **61** (1988) 1433;
I. Kostov, Mod. Phys. Lett. **A4** (1989) 217.

[19] E. Brézin, V. Kazakov, Al. B. Zamolodchikov, Nucl. Phys. **B338** (1990) 673.

PHASE TRANSITIONS IN ONE MATRIX MODELS

Gautam Mandal

Institute for Advanced Study
Olden Lane
Princeton, NJ 08540

Abstract: We discuss in detail the phase diagram of a matrix model with a sixth degree potential which describe the $k = 2$ and $k = 3$ multicritical models (pure gravity and Yang-Lee). We find that the usual (single-well) phase describing the $k = 2$ universality class has a finite (in the double scaling limit) tunnelling amplitude into other (multiple-well) phases, signalling instabilities of the pure gravity model. The Yang-Lee point can however be approached from within the single-well phase and is free from the problem of instabilities.

Introduction

In this talk we report on some results we found in an attempt to understand the nature and occurrence of phase transitions in the one matrix models. We started by looking at the "pure gravity" model described by the partition function,

$$Z_N = \int d\phi \exp[-\frac{1}{2}\mathrm{Tr}\phi^2 - \frac{g}{N}\mathrm{Tr}\phi^4], \tag{1}$$

where ϕ is an $N \times N$ Hermitian matrix. BPIZ [1] have given us the solution for this problem in the planar limit $N \to \infty$. Though the integral (1) is divergent for any $g < 0$, the BPIZ solution exists for negative g all the way upto $g_c = -1/48$ where it develops a non-analyticity of the form,

$$F \equiv -\log Z \propto (g - g_c)^{5/2} \tag{2}$$

One can ask the question: What exactly happens at this point? Is there a phase transition here in a statistical mechanics sense? If there is, what is the phase beyond g_c?

For $g < 0$, the potential in (1) is bottomless and hence Z_N is not well-defined. To cure this we add a small $\epsilon\phi^6$ ($\epsilon > 0$) term to the potential.

$$Z_N = \int d\phi \exp[-V(\phi)], \quad V(\phi) = \frac{1}{2}\mathrm{Tr}\phi^2 + \frac{g}{N}\mathrm{Tr}\phi^4 + \frac{\epsilon}{N^2}\mathrm{Tr}\phi^6 \tag{3}$$

As one knows, the non-analyticity of the BPIZ solution does not disappear for $\epsilon > 0$. The point P_2 ($g_c = -1/48, \epsilon = 0$) extends to a critical line $g_c(\epsilon)$ ending in a tricritical

Random Surfaces and Quantum Gravity
Edited by O. Alvarez *et al., Plenum Press, New York, 1991*

point P_3. We will call this line the BPIZ line (Figure 1). All points on this line have similar critical behaviour to P_2 in the sense that keeping ϵ fixed and approaching the BPIZ line from the right in Figure 1, the singularity in the free energy a la [1] goes as $(g - g_c(\epsilon))^{5/2}$ except at the point P_3 where the exponent changes from 5/2 to 7/3. In statistical parlance, the BPIZ line represents the universality class of the "pure gravity" model [2]. In the language of multicritical models, this line is called the $k = 2$ line, and P_3 represents the $k = 3$ tricritical point.

Can one understand this critical line as a line of phase transitions? After all, if one looks at how the potential V changes as one tunes g from 0 to negative values, one might expect a phase transition from a phase typical of a single-well potential to a phase typical of multiple-well potentials. Of course the self-interactions would change the effective potential somewhat but the qualitative features of the potential should probably remain the same. Is the BPIZ line then simply a transition from a single-well phase to a double-well phase? [3,4]

We describe first some numerical results where we look for possible phase transitions in various thermodynamic quantities as a function of g at different fixed values of ϵ numerically. If Z_N is written in terms of the eigenvalues λ_i of ϕ, $V(\phi)$ is replaced by $\sum V(\lambda_i)$ and the measure $d\phi$ with $\Pi_i d\lambda_i \Pi_{i<j}(\lambda_i - \lambda_j)^2$. The second factor is the Vandermonde determinant of the eigenvalues. The method we use consists of expanding this Vandermonde determinant as a polynomial in the eigenvalues and expressing the partition function and the Green's functions of the $N > 1$ problem in terms of its $N = 1$ counterparts. We do this algebra using $Mathematica^{TM}$ upto $N = 5$ and compute the $N = 1$ integrals numerically. We calculate the one-point function of energy $E(< \text{energy} >= \frac{1}{N^2} \sum_i < \frac{1}{2}\lambda_i^2 + \frac{g}{N}\lambda_i^4 + \frac{\epsilon}{N^2}\lambda_i^6 >)$ as a function of g at various fixed values of ϵ for $N = 1$ through $N = 5$. We find that $E(g)$ seems to undergo some kind of transition at g-values far less negative than $g_c(\epsilon)$ a la BPIZ (see 9 detail).

We mark these new g-values and plot them in Fig 1. We can see that the line of transition estimated by these points is to the right of the BPIZ line, at least for small values of ϵ. How does one understand this new line, given the fact that nothing special happens here according to [1]?

Let us reconsider the analysis of BPIZ on the planar limit in some detail. In [1] a function $\rho(\lambda)$ was introduced to describe the saddle-point density of eigenvalues. However, this function was assumed to have a single connected support. This is of course too restrictive, and in general one should allow for spectral densities with support $L \subset R$ where L is a union of n disconnected arcs $L_i = [a_i, b_i]$.

The function $F(\lambda)$ of [1] is now given by

$$F(\lambda) = \frac{1}{2}V'(\lambda) + f(\lambda)\sqrt{\Pi_{\text{arcs}}(\lambda - a_i)(\lambda - b_i)} \tag{4}$$

where $f(\lambda)$ is to be determined from the normalisation condition

$$F(\lambda) \to 1/\lambda + O(1/\lambda^3) \quad \text{as} \lambda \to \infty \tag{5}$$

Once we determine f the density ρ is given by

$$\rho(\lambda) = -f(\lambda)\sqrt{-\Pi_{\text{arcs}}(\lambda - a_i)(\lambda - b_i)} \tag{6}$$

What is the most general situation for a sixth degree potential? It is easy to show that the form of ρ and the demand that it be positive semidefinite limits the maximum number of arcs to three. For instance, if there are to be four arcs then since the radical in (6) changes sign every time λ hops to the next arc (as we move from the left of the real line, say) we need to have f also change sign three times. However since the radical $\sqrt{(\lambda^2 - a^2)(\lambda^2 - b^2)(\lambda^2 - c^2)(\lambda^2 - d^2)} \to \lambda^4$ as $\lambda \to \infty$, the maximum degree of f consistent with (5) is 1. Hence f cannot change sign thrice.

If we make the assumption that the arcs are symmetrically distributed around the origin (it would be an interesting possibility to consider other cases which would amount to a spontaneous violation of the reflection symmetry of the action) then we are left with the following three types of solutions,

1) one-arc solution, ρ has support $[-a, a]$

2) two-arc solution, ρ has support $[-c, -b] \cup [b, c]$

3) three-arc solution, ρ has support $[-c, -b] \cup [-a, a] \cup [b, c]$

The one- and two-arc solutions are uniquely determined by the conditions mentioned above, but the three-arc solution has one free parameter. The reason is that f is of the form,

$$f = A + B\lambda^2 \tag{7}$$

in this case. Condition (5) gives us four conditions in five quantities A, B, a, b and c. The free parameter represents the family of cases where the eigenvalues have only partially spilled from the central well of the potential to the outer ones.

Let us now explain the nature of solutions we find. Consider changing g from zero to negative values at a fixed ϵ. Figure 1 illustrates the results that one obtains. If we look for one-arc solutions and impose the requirement that $\rho(\lambda)$ has to be positive semidefinite, there is one unique solution, which ceases to exist when we reach the BPIZ line. When we check for other solutions, starting from $g = 0$ and decreasing g, up to a certain value of g the one arc solution is the only one that exists (the other solutions are unphysical, the roots for b, c being complex). As we continue to decrease g, we start seeing three-arc solutions with the outer arcs being infinitesimally small. This happens at the line L_2 in Figure 1. The one-arc solution still remains, of course, as [1] would predict. As we carry on, more and more three-arc solutions come into existence, with increasing size of the outer arcs and diminishing central arc. Let us denote by q the ratio of the central arc of a three-arc solution to the arc size of a single-arc solution exisiting at the same g-value. Roughly speaking, $1 - q$ represents the fraction of eigenvalues spilt out of the central well. We shall parametrise the family of three-arc solutions by q. It is easy to see that $q \to 1$ and $q \to 0$ represent the two extreme cases when the three-arc solutions go over to the one and two-arc solutions respectively.

To come back to the solutions, as we keep decreasing g we have an increasing q-band of coexisting solutions and at some point we start seeing double-well solutions as well. Now when we have different saddle point solutions existing together, we can determine which one is the most stable by looking at their free energies. It is easy (we used the numerical integration of $Mathematica^{TM}$) to compute the free energies of each of these solutions. The result shows that as we change g to more and more negative values (at fixed ϵ), the one-arc solution becomes less stable (having larger free energy) than multiple-arc solutions. The stablest solution makes a transtion from the one-arc solution through a sequence of three-arc solutions to the two-arc solution.

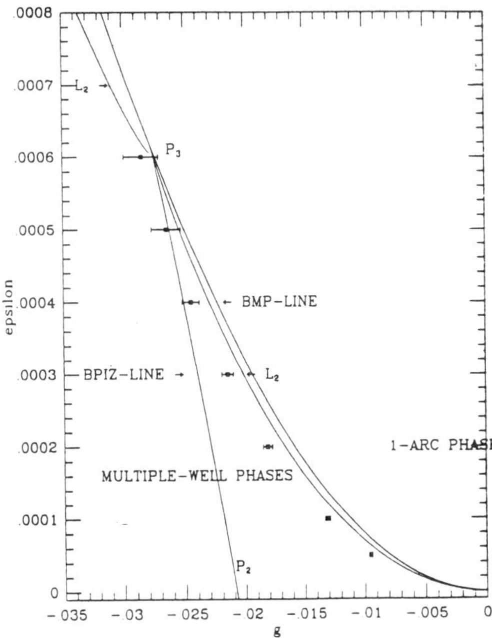

Fig 1. The phase diagram in the g-epsilon plane as determined from our analysis. The data points are the approximate location of the line of phase transition from the N=5 study. The curve L2, to the right of which multiple-arc phases do not exist, is found by the requirement that the the three-arc solution with q=1 exists with real b(=c).

What is the nature of the transition from the one-arc phase to the three-arc phases? Because of the q-band, the energy of the true ground state changes continously in contrast with what would have happened for a first-order transition between a single-well state and a double-well state. However, the lifetime of any q-state is infinite in the planar limit. Hence, in the planar limit the existence and nature of any thermodynamic singularity associated with this transition depends on the initial conditions imposed on the system.

Consequences for Random Surface Theories:

We have seen that as we tune g from zero to negative values, we encounter other phases long before we meet the BPIZ line. Of course it is possible in principle to try to stay on a metastable state (the one-arc state) and continue till it ceases to exist (like a supercooled system). The question is: is this state really metastable in the thermodynamic limit? If one takes the naive large-N limit then the answer is yes, because at any given value of g as we increase N the tunnelling amplitude for the eigenvalues (if initially prepared in the central well) decreases exponentially with N and in the limit $N = \infty$ the escape time goes to infinity so that the system equilibriates in the metastable state.

The situation is more subtle for the double-scaling limit where one takes g also close to $g_c(\epsilon)$ so that the renormalised coupling

$$x = (g - g_c(\epsilon))N^{4/5} \tag{8}$$

is kept constant. The system does know about $1/N^2$ fluctuations here which are likely to destablise the one-arc phase and lead it to the true ground state. We can estimate the lifetime of the one-arc phase in the double scaling limit when $\epsilon = 0$. Let us rescale the ϕ field so that the potential is of the form $V(\lambda) = N(\frac{1}{2}\lambda^2 + g\lambda^4)$. The time it takes an eigenvalue at the edge of the arc to move out is of $O(\exp \Delta F)$, where ΔF is the free energy barrier it has to cross before it can move out freely. We assume that all the other eigenvalues are fixed while the last one moves out. This places an upper bound on ΔF; in reality, the other eigenvalues will partially adjust to the perturbation caused by moving this last eigenvalue. We can ignore the change in the size of the arc caused by removing the last eigenvalue because that change is of $O((N\sqrt{g_c - g})^{-2/3})$, which is $O(N^{-2/5})$ in the double scaling limit. Let $F(\lambda)$ be the free energy of the system when the last eigenvalue is moved out to $\lambda\,(> a)$. Then

$$F(\lambda) - F(a) = V(\lambda) - V(a) - N\int_{-a}^{a} \rho(\mu)\ln|\lambda - \mu|d\mu + N\int_{-a}^{a} \rho(\mu)\ln|a - \mu|d\mu \qquad (9)$$

If we write λ as $a + \delta\lambda$, substituting the appropriate expression for $\rho(\mu)$ and a as functions of g leads to,

$$\frac{dF(\lambda)}{d\lambda} \sim N[\sqrt{g_c - g}\sqrt{\delta\lambda} - (\delta\lambda)^{3/2}] \qquad (10)$$

This can be integrated upto the maximum of free energy to give

$$\Delta F \sim N(g_c - g)^{5/4} \qquad (11)$$

In the double scaling limit, this is $O(1)$. Therefore the lifetime of the one arc phase in the double scaling limit is finite and the phase cannot be unambiguously defined. Since the nature of the singularities for the one-arc phase does not change for $\epsilon \neq 0$, we expect this result to carry over for general ϵ.

We see therefore that the BPIZ line cannot be approached on either side from the one-arc phase. Approaching the line from the right, it is *only* in the planar limit that the system can be made to be in the one-arc phase, so the double scaling region is not accessible. When we cross the BPIZ line from the left, the system remains in a two-arc phase; no singularity will be seen.

Interestingly, the $k = 3$ model is free from all these problems, as Figure 1 shows. The lines that correspond to tuning the cosmological constant are parabolas in the $g - \epsilon$ plane passing through the origin. (In the notation of [5] these lines correspond). to changing β and keeping all other parameters fixed; the origin corresponds to $\beta = \infty$ In Figure 1 the "BMP line" represents the parabola that passes through the $k = 3$ point P_3 [6]. As we can see, we can approach P_3 on this line from both sides staying in the one-arc phase. As [6] and [7] have shown, this makes available the use of both $x \to \pm\infty$ and determines a unique solution of the string equation.

We see therefore that the pure gravity universality class described in the double scaling limit by the Painleve equation has an inherent instability due to the existence of the ϕ multiple-well phases. Indeed the instanton action (11) which is finite in the double scaling limit denotes the amplitude of single eigenvalues to tunnel out of the central well, which causes the transition to the multiple well phase (this is in fact the origin of the oscillatory string solution found by [3]). It is interesting to note that the very phenomenon which gives us the first indication of non-perturbative effects in string theory also signals this instability of pure gravity (compare the works of [10] and [11] who reach simliar conclusions; in particular [11] shows that the exponential of the instanton action (11) precisely leads to the non-perturbative term in the Painleve solution with an imaginary coefficient). It is an important question whether it is possible to understand these non-perturbative effects in pure gravity from a well-defined theory with a bounded hamiltonian.

Acknowledgements: This work was done in collaboration with Gyan Bhanot and Onuttom Narayan[9]. We thank Itzhak Bars, Sumit Das, David Gross and Alexander Migdal for discussions. G.M.'s research was supported in part by a Department of Energy Grant No. DE-FG02-90-ER40542. It is a pleasure to thank the organisers of this conference for the opportunity to present this work.

REFERENCES

1. E. Brézin, C. Itzykson, G. Parisi and J. B. Zuber, Commun. Math. Phys. 59 (1978) 35.

2. See however alternative definitions of universality classes in O. Alvarez and P. Windey, LPTHE (Paris) preprint, February, 1990.

3. M. Douglas, N. Seiberg and S. Shenker, Rutgers University preprint, April, 1990.

4. L. Mollinari and E. Montaldi, I.N.F.N. (Milano) preprint 1990.

5. D. J. Gross and A. A. Migdal, Princeton University preprint, PUPT-1159, December, 1989.

6. E. Brézin, E. Marinari and G. Parisi, University of Rome preprint, ROM2F-90-09, February, 1990.

7. G. Moore, Yale University preprint, YCTP-P4-90.

8. E. Witten, Institute for Advanced Study preprint, IASSNS-HEP-90/37.

9. G.Bhanot, G.Mandal and O.Narayan, Phys. Lett. B251 (1990) 388.

10. J. Jurkiewicz, Niels Bohr Institute preprint, NBI-HE-90-21.

11. F. David, Saclay preprint, May 1990.

MATRIX MODELS OF 2D GRAVITY AND
ISOMONODROMIC DEFORMATION

Gregory Moore

Department of Physics and Astronomy
Rutgers University
Piscataway, NJ 08855-0849
and
Department of Physics
Yale University
New Haven, CT 06511-8167

1. Introduction

One of the principal goals of the theory of 2D gravity is making sense of the formal expression

$$Z(\mu, \kappa; t_i) = \sum_h \int_{MET_h} dg\, e^{\mu \int \sqrt{g} + \kappa \int R} Z_{QFT(t_i)}[g] \qquad (1.1)$$

where we integrate over metrics g on surfaces with h handles with a weight defined by the Einstein-Hilbert action (μ is the cosmological constant and κ is Newton's constant, or, equivalently, the string coupling) together with the partition function of some 2D quantum field theory, $QFT(t_i)$. The parameters t_i are coordinates on a subspace of the space of 2D field theories, or, equivalently, coordinates for a space of string backgrounds.

The expression (1.1) is of course extremely formal, especially when one includes the sum over topologies. Nevertheless rigorous definitions of this expression have been proposed using the methods of random matrix theory. Last year these concrete definitions led to a beautiful result for (1.1) for a special set of quantum field theories [1–6]. The answer is most simply expressed as a differential equation satisfied by

$$u(x; t_i) = \frac{\partial^2}{\partial x^2} Z \qquad (1.2)$$

Random Surfaces and Quantum Gravity
Edited by O. Alvarez *et al., Plenum Press, New York, 1991*

where x is the scaled string coupling. The differential equation satisfied by u for coupling to the (p, q) minimal conformal field theory [7] was most elegantly formulated by M. Douglas [5] in terms of a differential operator $L = D^q + u_{q-2}(x)D^{q-2} + \cdots u_0(x)$, where $D = d/dx$ and u_{q-2} is identified with $u(x)$, as the system of equations:

$$[L_+^{p/q}, L] = 1 \tag{1.3}$$

where the subscript $+$ keeps the differential operator part of a pseudodifferential operator. The equations for massive models interpolating between the minimal models have the form $\sum_p t_p [L_+^{p/q}, L] = 1$.

The specific case of the $(2p - 1, 2)$ models (corresponding to single hermitian matrix models) have been most intensively studied. In this case we have simply $L = D^2 + u(x)$ so the string equations are

$$\sum_j (j + \tfrac{1}{2})t_j R_j[u(x)] = x \tag{1.4}$$

where R_j are conserved densitites of the KdV hierarchy [3][4] .

Solutions to (1.4) satisfy several interesting properties. One remarkable result is that a solution to (1.4) , as a function of the t_j should satisfy KdV flow in the t_j [6] . The original argument of Banks, Douglas, Seiberg, and Shenker was proposed from the physical point of view using the matrix model integral, and makes certain implicit assumptions about boundary conditions. Later the issue of proper boundary conditions for physically acceptable solutions of (1.4) was clarified [8]. BMP argued that the existence of physically reasonable solutions depends on the parity of m, the largest index for which $t_m \neq 0$. Physical solutions should exist only for m odd, and in this case the asymptotics should fix a unique solution of (1.4) which is pole-free on the real axis. These arguments were supported by a numerical solution for the case $m = 3$. In [9] it was demonstrated, again numerically, that one cannot use KdV flow to define "pure gravity" (an $m = 2$ solution) by flowing from the well-defined $m = 3$ solution. Essentially, the solution develops a shock wave and there is no well-defined $t_2 \to \infty$ limit.

This paper is a review and continuation of [10] where the formalism of isomonodromic deformation and some related ideas were applied to the string equations. The purpose of our paper was threefold. First, as described in section three, the isomonodromic deformation formalism is well suited to proving rigorously the statements regarding the properties of KdV flow and existence and uniqueness of solutions mentioned above. The formalism applies equally well to the string equations associated with unitary-matrix models and with double-cut phases of the hermitian matrix models. Second, through the work of the Kyoto school   [12–15] it is known that the isomonodromic deformation formalism is closely related to the quantum field theory of free fermions in two-dimensional spacetime. This is extremely suggestive since hermitian matrix models can also be written in terms of nonrelativistic fermions. It is worth having a good understanding of any connection between these fermions since it might be indicative of a deep connection between nonperturbative 2D gravity and "quantum field theory on the spectral curve." These matters are discussed in sections two and four. Third, and more philosophically, interesting physics is usually related to interesting geometry, and this certainly ought to be the case for nonperturbative quantum gravity. We would like to know the underlying geometrical significance of the string equations. By this we do *not* mean the geometrical meaning of the terms in the asymptotic expansion of solutions to the string equations, for these have already been adequately understood from the point of view of topological field theory [16–20] . This third goal, which was our primary motivation in [10] , remains distant.

Closely related matters have been discussed in many recent papers. Of these we draw particular attention to   [21–24] where, among other things, the matrix models were shown to be equivalent to the infinite Toda chain - an important integrable system- even *before* taking the continuum limit.

2. Matrix Models and 2D Field Theory

In this section we outline how a "2D field theory on the spectral curve" may be seen to emerge from the matrix model integral. Our understanding of this phenomenon is woefully incomplete, but some relevant features may be seen already at this stage. Most importantly, the spectral parameter of inverse scattering theory is identified with the eigenvalue coordinate of the random matrix path integral. This point of view has also been emphasized in [23][24] (although some details are different).

2.1. The level-spacing problem

The clearest example of the phenomenon we are discussing can be seen already in the large N limit of hermitian matrix models. Consider the matrix model:

$$
\begin{aligned}
Z_N &= \int d^{N^2}\phi\, e^{-Ntr V(\phi)} \\
&= \int \prod d\lambda_i \Delta^2 e^{-N\sum_i V(\lambda_i)}
\end{aligned}
\tag{2.1}
$$

where $V(\lambda)$ is a polynomial in λ. and consider furthermore the probability $\tau(I;N)$ that no eigenvalue falls in the range $I = [\lambda_1, \lambda_2]$. It was shown in [25] that $\tau(a_1 - a_2) = \lim_{N\to\infty} \tau([a_1/N, a_2/N]; N)$ is the tau function for the isomonodromy problem related to the painlevé V equation. Since the context in which this was originally understood is (superficially) removed from matrix models we will show how it follows from that point of view.

Correlation functions with the measure (2.1) can be interpreted [26] [27] [6] as expectation values in a slater determinant of fermion one-body wavefunctions given by orthonormal functions:

$$
p_n(\lambda) = P_n(\lambda)e^{-\frac{N}{2}V(\lambda)}
$$

where P_n are orthonormal polynomials for the measure $d\lambda e^{-NV(\lambda)}$. As in [6] we may pass to second quantized wavefunctions:

$$\psi(\lambda) = \sum_{n=1}^{\infty} p_n(\lambda)a_n$$

$$\psi(\lambda)^\dagger = \sum_{n=1}^{\infty} p_n(\lambda)a_n^\dagger \qquad (2.2)$$

$$\{\psi^\dagger(\lambda), \psi(\lambda')\} = \delta(\lambda - \lambda')$$

where the ground state is the fermi sea with the first N levels filled. As argued in [6] the main contributions to correlation functions come from the neighborhood of the fermi level. For even potentials we can rewrite the recursion relation for orthonormal wavefunctions [1][2][3] in the form:

$$\lambda p_{2n}(\lambda) = \sqrt{r_{2n+1}}p_{2n+1} + \sqrt{r_{2n}}p_{2n-1}$$

$$\lambda p_{2n+1}(\lambda) = \sqrt{r_{2n+2}}p_{2n+2} + \sqrt{r_{2n+1}}p_{2n} \qquad (2.3)$$

By evaluating these at $\lambda = 0$ we see that if we expect a continuum limit for the orthonormal wavefunctions themselves in the neighborhood of $\lambda = 0$ we should define

$$p_{2n+1}(\frac{\lambda}{N}) = (-1)^n f_1(x, \lambda) \qquad p_{2n}(\frac{\lambda}{N}) = (-1)^n f_2(x, \lambda) \qquad (2.4)$$

where $x = n/N$. Assuming r_n has an expansion of the form $r_n = r(x) + \epsilon^2 r_1(x) + \cdots$ where $\epsilon = 1/N$ we find that (2.3) implies

$$f_1 = (r(x))^{-1/4}\sin\left[\lambda \int^x \frac{dx'}{\sqrt{r(x')}}\right]$$

$$f_2 = (r(x))^{-1/4}\cos\left[\lambda \int^x \frac{dx'}{\sqrt{r(x')}}\right] \qquad (2.5)$$

Since the dominant contributions of physical quantities come from the neighborhood of the fermi level ($x = 1$) the orthonormal wavefunctions become sines and cosines. These arguments can be checked explicitly using hermite functions in the case of a gaussian measure.

From the behavior of the 1-body wavefunctions it follows that $\hat{\psi}(\gamma; N) \equiv \frac{1}{\sqrt{N}}\psi(\gamma/N)$ has a smooth large N limit. For example the Darboux-Christoffel formula

$$\langle N|\psi^\dagger(\lambda_1)\psi(\lambda_2)|N\rangle = \sqrt{r_{N+1}}\frac{p_{N+1}(\lambda_1)p_N(\lambda_2) - p_{N+1}(\lambda_2)p_N(\lambda_1)}{\lambda_1 - \lambda_2} \qquad (2.6)$$

implies that

$$\langle N|\hat{\psi}^\dagger(\gamma)\hat{\psi}(\gamma')|N\rangle \to \frac{1}{\pi}\frac{\sin(\gamma - \gamma')}{\gamma - \gamma'} \qquad . \qquad (2.7)$$

On the operator level we define $p = (n - 2N)/2N$ and

$$
\begin{aligned}
a_1(p) &= \sqrt{N}(-1)^{n+N}\big(\hat{a}_{2n} - i\hat{a}_{2n+1}\big) \\
a_2(p) &= \sqrt{N}(-1)^{n+N}\big(\hat{a}_{2n} + i\hat{a}_{2n+1}\big)
\end{aligned}
\tag{2.8}
$$

where $\hat{a}_n \equiv a_{N+n}$. The sum over n becomes an integral $\int_{-1}^{\infty} dp$. Again, assuming that the main contributions come from the neighborhood of the fermi level we extend this to an integral over the entire p axis. Our main claim is therefore that $\hat{\psi}$ has a good large N limit and is given by

$$
\begin{aligned}
\hat{\psi}(\gamma) &= e^{i\gamma} \int_{-\infty}^{\infty} dp\, a_1(p)e^{i\gamma p} + e^{-i\gamma} \int_{-\infty}^{\infty} dp\, a_2(p)e^{-i\gamma p} \\
&= e^{i\gamma}\psi_1(\gamma) + e^{-i\gamma}\psi_2(-\gamma)
\end{aligned}
\tag{2.9}
$$

and that the fermi sea becomes the ground state defined by $a_i(p)|0\rangle = 0$ for $p > 0$ and $a_i^{\dagger}(-p)|0\rangle = 0$ for $p > 0$.

Now let us return to the problem of the level spacing. In terms of the orthonormal wavefunctions p_j we have [26][27]

$$
\begin{aligned}
\tau(I; N) &= det\left[\delta_{j,k} - \int_{\lambda_1}^{\lambda_2} d\lambda\, p_j(\lambda)p_k(\lambda)\right]_{0 \le j,k \le N-1} \\
&= \langle N| : exp\left(-\int_{\lambda_1}^{\lambda_2} \hat{\psi}^{\dagger}(\lambda)\hat{\psi}(\lambda)d\lambda\right) : |N\rangle
\end{aligned}
\tag{2.10}
$$

where the normal ordering puts ψ to the right of $\psi^{\dagger}$. In the $N \to \infty$ limit, taking $\lambda_i = a_i/N$, and $I = [a_1, a_2]$ we obtain:

$$
\tau(I) = \langle 0| : exp\left(-\gamma \int_I \hat{\psi}^{\dagger}\hat{\psi}\right) : |0\rangle
\tag{2.11}
$$

where $\gamma = 1$ and the normal-ordered exponential is evaluated by expanding in power series and point-splitting all the integrals. This expression with $\gamma = 2$ is a correlation function in the theory of the one-dimensional Bose gas, also known as the nonlinear Schrödinger theory, in the completely impenetrable case, and can be studied by the methods of [25][28] [29]. (Indeed, many integrable massive field theories have correlation functions related to painlevé equations [30].) In particular, $\tau(I)$ is the fredholm determinant $det\,(1 - K)$ where K is the kernel defined by (2.7) on the interval I and the integral operator $1 - K$ is in the infinite dimensional group of "completely integrable kernels" described in [28] . Following the general procedure described in [25][28][29] we define $\chi_1(z) = e^{iz}\psi_1(z)$ and $\chi_2(z) = e^{-iz}\psi_2(-z)$ so that $\hat{\psi} = \chi_1 + \chi_2$ and $\psi^{\dagger} = \chi_1^{\dagger} + \chi_2^{\dagger}$, and consider the "Baker-Akhiezer framing":

$$
\Psi_{\alpha\beta}(\lambda, \lambda') = \frac{\langle 0|\chi_\alpha^{\dagger}(\lambda)exp(-\int_I \chi^{\dagger}\chi)\chi_\beta(\lambda')|0\rangle}{\langle 0|exp(-\int_I \chi^{\dagger}\chi)|0\rangle}
\tag{2.12}
$$

As we will see in sec. 3.1 this matrix satisfies equations reminiscent of the Knizhnik-Zamolodchikov equation:

$$\frac{\partial}{\partial \lambda}\Psi = M_\lambda \Psi$$
$$\frac{\partial}{\partial a_i}\Psi = M_i \Psi \quad , \tag{2.13}$$

where M_λ, M_i are matrices which are rational in λ. We will see that the study of solutions of (2.13) gives information on the determinant $\tau(I)$.

2.2. Multi-Cut Solutions

We now turn to an example in which we take a double scaling limit. Consider a special class of multicritical potentials [31]

$$V_m'(\lambda) = k(m)\lambda^{2m+1}\left(1 - \frac{1}{\lambda^2}\right)^{1/2}\Big|_+ \tag{2.14}$$

where the subscript $+$ means we keep the polynomial part in an expansion about infinity and

$$k(m) = 2^{2m+1}\frac{(m+1)!(m-1)!}{(2m-1)!} \quad . \tag{2.15}$$

$V_m(\lambda)$ has the shape of a double well, and the bump at $\lambda = 0$ becomes progressively flatter as m increases. These potentials are characterized by the property that the eigenvalue distribution at tree level consists of two separate cuts, which, at the critical point, meet at the origin. These theories have very interesting phase transitions investigated in [32] [9][31] . The same critical behavior was discovered in the unitary-matrix ensembles [33] [34].

We consider again (2.3) , now assuming [32][9] that r_{2n} and r_{2n+1} have different scaling limits:

$$r_{2n} = r_c + a^{1/m}f(z) + a^{2/m}g(z) + \cdots$$
$$r_{2n+1} = r_c - a^{1/m}f(z) + a^{2/m}g(z) + \cdots \tag{2.16}$$

where, in the standard way, $x = \frac{2n}{N} = 1 - a^2(z - z_0)$ and $Na^{2+1/m} = 1$. The tree-level string equation is easily shown to be [31]

$$f^{2m} = \frac{-z}{2^{2m-1}(m+1)} \tag{2.17}$$

The critical behavior comes from the region $\lambda \cong \mathcal{O}(a^{1/m})$. By studying the recursion relation (2.3) in the neighborhood of $\lambda \cong 0$ we expect that the orthonormal wavefunctions

$$f^+(z,\lambda) \equiv j(a,\lambda)(-1)^k p_{2k}(a^{1/m}\lambda)$$
$$f^-(z,\lambda) \equiv j(a,\lambda)(-1)^k p_{2k+1}(a^{1/m}\lambda) \tag{2.18}$$

will have smooth limits, where we have allowed for a possible "wavefunction renormalization" j. The scaling which gives a smooth limit for the two-point function is $a^{1/2m}\psi(a^{1/m}\lambda) \to \psi(\lambda)$, so from (2.6)

$$\langle z_0|\hat{\psi}^\dagger(\lambda_1)\hat{\psi}(\lambda_2)|z_0\rangle = \frac{f^+(z_0,\lambda_1)f^-(z_0,\lambda_2) - f^+(z_0,\lambda_2)f^-(z_0,\lambda_1)}{\lambda_1 - \lambda_2} \quad . \quad (2.19)$$

This defines the kernel of a completely integrable integral operator so, following [25][28][29] we study the linear ODE's satisfied by $\vec{\psi} = (f^+(z,\lambda), f^-(z,\lambda))$. Indeed, the double-scaling limit of (2.3) becomes

$$\lambda\vec{\psi} = \sqrt{8}\left(-i\sigma_2\frac{d}{dz} - \sigma_1\frac{f}{4}\right)\vec{\psi} \quad . \quad (2.20)$$

Now consider the relation

$$\frac{d}{d\lambda}\vec{p} = N\big(V'(\lambda)_+\big)\vec{p} \quad (2.21)$$

where the subscript $+$ indicates that we keep only the upper triangular part of the matrix representation of the operator $V'(\lambda)$ in the basis of orthonormal wavefunctions. In the double scaling limit, with an appropriate choice of j and a slight redefinition of $\vec{\psi}$, the recursion relations become

$$\mathcal{L}\vec{\psi} \equiv \left(\frac{d}{dz} + \sigma_3\lambda + f\sigma_1\right)\vec{\psi} = 0 \quad (2.22)$$

$$\left(\frac{d}{d\lambda} + M_m(\lambda, f)\right)\vec{\psi} = 0 \quad (2.23)$$

where M_m is polynomial in λ and differential polynomial in f. We may find M_m explicitly as follows. Note that the commutator of M_m with $\mathcal{L}$ is just σ_3. In integrable systems theory the method of Zakharov and Shabat determines the space of matrices M which are polynomial in λ and differential polynomial in f and whose commutator with $\mathcal{L}$ is λ-independent. The vector space of such matrices is spanned by the matrices occuring in the Lax pairs for the modified KdV hierarchy. (See, e.g., [35] .)

The flatness condition following from (2.22) and (2.23) gives an ordinary differential equation for f. Comparing with the result (2.17) derived at tree level we find

$$M_m = (2m + 1)\left[\left(C_m + (V'_m/2\zeta) + \frac{z}{2m+1}\right)\sigma_3 + V_m\sigma_1 - (V'_m/2\zeta)i\sigma_2\right] \quad (2.24)$$

where

$$\begin{aligned} C_m &\equiv R_m + \zeta^2 R_{m-1} + \cdots \zeta^{2m} R_0 \\ V_m &\equiv \zeta S_{m-1} + \zeta^3 S_{m-2} + \cdots \zeta^{2m-1} S_0 \\ S_m &\equiv fR_m - \tfrac{1}{2}R'_m \end{aligned} \quad (2.25)$$

and the KdV potentials are evaluated for $u = f^2 + f'$. The compatibility conditions become the (2-cut) string equation $S_m - \frac{z}{2m+1}f = 0$ [33][34] . (Of course, the above argument is a variant of Douglas' original argument.)

More generally, by adding the multicritical potentials $V = \sum_\ell t_\ell a^{(2m-2\ell)/m}V_\ell$ we simply replace

$$M_m \to \sum_\ell t_\ell M_\ell \tag{2.26}$$

which yields the more general string equations:

$$\sum t_\ell S_\ell + \frac{1}{2m+1}zf = 0 \tag{2.27}$$

By arguments analogous to those above we find that the orthonormal wavefunctions must also satisfy linear equations in the variables t_ℓ:

$$\left(\frac{d}{dt_\ell} + \left(\zeta C_\ell + \tfrac{1}{2}V_\ell'\right)\sigma_3 + \left(\zeta V_\ell + S_\ell\right)\sigma_1 - \tfrac{1}{2}V_\ell'i\sigma_2\right)\vec{\psi} = 0 \tag{2.28}$$

The compatibility of (2.22) with (2.28) now shows that the solution f to (2.27) , as a function of t_ℓ satisfies mKdV flow

$$\frac{\partial f}{\partial t_\ell} = \frac{\partial}{\partial x}S_\ell[f] \tag{2.29}$$

while the solution f to (2.27) defines a self-similar solution to the flow.

As in the case of the level-spacing problem we can describe the partition function in terms of a fermion correlator. If we change the potential by changing $t_\ell \to t_\ell + \delta t_\ell$ then, for N finite we may write:

$$\frac{Z(t_\ell + \delta t_\ell)}{Z(t_\ell)} = \langle N|exp\left\{-\int_{-\infty}^{\infty}\sum_\ell \delta t_\ell a^{(2m-2\ell)/m}V_\ell(\lambda)\psi^\dagger\psi(\lambda)\right\}|N\rangle \tag{2.30}$$

By analytic continuation we can drop the subscript $+$ in (2.14) and, for small λ, $V_\ell(\lambda) \sim ik(\ell)\lambda^{2\ell+1}/(2\ell+1)$. Changing variables $\lambda \to a^{1/m}\lambda$ and taking the continuum limit gives

$$\frac{Z(t_\ell + \delta t_\ell)}{Z(t_\ell)} = \left\langle : exp\left\{-i\int_{-\infty}^{\infty}\sum_\ell \frac{k(\ell)}{2\ell+1}\delta t_\ell\lambda^{2\ell+1}\psi^\dagger\psi(\lambda)\right\} : \right\rangle \tag{2.31}$$

Because we interchanged limits several times (2.31) must be regarded as heuristic.

2.3. Single-Cut solutions

We now consider the phase of the matrix model in which the eigenvalue distribution forms a single cut. The critical behavior arises from the integration near a nonzero value of $\lambda = \lambda_c$, and for the mth multicritical point we have the scaling behavior $R_n \to r_c + a^{2/m}u(z)$ for $n/N = 1 - a^2(z - z_0)$, with $Na^{2+1/m}$ fixed [1][2][3] .

In this case it is natural to assume that the orthonormal wavefunctions have a limit as $a \to 0$:

$$p_n(\lambda_c + a^{2/m}\lambda) \to a^{-1/2m}p(z,\lambda) \qquad .$$

The recursion relation becomes

$$\left(\frac{d^2}{dz^2} + u(z)\right)p(z,\lambda) = \lambda p(z,\lambda) \tag{2.32}$$

so the double scaling limit of the orthonormal wavefunctions defines a Baker-Akhiezer function [1]. Similarly, in the limit we have the quantum field

$$a^{1/m}\psi(\lambda_c + a^{2/m}\lambda) \to \hat{\psi}(\lambda) = \int dz\, a(z)p(z,\lambda)$$

and the two point function is simply

$$\langle z_0|\hat{\psi}^\dagger(\lambda_1)\hat{\psi}(\lambda_2)|z_0\rangle = \sqrt{r_c}\frac{-p'(z_0,\lambda_1)p(z_0,\lambda_2) - p'(z_0,\lambda_2)p(z_0,\lambda_1)}{\lambda_1 - \lambda_2} \qquad .$$

$$\equiv K_m(\lambda_1,\lambda_2)$$

Following the procedure of [25][28][29] as before we consider the linear ODE satisfied by $\vec{\psi} = (p'(x,\lambda)\ p(x,\lambda))$:

$$\mathcal{L}\vec{\psi} \equiv \left[-\frac{d}{dx} + \begin{pmatrix} 0 & \lambda+u \\ 1 & 0 \end{pmatrix}\right]\vec{\psi} = 0 \qquad . \tag{2.33}$$

From the matrix model it is clear that

$$\left(\frac{d}{d\lambda} + N_m(\lambda,u)\right)\vec{\psi} = 0$$

where N_m is polynomial in λ and differential polynomial in u. The commutator of $\mathcal{L}$ with N_m is $\begin{pmatrix} 0 & 1 \\ 0 & 0 \end{pmatrix}$ so, by the method of Zakharov-Shabat, N_m is a linear combination of Lax pairs for the KdV hierarchy. The ℓth KdV flow can be written [35] as the compatibility condition $[2\partial/\partial t_\ell + \mathcal{P}_\ell, \mathcal{L}] = 0$, where the $sl(2)$ matrix

$$\mathcal{P}_\ell \equiv \begin{pmatrix} A_\ell & B_\ell \\ C_\ell & -A_\ell \end{pmatrix} \tag{2.34}$$

may be expressed in terms of the conserved densities R_ℓ of KdV flow [36] via

$$C_\ell = R_\ell + \lambda R_{\ell-1} + \cdots + \lambda^{\ell-1}R_1 + \lambda^\ell R_0$$
$$A_\ell = \tfrac{1}{2}C_\ell' \tag{2.35}$$
$$B_\ell = (\lambda + u)C_\ell - A_\ell'$$

---

[1] We take the double scaling limit of orthonormal wavefunctions explicitly, for the case of a gaussian ensemble, in appendix A.

Comparing with the tree-level equations we learn that if we define

$$\mathbb{P} = -\frac{d}{d\lambda} - \tfrac{1}{2}\sum_j (j + \tfrac{1}{2})t_j \mathcal{P}_{j-1} + \begin{pmatrix} 0 & x - (\sum_j (j + \tfrac{1}{2})t_j R_j) \\ 0 & 0 \end{pmatrix} \qquad (2.36)$$

(where $\mathcal{P}_{-1} = 0$) then the orthonormal wavefunctions satisfy

$$\mathbb{P}\vec{\psi}(\lambda, x, t_j) = 0$$
$$\mathcal{L}\vec{\psi}(\lambda, x, t_j) = 0 \qquad (2.37)$$
$$(2\frac{d}{dt_j} + \mathcal{P}_j)\vec{\psi}(\lambda, x, t_j) = 0$$

Compatibility of the first and last pairs gives the massive $(2l - 1, 2)$ and KdV equations, respectively. The compatibility conditions for the first and third equations then follow from the string and KdV equations. Similar considerations apply to the (p, q) equations.

If we perturb around the multicritical point we add the potential [3][4]

$$\delta V = N \sum \delta t_\ell (\lambda - \lambda_c)^{\ell + 1/2} a^{(2m - 2\ell)/m} \qquad (2.38)$$

As in the derivation of (2.31) the partition function becomes

$$\langle z_0 | : exp\left\{ \int_{-\infty}^{\infty} \psi^\dagger(\lambda)\psi(\lambda) \sum \delta t_\ell \lambda^{\ell + 1/2} d\lambda \right\} : | z_0 \rangle \qquad (2.39)$$

Thus, as a function of the δt_ℓ the partition function is simply the fredholm determinant $det(1 - K)$ for the kernel defined by

$$K(\lambda_1, \lambda_2; \delta t_\ell) = \sqrt{\vartheta(\lambda_1)} K_m(\lambda_1, \lambda_2)\sqrt{\vartheta(\lambda_2)}$$
$$\vartheta(\lambda) = \sum \delta t_\ell \lambda^{\ell + 1/2} \qquad (2.40)$$

The convergence of the integrals is rather delicate. The only case we can analyze explicitly is the "topological point" $m = 1$ in which case the relevent integrals are conditionally convergent.

2.4. Relation to other work

The emergence of a quantum field theory on the spectral curve has been discovered in several different guises in recent investigations into matrix models. In [37] [38] [39] it is shown that one may derive a quantum field theoretical representation of the $d = 1$ matrix model involving fields $\phi(t, \lambda)$ where t is the 1-dimensional time and λ is the eigenvalue coordinate of the string. This might be the $d = 1$ version of the phenomena discussed for $d < 1$ in this paper, and it would be extremely interesting to relate the $d = 1$ theories to the $d < 1$ theories by some analog of the Feigin-Fuks construction. On the other hand, we have analytically continued to complex λ, while λ is kept real in [37][38][39] .

There has also been a good deal of fuss over the discovery that the partition function of the matrix model is annihilated by operators forming a subalgebra of the Virasoro algebra [40] [20][23][24] . We will see in section 4.3 that the 2D quantum field theory on the spectral curve is related to a 2D conformal field theory, thus making the appearance of a Virasoro algebra completely natural [2]. Still, the relation between the two formalisms remains to be clarified.

It would be extremely interesting to understand how the field theory on the spectral curve arises from the sum over topologies of conformal-field-theoretic correlators. Some fascinating speculations along these lines are described in [41] [23] .

3. Isomonodromic Deformation

In the previous section we saw the appearance of a 2D quantum field theory on the spectral curve, and interpreted the string equations and KdV flow as compatibility conditions for first order linear differential equations satisfied by correlation functions in the theory. These equations are transport equations for a flat connection. Similar equations, for example null vector equations or the Knizhnik-Zamolodchikov equation, appear in conformal field theory. These equations depend on parameters, e.g. the moduli of some Riemann surface, and have interesting monodromy. Typically, deforming the parameters leaves the monodromy unchanged. In conformal field theory this is often expressed in terms of the flatness of the Friedan-Shenker vector bundles over moduli space [42]. This is an example of isomonodromic deformation. In this section we will see that isomonodromic deformation is the key behind the linear systems occuring in all the matrix model problems discussed above.

3.1. The level-spacing problem

We indicate the origin of (2.13). The monodromy of the two-point function (2.12) as a function of λ gives information on the partition function (2.11) [25] [3]. Because of the singular terms in the ope, $\chi_\alpha^\dagger$ has a nontrivial exchange algebra with $exp \int_I \chi^\dagger \chi$ so Ψ has monodromy as λ is continued around the endpoints of I. Moreover, by locality of the ope the monodromy matrix is unchanged as we deform the parameters a_i in the complex plane. Finally, the meromorphic matrix $d\Psi\Psi^{-1}$ where d is exterior differentiation in λ, λ', a_i is determined by its singularities at $\lambda = a_i, \lambda' = a_i, \lambda = \lambda'$ which are in turn dictated by the singular terms in the ope. In this way we can derive linear differential equations in λ, a_i which are analogous to the Knizhnik-Zamolodchikov

---

[2] Essentially the same remark was made in a pretty paper by A. Mironov and A. Morozov [24] .

[3] For readers of [25] one can identify, for example, $R_I^{\pm}(x, x'; \xi)$ with the two point function of $\hat{\chi}_i^\dagger(x)$ with $\hat{\psi}(x')$, and so on.

equations, obtained here as isomonodromic deformation conditions. Moreover, by studying the compatibility conditions for these linear equations we obtain nonlinear constraints which allow us to derive information on the partition function, which is simply the tau function for isomonodromic deformation (see below). Therefore, since (2.22)(2.23) and (2.37) are analogues of (2.13) we would like to interpret them as isomonodomic deformation conditions. This requires an enrichment of our notion of monodromy.

3.2. Stokes phenomenon

The isomonodromic deformation method focuses on the monodromy of the solutions of the equation in λ. Since (2.23)(2.37) is of the form $\frac{d}{d\lambda}\Psi = \mathcal{A}(\lambda)\Psi$ where $\mathcal{A}$ is polynomial in λ it would appear that there can be no monodromy. In fact, the singularity at infinity induces a kind of monodromy in the form of stokes phenomenon.

To put stokes phenomenon in perspective let us consider a general differential equation

$$\frac{d}{d\lambda}\Psi = \mathcal{A}(\lambda)\Psi \tag{3.1}$$

where $\mathcal{A}$ is meromorphic in λ. Near a regular singular point λ_0, where $\mathcal{A}$ has a pole, we may write a formal solution to the differential equation as

$$\Psi(\lambda) = \hat{\Psi}(\lambda)e^{M\ log(\lambda-\lambda_0)} \tag{3.2}$$

where $\hat{\Psi}$ is a formal power series in $\lambda - \lambda_0$. If we want true solutions corresponding to this formal solution we first choose angular regions of λ_0 in which we can define a branch of the logarithm. The formal and true solutions coincide in the angular regions [4], but upon comparison in the overlaps of regions we conclude that the true solutions differ by the monodromy $e^{2\pi i M}$.

Stokes phenomenon appears when we try to find solutions to (3.1) and $\mathcal{A}$ has an irregular singular point, i.e., a pole of order larger than one. For simplicity assume that

$$\mathcal{A}(\lambda) = \frac{A_{-r}}{(\lambda - \lambda_0)^{r+1}} + \cdots$$

with A_{-r} diagonalizable. In this case it can be shown [43] that there is a formal solution

$$\Psi = \hat{\Psi}e^{T(\lambda-\lambda_0)} \tag{3.3}$$

with

$$T = \frac{D_{-r}}{(\lambda - \lambda_0)^r} + \cdots \frac{D_{-1}}{(\lambda - \lambda_0)} + M\ log\ (\lambda - \lambda_0)$$

where the D_i are diagonal and commute with the "formal monodromy" M. In general $\hat{\Psi}$ is only an asymptotic series. It nonetheless has nontrivial analytic meaning since

---

[4] It is a nontrivial property of differential equations with regular singularities that $\hat{\Psi}$ is in fact a convergent power series.

there exist angular sectors in the neighborhood of the singular point in which there exist true solutions of (3.1) which are asymptotic to the formal solution (3.3) . The regions can be found by considering the nature of the essential singularity e^T. Notice that this is exponentially growing and decaying in the angular sectors where $\Re(\lambda-\lambda_0)^r$ is positive and negative, respectively. Typically a true solution asymptotic to the formal solution can only be defined in a sector of angular width π/r which contains regions of both growth and decay. Therefore if we compare two such solutions which are defined on regions with a nontrivial overlap they may differ by right-multiplication by a constant matrix. Such matrices are called stokes matrices, and, labelling the sectors by Ω_k we obtain a set of stokes matrices S_k associated with the differential equation.

According to [44][12–15] the stokes matrices are a generalization of the monodromy of the differential equation. This does not mean that a solution to the differential equation cannot be analytically continued to a single valued solution in an entire neighborhood of the singular point. The point is, such a single-valued solution has the "wrong" asymptotic behavior in all but two sectors.

A relevant example is provided by:

$$W' = \left(2\zeta^2\sigma_3 - \frac{1}{2\zeta}\sigma_1\right)W \qquad . \tag{3.4}$$

This is equivalent to the equation in λ in (2.37) for the "topological" case $\ell = 0$. The equation (3.4) has an irregular singularity of order 3 at infinity, and $T = \frac{2}{3}\zeta^3\sigma_3$ so there are six stokes sectors defined by the rays $\theta = \pm\frac{\pi}{6}, \pm\frac{\pi}{2}, \pm\frac{5\pi}{6}$. We can solve (3.4) exactly in terms of Airy functions. For example, defining $Ai_1(z) \equiv Ai(ze^{-2\pi i/3})$ we may write the fundamental solution, valid in the sector $\Omega_0 \cup \Omega_1 = \{\zeta| -\frac{\pi}{6} < arg\ \zeta < \frac{\pi}{2}\}$,

$$W_1 = \sqrt{\frac{\pi}{\zeta}} \begin{pmatrix} Ai_1'(\zeta^2) + \zeta Ai_1(\zeta^2) & Ai'(\zeta^2) + \zeta Ai(\zeta^2) \\ Ai_1'(\zeta^2) - \zeta Ai_1(\zeta^2) & Ai'(\zeta^2) - \zeta Ai(\zeta^2) \end{pmatrix} \begin{pmatrix} e^{-i\pi/6} & \\ & -1 \end{pmatrix} \tag{3.5}$$

whereas in the sector $\Omega_1 \cup \Omega_2 = \{\zeta|\frac{\pi}{6} < arg\ \zeta < \frac{5\pi}{6}\}$ we must use:

$$W_2 = \sqrt{\frac{\pi}{4\zeta}} \begin{pmatrix} -i(Ai'(\zeta^2) + \zeta Ai(\zeta^2)) & -(Ai'(\zeta^2) + \zeta Ai(\zeta^2)) \\ -i(Ai'(\zeta^2) - \zeta Ai(\zeta^2)) & -(Ai'(\zeta^2) - \zeta Ai(\zeta^2)) \end{pmatrix}$$

$$+ \sqrt{\frac{\pi}{4\zeta}} \begin{pmatrix} Bi'(\zeta^2) + \zeta Bi(\zeta^2) & i(Bi'(\zeta^2) + \zeta Bi(\zeta^2)) \\ Bi'(\zeta^2) - \zeta Bi(\zeta^2) & +i(Bi'(\zeta^2) - \zeta Bi(\zeta^2)) \end{pmatrix} \tag{3.6}$$

These two solutions are single-valued and holomorphic in the region $arg\ \zeta \neq \pi$ but are only asymptotic to the formal solution in the angular regions described above. Moreover, they are related by $W_2 = W_1 S_1$ where the stokes matrix is easily seen to be

$$S_1 = \begin{pmatrix} 1 & i \\ 0 & 1 \end{pmatrix} \qquad . \tag{3.7}$$

Similarly, from standard formulae one can find the the stokes matrices for the other sectors.

We now show that the linear ODE's of sec. 2 are isomonodromic deformation conditions. The differential equation

$$\frac{d}{d\lambda}\Psi = \sum t_\ell M_\ell \Psi \tag{3.8}$$

depends on parameters $z, t_j, f, f_z, f_{zz}, \ldots$, where, for the moment, we consider $f, f_z, f_{zz}, \ldots$ to be independent quantities. We will see in section four that z, t_j may be thought of as generalized moduli so it is natural to ask what conditions $f, f_z, f_{zz}, \ldots$ must satisfy if the monodromy(=stokes data) is to be independent of the moduli(=z, t_j). We may answer this question as follows.

As we vary z, t_j we obtain a family of invertible matrix solutions $\Psi(\lambda, z, t_j)$ to (3.8) so we may consider the quantity $\frac{\partial \Psi}{\partial z}\Psi^{-1}$, which is a rational matrix in λ. The only singularities can be at $\lambda = \infty$, so we need only know the behavior of a solution to (3.8) in this limit. Therefore we perform an asymptotic solution of (3.8) by relating the matrix elements of (2.24) to the resolvent $R(x, \lambda^2)$ of the Schrödinger operator $L = D^2 + u$. The result is that $\Psi \sim \hat{\Psi} e^T$ where

$$T = \tfrac{1}{2}\left(\sum_\ell \frac{2m+1}{2\ell+1} t_\ell \lambda^{2\ell+1}\right)\sigma_3$$

and

$$\hat{\Psi} = 1 + \frac{\hat{\Psi}_1}{\lambda} + \frac{\hat{\Psi}_2}{\lambda^2} + \cdots$$

$$\hat{\Psi}_1 = \frac{-if}{2}\sigma_2 + H\sigma_3$$

where $H' = \tfrac{1}{2}f^2$. Since the stokes data is unchanged under deformation of z we can compute $\partial_z \Psi \Psi^{-1}$ by substituting the asymptotic expansion and we find the condition (2.22). From this condition it follows that, as functions of z, f_z must be the derivative of f etc., and moreover $f(z)$ must satisfy the string equation. Similarly, one may compute $\frac{\partial}{\partial t_\ell}\Psi\, \Psi^{-1} mod(1/\lambda)$ to obtain the linear condtion (2.28) .This shows that if one considers a solution to (2.27) as a function of the t_ℓ then, if the stokes data are fixed, f must satisfy mKdV flow.

3.4. Asymptotic analysis of the PI family

A very similar computation can be carried out for the λ equation in (2.37) . There is one (important) technical change. Since the highest power of λ does not multiply an invertible matrix one must make a transformation [13] [45] $\lambda = \zeta^2$ and

$$\Psi(\lambda) = \zeta^{1/2}\begin{pmatrix} 1 & 1 \\ 1/\zeta & -1/\zeta \end{pmatrix} W(\zeta)$$

$$W(\zeta) = \zeta^{-1/2}\tfrac{1}{2}\begin{pmatrix} 1 & \zeta \\ 1 & -\zeta \end{pmatrix}\Psi(\lambda) \tag{3.9}$$

The leading singularity in the equation for W now multiplies σ_3 and $W \sim \hat{W}e^T$ where

$$T = \left(-\frac{1}{4}\sum_{j=1} t_j \zeta^{2j+1} + \zeta x\right)\sigma_3$$

$$\hat{W} = 1 + \frac{H_1}{\zeta}\sigma_3 + \frac{u}{4\zeta^2}\sigma_1 + \mathcal{O}(1/\zeta^3) \tag{3.10}$$

and $H_1' = \frac{1}{2}u(x)$. Again, this may be proved by relating the matrix elements in the differential equation to the resolvent of the Schrödinger operator $L = -D^2 + u$. As before, if we deform x keeping the stokes data fixed then we can evaluate the expression $\frac{\partial \Psi}{\partial x}\Psi^{-1}$ from its asymptotics. (In this case we must take into account the regular singularity at zero.) The result is just:

$$\frac{\partial \Psi}{\partial x}\Psi^{-1} = \begin{pmatrix} 0 & \lambda + u \\ 1 & 0 \end{pmatrix} \tag{3.11}$$

and comparing with section 2.3 we see that this means $u(x)$ must satisfy the string equation. Similarly we can ask for the condition that deformations in t_j keep the stokes data fixed, and we find remaining linear equations in (2.37) so that $u(x; t_j)$ satisfies KdV flow.

The string equation and KdV flow are compatible equations. This does not mean that, as we change the t_j a solution to the string equation automatically satisfies KdV flow. Nevertheless, we show in sec. 3.6 that the physical asymptotic conditions on $u(x)$ fix the stokes data uniquely, thus proving that $u(x; t_j)$ satisfies KdV flow. The following argument for KdV flow has been proposed in [21][22][23] . In the matrix model, before the continuum limit is taken, the jacobi matrix representing multiplication by λ in the space of orthogonal polynomials satisfies toda flow. It was proposed some time ago [46] that the continuum limit of toda flow should give (m)KdV flow.

3.5. τ functions

One of the beautiful results of the Kyoto school [12–15] is the definition of the τ function for isomonodromic deformation, which motivated the perhaps better known tau function of the KP hierarchy. Applying [12][13] to our case gives a closed one-form

$$\omega = Res_{\zeta=\infty} tr\left[\left(\hat{W}^{-1}\frac{d\hat{W}}{d\zeta}\right)dT\right] \tag{3.12}$$

on the space of deformation parameters. Since ω is closed one can define (locally) the *tau function* via $\omega = d(log\tau)$. Substituting (3.10) into (3.12) we get $\frac{\partial}{\partial x}(log\tau) = -2H_1$. On the other hand, (3.10) implies that $H_1' = u/2$ hence

$$u(x; t_j) = -\frac{d^2}{dx^2}log\tau \tag{3.13}$$

so the partition function of the matrix model (whose logarithm is the partition function of 2D gravity) is simply the tau function for isomonodromic deformation. Since the tau function is holomorphic [47] we see that all the string equations have the painlevé property: the *only* singularities of a solution u to the string equations are second order poles. This result may be easily extended to the entire hierarchy of (p,q) string equations.

Applying the definition (3.12) to the asymptotic expansion for the PII family gives

$$f^2 = 2H' = \frac{\partial^2}{\partial z^2} log\ \tau \qquad . \tag{3.14}$$

On the other hand, computing the connected two-point function of $tr\phi^2$ one can show [33][9][31]

$$\langle tr\phi^2\ tr\phi^2 \rangle_c = R_N \left(R_{N+1} + R_{N-1} \right) \rightarrow \frac{1}{8} - a^{2/m} f^2 \tag{3.15}$$

so that for this universality class we also can identify the partition function with the tau function.

3.6. Stokes matrices for the PI family

In order to obtain physical solutions to the string equation we need to choose proper boundary conditions. In the isomonodromic deformation literature it is shown that the initial conditions may be taken to be the stokes matrices of the associated linear problem. In this section we derive the stokes data for the PI family for the "physical" solutions to the string equations.

The issue of the correct choice of boundary conditions for a physically acceptable solution to the string equation is dictated by the matrix model integral [8][48]. For example, in the case of the $m = 3$ member of the PI family, i.e., $R_3[u(x)] = x$, the physical asymptotics are given by $u \sim \mathcal{O}(x^{1/3})$ as $x \rightarrow \pm\infty$. Since each asymptotic condition fixes two boundary conditions we expect that these asymptotics uniquely specify the solution. From the numerical integration in [8] it appears that the solution is pole-free.

We will show that the above asymptotics uniquely determine the stokes parameters in the monodromy problem associated to $R_3 = x$. Since one can reconstruct a solution from the stokes data (via the star operator, see below) it follows that the solution is unique. This solution is real for all x.

We investigate the direct monodromy problem following closely the treatment in [49] [45]. After the transformation (3.9) our equation can be written as

$$\frac{dW}{d\zeta} = \left[(B + \zeta^2 C + \Delta)\sigma_3 - (B - \zeta^2 C + \Delta)i\sigma_2 + (2\zeta A - \frac{1}{2\zeta})\sigma_1 \right] W \tag{3.16}$$

where

$$C \equiv -\tfrac{1}{2}\sum_j (j + \tfrac{1}{2})t_j C_{j-1} \qquad\qquad \Delta = x - \sum (j + \tfrac{1}{2})t_j R_j$$

$$A = \tfrac{1}{2}C' \qquad\qquad B = (\lambda + u)C - A'$$

Equation (3.16) has an irregular singularity of order $2\ell + 3$ at infinity and a regular singularity at the origin.

Consider first some general properties of the stokes matrices. If ℓ is the largest index with $t_{\ell+1} \neq 0$ there will be $4\ell + 6$ stokes sectors Ω_k each containing a unique ray $\theta = \frac{\pi}{4\ell+6}(2k - 1)$, $k = 0, \ldots 4\ell + 5$ along which $cos[(2\ell + 3)\theta] = 0$, thus we may take neighborhoods of infinity defined by:

$$\Omega_k \equiv \{\zeta | \frac{\pi}{4\ell + 6} + \frac{\pi}{2\ell + 3}(k - 2) < arg\zeta < \frac{\pi}{4\ell + 6} + \frac{\pi}{2\ell + 3}k\} \tag{3.17}$$

for $k = 0, \ldots, 4\ell + 5 \bmod 4\ell + 6$. On the overlap $\Omega_k \cap \Omega_{k+1}$ the two solutions W_{k+1} and W_k asymptotic to $\hat{W}e^T$ must be related by stokes matrices $W_{k+1} = W_k S_k$. The asymptotic expansion constrains the S_k to be of the form

$$S_{2k} = \begin{pmatrix} 1 & 0 \\ s_{2k} & 1 \end{pmatrix} \qquad S_{2k+1} = \begin{pmatrix} 1 & s_{2k+1} \\ 0 & 1 \end{pmatrix} \qquad . \tag{3.18}$$

The stokes parameters s_i are not all independent but satisfy:

$$s_{k+2\ell+3} = s_k \tag{3.19}$$

$$S_1 \cdots S_{2\ell+3} = i\sigma_1 \tag{3.20}$$

$$s_{2\ell+3-k} = -\bar{s}_k \qquad \text{if u is real} \tag{3.21}$$

We may prove these properties as follows. For (3.19) we use the symmetry of equation (3.16) to conclude that $W_{k+2\ell+3}(\zeta) = \sigma_1 W_k(-\zeta)\sigma_1$ and hence that $S_{k+2\ell+3} = \sigma_1 S_k \sigma_1 = S_k^{tr}$. For (3.20) we remark that the original equation in λ is regular throughout the λ plane. The solution is simply $Pexp \int^\lambda A(\lambda')d\lambda'$ for an appropriate matrix A hence the only singularities in Ψ can occur at infinity. Thus, near $\zeta = 0$ we have

$$W(\zeta) \cong \tfrac{1}{2}\zeta^{-1/2} \begin{pmatrix} 1 & \zeta \\ 1 & -\zeta \end{pmatrix} \tag{3.22}$$

By (3.22) , if we analytically continue W_1 from Ω_1 then $W_1(-\zeta) = i\sigma_1 W_1(\zeta)$ from which we obtain (3.20) . Finally if u is real then all the coefficients of powers of ζ in (3.16) are real so that $\bar{W}_k(\zeta) = W_{-k}(\zeta)$ implying (3.21) .

Since the determinant of (3.20) is automatically satisfied, (3.20) only imposes three independent constraints on the stokes matrices so we have $2\ell + 3 - 3 = 2\ell$ independent stokes parameters. Note that this is the number of initial conditions in the string equation. In fact, the two sets of parameter spaces may be regarded as the same [11][50].

Now we simplify (3.16) using the physical asymptotics. Consider (3.16) as $x \to \pm\infty$ first in the case where $t_j = 0$ for $j \neq \ell + 1, 0$. Since $R_\ell = \kappa_\ell u^\ell + \cdots$ with

$$\kappa_\ell = (-1)^\ell \frac{(2\ell - 1)!!}{2^{\ell+1}\ell!}$$

perturbation theory predicts that the solution to the string equation satisfies

$$u \sim \left(-1/2\kappa_{\ell+1}\right)^{1/(\ell+1)} x^{1/(\ell+1)} \qquad \ell \quad \text{even}, \quad x \to \pm\infty$$
$$u \sim \pm\left(1/2\kappa_{\ell+1}\right)^{1/(\ell+1)}(-x)^{1/(\ell+1)} \qquad \ell \quad \text{odd}, \quad x \to -\infty \tag{3.23}$$

In fact, perturbation theory tells us to take the $+$ root for u in the case of odd ℓ, but we may easily examine both cases at once. Rescaling variables and only keeping leading order terms (3.16) becomes

$$\frac{dW}{d\xi} = \tau\left[(2\xi^2 \pm 1)p_\ell^\pm(\xi)\sigma_3 \mp p_\ell^\pm(\xi)i\sigma_2 - \frac{1}{2\xi\tau}\left(\frac{8\ell\kappa_{\ell+1}}{\ell+1}\xi^2 p_\ell(\xi) + 1\right)\sigma_1 + \mathcal{O}(1/\tau^2)\right]W \tag{3.24}$$

where $p_\ell^\pm(\xi) = \sum_{p=0}^\ell (\pm 1)^p \kappa_p \xi^{2\ell-2p}$. For ℓ odd we obtain this equation with $x \to -\infty$, where:

$$\zeta = \left(-x/2\kappa_{\ell+1}\right)^{1/(2\ell+2)}\xi \qquad\qquad \tau = \left(-x/2\kappa_{\ell+1}\right)^{(2\ell+3)/(2\ell+2)} \qquad .$$

For ℓ even we obtain this equation with $\pm x \to +\infty$, where

$$\zeta = \left(\pm x/(-2\kappa_{\ell+1})\right)^{1/2\ell+2}\xi \qquad\qquad \tau = \left(\pm x/(-2\kappa_{\ell+1})\right)^{(2\ell+3)/(2\ell+2)} \qquad .$$

When $t_j \neq 0$ for $j \leq \ell$ we scale $\zeta^2 = u_0\xi^2$ where u_0 is the solution of the tree-level equation, so $\tau = |u_0|^{\ell+3/2}$. Expression (3.24) still holds but $p_\ell^\pm(\xi)$ receives corrections of order $\mathcal{O}\left(t_j\tau^{-2(\ell+1-j)/(\ell+1)}\right)$. These will not affect our argument below.

The evaluation of the stokes matrices is carried out by doing a WKB analysis in the $\tau \to \infty$ limit as in [11]. Thus true solutions to (3.24) are asymptotic as $\tau \to \infty$ to the WKB ansatz:

$$W^{WKB} \sim T\exp\left[\tau \int^\zeta \Lambda(\zeta')d\zeta'\right] \tag{3.25}$$

where T diagonalizes (3.24) to $\Lambda = \mu\sigma_3$. There are several WKB solutions in different regions defined by the turning points and conjugate stokes lines. The turning points are simply the roots ξ_i of μ, and the conjugate stokes lines are the lines defined by the vanishing real part:

$$\Re \int_{\xi_i}^\xi \mu(\xi')d\xi' = 0 \tag{3.26}$$

The general procedure for finding the stokes matrices is described in [11]. A simple consequence of this procedure allows us to obtain certain necessary conditions on the stokes matrices which, in some cases, fix the parameters uniquely. The main observation is that if, at a turning point which is a root of p_ℓ the conjugate stokes lines form three large regions each abutting an open region at infinity then the stokes matrix for the transition function associated with the middle region $\tilde\Omega_2$ is trivial [10] .

Therefore we describe the stokes lines. The lines for the case $\ell = 2$ have the form:

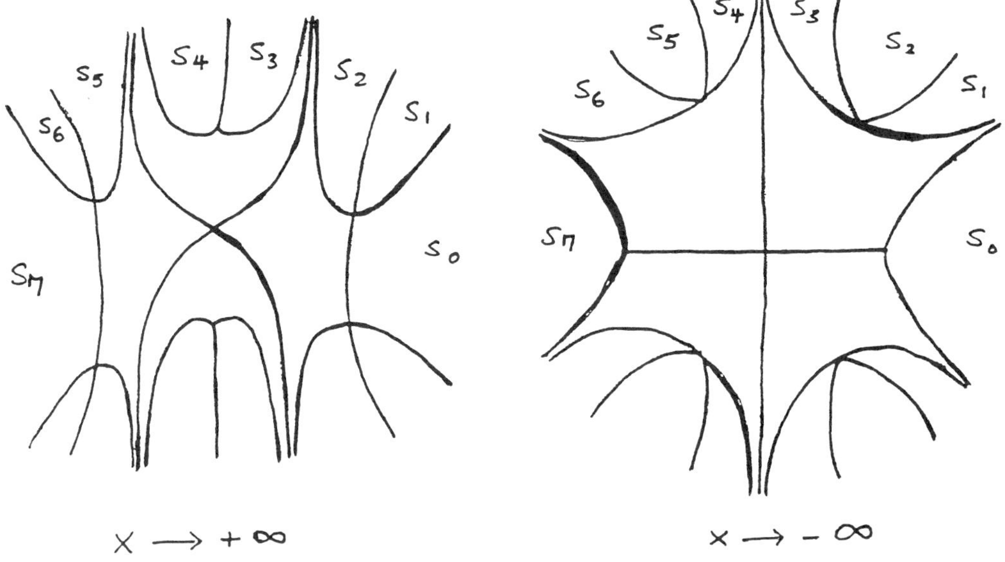

The limit $x \to +\infty$ gives $s_1 = s_6 = 0$ and the limit $x \to -\infty$ gives $s_2 = s_5 = 0$. From the monodromy constraints we get $s_0 = s_3 = s_4 = i$. Unfortunately, if we move on to higher ℓ the configuration of stokes lines becomes too complicated to use our observation immediately to set half the stokes parameters to zero. Nevertheless, the pattern for small ℓ leads to a natural guess for the stokes parameters in the general case. For ℓ even, comparing the constraints from the physical asymptotics at either end of the axis should fix two disjoint sets of stokes parameters. We expect that $s_1 = s_2 = \cdots = s_\ell = 0$ while $s_{\ell+1} = s_{\ell+2} = s_{2\ell+3} = i$ so that the solution is unique, and the stokes data is always concentrated on the wedges abutting the x and y axes in the $\zeta \ (=\sqrt{\lambda})$ plane. Similar considerations hold for ℓ odd. In particular, we expect (and can prove for the case $\ell = 1$) that (3.23) forces $s_0 = s_2 = \cdots = s_{\ell-3} = s_{\ell-1} = 0$, $s_{\ell+1} = s_{\ell+2} = i$.

We have not used the reality constraints in an essential way. Repeating the analysis for the triply truncated solution, characterized by $u \sim +(-x)^{1/2}$ for $x \to -\infty$ and $u \sim \pm i x^{1/2}$ for $x \to \infty$, one easily shows that in this case the constraints $s_2 = s_3 = i$, $s_0 = s_5 = 0$, $s_1 + s_4 = 0$ are supplemented by $s_1 = 0$ or $s_4 = 0$, depending on the sign of the imaginary part. In either case we confirm that the solution is unique.

An analog of the BMP solutions exists for the PII family [51] [52]. In this case physical asymptotics specify that $u(x)$ grows algebraically at one end of the axis and decays exponentially at the other end [9][52] . Applying the above techniques one finds that in this case (at least for the first two members of the PII family) the stokes data is concentrated entirely on the y-axis, but the necessary conditions leave a single

175

parameter undetermined. In this case, however, the inverse monodromy problem is equivalent to an integral equation (the self-similar Gelfand-Levitan-Marchenko equation) which can be examined directly. This is done in [52] where it is shown that the physical solution is unique.

3.7. KdV orbits are disconnected

Since KdV flow is compatible with the string equations we can consider the KdV orbits of solutions of the string equation. These orbits are parametrized by the largest index ℓ for which $t_\ell \neq 0$. We may wonder whether the orbits of physical solutions are connected. Certainly they are formally connected. For example, consider the equation

$$\tfrac{1}{2}(m_1 + \tfrac{1}{2})R_{m_1} + \tfrac{1}{2}(m_2 + \tfrac{1}{2})TR_{m_2} = x \tag{3.27}$$

with $m_1 > m_2$. As discussed in [6][53] if one scales a solution $u(x;T)$ to (3.27) using $v(y;T) = T^{2/(2m_2+1)}u\big(T^{1/(2m_2+1)}y;T\big)$, then the large T limit $v(y) = \lim_{T\to\infty} v(y;T)$ must be a solution of the lower order string equation $\tfrac{1}{2}(m_2 + \tfrac{1}{2})R_{m_2}[v] \sim x$, *provided the limit exists.* The existence of this limit is a very delicate issue, and in [9] numerical evidence is presented that the flow from $m = 3$ to $m = 2$ does not exist.

Since physical asymptotics fixes the stokes data, and since KdV flow is isomonodromic, we are in a position to investigate analytically the result in [9] . Suppose the $T \to \infty$ limit does exist. Then we can scale $\zeta \to T^{-1/(2m_2+1)}\zeta$ in (3.16) to obtain an equation with a smooth $T \to \infty$ limit. Since solutions can in principle be obtained from the path-ordered exponential of the "gauge field" on the rhs of (3.16) , solutions to (3.16) will also have smooth $T \to \infty$ limits. Moreover, from the asymptotics in ζ we see that the coefficients have smooth $T \to \infty$ limits and in fact approach the asymptotics of the lower order equation. Thus, fundamental solutions smoothly approach fundamental solutions for the lower order equation, although they will be defined on small regions of angular width $\frac{2\pi}{4m_1+2}$ and hence only define part of a fundamental solution for the lower order equation which is defined on the larger regions of angular width $\frac{2\pi}{4m_2+2}$. Because of this we find two rules governing flows:

1. A large region associated with a trivial stokes matrix for the "m_2 equation" cannot contain a small region with a nontrivial stokes matrix for the "m_1 equation."

2. A large region associated with a nontrivial stokes matrix for the "m_2 equation" must contain at least one small region with a nontrivial stokes matrix for the "m_1 equation."

Note that for even and odd ℓ the nontrivial stokes data disagrees on the real axis. Hence, by rule 1 it is impossible to use KdV flow to go from an even ℓ to an odd ℓ model, confirming the result of [9] . Note that flow from an even ℓ to a smaller even ℓ is consistent with rules 1 and 2. An alternative argument for the nonexistence of certain flows is given in [48] .

3.8. Stokes phenomenon and the eigenvalue distribution

The above WKB analysis is very similar to the work of F. David [48], who studies
carefully the saddle-point eigenvalue distribution, taking into account global stability
criteria. The reason for the similarity is that, at finite N the eigenvalue density $\rho_N(\lambda)$
and the fermion two-point functions are connected through

$$
\begin{aligned}
\rho_N(\lambda) &= \frac{1}{N}\langle N|\psi^\dagger\psi(\lambda)|N\rangle \\
&= \frac{\sqrt{r_N}}{N}\left(p_N\frac{\partial}{\partial\lambda}p_{N+1} - p_{N+1}\frac{\partial}{\partial\lambda}p_N\right)
\end{aligned}
\tag{3.28}
$$

and in the double scaling limit we therefore have

$$
\rho_N(\lambda_c + a^{2/m}\lambda) \longrightarrow a^{2-1/m}\left[\left(\frac{\partial}{\partial z}p\right)\left(\frac{\partial}{\partial\lambda}p\right) - p\left(\frac{\partial}{\partial z}\frac{\partial}{\partial\lambda}p\right)\right] \ .
\tag{3.29}
$$

The eigenvalue distribution vanishes algebraically along a cut $\mathcal{C}$ in the λ plane, ending
at λ_c and vanishes exponentially beyond λ_c. This has important consequences for the
large λ asymptotics of $p(z,\lambda)$. Introducing a second solution of (2.32), $q(z,\lambda)$ so that
p, q have unit wronskian we may write a matrix solution to (2.37)

$$
\Psi = \begin{pmatrix} \partial_z p & \partial_z q \\ p & q \end{pmatrix}
\tag{3.30}
$$

According to (3.25) the large λ asymptotics of Ψ have the form

$$
\Psi \sim \frac{1}{\lambda^{1/4}}\begin{pmatrix} \lambda^{1/2} & -\lambda^{1/2} \\ 1 & 1 \end{pmatrix}\left(1 + \mathcal{O}(\lambda^{-1/2})\right)exp\left[-\sigma_3\int^\lambda \mu(\lambda')d\,\lambda'\right]N
\tag{3.31}
$$

where $\mu^2 = A^2 + BC$ and N is an undetermined constant matrix. Note that the rhs
of (3.29) is the 21 matrix element of $\Psi^{-1}\partial_\lambda\Psi$. Therefore, just beyond λ_c along the
cut $\mathcal{C}$, p must decrease exponentially, and we can take $N = 1$. On the other hand,
along the cut $\mathcal{C}$ the algebraic behavior of ρ implies that p is not pure exponential but
instead is a cosine or a sine. This asymptotic behavior requires a different matrix

$$
N \propto \begin{pmatrix} 1 & 1 \\ 1 & -1 \end{pmatrix}
\tag{3.32}
$$

Thus, the exponential vanishing of the eigenvalue density outside of $\mathcal{C}$ and the algebraic vanishing of the density along $\mathcal{C}$ implies that the Baker Akhiezer functions must
exhibit stokes phenomenon. Moreover, it follows that the eigenvalue density in the
scaling region of λ_c is given by $-i\mu(\lambda)$. In particular, using the tree level approximation to μ we may obtain the tree level effective potential $Re[G(\lambda)]$ for eigenvalues
from

$$
G'(\lambda) = C[u,\lambda]\sqrt{\lambda + \overline{u}}
\tag{3.33}
$$

where we use the tree level approximation for C, u. Thus, David's stability sectors
coincide with the sectors (3.26) around the point $\xi_i = \lambda_c^{1/2}$ and his global stability
condition is essentially the condition that the integration over eigenvalues must not
pass through a region in which the double scaling limit of the orthogonal polynomials
has exponential growth.

4. Isomonodromy and Free Fermions

In this section we will interpret the isomonodromy problem connected with the string equations in conformal-field-theoretic terms. Our paradigm will be the solution of the Riemann-Hilbert problem for the case of regular singular points given ten years ago by the Kyoto school [54]. We review their construction first, in the light of subsequent developments in CFT. Then we consider the case of irregular singular points. Developing further some work of Miwa [55], we find that the theory of irregular singular points can be included at the expense of the introduction of a new kind of operator. We are extending conformal field theory by expanding the class of functions admitted in the theory from analytic functions with algebraic singularities to analytic functions with essential singularities. This is reflected in the need to expand the class of operators from twist operators to star operators.

4.1. Regular Singular Points

The basic idea of [54] is that the solution to an $m \times m$ matrix differential equation

$$\frac{d\Psi}{dz} = A(z)\Psi(z) \tag{4.1}$$

may be characterized uniquely by its monodromy properties. More precisely, suppose A has only simple poles at points a_ν and the residue can be diagonalized to L_ν. Then the matrix Ψ can be uniquely characterized by the requirement that

(i.) $\Psi(z_0) = 1$

(ii.) $\Psi(z)$ is holomorphic and invertible in $z \in \mathbb{P}^1 - \{a_1, \ldots, a_n\}$

(iii.) $\Psi(z) = \hat{\Psi}^{(\nu)}(z)e^{L_\nu \log(z - a_\nu)}$ for, $z \cong a_\nu$ where $\hat{\Psi}^{(\nu)}$ is holomorphic and invertible in a neighborhood of a_ν.

Conversely, any such matrix defines a rational matrix $A = \Psi_z \Psi^{-1}$ with at most simple poles. Thus, if one can construct appropriate "twist operators" φ_i such that the correlation function

$$\Psi_{\beta\alpha}(z_0; z) = (z_0 - z)\frac{\langle \bar{\psi}_\beta(z_0)\psi_\alpha(z)\varphi_n(a_n) \cdots \varphi_1(a_1) \rangle}{\langle \varphi_n(a_n) \cdots \varphi_1(a_1) \rangle} \tag{4.2}$$

has the correct monodromy properties, then it must be a solution of (4.1). Thus we have reduced the global Riemann-Hilbert problem to the *local* problem of finding conformal fields $\varphi_L(a)$ with the operator product expansion

$$\psi_\alpha(z)\varphi_L(a) \sim \left[\mathcal{O}^0_{L,\alpha}(a) + (a - z)\mathcal{O}^1_{L,\alpha}(a) + \cdots\right](z - a)^{L_\alpha} \tag{4.3}$$

The construction of these operators proceeds by choosing a basis of curves γ_ν circling once around a_ν and generating the fundamental group $\pi_1(\mathbb{P}^1 - \{a_\nu\}; z_0)$ and defining:

$$\varphi_M(a) = exp\left[\int_c^a tr\left\{log(M)J(y)\right\}\frac{dy}{2\pi}\right] \tag{4.4}$$

where $J_{\beta\alpha} = \bar{\psi}_\beta\psi_\alpha$ and $\mathcal{C}$ is a contour (a branch cut for Ψ) emanating from a. Consider now (4.2) with such operators inserted. As we analytically continue z around a the simple pole in the ope of ψ with J gives rise to the monodromy $\psi_\alpha \to \psi_\gamma M_{\gamma\alpha}$ in the Fermi field. Therefore (4.2) solves conditions (i-iii).

Let us now consider isomonodromic deformation of (4.1). It is clear from locality of the ope that changing the a_ν leaves the monodromy data unchanged. Necessary and sufficient conditions for isomonodromic deformation are given by:

$$\frac{\partial \Psi}{\partial z_0} = -\sum \frac{A_\nu}{z_0 - a_\nu}\Psi$$
$$\frac{\partial \Psi}{\partial a_\nu} = \left(-\frac{A_\nu}{z - a_\nu} + \frac{A_\nu}{z_0 - a_\nu}\right)\Psi \tag{4.5}$$

The compatibility conditions for these linear equations give the nonlinear Schlesinger equations [5]. From (4.2) we see that the linear equations of isomonodromic deformation theory should be thought of as transport equations on moduli space, analogous to the Knizhnik-Zamolodchikov equations, so that the theory of isomonodromic deformation for regular singular points fits nicely into the framework of Friedan-Shenker modular geometry.

The tau function associated to the deformation parameters a_ν is defined to be [12][13][14]

$$d \, log \, \tau(a_1, \ldots, a_n) = -\sum_\nu Res_{z=a_\nu} tr\left[(\hat{\Psi}^{(\nu)})^{-1}\frac{\partial \hat{\Psi}^{(\nu)}}{\partial z} d\left(log(z - a_\nu)L_\nu\right)\right] \tag{4.6}$$

where the d is a differential in the parameters a_ν. In fact the τ function is given by $\tau = \langle\varphi_1(a_1)\ldots\varphi_n(a_n)\rangle$ [12][13][14]. We will now rederive this using general principles of conformal field theory [6].

We have normalized (4.2) so that it is equal to $\delta_{\alpha\beta}$ at $z = z_0$. Taking the operator product expansion as $z \to z_0$ and matching this with an expansion of a solution to (4.1) around z_0 we find

$$\frac{\langle J_{\beta\alpha}(z_0)\varphi_n(a_n)\cdots\varphi_1(a_1)\rangle}{\langle\varphi_n(a_n)\cdots\varphi_1(a_1)\rangle} = -A(z_0)_{\beta\alpha}$$
$$\frac{\langle T(z_0)\varphi_n(a_n)\cdots\varphi_1(a_1)\rangle}{\langle\varphi_n(a_n)\cdots\varphi_1(a_1)\rangle} = \tfrac{1}{2}tr A^2(z_0) \tag{4.7}$$

where T is the stress energy tensor. Since L_{-1} always takes a derivative with respect to position we have

$$\frac{\partial}{\partial a_\nu}log[\langle\varphi(a_1)\cdots\varphi(a_n)\rangle] = \oint_{a_\nu} dz_0 \frac{\langle T(z_0)\varphi_n(a_n)\cdots\varphi_1(a_1)\rangle}{\langle\varphi_n(a_n)\cdots\varphi_1(a_1)\rangle} = \oint_{a_\nu} dz_0\tfrac{1}{2}tr A^2(z_0)$$

---

[5] For an appropriate choice of matrices, for example, these equations reduce to PVI [13].

[6] Exactly the same derivation was given 3 years ago by V. Knizhnik, [41], ch. 4. I was unaware of this work when I published [10].

On the other hand, substituting the local expansion $\Psi = \hat{\Psi}(z-a)^L$ we get

$$\hat{\Psi}^{-1}\hat{\Psi}_z + L/(z-a) = \hat{\Psi}^{-1}A\hat{\Psi} \quad . \tag{4.8}$$

Squaring this equation we find

$$\frac{\partial}{\partial a}(\log \tau) = Res\ tr\left(\hat{\Psi}^{-1}\hat{\Psi}_z\frac{L}{z-a}\right) = \tfrac{1}{2}Res\ tr\ A^2 \quad , \tag{4.9}$$

and hence the tau function is simply the correlation function of twist operators.

4.2. Irregular Singular Points

Let us now attempt to repeat the previous discussion for the case of a differential equation (4.1) where A is rational but can have irregular singularities. Our treatment is the same in spirit as the discussion of Miwa [55] , although there are some differences of detail.

Recall that at an irregular singular point we divide up a neighborhood of the point into sectorial domains, each containing a fundamental solution with asymptotics

$$\Psi \sim \left(\sum_{l\geq 0}\hat{\Psi}^{(l)}(z-a)^l\right)(z-a)^L e^{T(z-a)} \tag{4.10}$$

where L and

$$T(z-a) = \sum_{i=1}^{r}\frac{t_r}{(z-a)^r}$$

are diagonal, and $\hat{\Psi}^{(0)}$ is invertible. In particular, the analytic continuation of Ψ_1 will have the asymptotic expansion

$$\Psi_1 \sim \left[\sum_{l\geq 0}\hat{\Psi}^{(l)}(z-a)^l\right](z-a)^L e^T(S_1\cdots S_{k-1})^{-1} \tag{4.11}$$

in the sector Ω_k. A solution to the differential equation can be uniquely characterized by the conditions $(i)-(iii)$ above except that (iii) must be replaced by the requirement that Ψ have the asymptotic expansions (4.11) . Assume there is only one irregular singular point and define $\tilde{\Psi} \equiv \Psi_1 e^{-T(z-a)}$. We now search for quantum field operators $V_{S,T,L}(a)$, which we call "star" operators, such that

$$\tilde{\Psi}_{\beta\alpha}(z_0;z) = (z_0 - z)\frac{\langle\bar{\psi}_\beta(z_0)\psi_\alpha(z)V_{S_1,t_1,L_1}(a_1)\cdots\rangle}{\langle V_{S_1,t_1,L_1}(a_1)\cdots\rangle} \tag{4.12}$$

Evidently, a star operator is characterized by its operator product expansion with ψ, $\bar{\psi}$, e.g., for $z \in \Omega_k$ we have

$$\psi_\alpha(z)V_{S,T,L}(a) \sim \left[\mathcal{O}_\gamma^0(a) + (a-z)\mathcal{O}_\gamma^1(a) + \cdots\right](z-a)^{L_\gamma}\left[e^T(S_1\cdots S_{k-1})^{-1}e^{-T}\right]_{\gamma\alpha} \tag{4.13}$$

From this description it looks very unlikely that star operators exist [7]. We now give at least a formal construction of these operators.

Consider a ray C emanating from a point a. Consider the product of operators

$$exp\Big[\int_C^a dy\psi_\alpha(y)M_{\alpha\beta}(y)\bar\psi_\beta(y)\Big]\psi_\alpha(z)$$

where M is some matrix defined along the line. If we analytically continue in z through the curve C and compare with the other operator ordering it is a simple consequence of Cauchy's theorem and the operator product expansion that we have the exchange algebra:

$$exp\Big[\int_C^a dy\psi_\alpha(y)M_{\alpha\beta}(y)\bar\psi_\beta(y)\Big]\psi_\alpha(z-\epsilon) =$$

$$(4.14)$$

$$\psi_\gamma(z+\epsilon)(e^{M(z)})_{\gamma\alpha}exp\Big[\int_C^a dy\psi_\alpha(y)M_{\alpha\beta}(y)\bar\psi_\beta(y)\Big]$$

where z is a point on C and $z \pm \epsilon$ are points above and below z. Thus, defining $\mathcal{S}_k \equiv e^T S_k e^{-T}$ we may define, at least formally,

$$V_{S,T,L}(a) = \varphi_L(a)\prod_k exp\Big[\int_{C_k}^a tr\Big\{(log\mathcal{S}_l(y))J(y)\Big\}dy\Big] \qquad (4.15)$$

where we choose contours C_k in Ω_k such that the matrix $\mathcal{S}_k$ approaches the identity rapidly. In [55] Miwa obtained formulae for star operators using a slightly different formalism. Comparing his formulae in terms of expansions evaluated by Wick's theorem we obtain the same result. As shown in [55] contours of integration can be defined so that for small enough stokes data the integrals make sense, thus giving a more precise definition to the star operator.

It follows from locality of the operator product expansion that the differential equation satisfied by (4.12) has the property that the monodromy data S, L are preserved if we vary the parameters a_i, t_i. In strict analogy with the case of regular singularities, the tau function for isomonodromic deformation with irregular singular points is given by the correlation function of star operators [55] . One may give a formal argument for this following steps analogous to those leading from (4.7)to (4.9). The analog of (4.7) is

$$\frac{\langle J_{\beta\alpha}(z_0)V_{S_1,t_1,L_1}(a_1)\cdots\rangle}{\langle V_{S_1,t_1,L_1}(a_1)\cdots\rangle} = -A(z_0) + T'(z_0)$$

$$(4.16)$$

$$\frac{\langle T(z_0)V_{S_1,t_1,L_1}(a_1)\cdots\rangle}{\langle V_{S_1,t_1,L_1}(a_1)\cdots\rangle} = \tfrac{1}{2}tr(A - T')^2(z_0)$$

---

[7] For example, it is often claimed that operator product expansions in CFT are convergent. Note that (4.13) is only asymptotic. The reason for this is ultimately to be found in the fact that the string coupling has become dimensionful [2]. Note that it is x and the masses t_j which multiply the terms giving the essential singularity at infinity.

where $\mathcal{T}$ is the stress-energy tensor. Using formal manipulations with the ope one can show that

$$-Res_{z_0=a}tr\,\delta T(z_0)A(z_0) = \frac{\sum_k \int_{C_k}^a dy\,tr\left(\delta T(y)\frac{\delta}{\delta T(y)}\right)\langle V_{S,T,L}\cdots\rangle}{\langle V_{S,T,L}(a)\cdots\rangle} \tag{4.17}$$

Putting together (4.16) and (4.17) we then find

$$\frac{d}{da}\,log\,\langle V\cdots V\rangle = \tfrac{1}{2}Res_{z_0=a}tr\,A^2(z_0) = \frac{d}{da}\,log\,\tau \tag{4.18}$$

where the second equality follows from an argument similar to the case of regular singularities. Similarly, one can show that the dependence of $log\,\tau$ and $log\langle V\cdots V\rangle$ on other parameters is the same.

4.3. τ functions for 2D gravity

As a special case of the above formalism we can express the solution u of the string equations in terms of a fermion correlation function. We represent the solutions W to (3.16) and Ψ to (3.8) as fermion two-point functions in the presence of star operators. For the τ function of the PII family we may define $t(y) \equiv \tfrac{1}{2}\sum_{\ell\leq m}\frac{2m+1}{2\ell+1}t_\ell\lambda^{2\ell+1}$, so that $\tau(t_\ell) = \langle V(\infty; s_k, t_\ell)\rangle$ where

$$V(a; s_k, t_\ell) = \prod_k exp\left[s_{2k+1}\int_{C_{2k+1}}^a e^{2t(y)}\psi_1\bar{\psi}_2(y)\right] \\ exp\left[s_{2k}\int_{C_{2k}}^a e^{-2t(y)}\psi_2\bar{\psi}_1(y)\right] \tag{4.19}$$

For the PI family we have a fermion twist operator at the origin (from the regular singularity in (3.16)). If we bosonize the two fermions $e^{i\phi_i} = \psi_i$ then the twist operator at the origin is just $e^{i\omega/\sqrt{2}}$ where $-i\sqrt{2}\partial\omega = \bar{\psi}\sigma_3\psi$. The τ function is now

$$\tau(t_\ell) = \langle e^{-i\omega(\infty)/\sqrt{2}}V(\infty; s_l, t_\ell)e^{i\omega(0)/\sqrt{2}}\rangle \tag{4.20}$$

where the star operator is the same as in the PII case but $t(y) = -\tfrac{1}{4}\sum t_j\zeta^{2j+1}$ and the stokes data is nonzero for sectors along the x,y axes. The bosonized currents $\bar{\psi}_2\psi_1, \bar{\psi}_1\psi_2 = e^{\pm i\sqrt{2}\omega}$ involve only one scalar field ω so we have a correlator in a $c=1$ system. The expression (4.20) may be written explicitly using Wick's theorem. The contour integrals diverge near $y=0$ but this divergence may be regulated and is t_ℓ-independent.

These expressions are, of course, rather formal. It would be worthwhile making rigorous sense of them since, at least formally, they make transparent some interesting properties of the 2D gravity partition function. For example, a standard corollary of the operator formalism is that a τ function satisfies certain Virasoro constraints. Applying (4.16) to this case with A, T obtained from (3.16)(3.8) we find:

$$L_n\tau = \frac{1}{4}\delta_{n,0}\tau \qquad n \geq -1 \qquad \text{for PI}$$
$$L_n\tau = 0 \qquad\qquad n \geq -1 \qquad \text{for PII} \tag{4.21}$$

Similarly since commutation with $J = \bar{\psi}\sigma_3\psi$ rotates the fermions $\psi_{1,2}$ oppositely we may imagine that there is an identity like

$$e^{\int_C \left(t^{(2)}(y)-t^{(1)}(y)\right) J(y)dy} V(a;s,t_\ell^{(1)}) e^{-\int_C \left(t^{(2)}(y)-t^{(1)}(y)\right) J(y)dy} = V(a;s,t_\ell^{(2)}) \tag{4.22}$$

where C surrounds a. Hence we would have

$$\tau(t_\ell) = \langle t_\ell - \bar{t}_\ell | \Omega_{\bar{t}} \rangle \tag{4.23}$$

where $|\Omega_{\bar{t}}\rangle$ is the state created by the star operator at $\bar{t}$, and $\langle t_\ell - \bar{t}_\ell|$ is a coherent state for the scalar ω where only the odd oscillator modes are excited. (This follows since $t(y)$ involves only odd powers of y.) Combining with (4.21) we may obtain, formally, expressions similar (but not equal) to those in [40][20][23][24] .

Finally, let us compare with the fermion formalism of the matrix model described in section 2. There we found that the partition function at couplings t_ℓ can be expressed in terms of a correlation function in the ground state for couplings $\bar{t}_\ell$ according to:

$$\tau_{PI} = \left\langle exp\left(\int_{\mathbf{R}} \sum_\ell (t_\ell - \bar{t}_\ell)\lambda^{\ell+1/2}\psi^\dagger\psi(\lambda)d\lambda\right) \right\rangle$$
$$\tau_{PII} = \left\langle exp\left(i\int_{\mathbf{R}} \sum_\ell (t_\ell - \bar{t}_\ell)\lambda^{2\ell+1}\psi^\dagger\psi(\lambda)d\lambda\right) \right\rangle \tag{4.24}$$

While heuristic, these expressions are strikingly similar to (4.23) and could explain why, for matrix model asymptotics, the stokes data is nontrivial, and concentrated along the y axis, for the PII family and along the x, y axes for the PI family.

5. Grassmannians, Krichever's construction, and all that

When the connection to the KdV hierarchy was discovered in 2D gravity the theory of the KdV hierarchy, as presented in [56] [57] [58] , was already familiar to string theorists. Indeed, one of the main points of the so-called operator formalism[8] is the equivalence of the   Krichever theory with the theory of free fermions on an algebraic curve. In the previous section we related the τ function of 2D gravity to a correlation function of free fermions. In this section we will see that because of stokes phenomenon we require an extension of the the theory in [56][57][58] .

---

[8] Representative papers include [59] [60] and references therein.

5.1. Quasiperiodic KdV flow and isomonodromy

We begin by reviewing a remark of M. Jimbo and T. Miwa that the tau function of the quasiperiodic solutions to the KdV equations is a special case of the isomonodromic tau function [13].

Recall that quasiperiodic KdV flow is straightline motion along $Pic_{g-1}(X)$ for a riemann surface X of genus g, and, fixing an origin $\mathcal{L}_0$ for Pic_{g-1} the tau function is just

$$\tau(\mathcal{L}) = \frac{Det\ \bar{\partial}_{\mathcal{L}}}{Det\ \bar{\partial}_{\mathcal{L}_0}} \quad . \tag{5.1}$$

If $\mathcal{L} \otimes \mathcal{L}_0^{-1}$ has divisor $P_1 + \cdots P_g - Q_1 - \cdots - Q_g$ then by the insertion theorem [61] we have

$$\tau(\mathcal{L}) = \langle \psi(P_1) \cdots \psi(P_g) \bar{\psi}(Q_1) \cdots \bar{\psi}(Q_g) \rangle_{(X, \mathcal{L}_0)} \quad . \tag{5.2}$$

Choosing a point P_∞ "at infinity" and a local coordinate $1/z$ near P_∞ the Baker function is essentially

$$\langle \bar{\psi}(P_\infty) \psi(z) \psi(P_1) \cdots \psi(P_g) \bar{\psi}(Q_1) \cdots \bar{\psi}(Q_g) \rangle_{(X, \mathcal{L}_0)} \quad . \tag{5.3}$$

Now suppose $\pi : X \to \mathbb{P}^1$ is an m-fold branched covering. As is well-known from the theory of orbifolds [62–65] [41] we can represent one weyl fermion on X by m weyl fermions $\psi_\alpha, \bar{\psi}_\alpha$ on $\mathbb{P}^1$, where $\alpha = 1, \ldots, m$ labels the sheets, in the presence of twist operators at the branch points. For example, denoting the branch points on $\mathbb{P}^1$ by b_i we have the twist field correlator $Det\ \bar{\partial}_{\mathcal{L}_0} = \langle \prod \varphi_i(b_i) \rangle$. Similarly, if P_i, Q_i lie on branches α_i, β_i, respectively then the Baker function descends to the "Baker framing"

$$\tilde{Y}_{\alpha\beta}(\lambda) = \langle \bar{\psi}_\alpha(\infty) \psi_\beta(\lambda) \psi_{\alpha_1}(\pi(P_1)) \cdots \psi_{\alpha_g}(\pi(P_g)) \bar{\psi}_{\beta_1}(\pi(Q_1)) \cdots \bar{\psi}_{\beta_g}(\pi(Q_g)) \prod_i \varphi_i(b_i) \rangle_{\mathbb{P}^1}$$
$$\tag{5.4}$$

where $\lambda = \pi(z)$. (The actual Baker framining Y differs from $\tilde{Y}$ by an invertible diagonal matrix with an essential singularity at infinity of the form $\sim exp(\sum t_j z^j)$.) Regarding the fermion insertions as special cases of twist operators and following the reasoning of the previous section it follows that Y satisfies a linear ODE in λ with regular singularities at $\pi(P_i)$. Clearly we have isomonodromic deformation in these parameters, so the isomonodromic τ function is just

$$\left\langle \psi_{\alpha_1}(\pi(P_1)) \cdots \psi_{\alpha_g}(\pi(P_g)) \bar{\psi}_{\beta_1}(\pi(Q_1)) \cdots \bar{\psi}_{\beta_g}(\pi(Q_g)) \prod_i \varphi_i(b_i) \right\rangle_{\mathbb{P}^1} \quad , \tag{5.5}$$

but this is the same as (5.2) which is the tau function of the quasiperiodic KdV equations.

From this discussion we conclude that the required generalization of the operator formalism is the generalization from twist operators to star operators. In the next section we will arrive at the same conclusion from a different point of view.

One way to understand better the geometrical meaning of star operators is to investigate directly the required generalization of the Burchnall-Chaundy-Krichever theory through which one associates a riemann surface with a line bundle to a solution of the stationary KdV equations.

Recall that in the quasiperiodic theory we have $[P, L] = 0$ where $L = D^2 - u(x)$ and $P = \sum t_j L^{(2j+1)/2}|_+$. By simultaneously diagonalizing P, L we find that P, L must be algebraically related $P^2 - Q(L) = 0$ for some polynomial Q, defining a hyperelliptic curve Σ. Indeed, in terms of A, B, C defined in (2.35) we have the curve $\Sigma = \{(\mu, \lambda)|\mu^2 = A^2 + BC\}$. The simultaneous eigenfunction satisfies $L\psi(\lambda, x) = \lambda\psi(\lambda, x)$, i.e., is a Baker-Akhiezer function and is a section of a line bundle $\mathcal{L} \to \Sigma$. KdV flows starting from $u(x)$ are elegantly described as straightline flows of $\mathcal{L}$ along the Jacobian of Σ. Note especially that the parameters t_j are coordinates of hyperelliptic moduli space and are *unchanged* under KdV flow.

In 2D gravity we have the equation $[P, L] = 1$. This means we cannot diagonalize P, L simultaneously. What should we do? One idea [9] is that we should try to define a noncommutative spec in the spirit of noncommutative geometry, but this has not yet been pursued very far. An easier route is to consider the equation $[P, L] = \hbar$ and try to understand the $\hbar \to 0$ limit [10][66] [67] . Although the details in these papers are very similar the authors draw rather different conclusions. We now briefly sketch some salient points of these papers.

Consider a family of solutions $u_\hbar$ to the string equations $\sum(j + \frac{1}{2})t_j R_j = \hbar x$, where $t_j = t_j^{(0)} + \hbar t_j^{(1)} + \cdots$. By naive scaling one might think that the limit $\hbar \to 0$ corresponds to the limit $x \to 0$ of a solution at $\hbar = 1$ but this ignores possible $\hbar$ dependence of the boundary conditions. Using the analysis of Boutroux (and its extension to the higher order string equations) one may show that it is possible to choose boundary conditions so that $u_\hbar = u^{(0)} + \hbar u^{(1)} + \cdots$ and $u^{(0)}$ is a nontrivial solution of the stationary KdV equations (e.g. it is a Weierstrass $\wp$ function in the analysis of Boutroux).

In [10] we attempted to generalize the BCK theory by reinterpreting the equation $[P, L] = 0$ as equations on the matrix Baker-Akhiezer function. It was shown that in the case $\hbar \neq 0$ the equations generalize directly and are the first two compatibility equations of (2.37) . In particular the Baker-Akhiezer function exhibits stokes' phenomenon (as we saw in section 3.6). Any geometrical interpretation must take account of this fact. One attempt, in terms of framings of push-forwards of Krichever's line bundle, was made in [10]. An interesting consequence of this proposal is that the parameters $t_j^{(0)}$ and $t_j^{(1)}$ defined by $t_j = t_j^{(0)} + \hbar t_j^{(1)} + \cdots$ are, respectively, coordinates for hyperelliptic moduli space and the jacobian of the curve $\Sigma(t_j^{(0)})$. This reconciles the peculiar fact that in the stationary case KdV flow is a flow in $t_j^{(1)}$ while in the $\hbar \neq 0$ case the flow is in t_j.

---

[9] which also occured to C. Itzykson, C. Vafa, and possibly others

In [66][67] the equation $\mu^2 = A^2 + BC$ is taken to define a riemann surface Σ. The recursion relations for kdv potentials imply that $\frac{\partial}{\partial x}(A_\ell^2 + B_\ell C_\ell) = -2R'_{\ell+1}C$. Therefore, in the stationary case the curve Σ is independent of x (indeed, x parametrizes a direction along $Jac(\Sigma)$). When $\hbar \neq 0$ we have instead

$$\frac{\partial}{\partial x}\left(A^2 + BC[\lambda, u_\hbar]\right) = \hbar C[\lambda, u_\hbar] \tag{5.6}$$

so that the curve Σ now depends on x, t_j, and we have a family of Riemann surfaces. Now, *if* the standard BCK theory applied then we could identify [56]

$$u(x; t_j) = 2\frac{\partial^2}{\partial x^2} log \; \vartheta(\vec{v}x + \vec{w}|\tau) \tag{5.7}$$

where τ itself depends on x, t_j. In fact, substitution into (5.6) then yields a nontrivial transcendental equation for the moduli as a function of x, t_j [66][67] .

Unfortunately, it is not clear that solutions of the type (5.7) will have physically relevent asymptotics. For example, the transcendental equations in [66][67] might not have solutions. (It seems that, at best, the curve must be completely degenerate.) There is also a technical snag in applying the BCK theory to the family of curves $\Sigma(x)$. Since solutions to (2.37) for physical asymptotics have nontrivial stokes matrices the expansion of the Baker-Akhiezer function in $1/\lambda^{1/2}$ (or, equivalently, the expansion in $\hbar$) is only asymptotic while in standard kdv theory it is important that the expansion be convergent. This technicality also means that the potential $u(x; t_j)$ and the associated τ function is outside the class of functions $\mathcal{C}$ [58] for which the Segal-Wilson theory is valid. Nevertheless, the interpretation of the tau function of 2D gravity in terms of star operators, while formal, suggests that the Grassmannian picture of KdV flows should carry over more or less unchanged. Defining precisely the enlarged Grassmannian required to incorporate the new class of KdV flows encountered in 2D gravity is an interesting open problem.

Acknowledgements

In addition to those acknowledged in [10] I would like to thank L. Alvarez-Gaumé, T. Banks, F. David, C. Gomez, C. Itzykson, B. McCoy, E. Martinec, T. Miwa, H. Neuberger, G. Segal, S. Shatashvili, S. Shenker, M. Staudacher, E. Verlinde, and E. Witten for conversations. I am also grateful to the organizers of the Yukawa International Seminar in Kyoto, the Cargèse Workshop on Random Surfaces and 2D Gravity, and the Trieste conference on Topological Methods in Quantum Field Theories for the opportunity to present this material. I thank the Rutgers Dept. of Physics for hospitality while this paper was completed. This work was supported by DOE grants DE-AC02-76ER03075, and DE-FG05-90ER40559, and by a Presidential Young Investigator Award.

Appendix A. The double scaling limit of Hermite functions

In the case of a gaussian matrix potential $e^{-N\ tr\phi^2}$ the orthonormal wavefunctions are simply

$$p_n(\lambda) = \frac{N^{1/4}}{2^{n/2}\pi^{1/4}\sqrt{n!}} H_n(\sqrt{N}\lambda)e^{-N\lambda^2/2} \tag{A.1}$$

where H_n is a Hermite polynomial, and has the integral representation

$$H_n(x) = \frac{2^n}{\sqrt{\pi}} \int_{-\infty}^{\infty} (x+it)^n e^{-t^2} dt \tag{A.2}$$

If we let $n/N = 1 - a^2 z$, then we can estimate the integral by stationary phase approximation. At $\lambda = \lambda_c = \sqrt{2}$ the two saddle-points coalesce, and, expanding about the saddle point to third order we find

$$p_n\left(\sqrt{2}(1+a^2\lambda)\right) \sim a^{-1/2} \int_{-\infty}^{\infty} e^{-it(\lambda+z)-it^3/3} dt \tag{A.3}$$

up to numerical constants. Thus the double scaling limit of the Hermite functions are Airy functions, which satisfy (2.37) for the case $\ell = 0$. Using (A.3) one may easily reproduce the result of Brézin and Kazakov for the double scaling limit of the gaussian model resolvent [1] .

References

[1] E. Brézin and V. Kazakov, "Exactly solvable field theories of closed strings," Phys. Lett. **B236**(1990)144.

[2] M. Douglas and S. Shenker, "Strings in less than one dimension," Rutgers preprint RU-89-34.

[3] D. Gross and A. Migdal, "Nonperturbative two dimensional quantum gravity," Phys. Rev. Lett. **64**(1990)127.

[4] D. Gross and A. Migdal, "A nonperturbative treatment of two-dimensional quantum gravity," Princeton preprint PUPT-1159(1989).

[5] M. Douglas, "Strings in less than one dimension and the generalized KdV hierarchies," Rutgers preprint RU-89-51.

[6] T. Banks, M. Douglas, N. Seiberg, and S. Shenker, "Microscopic and macroscopic loops in non-perturbative two dimensional gravity," Rutgers preprint RU-89-50.

[7] A.A. Belavin, A.M. Polyakov, A.B. Zamolodchikov, Nucl. Phys. **B241**(1984)333.

[8] E. Brezin, E. Marinari, and G. Parisi, "A Non-Perturbative Ambiguity Free Solution of a String Model," ROM2F-90-09.

[9] M. Douglas, N. Seiberg, and S. Shenker, "Flow and instability in quantum gravity," Rutgers preprint, RU-90-19.

[10] G. Moore, "Geometry of the string equations," Yale preprint YCTP-P4-90.

[11] A. Its and V. Yu. Novokshenov, *The Isomonodromic Deformation Method in the Theory of Painlevé Equations,* Springer Lect. Notes Math. 1191.

[12] M. Jimbo, T. Miwa, K. Ueno, "Monodromy Preserving Deformation of Linear Ordinary Differential Equations with Rational Coefficients," Physica **2D**(1981)306.

[13] M. Jimbo and T. Miwa, "Monodromy Preserving Deformation of Linear Ordinary Differential Equations with Rational Coefficients. II," Physica **2D**(1981)407.

[14] M. Sato, T. Miwa, and M. Jimbo, "Aspects of Holonomic Quantum Fields Isomonodromic Deformation and Ising Model," in *Complex Analysis, Microlocal Calculus and Relativisitic Quantum Theory,* D. Iagolnitzer, ed., Lecture Notes in Physics 126.

[15] M. Jimbo, "Introduction to Holonomic Quantum Fields for Mathematicians," Proc. Symp. in Pure Math. **49**(1989)part I. 379.

[16] E. Witten, "On the structure of the topological phase of two dimensional gravity," preprint IASSNS-HEP-89/66.

[17] J. Distler, "2D quantum gravity, topological field theory and multicritical matrix models," princeton preprint PUPT-1161.

[18] R. Dijkgraaf and E. Witten, "Mean Field Theory, Topological Field Theory, and Multi-Matrix Models," IASSNS-HEP-90/18;PUPT-1166.

[19] E. Verlinde and H. Verlinde, "A Solution of two dimensional topological quantum gravity," preprint IASSNS-HEP-90/40.

[20] R. Dijkgraaf, H. Verlinde, and E. Verlinde, Princeton preprint PUPT-1184.

[21] E. Martinec, "On the origin of integrability in matrix models," Chicago preprint, EFI-90-67.

[22] E. Witten, "Two dimensional gravity and intersection theory on moduli space," IAS preprint, IASSNS-HEP-90/45.

[23] A. Gerasimov, A. Marshakov, A. Mironov, A. Morozov, and A. Orlov, "Matrix Models of 2D gravity and Toda Theory," P.N. Lebedev Institute preprint, July 1990.

[24] A. Mironov and A. Morozov, "On the origin of virasoro constraints in matrix models: lagrangian approach," P.N. Lebedev Institute preprint, July 1990.

[25] M. Jimbo, T. Miwa, Y. Mori and M. Sato, "Density Matrix of an Impenetrable Bose Gas and the Fifth Painlevé Transcendent" Physica **1D** (1980)80.

[26] M. L. Mehta, *Random Matrices* Academic Press,1967.

[27] See sec. 10.3 in C. Itzykson and J.-M. Drouffe, *Statistical Field Theory,* vol. 2, Cambridge Univ. Press. 1989.

[28] A.R. Its, A.G. Izergin, V.E. Korepin, and N.A. Slavnov, "Differential equations for quantum correlation functions," preprint.

[29] A.R. Its, A.G. Izergin, and V.E. Korepin, "Temperature correlators of the impenetrable bose gas as an integrable system," ICTP preprint IC/89/120.

[30] E. Barouch, B.M. McCoy, and T.T. Wu, Phys. Rev. Lett. **31** (1973)1409; C.A. Tracy and B.M. McCoy, "Neutron scattering and the correlation functions of the Ising model near T_c," Phys. Rev. Lett. **31**(1973)1500; T.T. Wu, B.M. McCoy, C.A. Tracy, and E. Barouch, Phys. Rev. **B13**(1976)316.

[31] Č. Crnković and G. Moore,"Multi-Critical Multi-Cut Matrix Models," Yale preprint YCTP-P16-90.

[32] G.M. Cicuta, L. Molinari, and E. Montaldi, Mod. Phys. Lett. **A1**(1986)125.

[33] V. Periwal and D. Shevitz, "Unitary-Matrix Models as Exactly Solvable String Theories," Phys. Rev. Lett. **64**(1990)1326.

[34] H.Neuberger, Nucl. Phys. **B340**(1990)703.

[35] Drinfeld and Sokolov, "Equations of Korteweg-de Vries type and simple Lie algebras," Sov. Jour. Math. (1985)1975, section 3.8.

[36] I.M. Gelfand and L.A. Dickii, "Asymptotic Behavior of the Resolvent of Sturm-Liouville Equations and the Algebra of the Korteweg-De Vries Equations," Russian Math Surveys, **30**(1975)77.

[37] S.R. Das and A. Jevicki, "String field theory and physical interpretation of d=1 strings," Brown preprint BROWN-HET-750.

[38] A.M. Sengupta and S.R. Wadia, "Excitations and interactions in d=1 string theory," Tata preprint.

[39] D. Gross, Lectures at the Cargèse workshop and at CERN, June 1990.

[40] M. Fukuma, H. Kawai, and R. Nakayama, Univ. of Tokyo preprint UT-562.

[41] V.G. Knizhnik, "Multiloop amplitudes in the theory of quantum strings and complex geometry," Usp. Fiz. Nauk. **159**(1989)401. English translation: Sov. Phys. Usp. **32**(1989)945, section 12.

[42] D. Friedan and S. Shenker, "The analytic geometry of two-dimensional conformal field theory," Nucl. Phys. **B281** (1987)509; D. Friedan, "A new formulation of string theory," Physica Scripta T **15**(1987)72.

[43] W. Wasow, *Asymptotic Expansions for Ordinary Differential Equations*, Interscience, 1965.

[44] H. Flaschka and A. Newell, "Monodromy and Spectrum-Preserving Deformations I," Comm. Math. Phys. **76**(1980)65.

[45] A. Kapaev, "Asymptotics of solutions of the Painlevé equation of the first kind," Differential Equations, **24**(1988)1107.

[46] J. Moser, "Finitely many mass points on the line under the influence of an exponential potential–an integrable system," in Lecture Notes in Physics **38**, J Moser, ed., p. 467.

[47] T. Miwa, "Painlevé property of monodromy preserving deformation equations and the analyticity of τ functions," Publ. Res. Inst. Math. Sci. **17** (1981)703.

[48] F. David, "Phases of the large N matrix model and non-perturbative effects in 2d gravity," Saclay preprint SPhT/90-090.

[49] See reference [11], especially, chapter 5.

[50] A.R. Its and V. Yu. Novokshenov, "Effective Sufficient Conditions for the Solvability of the Inverse Problem of Monodromy Theory for Systems of Linear Ordinary Differential Equations," Funct. Anal. and Appl. **22**(1988)190.

[51] S.P. Hastings and J.B. McLeod, "A boundary value problem associated with the second painlevé transcendent and the Korteweg-de Vries equation," Arch. Rat. Mech. and Anal. **73**(1980)31.

[52] Č. Crnković, M. Douglas, and G. Moore, "Physical solutions for unitary-matrix models," Yale preprint YCTP-P6-90.

[53] Č. Crnković, P. Ginsparg, and G. Moore, "The Ising model, the Yang-Lee edge singularity, and $2D$ quantum gravity," Phys. Lett. **237B**(1990)196.

[54] M. Sato, T. Miwa, M. Jimbo, "Holonomic Quantum Field Theory II," Publ. RIMS **15**(1979)201.

[55] T. Miwa, "Clifford operators and Riemann's monodromy problem," Publ. Res. Inst. Math. Sci. **17**(1981)665.

[56] Dubrovin, Matveev, and Novikov, "Non-Linear Equations of Korteweg-De Vries Type, Finite-Zone Liner Operators, and Abelian Varieties," Russian Math Surveys, **31** (1976)59.

[57] E. Date, M. Jimbo, M. Kashiwara, and T. Miwa, "Transformations Groups for Soliton Equations," I. Proc. Japan Acad. **57A**(1981)342; II. Ibid., 387;III. J. Phys. Soc. Japan **50**(1981)3806;IV. Physica **4D**(1982)343;V. Publ. RIMS, Kyoto Univ. **18**(1982)1111;VI. J. Phys. Soc. Japan **50** (1981)3813;VII. Publ RIMS, Kyoto Univ. **18**(1982)1077.

[58] G. Segal and G. Wilson, "Loop Groups and Equations of KdV Type," Publ. I.H.E.S. **61**(1985)1.

[59] L. Alvarez-Gaumé, C. Gomez, G. Moore, and C. Vafa, Nucl. Phys. **B303**(1988)455.

[60] E. Witten, "Conformal field theory, Grassmanians, and algebraic curves," Commun. Math. Phys. **113**(1988)189.

[61] L. Alvarez-Gaumé, J.-B. Bost, G. Moore, P. Nelson, and C. Vafa, Phys. Lett. **178B**(1986)41;Commun. Math. Phys. **112**(1987)503.

[62] Al. B. Zamolodchikov, "Conformal scalar field on the hyperelliptic curve and critical Ashkin-Teller multipoint correlation functions," Nucl. Phys. **B285**(1987)481.

[63] M. Bershadsky and A. Radul, "Conformal field theories with additional Z_N symmetry," Int. Jour. of Mod. Phys. **A2**(1987)165.

[64] L. Dixon, D. Friedan, E. Martinec and S. Shenker, "The conformal field theory of Orbifolds," Nucl. Phys. **B282**(1987)13.

[65] S. Hamidi and C. Vafa, "Interactions on Orbifolds," Nucl. Phys. **B279**(1987)465

[66] S.P. Novikov, "On the equation $[L, A] = \epsilon \cdot 1$," preprint.

[67] I. Krichever, "On heisenberg reltions for the ordinary linear differential operators," ETH preprint.

THE STRENGTH OF NONPERTURBATIVE EFFECTS IN STRING THEORY

Stephen H. Shenker

Department of Physics and Astronomy
Rutgers University
Piscataway, NJ 08855-0849

Abstract: We argue that the leading weak coupling nonperturbative effects in closed string theories should be of order $exp(-C/\kappa)$ where κ^2 is the closed string coupling constant. This is the case in the exactly soluble matrix models. These effects are in principle much larger than the $exp(-C/\kappa^2)$ effects typical of the low energy field theory. We argue that this behavior should be generic in string theory because string perturbation theory generically behaves like $(2g)!$ at genus g.

Nonperturbative effects are crucial ingredients in any attempt to describe the real world by string theory. Vacuum selection, supersymmetry breaking, and the vanishing of the cosmological constant are all examples of issues that must be addressed in an intrinsically nonperturbative way. As of yet there has been no real progress in understanding such phenomena in the critical superstring from a fully string theoretic point of view. The theory has not even been formulated nonperturbatively.

In the last year, though, substantial progress has been made in understanding nonperturbative phenomena in simple models of string theory corresponding to string propagation in less than [1] [2] or equal to one [3] dimension.[1] These systems are formulated in a nonperturbative way as matrix models [4] and have been shown to be related to topological field theory [5].

Such simple models have played an important role in theoretical physics. One need only remember how much was learned about scaling, universality, and the operator product expansion from the two dimensional Ising model, and about confinement

---

[1] It may well be more correct to interpret the Liouville field as a dimension and think of these theories as living in less than or equal to two dimensions.

Random Surfaces and Quantum Gravity
Edited by O. Alvarez *et al., Plenum Press, New York, 1991*

and the $U(1)$ problem from the Schwinger model. In order to generalize from these models it was crucial to identify which of their properties were *generic*, and not a consequence of the special simplicity that allowed them to be solved. The purpose of this note is to point out one such property in these simple string theories.

These simple string theories all have leading weak coupling nonperturbative effects of magnitude $exp(-C/\kappa)$ where C is a numerical constant and κ^2 is the closed string coupling constant, i.e., genus g amplitudes carry a factor of κ^{2g-2}. It is this property, we claim, that is generic in string theory. Note that the size of these effects is in principle much larger than one would expect from a low energy field theoretic approximation where leading nonperturbative effects would have the characteristic $exp(-C/\kappa^2)$ form.

We begin by reviewing nonperturbative phenomena in the matrix models. In the original exact solution of the one matrix model [1] the specific heat in the properly scaled continuum limit was expressed as a solution of a nonlinear ordinary differential equation–the string equation. Every derivative in the string equation is accompanied a factor of κ, the scaled version of $1/N$. For small κ the leading nonperturbative effects can be found by linearizing the equation around a reference $\kappa = 0$ algebraic solution. Because each derivative carries a κ, a WKB solution to the linear problem will be of the form $exp(-f(x)/\kappa)$, displaying the κ dependence described above.

For example, the specific heat $u(x = 1)$ of the $m = 2$ pure gravity one matrix model with even potential is described by the Painlevé I equation

$$u^2 - \frac{\kappa^2}{3}u'' = x.$$ (1)

Linearizing around the algebraic behavior $u = x^{1/2}$ for $x > 0$, we find a WKB solution to the homogeneous equation whose exponential dependence is of the form

$$u_{lin} \sim exp(-\frac{4\sqrt{6}}{5\kappa}x^{5/4}) \ .$$ (2)

The small imaginary part of the "triply truncated" solution to (1) that David [6] [7] showed describes the analytically continued $m = 2$ integral will be of this form. This exponentially small imaginary part reflects the nonperturbative instability of the model due to its unbounded potential.

Another example of leading nonperturbative effects in the one matrix models occurs in the flow [8] from the well defined $m = 3$ theory [9] to the $m = 2$ theory. The string equation here is

$$[u^3 - \kappa^2 uu'' - \frac{\kappa^2}{2}(u')^2 + \frac{\kappa^4}{10}u''''] + T_2[u^2 - \frac{\kappa^2}{3}u''] = x \ .$$ (3)

T_2 is the scaling field describing the flow to the $m = 2$ theory. The $\kappa = 0$ algebraic equation is

$$u^3 + T_2 u^2 = x.$$ (4)

The matrix integral tells us that the correct solution to expand around is the purely real solution that has a jump discontinuity at $x = \frac{4}{27}T_2^3 \equiv x_0$. When κ is finite this jump gets smoothed into a nonperturbatively sharp boundary layer that is another signature of the instability of the $m = 2$ theory. The leading exponential precursor to the boundary layer $(x > x_0)$ is given by linearizing in the neighborhood of the jump and is of the form

$$exp(\frac{-C(x - x_0)}{\kappa}) \ , \quad C = ((10 - \sqrt{10})T_2/3)^{\frac{1}{2}} \ . \tag{5}$$

The matrix integral provides us with a natural explanation for leading nonperturbative effects in terms of auxiliary saddle points. In terms of eigenvalues the one matrix integral is

$$\int d\lambda_1 \dots d\lambda_N exp(-S) \tag{6}$$

where S is given by

$$S = -\sum_{i \neq j} log(\lambda_i - \lambda_j)^2 + \sum_i NV(\lambda_i/\sqrt{N}) \ . \tag{7}$$

The $m = 2$ critical point which we will discuss first can be realized with the simple potential $V(\lambda) = \lambda^2/2 - \alpha\lambda^4$. Each sum in (7) is of order N^2 and so at large N a saddle point exists [10]. The lowest action and so perturbative saddle corresponds to all the eigenvalues in the well around $\lambda = 0$. This saddle is only locally stable, the eigenvalues would rather be at ∞, the instability discussed above.

The lowest action saddle that describes eigenvalues leaving the well is made by moving just one eigenvalue to the local maximum in the effective potential formed from V and the coulomb repulsion of the remaining eigenvalues (whose positions are essentially unchanged for large N). The key point is that this saddle corresponds to the movement of just one eigenvalue out of N so its change in action is proportional to N, not N^2. This means that the imaginary part of the integral is proportional to e^{-CN} and not the e^{-CN^2} we might have expected. In the scaled continuum limit N becomes $1/\kappa$ and the action of this saddle should become universal. The numerical value of this action should just match the value in (2) with $x = 1$. David [7] has recently verified that this is the case.

In the flow from $m = 3$ to $m = 2$, the effective potential for the last eigenvalue develops a secondary minimum for x near x_0 as T_2 is turned on. This minimum is above the eigenvalue filling level as long as $x > x_0$. The leading nonperturbative correction in this region (5) should correspond to one eigenvalue occupying the secondary minimum.

These one eigenvalue saddle points are simple examples of string instantons. Because they involve motion of only one out of N eigenvalues they have anomalously low action and so produce anomalously large effects.

The $D = 1$ model [3] also has $exp(-C/\kappa)$ nonperturbative effects. This model is formulated as one dimensional quantum mechanics of N noninteracting fermions with $\hbar$ equal to $1/N$ in a potential V like the one discussed above. The leading nonperturbative effects here correspond, as noted in the original papers, to tunneling out of the metastable well. Tunneling effects go like $exp(-C/\hbar) = exp(-CN)$. Here

C is the barrier penetration factor or instanton action. Again, the crucial point is that leading effects come from just one eigenvalue tunneling so the instanton action is not proportional to N. In the properly scaled continuum limit the tunneling effects for the eigenvalues at the top of the fermi sea become of order $exp(-C/\kappa)$.

A particularly interesting one dimensional model is the one formulated by Marinari and Parisi [11] that describes a string propagating in one super dimension. Its matrix model realization is one dimensional supersymmetric matrix quantum mechanics. For a cubic superpotential the Hamiltonian in the 0 fermion number sector becomes a standard N decoupled eigenvalue problem (where the eigenvalues are to be viewed as fermions of a different kind) in a fourth order potential. At the critical point $(\alpha = \alpha_c \equiv 0)$ around which the model scales the potential has a cubic inflection point. The region near the inflection point dominates the scaling limit so we can model the potential as $V(\lambda) = \lambda^3 - \alpha\lambda$. A small secondary minimum exists for $\alpha > 0$. The leading nonperturbative effects in this model, including supersymmetry breaking, are presumably driven by instantons. The fermi level at $N = \infty$ is at the secondary minimum, and so the appropriate instanton describing the behavior of the top eigenvalue (whose effect will be leading) is the one beginning from the fermi level in the main well and ending at the secondary minimum [12]. Its action is (for $H = p^2 + V$)

$$\frac{S_{instanton}}{\hbar} = \frac{4 \cdot 3^{1/4}}{5}\alpha^{5/4} N = -\frac{4 \cdot 3^{1/4}}{5\tilde{\kappa}} \tag{8}$$

where $\tilde{\kappa} = \alpha^{5/4} N$ is the $m = 2$ scaled coupling constant appropriate to this model. Instanton effects will be $exp(-C/\tilde{\kappa})$ here.[2]

At this workshop Parisi [14] has presented a calculation of supersymmetry breaking. The nonzero vacuum energy is just the escape rate in the Langevin evolution of eigenvalues in the $m = 2$ one matrix potential given by (7) . The escape rate is dominated by the action of the one eigenvalue saddlepoint discussed above and so the exponential dependence of the vacuum energy is just given by (2) with $x = 1$. The original universality arguments for the $m = 2$ string equation establish the universality of this effect. Normalizing coupling constants, we find that $\tilde{\kappa}$ in (8) is related to κ in (2) by $2^{\frac{1}{2}}\kappa = 3^{\frac{1}{4}}\tilde{\kappa}$. Comparing (2) and (8) we see that the nonvanishing of the vacuum energy is a two instanton effect, as is standard in supersymmetric theories.

Another signature of the $exp(-C/\kappa)$ effects in these models is the large order behavior of their perturbation theory. Writing the perturbation expansion of, say, the free energy F as $\kappa^2 F = \sum_g \kappa^{2g} A_g$, these models all have asymptotic behavior

$$A_g \sim C^{-2g}(2g)! \tag{9}$$

---

[2] In a recent interesting preprint Karliner and Migdal [13] have given an extensive analysis of the Marinari-Parisi model. They have shown by a rescaling of H with cubic plus linear potential that the model scales for arbitrary coupling and have demonstrated how to use the Gelfand-Dikii differential equation for the resolvent to determine its behavior.

as $g \to \infty$. The relation of this to nonperturbative effects is perhaps most simply described by the Borel transform[3]

$$B(t) = \sum_g \kappa^{2g} t^{2g} \frac{A_g}{(2g)!} \tag{10}$$

and its inverse

$$\kappa^2 F(\kappa) = \int_0^\infty dt e^{-t} B(t) \ . \tag{11}$$

Singularities in the Borel t plane on the positive real axis create ambiguities in the reconstruction of F from B. Singularities at t_0 are related to nonperturbative effects in the physical quantity F of magnitude e^{-t_0}. The large order behavior (9) implies that the nearest singularity to the origin is at $|t_0| = C/\kappa$ giving nonperturbative effects of size $exp(-C/\kappa)$. This presumably sets the scale for other singularities in B at $|t| \sim 1/\kappa$ (Of course there can be other singularities much further away from the origin giving rise to much weaker nonperturbative effects).[4]

This situation should be contrasted with that in field theory where in the loop expansion with κ^2 as the loop counting parameter the perturbation series $\sum_\ell \kappa^{2\ell} A_\ell$ typically has asymptotics $A_\ell \sim C^{-\ell} \ell!$ (*not* $(2\ell)!$). The appropriate Borel transform is $B(t) = \sum_\ell \kappa^{2\ell} t^\ell A_\ell / \ell!$. It has a leading singularity at $|t_0| = C/\kappa^2$ producing the usual $exp(-C/\kappa^2)$ nonperturbative effects of field theory.

It is well known that large order behavior can be described by instanton techniques. The one eigenvalue instantons discussed above are the source of this $(2g)!$.[5] The connection between supersymmetry breaking in the Marinari-Parisi model and the one eigenvalue saddle point that controls large order behavior in the $m = 2$ one matrix model (and hence in certain quantities in the Marinari-Parisi model) shows that the same instanton effects control supersymmetry breaking and large orders in perturbation theory in the Marinari-Parisi model.

In the exactly soluble models there are far more powerful techniques available to study nonperturbative effects than the asymptotics of perturbation theory. This is not the case in more complicated string theories like the critical strings where complete nonperturbative formulations do not yet exist. In such systems large order behavior can provide the first glimpse into their nonperturbative structure. [6]

The point of this note is to argue that the asymptotic behavior of perturbation theory in all closed string theories should be as in (9) and so leading nonperturbative effects should be of order $exp(-C/\kappa)$.

---

[3] Alternatively, one can study the dispersion relation connecting the discontinuity across the cut in the κ plane to large order behavior.

[4] Ginsparg and Zinn-Justin [15] have shown the Borel summability of the m odd one matrix models.

[5] This has been discussed in the $D = 1$ model by Ginsparg and Zinn-Justin [3] .

[6] The first work on the large order behavior of string perturbation theory was done by Gross and Periwal [16]. They showed that the series for the $D = 26$ bosonic string diverges and is not Borel summable by giving a $g!$ lower bound with positive coefficients.

The basic reason for the behavior (9) is the large volume of the moduli space of closed Riemann surfaces of genus g, $\mathcal{M}_g$, that one integrates over to calculate genus g perturbative amplitudes. This volume can be estimated by dividing $\mathcal{M}_g$ into cells[17] each with volume depending at most exponentially with the genus. The natural way to do this for moduli spaces of surfaces with at least one puncture is to use the Feynman diagrams of Witten's open string field theory [18] that produce a triangulation of the moduli space [19]. Diagrams (cells) for moduli space of genus g surfaces with n punctures, $\mathcal{M}_g^n$, are made up of $v = 4g + 2n - 4 = -2\chi$ cubic vertices. The number of such diagrams can be counted by large N matrix techniques [19] and is just the coefficient of N^n in the large N matrix gaussian expectation value

$$\frac{1}{v!} < \frac{tr M^3}{3} \frac{tr M^3}{3} \cdots \frac{tr M^3}{3} >_c \tag{12}$$

where there are v vertices, the $v!$ makes the vertices indistinguishable and the $\frac{1}{3}$'s account for a cyclic symmetry. Note that this enumeration of surfaces is dual to the one usually considered in matrix models [4] . The total number of diagrams in an open string field theory at order $\kappa^{-\chi}$ is given by (12) with $N = 1$ (to count all numbers of punctures equally). This is just zero dimensional ordinary ϕ^3 field theory and the result is clearly $\sim C^\chi(-\chi)!$. This leads to nonperturbative effects of order $exp(-C/\kappa)$. We expect such effects in open string theories because κ is the open string coupling constant and the theory can be formulated as a simple string field theory.

For closed string theories we want to enumerate the number of diagrams at genus g with a given number of punctures n, i.e., the term of order N^n in (12) . The techniques of [20] [21] allow the direct evaluation of this number. For simplicity we specialize to the case $n = 1$. The answer is,[7] asymptotically for large g,

$$Number\ of\ Cells(\mathcal{M}_g^1) \sim C^{-2g}(2g)!\ . \tag{13}$$

This is not surprising in light of the above results for open strings $(-\chi \sim 2g)$. The only thing that needs to be checked is that diagrams with just one puncture do not make up a vanishingly small fraction of all diagrams.

We do not expect string integrands to get anomalously large or small except perhaps at the compactification divisor where divergences may appear. We imagine working in a finite theory or one that is cut off in some manner so we may ignore this. The integration domain of the moduli in each cell is not too complicated, e.g., each cell is contractible, nor should it be unusually small. Therefore we expect the integral over each cell to be $\sim C^{-2g}$. Combining with (13) we have the estimate for a genus g string amplitude A_g coming from once punctured moduli space

$$A_g \sim C^{-2g}(2g)! \tag{14}$$

---

[7] This is just the leading term in α at a given order in $1/N$ in a one matrix (M^3) model and was evaluated for this reason (for an M^4 theory) in [20] . This demonstrates $(2g)!$ behavior in the unscaled one matrix theory.

as claimed above.

We now try to sharpen these arguments in the specific case of the $D = 26$ critical bosonic string in flat space. The vacuum amplitude V_g for the closed string can be written [22]

$$V_g = \int_{\mathcal{M}_g} d\mu_{WP} Z(2) Z'(1)^{-13} \tag{15}$$

where $Z(s)$ is the Selberg zeta function that describes regulated functional determinants and $d\mu_{WP}$ is the Weil-Peterson volume form. The integrand is manifestly positive everywhere and so $V_g > 0$. Penner [23] has shown that the Weil-Peterson volume of each cell in the triangulation of $\mathcal{M}_g^1$ is bounded below by C^{-2g}. This result combined with (13) gives the rigorous bound

$$\int_{\mathcal{M}_g^1} d\mu_{WP} > C^{-2g}(2g)! \quad . \tag{16}$$

It is very plausible, based on degeneration arguments for example, that this bound is true for the no puncture case, $\mathcal{M}_g$, as well,[8] although this has not yet been shown rigorously.

The Z function part of the integrand has been bounded by Gross and Periwal [16] in their original work on large order behavior.[9] They showed that away from the compactification divisor $Z(2)Z'(1)^{-13} > C^{-2g}$. Near the divisor the integrand is dominated by the tachyon double pole divergence. We cut off the integral by, say, replacing $Z(2)Z'(1)^{-13}$ by a genus independent constant whenever it exceeds that constant. Combining this with our information on the Weil-Peterson volume we arrive at the plausible bound

$$V_g > C^{-2g}(2g)! \quad . \tag{17}$$

Since away from the compactification divisor there is no reason for the integrand in (15) to become large we expect (17) to be a reliable estimate rather than a bound. The superstring case is more subtle to analyze because of potential cancellations. Nonetheless the expectation is that for certain quantities the $(2g)!$ will continue to hold. This is the case in the Marinari-Parisi supersymmetric model. We also expect this behavior to hold in non-critical strings, both ordinary and supersymmetric [26]. One effect of the Liouville functional integral will be to provide the power of the cosmological constant that is absorbed in defining the continuum string coupling constant in these theories.

We are proposing that this $(2g)!$ behavior is a basic signature of closed string theories, much as the $\ell!$ behavior in the loop expansion is a signature of particle

---

[8] This volume can be written as an intersection and evaluated using topological field theory [24] [25]. This may provide a simple way of relating the volumes of $\mathcal{M}_g$ and $\mathcal{M}_g^1$. I thank Ed Witten for this suggestion and for a number of other helpful remarks.

[9] I thank Mike Douglas for pointing this out to me.

theories. Roughly speaking it emerges because two open string vertices are required to make a closed string vertex and open strings are described by simple string field theories that have field theoretic large order behavior. The $(2g)!$ property is an obvious obstruction to building a simple covariant closed string field theory. It seems that any such theory will necessarily have complicated, coupling constant dependent interactions to build up the required large order behavior. Aspects of this problem have already been encountered at low genus by a number of workers [27].

As stressed above, the $(2g)!$ property indicates that the leading nonperturbative effects in closed string theory are of magnitude $exp(-C/\kappa)$. These effects are, for small enough κ, much larger than those found in a low energy effective field theory analysis of the closed string. The loop counting parameter in those theories is κ^2 so these effects would have the typical field theoretic magnitude $exp(-C/\kappa^2)$. Of course in the critical strings the coupling constant is another field in the theory, the dilaton, whose magnitude is conjecturally set by the dynamics of of the theory. The question of which effects are larger is then a dynamical one. The main point we want to make here is that there are new, intrinsically stringy, nonperturbative effects that must be understood in any study of the dynamics of string theory. The residue of these effects in the low energy effective field theory is, of course, of particular interest.

We have explained earlier that the $exp(-C/\kappa)$ effects in the matrix models can be understood as the signature of one eigenvalue instantons. It will be important to understand if a similar phenomenon holds in a more general setting. As a first step in this direction we can examine the $D = 1$ model which can be reformulated as a kind of field theory [28]. The field here is just the eigenvalue density as a function of λ and time, $\rho(\lambda, t)$. The one eigenvalue instanton in terms of ρ is the tree level eigenvalue distribution plus a delta function of strength $1/N$ that splits off from its edge, executes the instanton trajectory and then rejoins the tree level distribution. This is a peculiar ρ field configuration, although it is natural in terms of the original eigenvalues. It seems that the field likes to fall into pieces of size $1/N$, allowing the existence of anomalously low action instantons. We must understand what properties of the action for ρ allow such a singular stationary point. Clearly N must be involved to set the scale of the delta function. These issues can also be addressed from the point of view discussed by Tom Banks at this workshop [29].

In the question period after this presentation Spenta Wadia made the interesting remark that known results about Yang-Mills theory show the presence of analogous anomalously large nonperturbative phenomena. In the large N topological expansion, diagrams of genus g are weighted by $(1/N^2)^{g-1}$. Instanton effects are of magnitude $exp(-C/e^2)$ where e^2 is the Yang-Mills coupling constant. In the large N limit $\tilde{e}^2 = e^2 N$ is held fixed so instanton effects are order $exp(-CN/\tilde{e}^2)$ [30]. This is just the phenomenon discussed above. We can estimate the large order behavior in the $1/N$ expansion by making the usual assumption that it will be independent of the dimension in which the theory is defined. Two dimensional Yang-Mills theory (with a lattice cutoff) is just the one unitary matrix model. In the scaled continuum limit [31] this model has $(2g)!$ behavior, as we have come to expect. Presumably the unscaled limit appropriate for Yang-Mills theory does as well.

The underlying phenomenon in this example is that the Yang-Mills instanton of lowest action occupies just one $SU(2)$ subgroup of the $SU(N)$ gauge group and so its action does not scale with N. Again there is an indication of a string field falling into pieces of order $1/N$. We hope that further exploration of this phenomenon will cast light on the nonperturbative dynamics of string theory.

Acknowledgements

I would like to thank E. Brézin, F. David, P. Ginsparg, D. Gross, J. Horne, V. Kazakov, C. Lovelace, E. Martinec, G. Moore, H. Neuberger, G. Parisi, R. Penner, H. Verlinde, S. Wadia, E. Witten, S. Wolpert, and J. Zinn-Justin for valuable discussions. I am especially grateful to T. Banks, M. Douglas, and N. Seiberg for their crucial insights.

This note is an extended version of the second part of my presentation at the Cargèse Workshop on Random Surfaces, Quantum Gravity and Strings, May 28-June 1, 1990. I want to thank O. Alvarez, E. Marinari, and P. Windey for organizing an unusually stimulating conference.

This research was supported in part by grant DE-FGO5-90ER40559.

References

[1]   E. Brézin and V. Kazakov, Phys. Lett. B236 (1990) 144; M. Douglas and S. Shenker, Nucl. Phys. B335 (1990) 635; D. Gross and A. Migdal, Phys. Rev. Lett. 64 (1990) 127.

[2]   E. Brézin, M. Douglas, V. Kazakov and S. Shenker Phys. Lett. B237 (1990) 43; C. Crnković, P. Ginsparg and G. Moore Phys. Lett. B237 (1990) 196; D. Gross and A. Migdal, Phys. Rev. Lett. 64 (1990) 717; T. Banks, M. R. Douglas, N. Seiberg and S. Shenker, Phys. Lett. B238 (1990) 279; D. Gross and A. Migdal, Nucl. Phys. B340 (1990) 333; M. Douglas, Phys. Lett. B238 (1990) 176; P. Di Francesco and D. Kutasov, Princeton preprint PUPT-1173 (1990).

[3]   E. Brézin, V. Kazakov and Al. Zamolodchikov, Ecole Normale preprint, December 1989, LPS-ENS 89-182, Nucl. Phys. B (in press); P. Ginsparg and J. Zinn-Justin, Phys. Lett. B240 (1990) 333; D. Gross and N. Miljković, Phys. Lett. B238 (1990) 217; G. Parisi, Phys. Lett. B238 (1990) 209; G. Parisi, Phys. Lett. B238 (1990) 213; D. Gross and I. Klebanov, Princeton preprint, March 1990, PUPT-1172.

[4]   J. Ambjørn, B. Durhuus and J. Fröhlich, Nucl. Phys. B257 (1985) 433; F. David, Nucl. Phys. B257 (1985) 45; V. Kazakov, Phys. Lett. B150 (1985) 282; V. Kazakov, I. Kostov and A. Migdal Phys. Lett. B157 (1985) 295.

[5]   E. Witten, Nucl. Phys. B340 (1990) 281; J Distler, Princeton preprint, 1990 PUPT-1161; R. Dijkgraaf and E. Witten, IAS preprint, February 1990; E. Verlinde and H. Verlinde, IAS preprint, April 1990; R. Dijkgraaf, E. Verlinde and H. Verlinde, IAS preprint, May 1990.

[6]   F. David, Mod. Phys. Lett. A5 (1990) 1019.

[7] F. David, Saclay preprint, July 1990, SPhT/90-090.

[8] M. Douglas, N. Seiberg, and S. Shenker, Rutgers preprint, April 1990, Phys. Lett. B (in press). For rigorous results see G. Moore, Yale preprint, April 1990.

[9] E. Brézin, E. Marinari and G. Parisi, Phys. Lett. B242 (1990) 35.

[10] E. Brézin, C. Itzykson, G. Parisi and J. Zuber, Comm. Math. Phys. 59 (1978) 35.

[11] E. Marinari and G. Parisi, Phys. Lett. B240 (1990) 375.

[12] T. Banks, N. Seiberg and S. Shenker, unpublished.

[13] M. Karliner and A. Migdal, Princeton preprint, July 1990.

[14] G. Parisi, presentation at the Cargèse workshop.

[15] P. Ginsparg and J. Zinn-Justin, presentation at the Cargèse workshop.

[16] D. Gross and V. Periwal, Phys. Rev. Lett. 60 (1988) 2105.

[17] The link between the (2g)! growth of perturbation theory in the matrix models and the number of cells in the moduli space was first made by Douglas and the author in [1] .

[18] E. Witten, Nucl. Phys. B268 (1986) 253; S. B. Giddings, E. Martinec and E. Witten, Phys. Lett. B176 (1986) 362.

[19] J. Harer and D. Zagier, Inv. Math. 185 (1986) 457; R. Penner, Comm. Math. Phys. 113 (1987) 299; J. Diff. Geom. 27 (1988) 35.

[20] D. Bessis, C. Itzykson and J.-B. Zuber, Adv. Appl. Math. 1 (1980) 109.

[21] C. Itzykson and J.-B. Zuber, Saclay preprint, January 1990, SPhT/90-004.

[22] E. D'Hoker and D. Phong, Nucl. Phys. B269 (1986) 205.

[23] R. Penner, Institut Mittag-Leffler preprint no. 10, 1989.

[24] E. Witten, IAS preprint, May 1990.

[25] J. Horne, Princeton preprint June 1990, PUPT-1185.

[26] D. Kutasov and N. Seiberg, Rutgers preprint, July 1990.

[27] M. Kaku, in *Functional Integration, Geometry and Strings*, Z. Haba and J. Sobczyk, eds. Berlin (1989); T. Kugo, H. Kumitomo and K. Suehiro, Phys. Lett. B226 (1989) 48; M. Saadi and B. Zweibach, Ann. Phys. 192(1989)213.

[28] S. Das and A. Jevicki, Brown preprint, 1990; J. Polchinski, Texas preprint, 1990, UTTG-15-90.

[29] T. Banks, presentation at the Cargèse workshop.

[30] E. Witten, Nucl. Phys. B156 (1979) 269.

[31] V. Periwal and D. Shevitz, Phys. Rev. Lett. 64 (1990) 1326.

APPROXIMATING THE CRITICAL STRING MEASURE USING DYNAMICALLY TRIANGULATED SURFACES

Dirk-Jan Smit

Department of Theoretical Physics
University of California
Berkeley, CA 94720 USA

1. Introduction

The recent formulation of 2D-gravity and string theory based on summations over dynamically triangulated random lattices [1, 2] (referred to as the DT formulation), reveals an interesting thermodynamical limit. In this limit certain singularities (i.e. critical exponents) appear, signalling a phase transition beyond which one may hope for a non-perturbative description of the theory.

In certain special situations one can compare the critical exponents with the ones found in a continuum formulation of the conformal field theory coupled to gravity as described in [3]. So far the only (partially) explicit verifications are for $c < 1$ theories in which the continuum theory and the DT formulation are in the same universality class. For these theories the DT formulation turns out to be equivalent with a quantum mechanical formulation of a suitably chosen matrix model which turns out to be exactly solvable [4].

For more general theories a matrix model has not been very useful so far to gain some insight in the thermodynamical limit. It therefore remains to be seen whether for example the DT approach to critical string theory as proposed in [1] indeed agrees with the results one obtains from the continuum theory after a suitable gauge-fixing procedure.

In the first part of this talk we will establish an algebraic formulation of the DT approach which may be useful to study the problem in what sense the DT formulation approximates to continuum path integral formulation of critical string theory. The need for such a formulation has already been given in [5]. For this we will make use of a theorem proved in [6] implying that there is a *one-to-one* correspondence between a dynamical triangulation (i.e. a graph specified by an adjacency matrix G_{ij} which takes only the values zero or one, so that all sites are of the same fixed length) and special Riemann surfaces specified by a set of polynomial equations with only *integer* coefficients.

Random Surfaces and Quantum Gravity
Edited by O. Alvarez *et al., Plenum Press, New York, 1991*

Such surfaces are said to be defined over the number field $\overline{\mathbf{Q}}$, the algebraic closure of the field of rationals $\mathbf{Q}$. We will give an informal, intuitive, description of this theorem and present an explicit example.

The main point of this note is discussed in the second part in which we show that as a consequence of the above relation with algebraic points, the sum over all adjacency matrices defining triangulations of a fixed area is really a sum over the K-rational points in the moduli space of a given genus g Riemann surface. We will give an explicit expression for this summation in the DT approximation of critical string theory. The Polyakov critical string partition function in the DT approach [1] reads

$$
Z = \int \prod_i DX_i \sum_{G_{kl}} \exp\left(-\tfrac{1}{2} \sum_{k,l} G_{kl}(X_k - X_l)^2 + N \cdot \text{const.} \right),
\tag{1.1}
$$

where X_i, $i = 1, \cdots, 26$ denotes the string coordinate in a 26 dimensional Euclidean space and N denotes the number of triangles. The sum over G_{ij} is for a fixed area of triangulation. We will show that, after integrating out the X variables, (1.1) is equivalent to the so-called *modular height function* $h(C_K)$ on the moduli space summed over the K-rational points C_K in moduli space. (K-rational points correspond to Riemann surfaces specified by polynomial equations whose solutions and coefficients are in a finite extension K of the field $\mathbf{Q}$.) The modular height function can be seen as a (logarithmic) distance function measuring the distance to the nearest K rational point in moduli space from a given point C. More precisely: the result will turn out to be that (1.1) (considered for a fixed number of triangles N) is equal to

$$
Z = \lim_{K \to \overline{\mathbf{Q}}} \sum_{C \in \mathcal{M}_{g,K}} \exp(h(C)) \cdot \Lambda_{g,K},
\tag{1.2}
$$

where C_K is a constant depending only on the degree of the extension K. This expression for the string partition function was already introduced in in [10, 9]. The label K attached to $\mathcal{M}_g$ indicates that we consider only those points in the moduli space which are K-rational. The degree of the extension roughly specifies the number of triangles, i.e. the area.

The continuum limit, in which $N \to \infty$, corresponds to considering all $\overline{\mathbf{Q}}$-algebraic points in $\mathcal{M}_g$. In this limit one expects to get a measure on the moduli space of a given genus g Riemann surface. The question is whether this measure corresponds to the Polyakov measure [7] which corresponds to the square of the Mumford form on moduli space obtained after appropriate gauge-fixing the continuum theory [8]. This,indeed is a subtle point since the $\overline{\mathbf{Q}}$-algebraic points are very likely badly distributed in the moduli space for high genus surfaces. This poses the question if one should introduce a non-trivial weight for every point in the sum (1.2). In the DT approach one weighs every adjacency matrix with weight one, however, given the irregularity in the distribution of algebraic points in generic genus moduli spaces it is far from obvious that the summation over adjacency matrices indeed approximates the continuum path integral on the moduli space.

To study this problem one needs explicit knowledge of the height function. For low genus the distribution problem does not exist. In fact for genus $g = 1,2$ the height function can be computed explicitly [11, 9] from which it follows that (1.2) (with all the weights to be unity) indeed converges to the known expression for the Polyakov measure [13].

The distribution problem is well known in the mathematical literature. Its relevance for discrete approximation to string theory was already discussed in [10, 9] and more recently, in the context of 2D gravity in [13].

2. Equilateral Triangulations and Algebraic Surfaces

The algebraic points in moduli space have a nice geometrical interpretation in terms of triangulated surfaces due to a result of Voedvodski and Shabat [6]. Consequently only certain 'special' Riemann surfaces can be described by adjacency matrices in the limit $N \to \infty$ (N the number of triangles), namely exactly those that are $\overline{\mathbf{Q}}$-algebraic. More precisely: to each equilateral triangulation one can assign a unique complex structure compatible with the adjacency matrix. The polygon (consisting of only equilateral triangles[1]) specified by the adjacency matrix equipped with this complex structure can be shown to be equivalent with an algebraic Riemann surface. Furthermore, *all* $\overline{\mathbf{Q}}$-surfaces can be obtained in this way.

As we shall show this implies that the original sum over adjacency matrices in (1.1) for critical string theory is indeed given by the modular height function summed over algebraic points in the moduli space. The question whether the sum over adjacency matrices agrees with the continuum formulation of the theory thus becomes essentially the problem of how the algebraic points are distributed in the moduli space.

Let us begin with presenting a geometrical interpretation of an algebraic point in $\mathcal{M}$. The starting point is a theorem by Belyi [14]. It states that a complete algebraic curve X over a field of characteristic zero (e.g. $\mathbf{C}$), can be defined over $\overline{\mathbf{Q}}$ *if and only if* there exists a covering map

$$\phi : X \longrightarrow \mathbf{P}^1 \tag{2.1}$$

which is holomorphic outside three points (e.g. $\{0, 1, \infty\}$).

In practice it is not so easy to find examples of such functions. In [6] it is shown that such functions are intimately related with functions depending on a period matrix τ and a complex coordinate which for certain special values of τ are functions on the complex plane that give rise to a polygon consisting exclusively of equilateral triangles. This polygon is holomorphically equivalent to the Riemann surface characterized by τ. That is, the required function 'transfers' the canonical complex structure (inherited from $\mathbf{C}$) on the interior of each triangle to the polygon such that it defines a complex structure on a two dimensional surface characterized by the period matrix τ. This complex structure on X is called the equilateral complex structure. The result is stated in the following

Theorem [6]

Let X be a complete nonsingular complex algebraic curve. It is defined over $\overline{\mathbf{Q}}$ if and only if there exists a simplicial scheme for which X is biholomorphically equivalent to the Riemann surface X with the equilateral complex structure.

---

[1] Although everything we say is true for arbitrary triangulations in which the length of the sites is *not* a dynamical variable, we will restrict ourselves to equilateral triangles.

The proof of the theorem relies for a large extend on the construction of the equilateral complex structure. Using conformal mappings, it is explained how in principle the required Belyi function ϕ can be constructed. We will explain below how one obtains from a given set of equilateral triangles a unique Riemann surface of given genus. Instead of reviewing the abstract procedure by which one shows how the complex structure is obtained, we show this explicitly in an example which turns out to be useful in the rest of this section.

In order to relate the combinatorial data of a set of triangles with a compact two dimensional surface one introduces a so-called simplicial scheme S which consists of a finite set $J_0(S)$ which will correspond to the vertices and two collections $J_1(S)$ and $J_2(S)$ consisting of respectively two-element and three-element subsets of $J_0(S)$. $J_1(S)$ respectively $J_2(S)$ correspond to the set of edges respectively faces. For consistency one requires that all two element subsets of $J_2(S)$ are contained in $J_1(S)$. Out of S one then constructs a polyhedron in a real Euclidean vector space of dimension equal to the order of the set $J_0(S)$. S is identified with the coordinate vectors on this vector space and the convex hulls of the images of $J_1(S)$ and $J_2(S)$ form the edges and faces of this polyhedron. Such a polyhedron is called the realization of S and denoted by $|S|$.

One next introduces a *flag-set* $\mathcal{F}(S)$ associated with S which allows one to identify triangles on the polyhedron. It is defined as the set of tuples

$$\mathcal{F}(S) = \{\alpha, \beta, \gamma \in J_0(S) \times J_1(S) \times J_2(S) | \alpha \subset \beta \subset \gamma\}. \tag{2.2}$$

One also introduces the projections

$$\pi_q : \mathcal{F}(S) \longrightarrow J_q(S), \qquad q = 0, 1, 2. \tag{2.3}$$

In order to study a complex structure on the faces of the polyhedron which extends to the edges and vertices such that $|S|$ becomes homeomorphic to a compact Riemann surface, it turns out to be useful to introduce a group action on $\mathcal{F}(S)$ which maps triangles into each other. One defines a group C_2 acting on $\mathcal{F}(S)$ by the following relations for its generators σ_p, $p = 0, 1, 2$ for all elements $F \in \mathcal{F}(S)$:

$$\sigma_0^2 = \sigma_1^2 = \sigma_2^2 = (\sigma_0 \sigma_2)^2 = 1, \tag{2.4}$$

having the following properties

1.
$$\pi_q(\sigma \circ F) = \pi_q(F) \Longleftrightarrow p \neq q;$$

2. C_2 must act transitively, so that the realization $|S|$ is connected;

3. there must be an element $O \in C_2$

$$O : \mathcal{F}(S) \longrightarrow \{\pm 1\}$$

such that

$$O(\sigma_q \circ F) = -O(F).$$

It is not hard to see that for a given S its realization is homeomorphic to a two dimensional surface if and only if there is a *unique* action of C_2 on $\mathcal{F}(S)$ satisfying the above properties.

Let us pause for a moment and see how this works in a simple example[2]. We will build a regular tetrahedron with vertices numbered from 1 to 4. (See figure 1).

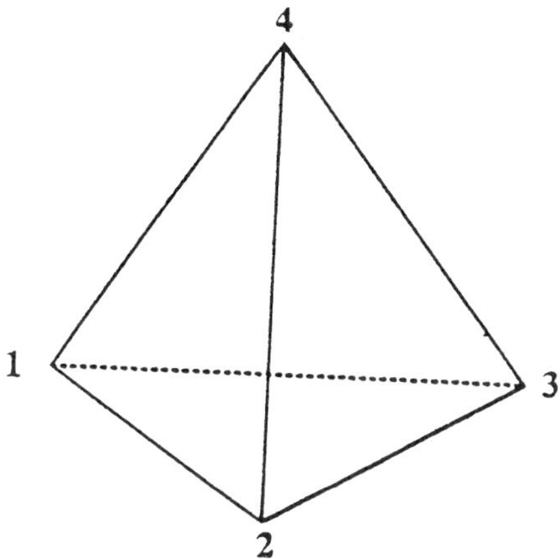

Figure 1
The tetrahedron.

So we have

$$
\begin{aligned}
J_0(S) &= \{1,2,3,4\} \\
J_1(S) &= \{\{1,2\},\{1,3\},\{1,4\},\{2,3\},\{2,4\},\{3,4\}\} \\
J_2(S) &= \{\{1,2,3\},\{1,2,4\},\{2,3,4\},\{1,3,4\}\}
\end{aligned}
$$

and S is realized in $\mathbf{R}^4$. The flagset $\mathcal{F}(S)$ is given by (3.2) and thus is of order $4 \cdot 3 \cdot 2 = 24$. The action of the group C_2 is easily determined from the definition of its generators. Take $F = (1,\{1,2\},\{1,2,3\}) \in \mathcal{F}(S)$ then

$$
\begin{aligned}
\sigma_0(F) &= (2,\{1,2\},\{1,2,3\}) & (2.5) \\
\sigma_1(F) &= (1,\{1,3\},\{1,2,3\}) & (2.6)
\end{aligned}
$$

These actions are natural; we have illustrated them in figure 2. The action of σ_2 is less automatic, since it adds a new vertex to the configuration; we have, (see figure 3),

$$
\sigma_2(F) = (1,\{1,2\},\{1,2,4\}) , \tag{2.7}
$$

---

[2] I thank B. Edixhoven for discussions on this point.

where the new vertex is labelled by 4.

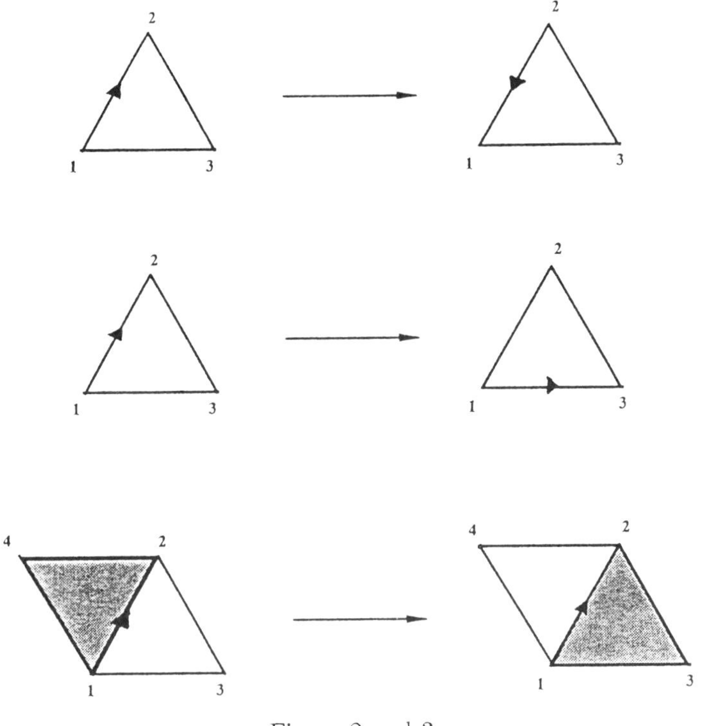

Figure 2 and 3
The action of C_2.

A priori, this may not be a unique vertex, however, in this example the action of σ_2 is *unique* since the order of the set

$$\{\gamma \in J_2(S) | \beta \in \gamma\} = 2.$$

Since all σ_i change the orientation of the triangle described by F, the above action of σ_i satisfies all the properties listed above. Observe that in addition to the relation (3.4) we have $(\sigma_0\sigma_1)^3 = 1$. This tells exactly how many triangles meet at each vertex. More generally: the orbits of the group generated by σ_i, $i = 1, 2$ correspond to the number of triangles meeting at each vertex. Thus we see that from a given S and a flagset $\mathcal{F}$ the action of C_2 gives a unique realization $|S|$.

To show that $|S|$ is homeomorphic to a Riemann surface, requires the construction of a complex structure. In [6] such a complex structure is constructed out of the canonical complex structure on each interior of a triangle inherited from $\mathbf{C}$. Of course, the extension to the edges and vertices will lead to a compatability condition for each of the complex structures on the faces. It turns out that there is a *unique* complex structure on $|S|$ which does this.

In order to describe this complex structure one constructs a slightly different realization of S in which one considers the set of triangles in the complex plane. Consider the set of equilateral triangles $T^{\pm}$ of which one side is along the interval $[0, 1]$ on the real axis in the complex plane and a vertex in the upper respectively lower half plane. Form the disconnected set of triangles

$$\mathcal{C}(S) = \left\{ z \times F \in \mathbf{C} \times \mathcal{F}(S) \mid z \in T^{\pm} \text{ for } O(F) = \pm 1 \right\}. \tag{2.8}$$

The group $\mathcal{C}_2$ acts on this set so its is natural to define two equivalence relations on this set. The first equivalence relation, R_1 says that two elements in $\mathcal{C}$ are equivalent whenever $F' = \sigma_2 F$ and $z = z' \in \mathbf{R}$. The other relation R_2 involves the action of σ_0 and σ_1. The group $\mathcal{C}_1$ generated by $\sigma_{0,1}$ acts as

$$\begin{aligned}
\sigma_0(z \times F) &= ((1 - z) \times (\sigma_0(F))), \\
\sigma_1(z \times F) &= ((e^{-i\mathcal{O}(F)\pi/3}) \times (\sigma_1(F))).
\end{aligned} \tag{2.9}$$

The quotient $\mathcal{C}(S)/R$ defined by the weakest equivalence the relation R generated by $R_{1,2}$, defines a space $X(S)$ on which the following complex structure inherited from $\mathbf{C}$ can be defined.

Let n be the number of triangles meeting at each vertex, i.e. for a given $F \in \mathcal{F}$, n is the smallest integer for which $(\sigma_1 \sigma_2)^n(F) = F$. The sum of the angles at each vertex thus equals $n\pi/6$. Let Z_ν be a holomorphic function with respect to the canonical complex structure (inherited from $\mathbf{C}$) on the interior of one the ν-the triangle, $\nu = 1, \cdots n$. Whenever a point P lies on an edge, say of the ν-th and $\nu + 1$-the triangles then

$$Z_\nu(P) = Z_{\nu+1}(P).$$

Thus it is clear that we have holomorphic coordinates on all points except possibly at the vertices. So let P be a vertex of a triangle Δ_ν. We may assume that $Z_\nu(P) = 0$, and that Δ_ν has an edge in common with the triangle $\Delta_{\nu+1}$. (We identify the ν with $\nu + 1$.) Let V_ν be the angle of Δ_ν at $Z_\nu = 0$. Then one may define a local coordinate Z at P by

$$Z|_{V_\nu} = Z_\nu(Q)^\mu e^{\mu(\sum V_\nu)} \tag{2.10}$$

where Q is point in Δ_ν and the sum runs over all angles meeting at the origin (with $V_0 = 0$). The number μ is the sum of the angles: $\mu = \frac{2\pi}{\sum V_\nu}$. Since all triangles are equilateral one may recast this into

$$Z|_{V_\nu} = \exp(2\pi i\nu/n)(Z_\nu)^{6/n} \tag{2.11}$$

This turns $X(S)$ into a compact Riemann surface. In fact this formula tells us that each open neighborhood of a vertex in $X(S)$ with the vertex itself removed is biholomorphic to a punctured disc. The complex structure on $X(S)$ is called the equilateral complex structure. Note that when $n = 6$, the triangulation describes a flat surface and the equilateral complex structure coincides to the canonical complex structure on $\mathbf{C}$. Namely, if $n = 6$, rotation over 180 degrees, (that is, taking $\nu = 3$) indeed corresponds to multiplication with -1. Another important observation is that changes of the lengths of the edges of the triangles does not change the conformal class of the discretized metric.

The theorem in [6] quoted above implies that each equilateral triangulation endowed with its equilateral complex structure defines a $\overline{\mathbf{Q}}$-algebraic Riemann surface and vice versa. For us the part of the theorem

$$\textit{Triangulation} \longrightarrow \textit{Riemann surface}$$

is most important, as we shall now discuss. It follows from the existence of a Belyi function. Namely, draw the medians in each of the triangles $T^{\pm} \in \mathcal{C}(S)$ so that one obtains 6 small rectangular triangles. Label these small triangles according to their orientation following from $\mathcal{O}(F)$. The images of these triangles on $X(S)$ are thus in a one-to-one correspondence with the flag set $\mathcal{F}(S)$. One now proceeds with constructing a (continuous) conformal transformation of $\mathcal{C}(S)$ onto $\mathbf{P}^1$ by mapping each rectangular triangle onto the upper half plane respectively lower half plane according to their orientation such that the three angles are sent to 0,1, ∞. This map is invariant under the equivalence relation R so that it determines, at least in principle, the Belyi function. Hence, to a given equilateral triangulation one can assign the equilateral complex structure, obtained from a simplicial scheme S, provided this surface is $\overline{\mathbf{Q}}$-algebraic. It is important to realize that this map is *independent* of the metrical properties, i.e. of the size of the edges of the triangles!

It is intuitively clear that the surface is necessarily algebraic: the only parameters entering the description of the map are the three branch points which are obviously algebraic. Thus we have assigned a to a given triangulation a branched cover of the sphere, which is holomorphic (with respect to the equilateral complex structure) outside the three branch point. The conclusion is that this branched cover describes an arithmetic surface. That is, every planar graph of arbitrary topology (generated from a potential in matrix model, say) corresponds to an arithmetic surface described as a multiple cover of the sphere, branched over exactly three points. Even more is true: *all* arithmetic surfaces can be obtained in this way.

3. The Sum over Algebraic Points in Moduli Space

Form the previous section we learn that only $\overline{\mathbf{Q}}$-*algebraic* Riemann surfaces can be described by the combinatorial data of triangulation specified by adjacency matrices $G_{ij} = 0, 1$. Consequently, the sum over adjacency matrices in the discretized critical string defines a measure on the moduli space which gets it support only from the $\overline{\mathbf{Q}}$-algebraic points. We would like to compare this measure with the string measure obtained in the continuum formulation of the theory.

Of course, to address this problem in full detail it is necessary to have explicit knowledge about the correspondence between a given triangulation and a K-rational point in moduli space. Unfortunately, we were unable to solve this problem in complete generality. For certain spacial cases, e.g. triangulations of a genus zero surface corresponding to the regular Platonic polyhedrons, the problem can be solved [15, 16].

In [16] a technique has been presented which in principle allows one to find an explicit geometrical description of an algebraic surface. In a few simple cases this technique leads to an explicit construction of a Belyi function. More importantly, however, is the fact that it reveals a relation between the volume measured by the string measure on the moduli space and the modular height function for $\overline{\mathbf{Q}}$-algebraic points. We will briefly discuss this technique here referring the reader to [16] and the references therein for more details.

Let us consider an arbitrary curve C of genus g (corresponding to a compact Riemann surface of genus g). We know that for certain special values of the period matrix describing the Jacobian it will be possible to define the curve over $\overline{\mathbf{Q}}$. We would like to get a condition on how this 'tuning' of the period matrix can be done a more geometrical way.

Let us assume that the curve C is uniformized by some finite index subgroup $\Gamma \subset PSL_2(\mathbf{R})$ acting on the upper half-plane:

$$C \simeq H/\Gamma. \tag{3.1}$$

The images of the parabolic elements in Γ in the upper half-plane define certain coordinates z_i, which we will identify later on with some of the vertices of a triangulation. In general Γ will be some Kleinian group [17]. For simplicity we will assume that Γ is a Fuchsian group. Consider also the affine curve X which is defined as the n-punctured sphere

$$X = \mathbf{C} - \{z_1, \ldots, z_n\}.$$

The surface X is uniformized by a group $\hat{\Gamma}$ which contains the group Γ as a finite index subgroup. (The cases where $\hat{\Gamma} \simeq \Gamma$ are precisely those surfaces that correspond to punctured spheres. In general one either has to consider a genus g Riemann surfaces with some punctures or a finite index subgroup of $\hat{\Gamma}$.)

The reason we introduce punctures is as follows. The punctures will be identified later with some of the vertices of a triangulation. Keeping the number of punctures fixed we would like to characterize the period matrix of the algebraic curve in the limit where the number of vertices goes to infinity, while the Euler number is kept fixed. In the limit it is intuitively clear that we obtain an arithmetic surface (i.e. a surface defined over the closure $\overline{\mathbf{Q}}$) of genus g. In the rest of this section we will consider *only* those arithmetic surfaces which correspond to n-punctured spheres. In this case one can, at least in principle, characterize the algebraic points in the moduli space of all n-punctured spheres in a geometrical way. So strictly speaking the argument presented below is only valid for n-punctured spheres. However, as we shall argue, the result we will obtain will be true for all arithmetic surfaces.

Let us consider the n-punctured sphere in more detail. The inverse $J^{-1}(z)$ of the map $J : H \to X$, is a multivalued analytic function on H of which the branches are transformed into each other under linear fractional transformations in $\hat{\Gamma}$. It is well known see e.g. [17] that the Schwarzian derivative $\mathcal{S}(J^{-1})$ defines a rational function T_X on X. In case $n = 3$, (i.e. X corresponds to a three punctured sphere) it reads

$$\mathcal{T} = \frac{1}{2}\frac{(1 - \nu_1^{-2})}{z^2} + \frac{1}{2}\frac{(1 - \nu_2^{-2})}{(z-1)^2} + \frac{a}{z} + \frac{b}{(z-1)}, \tag{3.2}$$

with the constants a and b given by

$$a = -b = \frac{1}{2}\left(1 - \nu_1^{-2} - \nu_2^{-2} - \nu_3^{-2}\right). \tag{3.3}$$

The numbers ν_i are the ramification indices of the points $0, 1, \infty$.

For a generic surface, having n punctures, the function T_X is given by the term $\mathcal{T}$ plus a term involving the so-called accessory parameters c_j, $j = 4, \ldots n$

$$T_X = \mathcal{T} + \sum_{j=4}^{n} \frac{w_j(w_j - 1)}{w(w-1)(w - w_j)}\left(\frac{(1 - \nu_j^{-2})}{2(w - w_j)} + c_j\right). \tag{3.4}$$

The accessory parameters are given as residues on X

$$c_j = \frac{1}{w_j(w_j - 1)} \mathrm{Res}_{w_j} w(w - 1) T_X(w) dw^2, \qquad j = 4, \cdots, n, \qquad (3.5)$$

where Res_{w_j} denotes the residue at $w = w_j$. Thus, the accessory parameters are uniquely determined by the singular points z_j and hence by the uniformizing group. In practice it turns out to be very difficult to compute the accessory parameters explicitly. However, even without explicit knowledge one can extract the following useful geometrical information [18]. It turns out that the accessory parameters correspond to real analytic functions on the space

$$W_n = \{(z_4, \ldots z_n) \in \mathbf{C}^{n-3} | z_i \neq 0 \text{ and } z_i \neq z_j \text{ for } i \neq j\}. \qquad (3.6)$$

Furthermore, it is known that there exists a holomorphic covering map Ψ

$$\Psi : \mathcal{T} \longrightarrow W_n \qquad (3.7)$$

where $\mathcal{T}$ is the Teichmüller space of n-punctured spheres. These facts are used in [18] to show that the variation of the accessory parameters with respect to the location of the punctures defines the Kähler potential for the Weil-Petersson metric on the moduli space of n-punctured spheres:

$$\frac{\partial c_i}{\partial z_j} = \frac{1}{2\pi} < \frac{\partial}{\partial z_i}, \frac{\partial}{\partial z_j} >= \lambda_{ij}, \qquad (3.8)$$

with λ_{ij} denoting the metric tensor of the Weil-Petersson metric. This result is useful, since we can now invoke some results of [19] to relate this with the string measure on moduli space.

But before we do this, we want to point out that the surface X will be algebraic if and only if the coordinates of the punctures on the sphere are algebraic integers. This then implies at the same time that the compact curve C is algebraic. We thus conclude that if we would know the dependence of the accessory parameters on the location of the punctures then with (3.8) we could compute the density of the Weil-Petersson metric at the algebraic point in the moduli space of n-punctured spheres.

Let us now see how this is related to the height of an algebraic point in the moduli space. First we recall that the Weil-Petersson metric defines a cohomology class in $H^2(\mathcal{M}_g, \mathbf{Q})$. In fact Wolpert showed that this class is equal, on the level of differential forms to the class of the bundle of 'holomorphic differentials on the Riemann surface' or more precisely to the dualizing sheaf of $\pi : X \to \mathcal{M}_g$, where X denotes the universal curve over the the moduli space of genus g Riemann surfaces:

$$\frac{1}{24\pi^2} \omega_{W.P} = c_1(\lambda), \qquad \lambda \equiv \det(\pi_* \omega_{X/\mathcal{M}_g}). \qquad (3.9)$$

Furthermore, we recall that the string measure M_g on $\mathcal{M}_g$ is given

$$M_g = \frac{\mu(s) \wedge \overline{\mu(s)}}{||s||^2}, \qquad (3.10)$$

with s a section of λ^{13} and μ the isometry (for the Quillen metric)

$$\mu : \lambda^{13} \simeq \Omega_{\mathcal{M}_g}, \tag{3.11}$$

with $\Omega_{\mathcal{M}_g}$ denoting the cotangent bundle of $\mathcal{M}_g$. The important fact is that this measure is in fact defined for the moduli space $\mathcal{M}_g$ defined over the integers $\mathbf{Z}$ [9].[3] Thus it makes sense to evaluate the measure at the $\overline{\mathbf{Q}}$ algebraic points in $\mathcal{M}_g$ and try to replace the integration over $\mathcal{M}_g$ by a summation over these points. One way to describe these points would follow from dependence of the accessory parameters on the location of the punctures. Let us assume that we have this knowledge, and evaluate the Weil-Petersson metric at those values for c_i corresponding to algebraic coordinates z_i. We will now argue that the sum over the contribution (3.8) from each algebraic point corresponds to the volume of the *lattice in the tangent space* of $\mathcal{M}_g$, of which the lattice points are the algebraic points, measured by the Weil-Petersson metric! We can be much more precise on this volume. Consider an arithmetic surface

$$\pi : C \longrightarrow S = \mathrm{Spec}(\mathcal{O}_K), \tag{3.12}$$

where K is a number field and $\mathcal{O}_K$ denotes its ring of integers (see [9] for notation). Let $P \in \mathcal{M}_g$ be the point in the moduli space over $\mathbf{Z}$ corresponding to (3.12), i.e. we have the following diagram

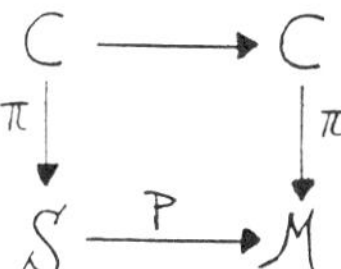

Denote by $T_{\mathcal{M}_{\mathbf{Z}}}$ the tangent bundle of $\mathcal{M}_g$. The 'pullback' $P^*T_{\mathcal{M}_g}$[4] defines a lattice in the vector space $P^*T_{\mathcal{M}_g} \otimes_{\mathbf{Z}} \mathbf{R}$. The measure M_g in (3.10) defines Hermitean metrics at infinity (i.e. at the induced line bundle over the infinite places of K) on the determinant $\det(P^*T_{\mathcal{M}_g})$. This metric defines a volume form on the vector space $P^*T_{\mathcal{M}_g} \otimes_{\mathbf{Z}} \mathbf{R}$. The volume we are interested in is thus

$$\mathrm{Vol}_K \left((P^*T_{\mathcal{M}_g} \otimes_{\mathbf{Z}} \mathbf{R}) / P^*T_{\mathcal{M}_g} \right). \tag{3.13}$$

The conclusion now is that this corresponds to summing the Weil-Petersson density over the algebraic points computed from the accessory parameters at the algebraic integers z_i. This technique is only applicable for n-punctured spheres.

There is, however, a way to characterize the the Weil-Petersson density at the algebraic points in a more abstract way using a Riemann-Roch formula for arithmetic surfaces [20]. It then follows that the volume is expressible in terms of the so-called modular height

---

[3]More precisely over the moduli stack over $\mathbf{Z}$.
[4]One should think of $P^*T_{\mathcal{M}_g}$ as an $\mathcal{O}_K$-module.

function which is related with the class of the dualizing sheaf of $\pi : C \longrightarrow \mathrm{Spec}(\mathcal{O}_K)$, namely

$$h(C) = \frac{1}{[K:\mathbf{Q}]} \deg \det(\pi_* \omega_{C/S}). \tag{3.14}$$

The volume is computed explicitly in [10, 21, 22]. The result is

$$Vol_K\left((P^*T_{\mathcal{M}_g} \otimes_{\mathbf{Z}} \mathbf{R})/P^*T_{\mathcal{M}_g}\right) = \exp(13h(C)) \times \Lambda_{g,K}, \tag{3.15}$$

with $\Lambda_{g,K}$ is a constant only depending on the genus and the number field K. The degree of the extension K controls roughly speaking the shape of the triangulation, i.e. the number of vertices necessary to describe a given genus g algebraic point [16]. Fixing K corresponds to keeping fixed the area of the triangulation. The volume (3.15) is thus the volume obtained in (1.1) after performing the summation over all adjacency matrices with fixed area!

Now, we require that in the limit to $\overline{\mathbf{Q}}$ the volume (3.15)'approximates' the volume as measured by the string measure (3.10), that is, it should have for example the same singularity behaviour at points corresponding to degenerate curves. This in principle requires the inclusion of nontrivial weights $\Delta(C)_K$ attached to each arithmetic curve C. As we argued in the introduction, for low genus it seems reasonable to expect that all weights can be safely chosen to be unity. Indeed, explicit calculations of the height function for smooth genus one and two [11, 9] show that one recovers the known expressions [12] for the string measure in these cases. For high genus it is far from obvious that one can still recover the string measure without introducing nontrivial weights, since the $\overline{\mathbf{Q}}$ algebraic points are presumably badly distributed in $\mathcal{M}_g$ for $g > 23$ [23]. This thus raises a nontrivial question on the effectiveness of the dynamical triangulation approach to string theory.

Leaving aside this potential problem for future work, we would like to illustrate briefly in conclusion the technique above in the example of the tetrahedron [16]. It turns out that when the four triangles are layed down in the complex plane and and applying the technique above, one finds that the Weierstrass $\mathcal{P}$-function identifies the edges and vertices such that a tetrahedron arises (see fig. (4)).

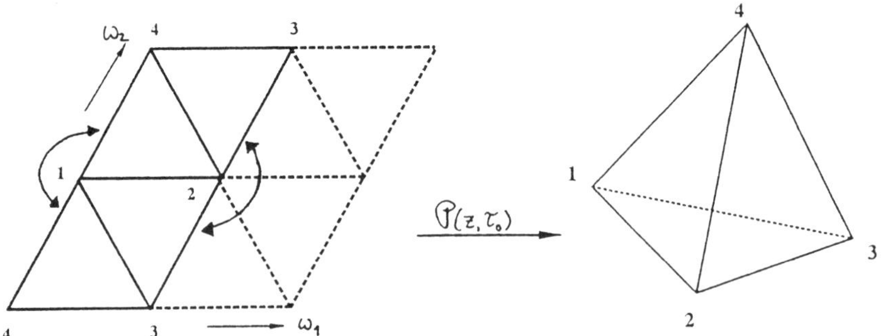

Figure 4

Glueing the tetrahedron using the Weierstrass $\mathcal{P}$-function.

Thus, using the functional equation satisfied by the Weierstrass function and its first derivative, one finds the equation for the elliptic curve

$$y^2 = x^3 + 1 \qquad (3.16)$$

which parametrizes a two-folded $g = 1$ cover of the sphere, with modular parameter $\tau = \tau_0 = \exp(2\pi i/6)$. In order to get the genus zero triangulation one restricts to only one sheet of the cover.

Vice versa, consider a genus one curve parametrized by

$$y^2 = x(x - 1)(x - \tau) \qquad (3.17)$$

with $\tau = c\tau_0$, where c is any rational number. It is well known see e.g. [24] that *all* tori parametrized by the modular parameter

$$\lambda(\tau) = \left(\frac{\vartheta_3(\tau)}{\vartheta_1(\tau)} \right)^4, \qquad \tau = c\tau_0, \qquad (3.18)$$

are algebraic since for those τ the parameter $\lambda(\tau)$ is an algebraic integer. The λ-points are dense and nicely distributed in the Teichmüller space of tori. The tetrahedron corresponds to $c = 1$, i.e. $\tau = \tau_0$. In general the tori described by other choices for c (not equal to $c = \frac{1}{2}, 2$) will require more than just four triangles. However, always four vertices will correspond to punctures. In the limit of infinite triangles one describes the set of all arithmetic four punctured spheres which are smoothly distributed in the moduli space of four punctured spheres.

Generically the height function of a given K-rational point gives the distance to the closest K-rational point. We will present elsewhere an explicit illustration of this fact for rational four punctured spheres.

Acknowledgements

I would like to thank O. Alvarez and P. Windey for many comments and for reading the manuscript. This work is supported by NSF grant PHY85-15857.

References

[1] V. Kazakov: Bilocal reguralization of models of random surfaces, Phys. Lett. **150B**, 282, (1985).
V. Kazakov, I. Kostov, A. Migdal: Critical properties of randomly triangulated planar random surfaces, Phys. Lett. **157B**, 295 (1985).

[2] F. David: Planar diagrams, two-dimensional lattice gravity and surface models, Nucl. Phys. **B257** [FS14], 45, (1985); *and:* A model of random surfaces with non-trivial critical behaviour, Nucl Phys. **B257** [FS14], 543, (1985).

[3] V. Kniznik, A. Polyakov, A. Zamolodchikov: Fractal structure of 2d-quantum gravity, Mod. Phys. Lett. A. **3** (1988), 819.

[4] E. Brezin and V. A. Kazakov: Exactly solvable field theories of closed strings, Phys. Lett **236B**, 144, (1990),
M. Douglas and S. Shenker: Strings in less than one dimension, Rutgers preprint RU-89-34. October 1989.
D. Gross and A. Migdal: Non-perturbative two dimensional quantum gravity, Phys. Rev. Lett. **64**, 127, (1990).

[5] Yu. Manin: Reflections on arithmetical physics, *in:* The proceedings of the Poiana-Brasov school on strings and conformal field theory, 1987.

[6] V. Voedvodski and G. Shabat: Equilateral triangulations of Riemann surfaces and curves over algebraic number fields, Sov. Math. Dokl. **39**, 38, (1989).

[7] A. Beilinson, Yu. Manin: The Mumford form and the Polyakov measure in string theory, Comm. Math. Phys. **107**, 359, (1986).

[8] A. Belavin. V. Kniznik: Algebraic geometry and the geometry of quantum strings, Sov. Phys. JETP **64**, 214, (1986).

[9] D.-J. Smit: String theory and the algebraic geometry of moduli spaces, Comm. Math. Phys. **114**, 645, (1988).

[10] K. Ueno: Bosonic strings and arithmetic surfaces, Kyoto University preprint December 1986.

[11] J. Silverman: Height and elliptic curves, *in:* Arithmetic geometry, G. Cornell, J. Silverman (eds.), Berlin Springer 1986.

[12] A. Morozov: Explicit formulae for one, two, three and four loop string amplitudes, Phys. Lett. **184B**, 173, (1977).

[13] A. Levin, A. Morozov: On the foundations of random lattices approach to quantum gravity, ITEP-preprint Moscow, 1990.

[14] G. Belyi: On Galois extensions of a maximal cyclotomic field, Math. USSR Izv. **14**, 247, (1980).

[15] C. Itzykson, private communication.

[16] D.-J. Smit: Summations over equilaterally triangulated surfaces and the critical string measure, Univ. Cal. preprint UCB-PTH-90/36, August 1990.

[17] I. Kra: Accessory parameters for punctured spheres, MSRI-preprint, Berkeley 1988, *and:* Automorphic forms and Kleinian groups, Benjamin Reading, Mass. 1972.

[18] P. Zograf and L. Takhtajan: Action of the Liouville equation is a generating function for the Weil-Petersson metric on the Teichmüller space, Funct. Ann. Appl. **19**, 219 (1985) *and:* On the Liouville equation, accessory parameters and the geometry of Teichmüller space for Riemann surfaces of genus zero, Math. USSR Sbornik **60**, 143 (1988).

[19] S. Wolpert: On the homology of the moduli space of stable curves, Invent. Math. **118**, 491 (1983).

[20] G. Faltings: Calculus on arithmetic surfaces, Ann. Math. **119**, 387, (1984).

[21] B. Edixhoven: The Polyakov measure and the modular height function, *in:* Proceedings of the Arbeitstagung on arithmetical algebraic geometry, Wuppertal 1987.

[22] D.-J. Smit: Algebraic and arithmetic geometry in string theory, *in:* Proceedings of the 16-th colloquium on group theoretical methods in physics, Lect. Phys. 313, 515, Springer New-York 1988.

[23] J. Harris and D. Mumford: On the Kodaira dimension of curves, Invent. Math. **67**, 23, (1982).

[24] Higher transcendental functions, Bateman Manuscript Project, eds. A. Erdélyi *et al.*, Vol. 4, McCraw-Hill, New-York, (1953).

MATRIX MODELS, STRING FIELD THEORY AND TOPOLOGY

Tom Banks

Department of Physics and Astronomy
Rutgers University
Piscataway, NJ 08855-0849

1. Introduction

This conference has been devoted to what might be called completely integrable string theories. The last year has seen a proliferation of exact solutions of simple string models (more properly: resummation of the perturbation series around simple solutions of the classical string equations of motion). These models all have one and two dimensional target spaces, a fact directly connected to their solubility. Indeed, it is easy to show that all of the matrix chain models can be rewritten in terms of free fermions which in some sense live on a two dimensional target space. The easiest way to demonstrate this is to write the Mehta et. al.[1] formula for the partition function for a matrix chain with general nearest neighbor transfer matrix[1]

$$Z = \int d^N \lambda_F d^N \lambda_I \Delta(\lambda_F)\Delta(\lambda_I)e^{U(\lambda_F)+U(\lambda_I)} \prod_{i=1}^{N} < \lambda_F^i | \prod_k \mathcal{T}_k | \lambda_I^i > \qquad (1.1)$$

$$\mathcal{T}_k(\lambda, \kappa) = e^{c\kappa\lambda + V_k(\lambda) + V_k(\kappa)} \qquad (1.2)$$

and use a Grassmann integral representation of the Vandermonde determinants:

$$\Delta(\lambda) = \int d^N \chi d^N \psi \;\; e^{[\chi_i \lambda_i^{j-1}\psi_{j-1}]} \qquad (1.3)$$

The integrals over eigenvalues factorize and one can redo the χ integrals to obtain:

$$Z = \int d^N \psi_F d^N \psi_I \;\; e^{[\psi_F^j D_{jk} \psi_I^k]} \qquad (1.4)$$

---

[1] Note that in these formulas, the function U which determines the boundary conditions is different from the functions V_k which appear in the transfer matrix.

Random Surfaces and Quantum Gravity
Edited by O. Alvarez et al., Plenum Press, New York, 1991

Here

$$D_{jk} = \int d\lambda_F d\lambda_I \quad e^{U(\lambda_F)+U(\lambda_I)} < \lambda_F | \prod_k T_k | \lambda_I > \lambda_F^j \lambda_I^k \tag{1.5}$$

The orthogonal polynomial method can be viewed as a way of diagonalizing the operator D. It is well known by now that in the continuum limit, the index space on which this matrix is defined becomes continuous, forming one of the dimensions in which the fermion fields live. The second is the "time" parameter along the chain. In the "time" continuum limit, the product of transfer matrices above becomes the time ordered exponential of a (generically time dependent) hamiltonian operator h. It is easy to show that the partition function can then be rewritten as[2]:

$$Z = \int d\overline{\psi}(t)d\psi(t)e^{\int_0^T \overline{\psi}(\partial_t + h(t))\psi} \tag{1.6}$$

with boundary conditions $\overline{\psi}(T), \psi(0) \in W$. Here W is the subspace of the one body Hilbert space spanned by the functions $\lambda^k e^U$ with $k \leq N - 1$. Thus all single chain matrix models can be reformulated in terms very similar to the Segal-Wilson[2] presentation of the KdV hierarchy, even before the worldsheet continuum limit is taken. This presentation of the matrix chain models also makes clear their relation to the fermion formulation of the $d = 1$ string.[3]

Rather than reviewing my own work on these soluble models (which has been adequately done at this conference by my collaborators) I decided to talk about the possible implication of matrix model ideas for the solution of the problems of unified string theory. My results in this direction can at best be characterized as work in progress, and at worst as rank speculation. Nonetheless, I felt that we were getting a little too bogged down in technicalities, and the usual mathematical richness of completely integrable systems, and that a speculative talk of this nature might be of some use.

The deficiencies of perturbative string theory have been expounded upon in countless papers over the last few years. Even the most elegant formulation of string perturbation theory in terms of conformal field theory/sigma model ideas cannot deal with the crucial issues of supersymmetry breaking which are necessary to a confrontation of string theory with the real world. On a more conceptual level, we do not know what, if anything, singles out a true quantum ground state from the myriad classical static solutions of superstring theory. Indeed, we do not have any real proof that the different string vacua truly correspond to solutions of the same theory. There is no convincingly crushing reply to those benighted souls who would prefer to think of different vacua as different theories. We cannot be sure that there are not many

---

[2] Actually this is true up to a factor which depends only on the dimension of the subspace W of the one body Hilbert space spanned by the functions $\lambda^k e^U$ with $k \leq N - 1$.

string theories (with the same world sheet gauge algebra) whose field configurations belong to disconnected topological sectors.

String field theory, which was supposed to give nonperturbative answers to many of these questions, has been a dismal failure. All current formulations of string field theory (which, with the exception of Witten's theory for open bosonic strings, are spectacularly ugly) rely on constructions which can only be defined rigorously in perturbation theory around a given vacuum state. They depend heavily on conformal invariance, which is not a sensible concept for off shell string field configurations. If one believes in the slogan according to which the configuration space of strings is "the space of all two dimensional field theories", one is practically forced to admit that a proper formulation of string field theory must introduce a world sheet cutoff.

Large N matrix models provide a nonperturbative cutoff formulation of string theory. They have already been demonstrated to be a powerful tool for studying nonperturbative questions about simple classical solutions of bosonic string theory. Among the questions which can be addressed is nonperturbative breaking of a (toy model of)target space supersymmetry. One cannot help being impressed by the difficulty of reproducing these results from a continuum field theory point of view. It seems clear that these models provide our best current hope for obtaining nonperturbative information about string theory.

There is however one sense in which the matrix model formulation is distinctly disappointing. Although completely nonperturbative, it nevertheless seems to be tied to particular perturbative string vacua. The nonperturbative results can be viewed as a particular method of resumming the perturbation series around a given vacuum, and one is led to ask about background independence: where are the other ground states? Shouldn't they be "seen" in a nonperturbative formulation of the theory?

The purpose of this paper is to provide the beginnings of an answer to this question. We will see that the answer has two parts; one topological and the other dynamical. We will argue that in the matrix model formulation, (and probably in any cutoff formulation of string field theory) the space of string configurations breaks up into topological sectors, loosely corresponding to the topology of the base space on which the matrices live. A more precise measure of the topology is perhaps the set of degrees of freedom in the cutoff world sheet theory defined by the matrix model. The topological classification of the space of string field configurations is unlike the topological classification of base manifolds in that it is hierarchical rather than exclusive. To give some simple examples: matrix models on a space of two points include those on a space of one point. Models in higher dimension contain those in lower dimension. A choice of the fundamental set of matrix variables, (and if they live on a continuous manifold, the topology of this manifold) fixes the highest point in the hierarchy that can be studied. This is the primary reason that the nonperturbative solution of matrix models on small target spaces cannot see all the vacua of the bosonic string. A given tree level vacuum of string theory is a continuum field theory. By itself it cannot determine the set of variables in the cutoff theory from which it was defined. This is because of the phenomenon of decoupling: lattice degrees of freedom can disappear in the continuum limit. A full nonperturbative formulation of string theory must specify the full space of cutoff variables. Matrix models on small target spaces keep only a few degrees of freedom of the full string theory, and vacua which require more degrees of freedom for their description cannot be seen even in the nonperturbative solutions of these models. In the last section of this paper we will describe a model which allows us to access all possible topological classes, and thus all possible perturbative vacua of the bosonic string.One might argue that only this model is the "true" nonperturbative formulation of bosonic string theory.Alternatively we must make a choice of topological sector, and define nonperturbative string theory in terms of a matrix model whose degrees of freedom do not allow us to describe all classical string vacua. In this hierarchical sense there are many different topological classes of nonperturbative string theory. Perhaps only some of them are well defined or unitary.[3]

The matrix model formulation of string dynamics is somewhat remote from previous attempts to understand string theory in terms of spacetime equations of motion. I would now like to discuss a reformulation of matrix model dynamics which clarifies the relation to previous approaches. The loop equations of the matrix model[4] can be reformulated as differential equations for the partition function, viewed as a function of all possible local[4] $U(N)$ invariant couplings of the given matrix degrees of freedom. These equations provide an alternative definition of the model. They strongly resemble the Schwinger Dyson equations of a nonlocal field theory containing an infinite number of multilocal fields $\Phi_n(x_1, t_1; ...; x_n, t_n)$ and we can write a formal solution of the equations in terms of a path integral over these fields.

The spacetime on which our fields are defined is quite novel. Spatial hypersurfaces are just the space on which the original matrices live, while the time component may be identified with the two dimensional lengths of loops. The source conjugate to Φ_n is a coupling in the matrix model to an operator which creates n boundaries on the world sheet with world sheet lengths $t_1, ..., t_n$ (in lattice units). The nth boundary is localized at the point x_n in embedding space.

The idea that the intrinsic size of the spatial universe should be viewed as a time variable is as old as quantum gravity[5]. In string theory the idea that the conformal factor should be identified with time in the embedding space was proposed by a number of authors[6], and the relation between these two points of view was discussed

---

[3] Of course we are assuming here that string theory is defined in terms of *some* matrix model. D. Gross and A. Migdal have emphasized that there is no particular reason to make this assumption. However, since there is no other nonperturbative definition of string theory at present, I feel that we have to make do with what is available.

[4] In the sense of world sheet locality. In matrix language this means that the action should involve only a single trace of some function of the matrices.

extensively in[7]. Here we find the relation between two dimensional geometry and external time flowing automatically from the formalism of matrix models: n-tuples of points in space-worldvolume(time) are the natural labels for the complete set of invariant couplings available to a given set of matrix degrees of freedom.

A byproduct of this point of view is a further clue to understanding the difference betweeen gauge theory strings and the string theories described by the Feynman diagrams of large N scalar field theories. In gauge theories the length of a string or Wilson loop is completely tied to the length of a particular path in the embedding space. In contrast, in scalar models, the loop length is an independent variable. Thus gauge theory strings live in the embedding spacetime of the gauge theory, while matrix model strings live in a spacetime of one higher dimension. There is no Liouville coordinate for gauge theory strings.

Our reformulation of matrix models as string field theory contains a surprise for those who expected to see string theory formulated as an integral over the couplings of two dimensional field theories. In our formulation, the matrix model couplings (which are in one to one correspondence with couplings on the cutoff world sheet) are sources for the dynamical fields Φ_n. Although I have only an incomplete understanding of this point I believe it has something to do with our confusion (in the continuum) of bare and renormalized world sheet couplings. Indeed, in the Koba-Nielsen formula for the generating functional for the scattering matrix, the bare couplings to vertex operators act as (on shell) sources: derivatives with respect to them generate the S-matrix which is a *connected* rather than a one particle irreducible Green's function.[5] On the other hand, renormalized couplings are solutions to the renormalization group equations and should clearly be thought of as fields rather than sources. In our cutoff formalism the distinction between field and source is somehow more clear.

The distinction between bare and renormalized couplings is the key to the dynamical part of the question of background independence. At first glance it would appear that if bare matrix model (= world sheet) couplings are freely specifiable sources in string field theory, then the theory does not single out a particular string vacuum state. The resolution of this difficulty is connected with the notion of taking the worldsheet continuum limit of a matrix model. We must tune the bare couplings to a point where the discrete surfaces of large N matrix perturbation theory become continuous. At tree level (leading order in the $\frac{1}{N}$ expansion) this tuning process can be accomplished in many ways, corresponding to possible conformally invariant quantum

---

[5] That these are indeed bare couplings follows from the fact that ultraviolet divergences are not subtracted out of the Koba Nielsen amplitudes. They correspond to internal on shell lines. One should also note that the couplings are more properly viewed as *asymptotic fields* than as sources. On shell sources vanish everywhere except at space time infinity. Asymptotic fields are the solutions of free field equations in the presence of these sources. The point of view expressed in this paragraph was developed in discussions with E. Martinec.

field theories that can be constructed by combining the Liouville field (whose emergence in matrix models is as yet only partially understood) and the available lattice degrees of freedom.

Most of these tree level solutions to the problem of taking the continuum limit are not satisfactory. For example, if one has a matrix field theory in 25 space dimensions, then one of the tree level solutions is the critical bosonic string in flat spacetime. As is well known, the perturbative tachyon and dilaton loop divergences in this theory violate conformal invariance. Thus the problem of an exact continuum limit is not necessarily solved in this vacuum state of the bosonic string. The dilaton divergences can perhaps be renormalized by the Fischler-Susskind mechanism[8], but this makes the existence of the continuum limit in a nonperturbative sense a key unanswered question.

The exactly soluble $c < 1$ matrix models provide an illuminating example of this sort of phenomenon. Here, there are many models whose continuum limits exist in each order of perturbation theory[6] However, all of the unitary[7] models of this sort suffer from a nonperturbative instability[9]. Their nonperturbative continuum limits do not exist.

The tachyon divergences (which Seiberg has argued will exist for any bosonic string vacuum with an infinite number of physical states) can not even be treated in this way. In Seiberg's interpretation[10] tachyon couplings in the tree level world sheet Lagrangian (which would be required in an implementation of the Fischler Susskind mechanism for tachyons) do not correspond to local operators. In the presence of such couplings the world sheet degenerates into branched polymers and has no smooth surface limit.

Thus it is our contention that background independence in string theory may be achieved in the nonperturbative continuum limit. The plethora of tree level string vacua is an illusion. Indeed it is even reasonable to suggest that there are *no* nonperturbative unitary continuum limits of bosonic string theory. As emphasized above,the results of Seiberg[10]show that all classical vacua of the bosonic string with an infinite number of physical states have tachyons, and thus suffer from the branched polymer instability. The only stable classical solutions are the minimal models, and all the unitary minimal theories suffer from a nonperturbative quantum instability. If this is indeed the case, then one is led to conjecture that a supersymmetric version of the matrix models might have a unique unitary nonperturbative continuum limit, thus achieving background independence. There are many obstacles remaining in the

---

[6] Perturbatively one seems to get a consistent continuum quantum theory for every classical vacuum in the $c < 1$ hierarchy.

[7] Here and throughout this talk, I will use the word unitary in the sense in which it is used in critical string theory. A model is unitary if (after gauge fixing) it can be written as a unitary field theory coupled to the Liouville field.

search for a super version of matrix models. They will not be overcome in the present paper.

It should be emphasized that the reformulations of matrix models that I discuss here have so far been of little use in obtaining analytic solutions of previously unsolved models. In low dimensional target spaces the loop equations provide an efficient method for solving string models. They generate the $\frac{1}{N}$ expansion in a simple way. This is not the case in higher dimensions. Even the large N limit of models with more than one spatial coordinate is described by a two dimensional field theory with an infinite number of gauge invariant degrees of freedom. We should not be surprised that it is difficult to solve. Rather than an analytical tool, I view the loop equations for general models as an analogue of the lattice formulation of gauge theories. They provide a regularized nonperturbative description of string theory which may be amenable to numerical or renormalization group techniques. They emphasive the spacetime physics of the string theory and provide tentative answers to conceptual questions (e.g. background independence). They represent a program rather than a solution.

Before concluding with the introduction I should note that the primary motivation for this work was the discovery by E. Verlinde, H. Verlinde, and R. Dijkgraaf[11] of a set of coupling constant differential equations for the integrable continuum limits of a large class of matrix models. In an attempt to understand the lattice origin of these equations, I realized that the matrix model loop equations could be rewritten in a form remarkably similar to those of[11].[8] The generalization to more complicated matrix models was then obvious. The precise correspondence between the equations of[11]and the continuum limit of the loop equation has been explored by Fukuma and collaborators[12].

2. Loop Equations and String Field Theory

We will start by deriving the loop equations for simple one-matrix models.[4]We consider a theory with partition function of the form

$$Z = \int dM e^{S} \tag{2.1}$$

with $S = N \sum_n Tr J_n (\frac{M}{\sqrt{N}})^n$. Define an orthogonal basis in the space of Hermitian N x N matrices : $Tr \lambda_a \lambda_b = \delta_{ab}$, and write any matrix as $M = \sum \Lambda_a \lambda_a$ The integral over M is then just an integral over the coefficients Λ_a. Note also that the basis satisfies the completeness relation $(\lambda_a)_{ij}(\lambda_a)_{kl} = \delta_{il}\delta_{jk}$.

---

[8] E. Brezin has informed me that he and Kazakov wrote down this form of the loop equations some time ago in an attempt to find the continuum limit of the matrix models.

Now write the identity

$$\int d\Lambda_a \partial_{\Lambda_a} (Tr\lambda_a(\frac{M}{\sqrt{N}})^p e^S) \tag{2.2}$$

Carrying out the differentiation we get

$$0 = \int \sum_{r=0}^{p-1} Tr(\lambda_a(\frac{M}{\sqrt{N}})^r \lambda_a(\frac{M}{\sqrt{N}})^{p-r-1})e^S + N \int Tr\lambda_a(\frac{M}{\sqrt{N}})^p Tr\lambda_a \sum n J_n(\frac{M}{\sqrt{N}})^{n-1} e^S \tag{2.3}$$

Using the completeness relation we see that both terms involve only traces of powers of M, so we can write them as derivatives with respect to the couplings J_n.

$$0 = \sum_{n=o}^{\infty} n J_n \frac{\partial Z}{\partial J_{n+p-1}} + 1/N^2 \sum_{r=0}^{p-1} \frac{\partial^2 Z}{\partial J_r \partial J_{p-r-1}} \tag{2.4}$$

It is a remarkable fact that the differential operators appearing in these equations satisfy the Virasoro algebra. This was first noticed by E. and H. Verlinde[11]in their derivation of the continuum version of these equations from topological field theory. Y. Matsuo[13] has pointed out the origin of this algebra in the single matrix model. Instead of considering the loop equations as describing the effect of arbitrary local $U(N)$ invariant perturbations of our matrix model, we can instead view the perturbations as arising from a change of variables in the matrix functional integral. The set of $U(N)$ covariant nonsingular, infinitesimal transformations of the matrix is simply $M \rightarrow M + F(M)$, where F is an analytic function. The algebra of these transformations is generated by the Virasoro operators $L_{-1}, L_0, L_1, L_2 \ldots$.

The source J_0 couples to the identity operator. It is convenient to write its contribution to these equations explicitly using the obvious identity $\frac{\partial Z}{\partial J_0} = N^2 Z$. In terms of the free energy defined by $F = \frac{1}{N^2} lnZ$, we obtain

$$\sum_{n=1}^{\infty} n J_n \frac{\partial F}{\partial J_n} + 1 = 0 \tag{2.5}$$

$$\sum_{n=1}^{\infty} (n+1) J_{n+1} \frac{\partial F}{\partial J_n} + J_1 = 0 \tag{2.6}$$

$$\sum_{n=1}^{\infty} n J_n \frac{\partial F}{\partial J_{n+q}} + 2 \frac{\partial F}{\partial J_q} + \sum_{r=1}^{q-1}(\frac{\partial F}{\partial J_r} \frac{\partial F}{\partial J_{q-r}} + \frac{1}{N^2} \frac{\partial^2 F}{\partial J_r \partial J_{q-r}}) \tag{2.7}$$

The last of these equations is actually an infinite sequence of equations, one for each positive integer q.

These equations may be viewed as world sheet Schwinger Dyson equations, which express local relations between world sheet operators . In the language of the renormalization group, they allow one to express redundant operators in terms of a complete set of local physical operators. The novelty of these relations from the worldsheet point of view is that they relate Riemann surfaces of different genera. In this they resemble the Fischler-Susskind[14] equations of critical string theory. The derivation of the continuum analogue of these relations[11]requires a great deal of clever manipulation, while in the matrix models they follow from the usual machinery of path integrals.

To see more explicitly that these equations are indeed world sheet equations of motion, take all the J_k for $k > k_0$ equal to zero. Then the second equation allows us to eliminate the operator represented by $tr M^{k_0-1}$ in terms of lower order operators, the first equation allows us to eliminate $tr M^{k_0}$, and the qth equation in the sequence allows us to eliminate $tr M^{k_0+q}$. If we write each of these operators as a formal sum over points on a discrete world sheet of fixed genus, then the linear first derivative terms in the equations are local relations. The other terms in the equation exhibit the nonlocality characteristic of wormhole interactions.

In fact, the loop equations determine all correlation functions of all operators, at least order by order in the $\frac{1}{N}$ expansion. To see this note that in each order in $\frac{1}{N}$, the free energy is an analytic function of all the couplings in the neighborhood of the Gaussian point where only J_2 is nonzero. The equations then become recursion relations which determine all the coefficients of the power series expansion of this function. Note however, that for any model with only a finite number of nonzero couplings, most of the operators are determined simply and locally on the world sheet (up to wormhole effects) in terms of the first few. The equations for this finite set of operators are difficult to solve except in perturbation theory. These are the "fundamental local field operators of the model".

A general formal solution of the loop equations may be obtained, as in ordinary field theory by Fourier transforming the linear equations for the partition function. These equations are partial differential equations whose coefficients depend only linearly on the variables J_n. Therefore the Fourier transformed equations are first order partial differential equations whose formal solution may be immediately written down.

There are however several subtleties in the derivation of these equations. The original matrix model integrals are well defined only for a restricted range of couplings. For example, if there are only a finite number of nonzero couplings, the largest one must be even and negative. We can define the partition function and correlation functions by analytic continuation for other values of the couplings. The question now is whether one can find a region in complex coupling space such that the partition function can be written as a Fourier transform

$$Z(J_n) = \int d\Phi_n \quad e^{N^2 \Sigma J_n \Phi_n} e^{N^2 S(\Phi)} \tag{2.8}$$

In addition one must require that the contour in Φ space be such that integration by parts does not produce any surface terms. We will assume that such a contour can be found. In this case, the differential equations for Z can be rewritten as constraints on S:

$$\frac{\partial S}{\partial \Phi_1} + \sum_{n=1}^{\infty} (n+1)\Phi_n \frac{\partial S}{\partial \Phi_{n+1}} = 0 \tag{2.9}$$

$$1 = \sum_{n=1}^{\infty} n \left(\frac{\partial S}{\partial \Phi_n} + \frac{1}{N^2} \right) \tag{2.10}$$

$$2\Phi_q + \sum_{r=1}^{q-1} \Phi_r \Phi_{q-r} = \sum_{n=1}^{\infty} n \Phi_{n+q} \frac{\partial S}{\partial \Phi_n} \tag{2.11}$$

These equations have the form

$$G^{ij}(\Phi) \frac{\partial S}{\partial \Phi^j} = D_j^i \Phi^j + C_{jk}^i \Phi^j \Phi^k \tag{2.12}$$

and bear a striking formal resemblance to the Wilson renormalization group equations proposed in[15], with G_{ij} playing the role of the Zamolodchikov metric. It remains to be seen whether this is more than a formal analogy.

A solution of the equations may be written down in the following way: choose a point Φ^* for which $G^{ij}(\Phi^*)$ is invertible. Let $\Phi(s)$ be any function for which $\Phi(0) = \Phi^*$ and $\Phi(1) = \Phi$.[9] Then

$$S[\Phi] = \int_0^1 ds \frac{d\Phi_i}{ds} G_{ij}(\Phi(s))[D_j^i \Phi^j(s) + C_{jk}^i \Phi^j(s) \Phi^k(s)] \tag{2.13}$$

These formulae are far from pretty, but this was probably inevitable. The genus g coefficients of closed string perturbation theory diverge like $2g!$, faster than the loop expansion of any field theory with a nonsingular Lagrangian. All attempts to find a closed string field theory for critical strings[16] have led to non polynomial Lagrangians which must be improved at each order in the genus expansion in order to cover moduli space. Our action for the Φ_n has a somewhat different disease. It has a zeroth order term which is the solution of

$$\frac{\partial S_0}{\partial \Phi_1} + \sum_{n=1}^{\infty} (n+1) \Phi_n \frac{\partial S_0}{\partial \Phi_{n+1}} = 0 \tag{2.14}$$

$$1 = \sum_{n=1}^{\infty} n \frac{\partial S_0}{\partial \Phi_n} \tag{2.15}$$

$$2\Phi_q + \sum_{r=1}^{q-1} \Phi_r \Phi_{q-r} = \sum_{n=1}^{\infty} n \Phi_{n+q} \frac{\partial S_0}{\partial \Phi_n} \tag{2.16}$$

and a term of order $\frac{1}{N^2}$ which satisfies the singular equations

$$\frac{\partial S_1}{\partial \Phi_1} + \sum_{n=1}^{\infty} (n+1) \Phi_n \frac{\partial S_1}{\partial \Phi_{n+1}} = 0 \tag{2.17}$$

---

[9] We should also choose this trajectory such that the metric is invertible all along it.

226

$$-\sum_{n=1}^{\infty} n = \sum_{n=1}^{\infty} n \frac{\partial S_1}{\partial \Phi_n} \tag{2.18}$$

$$0 = \sum_{n=1}^{\infty} n \Phi_{n+q} \frac{\partial S_1}{\partial \Phi_n} \tag{2.19}$$

In order to make sense of this equation we must cut it off at large loop lengths (n). Now, consider a classical solution $\Phi_c(0)$ of the zeroth order action. There may well be small perturbations (nominally of order $\frac{1}{N}$) of this solution which are more singular than it at large loop size. Because of the singular second term in the action they can give a finite contribution to the integral. As emphasized by Shenker[17] such configurations are precisely what is needed to give e^{-N} contributions to the partition function and a $2g!$ divergence of perturbation theory.

It is easy to see that the solutions of the classical equations of motion of our field theory reproduce the standard results of large N perturbation theory. We can write them down even though it is difficult to get an explicit expression for the action. Simply substitute the equation $\frac{\partial S}{\partial \Phi_n} = -J_n$ into the defining relations for S to obtain:

$$\sum n J_n \Phi_{n+q} = 2\Phi_q + \sum \Phi_r \Phi_{q-r} \tag{2.20}$$

If the Φ_n are written as the moments $\int d\lambda \rho(\lambda)$ of an eigenvalue density, then these equations are easily seen to be equivalent to the standard large N equation for the eigenvalue density.

These considerations can be extended to matrix models on a spatial manifold of any dimension d. We must simply include all possible invariant couplings in the matrix Lagrangian. These have the form

$$\int dx^1 \ldots dx^n \quad J_n(t^1, x^1 \ldots t^n, x^n) \quad Tr M^{t^1}(x^1) \ldots M^{t^n}(x^n) \tag{2.21}$$

where the J_n are cyclically symmetric functions of their arguments. The J_n may then be thought of as multilocal sources in a $d+1$ dimensional space time whose "time" coordinate is discrete.[10] If we take J_n to transform like a tensor product of volume forms and M like a scalar, then all of our equations will be covariant under spatial diffeomorphisms, but not under coordinate transformations that mix up the space and time directions. Matrix models appear to have chosen a sort of synchronous coordinate system in spacetime.

It should be emphasized that although this interpretation of loop length as an extra coordinate in the embedding space is consistent with the idea that the Liouville

---

[10] It is not clear from this description what the signature of the spacetime is or whether the discrete coordinate should be considered a timelike or spacelike variable. We call it time only because of previous identifications of the conformal factor as a time coordinate.

field in string theory be identified with time, its derivation is completely independent of that idea. We have written the single matrix model partition function as an expectation value in a field theory. When we attempt to do the same for matrix models on a d dimensional space, the field theory clearly lives in d+1 dimensions. The emergence of an extra embedding space dimension is seen from a completely different point of view in the work of Das and Jevicki, and Polchinski[18] on the so called d=1 string. It seems like an inescapable feature of the matrix model approach to string theory.

It is probably important to note that the emergence of the Liouville coordinate is a feature of scalar large N matrix models that will not be found in gauge theories. The invariants of gauge theories are Wilson loops which are tied to paths in the embedding space. There is no room for an extra length coordinate, the length of a path being specified already by the *a priori* metric on the embedding space. This may be part of the reason for the difference between gauge theory strings and Polyakov strings.

The awkwardness of having a discrete time variable and continuous space variables can be formally overcome by introducing the generating function of the matrix powers, e^{LM} where L is a continuous variable. General couplings then have the form

$$\int dL_i \quad d^D x_i \quad J(L_1, x_1; \ldots; L_n, x_n) \quad Tr e^{L_1 M(x_1)} \ldots e^{L_n M(x_n)} \tag{2.22}$$

Note that although both space and time coordinates are now continuous, L still plays a distinct role. Another important difference between the time and space coordinates has to do with spacetime locality. Although it is certainly incorrect and unnecessary to insist that the matrix field theory be microscopically local[11] we should probably choose the coupling functions to fall off faster than any power of spatial separation. It is not clear whether a similar restriction should be placed on the behavior of the J's as a function of either the discrete or continuous time coordinate.

3. THE CONTINUUM LIMIT

We have reformulated arbitrary matrix models in terms of a set of differential equations which may be formally solved by a field theoretic path integral. The connection of this formalism to conventional string theory is far from obvious. In order to see this connection we must take the continuum limit of our equations. In principle, this can be done by considering loop correlation functions. Questions about the continuum limit are equivalent to questions about the behavior of loop correlators for large loops in a theory with fixed cutoff. At the moment, the only systematic method available to study the continuum limit is based on the $\frac{1}{N}$ expansion and the

---

[11] If we want to obtain the standard sort of Gaussian string action on the discretized world sheet, we must certainly choose a nonlocal propagator for the matrix field theory. The real point is that, within limits, the precise degree of spacetime locality is *irrelevant* in the world sheet continuum limit.

interpretation of the terms in this expansion as two dimensional field theories on a random lattice. All known soluble examples can be equivalently formulated in terms of the continuum Liouville theory coupled to a two dimensional field theory in a generally covariant way (i.e. the full coupled field theory is a conformal field theory with central charge 26), and it is plausible that this is the general case. Thus we have a qualitative understanding of the continuum limit of random surfaces with the topology of a sphere. In the matrix model formulation of the theory, the loop equations connect spherical surfaces with surfaces of higher topology. Thus a "classical string background" is only the first approximation to the continuum limit of a solution of the loop equations. As in ordinary field theory, we have not really found the continuum limit until we have a proof that it exists nonperturbatively in the loop expansion.

The crucial missing feature of the loop equation approach to string theory is a formulation of the renormalization group in terms of the loop equations themselves. At present our only handle on the continuum limit of the loop equations is via the Feynman diagram expansion of matrix models and its connection with two dimensional random lattice field theory. This renormalization scheme is intrinsically perturbative. If this approach to string theory is to be successful a nonperturbative method of renormalizing the loop equations must be found. At present I can only report that several attempts fo find such a scheme have ended in failure.

One question that will have to be answered in the continuum limit is the fate of the "angle variables" of the matrix model. The loop equations of multimatrix models involve of order N^2 independent variables. However, for minimal models Douglas[19] has shown that one can understand all of the continuum operators in terms of $o(N)$ lattice operators. The angular degrees of freedom seem to become redundant in the continuum limit. Gross and Klebanov[20] have made similar claims about the compact $c = 1$ theory. If this decoupling of the angle variables is a general phenomenon then it may be possible to obtain a more tractable form of the loop equations near the continuum limit.

Finally, let us note the intriguing fact that the continuum limit of string theory can be thought of as a large time limit in the embedding space if we believe in the correspondence between loop length and time. Is it possible that this formal connection has anything to do with cosmology? Should we view the cutoff on the string worldsheet as finite in the real world, and negligible only at "late times"?

4. THE TOPOLOGY OF STRING FIELD SPACE

We have seen that a general matrix model may be written as a string field theory involving an infinite number of multilocal spacetime fields. It is clear that all models involving the same number of matrix variables may be continuously connected to each other by varying the sources in this field theory. In the continuum or long time limit, correlation functions will become independent of the sources or depend on only a few relevant couplings. In this limit we may also relate a matrix model with a given number of variables to one with fewer variables. In lowest order in the $\frac{1}{N}$ expansion the continuum limit is describable by a conformal field theory whose fields take values in the space on which the matrix variables live, extended by the Liouville mode. However, some of these fields may decouple at the fixed point because mass terms or other relevant couplings go to infinity. The standard procedure in critical string theory is to take the continuum limit at tree level. Fischler-Susskind divergences invalidate this procedure, but even more seriously, it may completely miss degrees of freedom. As an example, one may consider $c < 1$ strings. At tree level, these may be considered as Euclidean time dependent classical solutions of critical string theory. To do so one must consider them to be obtained by letting most of the two dimensional scalar fields of the bosonic string become massive and decouple, while one of them has a multicritical Landau Ginzburg potential. The standard matrix model perturbation series around these solutions is not the same as the one generated by thinking of them as solutions of the full critical string. The latter series would contain contributions from the degrees of freedom which decouple in the tree level continuum limit. Certainly one cannot expect to access the real solution of bosonic string theory without taking them into account.[12]

A completely controllable example of this phenomenon occurs within the context of integrable models. There are fixed points of the two matrix models which are completely equivalent to the one matrix model. Thus, in the language of critical string theory, pure two dimensional gravity coupled to the Lee-Yang edge field theory[13] is a classical solution of two matrix string field theory. There is no way to recover the two matrix solution from the exact continuum quantum mechanics of the one matrix problem, or from the cutoff theory with only one matrix. This is related to a sort of failure of decoupling theorems for theories in which we sum over the geometry and topology of space time. In ordinary field theory, the effect of heavy degrees of freedom can be incorporated in a series of local counterterms in the Lagrangian of the light degrees of freedom. In principle, accompanied by an appropriate definition of the contact terms in Green's functions of these nonrenormalizable operators (and possibly a prescription for resumming the series), this series allows us to recover the effects of the missing degrees of freedom. In this sense, one might by a stretch of the imagination think that string solutions in which a larger number of degrees of freedom are excited could be encoded in a choice of couplings in the one matrix model. In theories with sums over topology however, heavy degrees of freedom propagating in closed loops in spacetime can induce nonlocal couplings between light degrees of freedom. Any attempt to produce a one matrix model from a multimatrix model by integrating out

---

[12] It is possible to view this argument as evidence that the $c < 1$ models are *not* classical solutions of critical string theory. This should certainly be taken seriously but I will ignore it in what follows.

[13] We choose this example to avoid nonperturbative instabilities.

230

some subset of degrees of freedom[14] leads to terms of the form $Tr M^k Tr M^p$ etc. in the effective action. In world sheet terms, these are nonlocal couplings. It is in fact quite remarkable that elimination of the heavy degrees of freedom in flow between different multicritical points of the single matrix models does not produce such nonlocal effects. This I take as evidence for Polyakov's suggestion[21] that all of these theories can be described by the Liouville field with some higher derivative action. They have different numbers of particles but the same set of world sheet fields. Indeed, in any field theory with topology change, integrating out short distance degrees of freedom leads to nonlocal effective actions. This procedure does not however produce the most general possible nonlocal action, for a given set of fields we will only get a restricted class of actions. If all fields in the short distance theory are still present in the long distance theory, then essentially by construction, all the nonlocalities can be absorbed into the bare couplings of those fields. Thus in a theory with topology change, a flow between fixed points which does not induce nonlocality is evidence that the fixed points are described by the same set of local degrees of freedom.

Thus we see that the space of possible string fields (taken to be the dual space of the space of all possible couplings in all possible matrix models) has a sort of hierarchical topology determined (to lowest order in the $\frac{1}{N}$ expansion) by the renormalization group of two dimensional field theories. A lattice field theory with a given set of variables can flow in the continuum limit to one with fewer variables. Thus two such theories may be said to be continuously connected. But once we fix the set of two dimensional field variables (and the topology of the space they live in if they are continuous variables) we have cut ourselves off from all theories with more variables or a different topology. This restriction on the accessible string vacua is present both in the matrix formulation or in the formulation in terms of field theories on dynamically triangulated random surfaces. Thus it is not an artifact of the string loop expansion but exists nonperturbatively.

Consider for example theories of r lattice scalar fields which take values on a torus. Among the fixed points of the renormalization group on this space of theories are models which are equivalent to the $k = 1$ Wess-Zumino-Witten model for a simply laced group of rank r. However, this is a model which can also be written in terms of a number of scalar fields equal to the dimension of the group. And the renormalization group in this larger space of fields contains fixed points which cannot be accessed in the smaller theory space.

Any definition of the space of all two dimensional field theories in terms of a cutoff must cope with this topological question. And the whole question of spacetime topology change in string theory depends on the answer to it. It was first emphasized by V. Kaplunovsky that if we allow an arbitrary number of fields in our two dimensional field theory, topology change must occur. Any two manifolds can be embedded

---

[14] It should be said that it is not at all clear how to eliminate precisely a particular set of *world sheet degrees of freedom* by partial integration over matrices.

in a Euclidean space of high enough dimension, and in the renormalization group of theories whose target space is that Euclidean space, the fixed points corresponding to hypothetical conformally invariant sigma models on the two original spaces are continuously connected.

Is there an embedding theorem, similar to that for Riemannian manifolds, which states that any conformal field theory with central charge 26 can be found as the fixed point of a lattice theory with some maximal number of degrees of freedom? (The answer to this appears to be no, at least if we accept the idea that theories with Chodos-Thorn-Feigen-Fuks[22] terms should be accepted as time dependent solutions to string theory. These theories can have $c = 26$ and an arbitrarily large number of scalar fields, and so far as I know cannot be written in terms of a smaller number of degrees of freedom.However, it is not clear whether these are really acceptable string solutions.) If there is no finite dimensional embedding theory, a fundamental formulation of string theory would have to involve infinite component two dimensional fields.

Fortunately, there is a simpler way to access all perturbative string solutions. Klebanov and Susskind[23], and more recently Parisi[24] have shown that string theories with discrete target spaces can have worldsheet continuum limits that are identical with string theories on continuous spacetimes. This is also implicit in Zamolodchikov's rewriting of discrete series models as Landau Ginsburg theories[25] and has been emphasized by Martinec[26]in his discussion of $N = 2$ Landau Ginsburg models. The basic mechanism underlying this phenomenon is the Kosterlitz-Thouless transition[27]. When discrete variables are allowed to have large fluctuations, they can appear continuous. We can use this idea in the following way. Let us consider a matrix model with matrices defined on a one dimensional index set. For every integer we have a matrix $M(n)$. This is just like the infinite version of the matrix chain models of Mehta *et. al.*[1]. However, we now drop the awkward restriction to couplings which fall off with one dimensional separation and allow arbitrary interactions between the matrices. The set of allowed couplings is just $J(t_1, m_1; \ldots; t_n, m_n)$ $TrM^{t_1}(m_1) \ldots M^{t_n}(m_n)$, with no restriction on the J functions except cyclic symmetry and the positivity required in order for the matrix integral to be well defined. With appropriate choices of the coupling functions the $\frac{1}{N}$ expansion of this model will be given by a sum over random two dimensional surfaces embedded in a discrete space of arbitrary dimension and topology. Thus, at tree level, the set of fixed points for the loop equations of this general class of matrix models will include all possible bosonic string vacua.[15]

In the previous sections we advocated interpreting the correlation functions of the matrix models as correlation functions of fields in space time, with the loop length playing the role of time. In the present context, the awkward distinction we

---

[15] Actually, we do not know if it is possible to formulate asymmetric orbifolds, and other conformal field theories whose construction involves chiral fields, as limits of cutoff theories.

232

encountered between discrete time and continuous space disappears. If we now ask for the large time behavior of correlation functions we will be led in the large N approximation to an infinite number of different nontrivial limits, corresponding to conformal field theories (which include the Liouville mode in their space of degrees of freedom).

Our discrete system has fixed points corresponding to all possible theories of scalar fields coupled to gravity. These are obtained by first choosing the couplings J so that the matrix model lives on a lattice of fixed dimension and topology, and then taking the continuum limit from the correct side of the Kosterlitz Thouless point. Thus matrix models with a discrete infinite set of matrices can reproduce all possible perturbative vacua of the bosonic string. Below we will write down an infinite closed system of differential equations, which determines the dependence of the partition function on all the possible invariant couplings in the matrix model. Solutions of these equations which have nontrivial continuum (large time) limits, would represent nonperturbative solutions of bosonic string theory. The perturbative vacuum states are putative starting points for a systematic construction of these solutions.

As emphasized above, most of these vacua will not have sensible continuum perturbation expansions, because they contain tachyons. This is the case for any model with an infinite number of physical states.[10] The remaining discrete series of models is sensible in perturbation theory but all those which correspond to unitary conformal field theories coupled to gravity probably suffer from the nonperturbative instability discovered in[9]. The nonunitary models have only a finite number of physical states.

The problems can be cured only by letting the matrices depend on Grassmann variables. Such matrix models give rise to worldsheet theories whose "Hilbert space" has a fundamental form which is partly skew symmetric. The resulting Grassmann zero modes allow one to circumvent Seiberg's no go theorem[10], and these models have continuum limits at higher genus despite the fact that they have an infinite number of physical states. Their nonperturbative definition is under investigation.[28] It seems that Grassmann variables, if not spacetime supersymmetry, are necessary to the consistent definition of any string theory which has an infinite number of states at tree level. It is possible that spacetime supersymmetry is a necessary requirement for a consistent quantum mechanical interpretation of such Grassmann string theories.

At any rate, one can certainly write down a generalization of the equations we have studied to the case where matrices depend on Grassmann variables. Our spacetime field theory will then contain both commuting and anticommuting fields. There are many puzzles to be resolved, such as how many Grassmann variables to introduce, how to achieve the chiral world sheet variables that are necessary in all sensible superstring vacua, how to understand the connection between unitarity and spacetime supersymmetry etc. But if all these problems can be resolved, we will have a completely nonperturbative formulation of superstring theory, in terms of a set of differential equations and one might hope that the large time (loop length) limit of the solutions of these equations was unique, independent of the bare couplings, and corresponded to the real world.

For reference, let me record the bosonic version of the loop equations which determine all possible perturbative string vacua. Perhaps they will be helpful in the search for a consistent nonperturbative string theory. The basic couplings of the model have the form

$$\sum J_n(t^1, x^1 \ldots t^n, x^n) \quad Tr M^{t^1}(x^1) \ldots M^{t^n}(x^n) \tag{4.1}$$

The loop equations are now second order differential equations for the partition function in terms of these couplings. They are

$$\sum_{y_m; s_m} \Big(\sum_{l=1}^{m} \delta(x - y_l) \sum_{r=0}^{s_l - 1} J(y_1, s_1; \ldots; y_m, s_m)$$

$$\frac{\delta Z}{\delta J(y_1, s_1; \ldots; y_l, r; x_1, t_1; \ldots; x_n, t_n; y_l, s_l - r - 1; \ldots : y_m, t_m)} \Big) \tag{4.2}$$

$$+ \frac{1}{N^2} \sum_{j=1}^{n} \delta(x - x_j) \sum_{r=0}^{t_j - 1} \frac{\delta^2 Z}{\delta J(x_1, t_1; \ldots; x_j, r) \delta J(x_j, t_j - r - 1; \ldots; x_n, t_n)}$$

ACKNOWLEDGEMENTS

I would like to thank my colleagues N. Seiberg and S. Shenker for innumerable discussions about string theory and quantum gravity. I would also like to thank E. Verlinde, H. Verlinde, and R. Dijkgraaf for discussion of their work before publication, and Y. Matsuo for an important conversation. I would like to thank the organizers of the Cargese workshop, O. Alvarez, E. Marinari, P. Windey and Marie-France Hanseler, for a marvelous physics conference. The penetrating comments of J. Zinn-Justin on my presentation were also thoroughly appreciated. Finally I would like to thank E. Martinec for a number of enlightening conversations about perturbative and nonperturbative string theory. This work was supported in part by the United States Department of Energy under grant No. DE - FG05 - 90ER40559.

References

[1] S. Chadha, G. Mahoux, M.L. Mehta, J. Phys. A: Math. Gen. 14, 579, (1981).

[2] G. Segal, G. Wilson, Pub. Math. IHES 61, 5, (1985).

[3] E. Brezin, C. Itzykson, G. Parisi, J.B. Zuber, Comm. Math. Phys. 59, 35, (1978) ; S. Wadia, Cargese Lecture, this volume; H. Sengupta, S. Wadia, Tata preprint, TIFR TH-90-33, July 1990; D. Gross, Cargese Lecture, this volume.

[4] A.A. Migdal, Yu. Makeenko, Phys. Lett. 88B, 135, (1979); D. Bessis, C. Itzykson, J.B. Zuber, Adv. Appl. Math 1, 109, (1980); A.A. Migdal, Physics Reports 102C, 199, (1983).

[5] B.S. deWitt, Phys. Rev. 160, 1113, (1967).

[6] F. David, Mod. Phys. Lett. A3, 1651, (1988); S. Das, R. Naik, S. Wadia, Modern Physics Letters A4, 1033, (1989); J. Polchinski, Nucl. Phys. B324,123, (1989).

[7] T. Banks, *Physicalia* 12, 19, (1990); T. Banks and J. Lykken, Nucl. Phys. B331, 173, (1990).

[8] W. Fischler, L. Susskind, Phys. Lett. B173, 262, (1986).

[9] F. David, Saclay preprint SPhT/90-043, March 1990; M. Douglas, N. Seiberg, S. Shenker, Phys. Lett. B244, 381, (1990).

[10] N. Seiberg, *Notes on Quantum Liouville Theory and Quantum Gravity*, Rutgers preprint RU-90-29, Lectures given at the 1990 Yukawa International Seminar, Common Trends in Mathematics and Quantum Field Theories, and at the Cargese meeting, June 1990, this volume.

[11] E. and H. Verlinde, *A Solution of Topological Two Dimensional Gravity*, IASSNS-HEP-90/40,April 1990 ; R. Dijkgraaf, E. and H. Verlinde, *Loop Equations, Virasoro Constraints, and Nonperturbative Two Dimensional Quantum Gravity*, IASSNS-HEP-90/48, May 1990.

[12] M. Fukuma, H. Kawai, R. Nakayama, UT-562, KEK-TH-251, KEK preprint 90-27, May 1990.

[13] Y. Matsuo, private communication.

[14] W. Fischler, L. Susskind, Phys. Lett. B173, 262, (1986); B171, 383, (1986).

[15] T. Banks, E. Martinec, Nucl. Phys. B294, 733, (1987).

[16] M. Kaku, in *Functional Integration, Geometry and Strings*, Z. Haba and J. Sobczyk, eds. Berlin (1989); T. Kugo, H. Kunitomo and K. Suehiro, Phys. Lett. B226, 48, (1989); M. Saadi and B. Zweibach, Ann. Phys. 192, 213, (1989).

[17] S.Shenker, *The Strength of Nonperturbative Effects in String Theory*, Rutgers preprint RU-90-47, 1990 Cargese Lectures, this volume.

[18] S. Das, A. Jevicki, BROWN HET-750-April 1990; J. Polchinski, UTTG-15, 1990.

[19] M. Douglas, Phys. Lett. B238, 176, (1990).

[20] D. Gross, I. Klebanov, PUPT-90-1172, March 1990.

[21] A. Polyakov, Unpublished.

[22] A. Chodos, C. Thorn, Nucl. Phys. B72, 509, (1974); Feigen, Fuks, unpublished.

[23] I. Klebanov, L. Susskind, Nucl. Phys. B309, 175, (1988).

[24] G. Parisi, Phys. Lett. B238, 213, (1990).

[25] A.B. Zamolodchikov, Yad. Fiz. 46, 1819, (1987).

[26] E. Martinec, *Criticality, Catastrophes and Compactification*, U. Chicago preprint 89-0373, to appear in the Knizhnik memorial volume edited by L. Brink et. al.

[27] J.M. Kosterlitz, D.J. Thouless, J. Phys. C6, 1181, (1973).

[28] T. Banks, N. Seiberg, S. Shenker, *Never to be published.*

MULTIPOINT CORRELATION FUNCTIONS IN ONE-DIMENSIONAL STRING THEORY

Dmitri Boulatov

Laboratoire de Physique
Théorique et Hautes Energies
Université de Paris XI
Bâtiment 211, 91405 Orsay Cedex
France and Cybernetics Council
Academy of Sciences
ul. Vavilova, 117333 Moscow, U.S.S.R.

<u>Abstract</u>

The general closed expression is found for multipoint correlation
functions in the D = 1 matrix model which correspond to the scalar
amplitudes in one-dimensional string theory.

Correlation functions play an important role in quantum field theory.
It is known, for example, that they are connected via the Lagrange
transform to the effective action functional and, hence, contain full
information about the vacuum state of theory. Usually, the calculations
of the complete set of correlation functions is a complicated problem
but, in several simplest cases, it turns out to be solvable. A well-known
example is the famous Koba-Nielsen formula which gives the general
expression for multipoint correlation functions of scalar operators in the
theory of D = 26 bose strings.[1]

For a recent few years, the lattice approach to string theory has been
elaborated by many authors.[2] It often leads to significant simplifications
of analysis making it possible to get answers to questions which sometimes
hardly may be asked in continuous theory. One of the most puzzling features
of string theory is the presence of the gap for dimensions of the target
space from 1 to 25 where the theory probably does not exist in the usual
sense. At the critical dimension D = 1, string theory turns out to be
exactly solvable in the lattice framework.[3] In many respects this case is
its simplest realization.[4] It is reduced to the solution of the Schrödinger
equation for the U(N) rotator[5] :

$$\text{tr}\{ - \frac{1}{2} (\frac{\partial}{\partial \phi})^2 + \frac{1}{2} \phi^2 - \frac{g}{4N} \phi^4 \} \ \psi = N^2 E \psi \tag{1}$$

where ϕ is the $N \times N$ hermitean matrix ; ψ is the wave function. The free
energy of the ensemble of random surfaces is equal to the ground-state
energy per one degree of freedom E_0 (see Ref. 3 for details). After usual
decomposition ϕ into the diagonal $\lambda = \text{diag}(\lambda_1, \lambda_2, \ldots, \lambda_N)$ and the
unitary Ω parts ($\phi = \Omega \lambda \Omega^+$), the equation (1), in the U(N) symmetric sector,

Random Surfaces and Quantum Gravity
Edited by O. Alvarez *et al., Plenum Press, New York, 1991*

takes the form :

$$\sum_{i=1}^{N} \left\{ -\frac{1}{2} \frac{\partial^2}{\partial \lambda_i^2} + \frac{1}{2} \lambda_i^2 - \frac{g}{4N} \lambda_i^2 \right\} \Phi(\lambda) = N^2 E \Phi(\lambda) \tag{2}$$

where $\Phi(\lambda)$ is the completely antisymmetric wave function : $\Phi(\lambda) = \Delta(\lambda)\psi(\lambda)$; $\Delta(\lambda) = \Pi_{i<j}^{N} (\lambda_i - \lambda_j)$. Hence, the problem is reduced to the consideration of the system of N free fermions moving in the inverse W-potential. The $N \to \infty$ limit corresponds to fixation of the spherical topology of surfaces and, at the same time, makes the quasiclassical Thomas-Fermi approximation applicable.

Our aim is to obtain the general expression for multipoint correlators in the framework given by the equation (2). In the $N \to \infty$ limit, we have the equidistant spectrum of energy with all levels below the Fermi one filled by fermions. It is natural to identify this spectrum with the equidistant spectrum of anomalous dimensions in the continuous $D = 1$ theory.[6] But, in the lattice framework, it is impossible to distinguish between operators having the same dimension or introduce tensor operators. Hence, we have to consider universal quantities in order to be able to verify our predictions in the continuous framework. The set of such quantities (the amplitudes as we shall refer to them) can be defined as follows. Let us consider the connected correlation functions :

$$G(t_1, \dots, t_n) = \lim_{N \to \infty} \frac{1}{N} \ll O_1(t_1) \dots O_n(t_n) \gg \tag{3}$$

where $O_i(t_i) = \mathrm{tr}\, \lambda^{m_i}(t_i)$ are the Heisenberg operators ; m_i are integer numbers ; $\ll \dots \gg$ denotes the connected correlator :

$$\ll O_1 O_2 \gg = \langle O_1 O_2 \rangle - \langle O_1 \rangle \langle O_2 \rangle$$

$$\ll O_1 O_2 O_3 \gg = \langle O_1 O_2 O_3 \rangle - \langle O_1 O_2 \rangle \langle O_3 \rangle - \langle O_2 O_3 \rangle \langle O_1 \rangle$$

$$- \langle O_3 O_1 \rangle \langle O_2 \rangle + \langle O_1 \rangle \langle O_2 \rangle \langle O_3 \rangle \tag{4}$$

and so on.

$\langle \dots \rangle = \langle V | \dots | V \rangle$ is the standard vacuum mean value (the average with respect to the ground state of the hamiltonian (2)). $\langle O_1 \dots O_n \rangle = O(N^n)$, but correlators (3) are finite in the $N \to \infty$ limit and can be intepreted as the mean values in the ensemble consisting of the single spherical random surface attached at points t_1, t_2, $\dots$, t_n by operators O_1, O_2, $\dots$, O_n.

In terms of eigenvalues

$$\langle O_1(t_1) \dots O_n(t_n) \rangle = \langle V| \exp(-iHt_1) \sum_i^N \lambda_i^{m_i} \exp(iHt_1) | \psi \rangle$$

$$\sum_\psi \langle \psi | \dots | \psi \rangle \sum_\psi \langle \psi | \exp(-iHt_n) \sum_i^N \lambda_i^{m_i} \exp(iHt_n) | V \rangle \tag{5}$$

where H is the hamiltonian from the radial Schrödinger equation (2). $| \psi \rangle \sum_\psi \langle \psi |$ is the identity expansion in the space of eigenfunctions of (3). Since $\mathrm{tr}\, \lambda^m$ are U(N) invariants, the angular excitations of the hamiltonian (1) do not appear in the sums over intermediate states in (5).

We are interesting in the continuous limit when the coupling constant g in (2) tends to its critical value. In the quasiclassical approximation, the energy spectrum is equidistant and infinite in both up and down directions (if we hold the Fermi energy equal to 0). The vacuum $|V>$ is the state with all levels below the Fermi one filled by fermions. Hence, $\sum_\psi |\psi> <\psi|$ is the sum over all excitations of this one-dimensional Dirac sea. Such an unphysical system appeared because, from the formal point of view, we had the Plank constant depending on the number of fermions N : $\hbar \sim 1/N$, while $N \to \infty$.

Since we have the ensemble of free quasiclassical particles, all information about the puncturing operators $O_i(t_i)$ is contained in the quasiclassical one-particle matrix elements :

$$(\tilde{\lambda}^m)_k = \omega \int_0^{2\pi/\omega} dt \ \exp(-i\omega kt) \ \lambda^m(t) \tag{6}$$

which are the Fourier coefficients of powers of the trajectory of a classical particle moving periodically in the inverse W-potential with the frequency ω. It is convenient to write down the correlators (3) in the form :

$$G(t_1, \ldots, t_n) = \sum_{k_i \in Z} (\tilde{\lambda}^{m_1})_{k_1} \ldots (\tilde{\lambda}^{m_n})_{k_n} \exp(i\omega(k_1 t_1 + \ldots + k_n t_n)) \cdot A_{k_1 \ldots k_n} \tag{7}$$

The amplitudes $A_{k_1 \ldots k_n}$ do not depend on details of the definition of the model and, hence, are the universal quantities we are looking for. The most convenient definition of $A_{k_1 \ldots k_n}$ can be done in terms of the creating a_n^+ and deleting a_n operators :

$$a_n^2 = a_n^{+2} = 0 \quad ; \quad \{a_n, a_m^+\} = \delta_{nm}$$

$$A_{k_1 \ldots k_n} = << : \sum_{j_n \in Z} a_{j_n+k_n}^+ a_{j_n} : \ldots : \sum_{j_1 \in Z} a_{j_1+k_1}^+ a_{j_1} :>> \ . \tag{8}$$

Operators between colons in (8) are assumed to be normally ordered, i.e. they commute with each other. The vacuum is the state with all negative levels being occupied :

$$<V| = <\phi|\ldots a_{-n} \ldots a_{-2} \ a_{-1} \ ; \ |V> = a_{-1}^+ \ a_{-2}^+ \ldots a_{-n}^+ \ldots |\emptyset> \tag{9}$$

where $|\emptyset>$ is the "empty" vacuum. The sequence of indices k_i in (8) assumes the strict time ordering in (7) :

$$t_1 < t_2 < \ldots < t_{n-1} < t_n \ . \tag{10}$$

In order to drop this constraint, one has to sum over all permutations of the points t_i multiplying each term by the product of corresponding θ-functions. But, in what follows, the time order (10) will be implied.

Introducing the formal generating functions

$$a^+(Z) = \sum_{n \in Z} a_n^+ \ \bar{Z}^n \qquad a(Z) = \sum_{n \in Z} a_n \ Z^n$$

$$Z = \exp(it) \quad ; \quad \bar{Z} = \exp(-it) \ ; \tag{11}$$

we can write down the generating function for the amplitudes (8) as follows :

$$(Z_1, \ldots, Z_n) = <<:a^+(Z_n)a(Z_n): \ldots :a^+(Z_1)a(Z_1):>> \qquad (12)$$

$$A_{k_1 \ldots k_n} = \oint_{c_n} \frac{dZ_n}{2\pi i} \ldots \oint_{c_1} \frac{dZ_1}{2\pi i} \quad (Z_1, \ldots, Z_n) \; Z_1^{k_1-1} \ldots Z_n^{k_n-1} \qquad (13)$$

The paths of integration in (13) are concentric circles around the center-of-coordinate point ordered in the radial direction in accordance with the time ordering : $|Z_1| > |Z_2| > \ldots > |Z_n|$, provided $t_1 < t_2 < \ldots < t_n$.

The colons imply that

$$<V|:a^+(Z) \; a(Z): \; | V> = 0 \quad . \qquad (14)$$

In the original matrix model (1), the mean value

$$<V|tr \; \phi^m|V> \neq 0 \qquad (15)$$

for even m, but in the continuous limit, we are interesting in, the mean value (15) cannot be made finite in a selfconsistent way and, hence, the condition (14) implies the subtraction of infinite terms from all answers.

Actually, the integral representation (13) can be reduced to the following simple form (see the proof in Ref. 7) :

$$A_{k_1 \ldots k_n} = 2^{n-2} \oint_{c_n} \frac{dZ_n}{2\pi i} \ldots \oint_{c_1} \frac{dZ_1}{2\pi i} \; \frac{Z_1^{k_1} Z_2^{k_2} \ldots Z_n^{k_n}}{(Z_1-Z_2)(Z_2-Z_3) \ldots (Z_{n-1}-Z_n)(Z_1-Z_n)} .$$
$$(16)$$

After integration over Z_1, one can obtain the recurrence relation for the amplitudes :

$$A_{k_1,k_2,\ldots k_n} = 2\Theta(k_1)(A_{k_1+k_2,k_2,\ldots,k_n} - A_{k_2,\ldots,k_{n-1},k_1+k_n}) \qquad (17)$$

where $\Theta(x)$ is the step function : $\Theta(x) = 1$, if $x > 0$; $\Theta(x) = 0$, if $x \leq 0$. On the right hand side of Eqs. (17), the number of the arguments less than on the left hand side by 1. It is not difficult now using (16) and (17) to write down the expressions for several first amplitudes* :

$$A_{k_1} = 0 \quad ; \quad A_{k_1,k_2} = k_1 \Theta(k_1) \; \delta_{k_1,k_2}$$

$$A_{k_1,k_2,k_3} = 2\Theta(k_1) \left[(k_1+k_2) \Theta(k_1+k_2) - k_2 \Theta(k_2) \right] \delta_{k_1+k_2+k_3,0} =$$
$$= 2 \min(|k_1|, |k_3|) \; \Theta(k_1) \; \delta_{k_1+k_2+k_3,0}$$

$$A_{k_1,k_2,k_3,k_4} = 4\Theta(k_1) \left[(k_1+k_2+k_3) \Theta(k_1+k_2)\Theta(k_1+k_2+k_3) - (k_2+k_3) \right.$$
$$\left. \Theta(k_2) \; \Theta(k_2+k_3) + k_3 \Theta(k_3) (\Theta(k_2) - \Theta(k_1+k_2)) \right] \delta_{k_1+k_2+k_3+k_4,0}$$
$$(18)$$

There are the following general relations :

$$A_{k_1,\ldots,k_n} \geq 0 \quad ; \quad A_{\lambda k_1,\lambda k_2,\ldots,\lambda k_n} = \lambda \; A_{k_1,k_2,\ldots,k_n} \qquad (19)$$

---

* The two point correlation function was first calculated in Ref. 8.

$$A_{0,k_2, \ldots , k_n} = A_{k_1, \ldots , k_{n-1},0} = 0$$
$$A_{k_1,\ldots k_{i-1},0,k_{i+1},\ldots k_n} = 2A_{k_1,\ldots,k_{i-1},k_{i+1},\ldots,k_n} \cdot \tag{20}$$

The equations (2) have the following physical meaning. The condition $k_i = 0$ corresponds to integration over all positions of the i-th point which is equivalent, up to a numerical factor, to the amplitude with the less number of attached points.

In the other limit, when $k_1 \to + \infty$,

$$A_{k_1,k_2,\ldots,k_{n-1},k_n} \underset{k_1 \to \infty}{\sim} 2^{n-2} A_{k_1,-k_1} \tag{21}$$

and, when $k_i \geq \max(|k_1|, |k_2|, \ldots, |k_n|)$,

$$A_{k_1,\ldots,k_i,k_n} = 2\, A_{k_1,\ldots,k_{i-1},k_{i+1},\ldots k_{n-1},k_n+k_i} \cdot \tag{22}$$

Hence, excitations with very high energy behave like ones with zero energy : the amplitudes do not "feel" them in the intermediate states.

In principle, the results of this work can be verified in the framework of continuous theory, but it is not so easy to do. For the two point correlator, the correspondence of these two approaches was established,[9] but, in the general case, it remains to be done.

References

1. M.B. Green, J.H. Schwarz, and E. Witten, "Superstring theory", CUP (1986).
2. V.A. Kazakov, _Phys. Lett. B_, 150:282 (1985) ; F. David, _Nucl. Phys. B_, 257:45 (1985) ; J. Ambjorn, B. Durhuus, and J. Fröhlich, _Nucl. Phys. B_, 257:433 (1985).
3. V.A. Kazakov and A.A. Migdal, _Nucl. Phys. B_, 311:171 (1988).
4. V.A. Kazakov, this volum ; D. Gross, this volum.
5. E. Brezin, C. Itzykson, G. Parisi, and J.-B. Zuber, _Commun. Math. Phys._, 59:35 (1978).
6. V.G. Knizhnik, A.M. Polyakov, and A.B. Zamolodchikov, _Mod. Phys. Lett. A_, 3:819 (1988).
7. D.V. Boulatov, _Phys. Lett. B_, 237:202 (1990).
8. I.R. Kostov, _Phys. Lett. B_, 215:499 (1988).
9. Z. Yang, "Correlation functions in D = 1 random surfaces from effective field theory", University of Texas preprint UTTG-35-90.

THE PENNER MODEL AND D = 1 STRING THEORY

Jacques Distler

Joseph Henry Laboratories
Princeton University
Princeton, NJ 08544

Cumrun Vafa

Lyman Laboratory of Physics
Harvard University
Cambridge, MA 02138

After the beautiful contribution by Claude Itzykson to this volume, I'm a little embarrassed to add my own comments on the matrix model of Penner. Nevertheless, I am going to try to convince you that Penner's matrix model can in fact be interpreted as $D = 1$ string theory [1] (perhaps in a topological phase). Even if I fail in convincing you of this, I hope that you will agree that the Penner model is a useful laboratory for exploring some of the issues that arise in $D = 1$ string theory (such as the origin of the logarithmic scaling violation at low genus [1][2]) in a *much* simpler context.

Let us begin by recalling the problem that Penner [3], Harer and Zagier [4] and others were interested in solving, namely: *What is the orbifold Euler characteristic $\chi(\mathcal{M}_{g,n})$ of the moduli space of Riemann surfaces of genus g and n punctures?* Recall that moduli space is an orbifold, the quotient of a smooth variety by a non-freely acting group. The orbifold Euler characteristic is like the topological Euler characteristic, except that we weight the fixed-point sets by one over the order of the group that leaves them fixed.

This is what we want to calculate, and in simple cases, it is not too hard. Consider, for instance, $\mathcal{M}_{1,1} \cong \mathcal{M}_{1,0}$. The moduli space is, as we all know by now, the fundamental domain in the upper half plain. Topologically, this is a sphere minus the point at infinity. However there are orbifold points at $\tau = i$ and $\tau = e^{i\pi/3}$ which are fixed by automorphism groups of order 2 and 3, respectively. In addition, *every* point is fixed by the $\mathbb{Z}_2$ involution of the corresponding torus, so we should divide the answer by an overall factor of 2. Putting all this together, we obtain the well-known result

$$\chi(\mathcal{M}_{1,1}) \cong (\mathcal{M}_{1,0}) = \tfrac{1}{2}(2 - 3 + 1/2 + 1/3) = -1/12 \quad . \tag{1}$$

How can we calculate this number in the general case?

One way to approach the problem is to recall a theorem of Strebel [5]. Let η be a quadratic differential on Σ, which is holomorphic on $\Sigma \setminus \{p_i\}$ and has double poles with residues $-c_i$ (with c_i real, positive) at the p_i. This means that in local coordinates near one of the p_i, $\eta = -c_i \, dz^2/z^2$. Now such a quadratic differential defines a complex (singular) metric on Σ and hence defines a *foliation* of Σ as follows. At each point, we can ask, for what tangent vectors v is $\eta(v,v)$ real and positive? Call this direction "horizontal". Clearly if v is horizontal, so is $-v$. Drawing all the horizontal directions through each point on the surface gives us the desired foliation.

For instance, let us look in the neighbourhood of one of the poles of η. There

$$\eta = -c \, \frac{dz^2}{z^2} = -c \, \frac{(dr + ir\,d\theta)^2}{r^2}$$

so that $\eta(\partial/\partial\theta, \partial/\partial\theta) > 0$ and the horizontal leaves are concentric circles about the origin (Fig. 1).

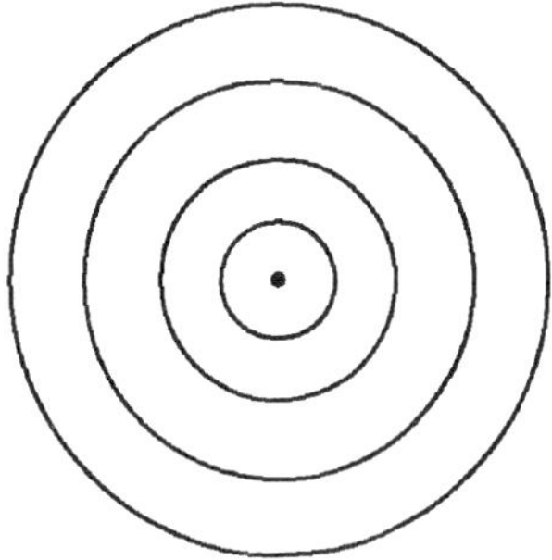

Fig. 1. Horizontal leaves near a pole of η.

Now let us look in the neighbourhood of a zero of η. Near a zero,

$$\eta \sim z\,dz^2 = re^{3i\theta}(dr + ir\,d\theta)^2$$

so there are three directions in which the radial tangent vectors are horizontal. At a simple zero of η, three singular leaves meet at a point. Similarly, at an mth order zero, $m + 2$ singular leaves meet (Fig. 2).

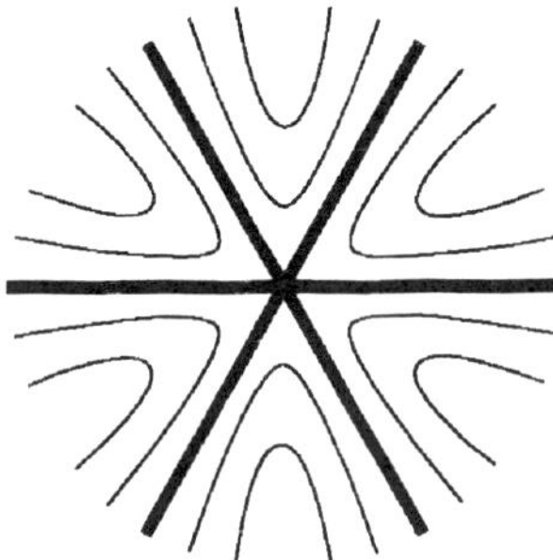

Fig. 2. Horizontal leaves near a 4th order zero of η.

Those leaves which do not come abruptly to an end at one of the zeroes of η are called *nonsingular*, and we can demand that the nonsingular leaves form closed curves, rather than, say, ergodically filling some region on Σ. if the nonsingular leaves form closed curves, then we say that η is *horocyclic*.

Strebel's Theorem says that given $\Sigma, \{p_i\}$ there is a unique (up to an overall scale) horocyclic quadratic differential with double poles at the p_i's, holomorphic on $\Sigma \setminus \{p_i\}$.

This gives us a way to associate a graph to every point in the moduli space $\mathcal{M}_{g,n}$: 1) Given $\Sigma, \{p_i\} \in \mathcal{M}_{g,n}$, find the corresponding Strebel differential η. 2) Then the singular leaves of η form a graph (whose vertices correspond to the zeroes of η).

Actually, we've done better than this. If move slightly in moduli space, by moving the p_i or deforming the surface, the topology of the graph we obtain doesn't change, so what we have really done is associate a graph to some *region* in moduli space. If you think about it, the only way the topology of the graph can change is when one of the propagators shrinks to zero length, corresponding to an nth and an mth order zero coalescing into an $(n+m)^{th}$ order zero, or the converse process. So what we actually have obtained is a *cell-decomposition* of the moduli space [6].

From this cell decomposition, it is easy to compute the Euler characteristic. Actually we are interested in the virtual Euler characteristic, which is obtained by weighting the cells by

$$\frac{(-1)^{d_\mathcal{F}}}{|G_\mathcal{F}|} \tag{2}$$

where $|G_\mathcal{F}|$ denotes the order of a stabilizer of the subgroup of the mapping class group which fixes the class of the fat graph. This in turn is precisely the symmetry of the fat graph itself. This explains in particular why we dropped the fat graphs corresponding to the boundaries of moduli space, as there we would have infinite order for $G_\mathcal{F}$ and they would not contribute to (2). As an example of (2) consider again the torus with one puncture. We have two fat graphs (Fig. 3) one with a symmetry of order 6 and the other with a symmetry of order 4 so we find the result

$$\chi_{1,1} = \frac{(-1)^2}{6} + \frac{(-1)^1}{4} = -\frac{1}{12}$$

just as in (1).

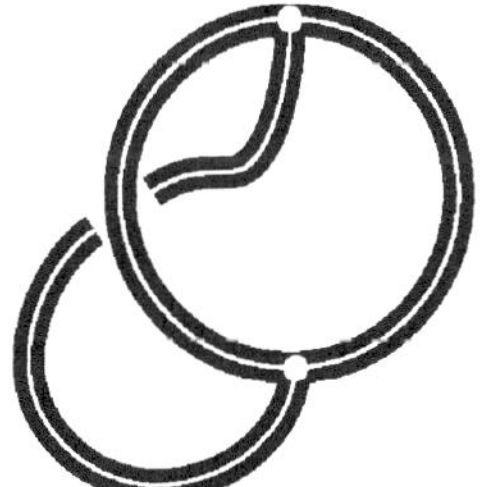 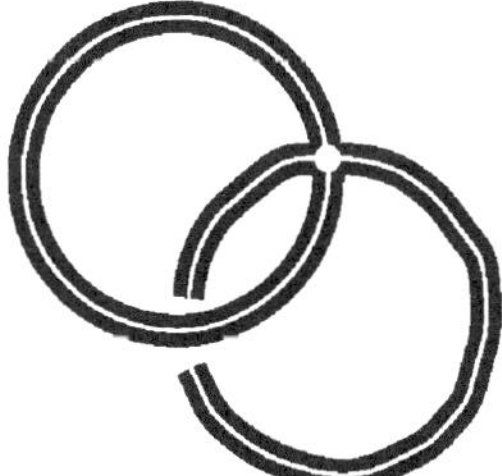

Fig. 3. Fat graphs for $\mathcal{M}_{1,1}$.

The two fat graphs in Fig. 3 correspond respectively to a cell of dimension 2 (generic values of τ in the fundamental domain) and a cell of dimension 1 (τ pure imaginary). As an aid to the reader's eye, we have redrawn them as the actual foliations of the corresponding tori in Fig. 4.

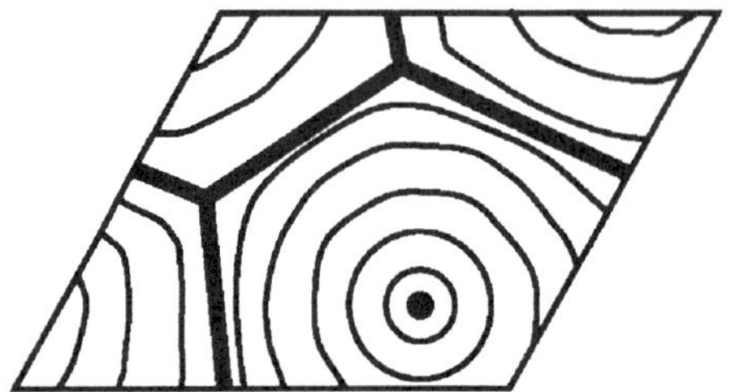 

Fig. 4. The corresponding foliations.

Following Penner [3] we would like to introduce a matrix model whose Feynman rules generate diagrams (fat graphs) with the weighting factor (2). Fat graphs are simply the Feynman diagrams of a large-N Hermitian matrix model in "double-line" notation. Consider the potential

$$V(\phi) = -\sum_{n=2}^{\infty} tr \frac{\phi^m}{m} = tr\,[log(1-\phi)+\phi] \quad .$$

The Feynman rules for expanding $\int d\phi e^{NtV(\phi)}$ weight each Feynman graph with v vertices, e edges and n traces (closed loops in the double-line notation) by the factor

$$\frac{(-1)^v N^{v-e+n} t^{v-e}}{s(G)}$$

where $s(G)$ is the symmetry factor of the graph. Now, the genus of the Riemann surface given by the fat graph is $2-2g = v-e+n$. The dimension of the cell in moduli space corresponding to a given fat graph is $d = dim\mathcal{M}_{g,n} + 3v - 2e$ (so that a graph with purely cubic vertices is a top-dimensional cell). Since $dim\mathcal{M}_{g,n} = 6g-6+2n$ is even, $d = v$ (mod 2), so each Feynman graph is weighted by

$$\frac{(-1)^v}{s(G)} N^{-(2g-2)} t^{-(2g-2+n)}$$

which is precisely what we want for (2). To calculate the Euler characteristic of $\mathcal{M}_{g,n}$, we simply have to sum all of the Feynman diagrams which contribute to a given order in N and t.

Before going further, let's ask what a continuum theory which computed the virtual Euler characteristic of moduli space would look like. That is, assume we have some continuum theory whose partition function on a genus g Riemann surface is χ_g, i.e., $F_g = \chi_g$. We can write the sum of the contributions of all genera as

$$F = \sum_g \chi_g \, \mu^{a(2-2g)} \tag{3}$$

where μ is the cosmological constant and a is determined from the conformal theory coupled to gravity [7] and is related to the string susceptibility by $\gamma_{str} = 2 - 2a$.

In such a theory, what would the n point function of the area operator look like? If the free energy is the Euler character of moduli space of Riemann surfaces, each time we insert an area element all we are doing is creating a puncture (which is why it is often called the *puncture operator* [8]) and it would be natural to expect that this should be related to Euler character of moduli space of Riemann surface of genus g with n punctures $\chi_{g,n}$. Let us see if this expectation is actually realized. In fact there is a relation between $\chi_{g,n}$ and χ_g which follows from a simple argument (see for example [3]):

$$\chi_{g,n} = \frac{(-1)^n (2g - 3 + n)!}{n!(2g - 3)!} \chi_g \quad . \tag{4}$$

It is straightforward to obtain the n point expectation value of the area operator, as that is simply a matter of differentiation of the free energy F with respect to μ, i.e.,

$$\frac{1}{n!} < AA...A > = \frac{1}{n!} \frac{\partial^n}{\partial \mu^n} F \quad . \tag{5}$$

In particular on a genus g surface we obtain

$$\frac{1}{n!} < AA...A >_g = \frac{1}{n!} \chi_g \, a(2-2g) \cdot (a(2-2g)-1) \dots (a(2-2g)-n+1) \, \mu^{a(2-2g)-n} \tag{6}$$

We thus see that if $a = 1$, this is in fact $\chi_{g,n}$, i.e., for $a = 1$ we have:

$$\frac{1}{n!} < AA...A > = \frac{1}{n!} \frac{\partial^n}{\partial \mu^n} F = \sum \chi_{g,n} \, \mu^{2-2g-n} \quad . \tag{7}$$

It is quite a surprise that the free energy F, defined by (3) with $a = 1$, automatically 'knows' about the relation (4) as it is able to reproduce it by differentiation. In fact this is the first sign that there might be a physical theory with the properties we are looking for. Along the way we have learned that $a = 1$. This value of a corresponds to vanishing string susceptibility, $\gamma_{str} = 0$, which, from the formulas of [7], suggests that we are dealing with a gravity theory with central charge $c = 1$.

If this were to be identified with a $c = 1$ model, we might well ask what happened to logarithmic corrections which are known to accompany $c = 1$ models in genus zero [1][2]. To answer this question we go back to expression (3) and notice that for genus zero we have to define F_0. This is somewhat ill-defined because the moduli space of the sphere has virtual dimension -6 due to the $SL(2,\mathbb{C})$ invariance of the sphere. However, to define it in a sensible way we resort to equation (7) and take that as the defining property of F_0. By taking enough punctures we can obtain a moduli space with positive virtual dimension, and hence a sensible Euler characteristic. In

particular the Euler character of the moduli space of a sphere with three punctures is
1/6 (it consists of a single point, except that the three punctures are indistinguishable
giving a factor of 1/3! for the virtual Euler characteristic). From (7) this implies that
the genus zero contribution to free energy satisfies

$$F_0{}'''(\mu) = \mu^{-1} \quad . \tag{8}$$

From this we solve for the genus zero contribution to F to be

$$F_0 = \tfrac{1}{2}\mu^2 log\mu + A\mu^2 + B\mu + C \tag{9}$$

where A, B, C are some undetermined constants of integration. We thus get a modification of the simple expression (3) in such a way that the genus zero contribution
is in fact not just μ^2 but actually accompanied by $log\mu$ as is expected for a theory at
$c = 1$!

I think that there is an important message here. The explanations that we have
heard at this conference of the origin of the logarithmic scaling violation at low genus
are, to my ears, rather baroque. On the other hand, the origin of the logarithmic
scaling violation (in this "hypothetical" continuum theory and in the continuum limit
of the Penner model, to be discussed below) is stunningly simple. The appearance of
logarithmic terms in the free energy at genus zero is a direct result of the $SL(2,\mathbb{C})$
invariance of the sphere! Rather than being some exotic target space phenomenon,
the logarithmic scaling violation is a symptom of the conformal killing symmetry of
the worldsheet.

Again at genus one, the torus with no punctures has $U(1) \times U(1)$ conformal killing
symmetries, and so the virtual dimension of moduli space is zero, rather than being
equal to the true dimension (two). Again we resort to (7) for the correct definition
of χ_1, and in particular by looking at the torus with one puncture, we find that the
contribution of the genus one surfaces to F is $-\tfrac{1}{12} log\mu$. This now finally completes
our precise definition of the free energy F, as there is no other Riemann surface with
conformal killing symmetries.

So far we have not given the explicit form of χ_g as it was not needed. All we
needed was the relation between $\chi_{g,n}$ and χ_g given by (4). However it is useful to give
an explicit form which was computed by Harer and Zagier [4]. In fact their method
of computation which involved integration over hermitian matrices was subsequently
generalized by Penner [3] to the full-fledged matrix model that is the focus of this
talk. The result they obtain for χ_g is

$$\chi_g = \frac{B_{2g}}{(2g)(2g-2)} \tag{10}$$

where B_{2g} is the 2g-th Bernoulli number. In order to write F in a compact form it is
convenient to introduce the ψ function:

$$\psi(\mu) = \frac{d}{d\mu} log\Gamma(\mu) = -\gamma + \sum_{k=0}^{\infty} \frac{1}{k+1} - \frac{1}{\mu+k} = log\mu - \frac{1}{2\mu} - \sum_{g=1}^{\infty} \frac{B_{2g}}{2g}\mu^{-2g} \quad . \tag{11}$$

Using this and equations (10)(11) and the definition of F given in (3) with the modifications discussed above for genus zero and one we finally obtain the final compact form of F:

$$F(\mu) = \int^{\mu} x\psi(x)\,\mathrm{d}x \quad . \tag{12}$$

It is quite surprising that the final form for F which *a priori* did not look like a simple function can be expressed in such a compact from. Note that the above form for F fixes the undetermined quadratic term up to the addition of a constant. Before addressing whether or not there is a gravity theory with this free energy we remark that as it stands (12) does not describe the free energy of a unitary theory as the contributions to it are not positive definite. However, if we replace $\mu \to i\mu$ then the free energy *is* positive definite (apart from sphere and torus which are logarithmic anyways), as we would expect from a unitary theory. The alternating signs in the free energy (12) for real μ is simply due to the fact that sign of the Euler character of moduli space of genus g surfaces is $-(-1)^g$ (reflecting the fact that most of the cohomology is coming from the middle dimension) as is indicated by the alternating signs of the Bernoulli numbers.

Returning to the Penner model, we can sum all the connected fat graphs to obtain the free energy $F = logZ$ of the matrix model,

$$F = \sum_{g,n} N^{2-2g} t^{2-2g-n} \chi_{g,n} \quad . \tag{13}$$

Penner used the matrix model techniques developed in [9] to compute in this way the coefficients $\chi_{g,n}$, and he found

$$\chi_{g,n} = \frac{(-1)^n(2g-3+n)!(2g-1)}{(2g)!n!} B_{2g} \quad . \tag{14}$$

This computation has recently been simplified by [10]. The detail of this method will not concern us as we will later give a different derivation of the free energy in the continuum limit.

So far we have only described the non-continuum theory. To obtain a continuum theory we wish to take $N \to \infty$, and adjust $t \to t_c$, in such a way that the perturbation series diverges and the contribution to F_g is dominated by surfaces composed of many little pieces, *i. e.* by n, the number of plaquettes, $\to \infty$. Given the simple form of (14), the sum can be simply done

$$F_g = \sum_{n=0}^{\infty} N^{2-2g} t^{2-2g-n} \chi_{g,n}$$

$$= \frac{B_{2g}}{2g(2g-2)} (Nt)^{2-2g} \sum_n (-t)^n \binom{2g-3+n}{n} \quad .$$

Clearly, the sum diverges as $t \to -1$. The double scaling limit is to take $N \to \infty$, $t \to t_c = -1$, holding $\mu = (t_c - t)N$ fixed. In this limit, we simply find that for $g > 1$

$$F_g = \frac{B_{2g}}{2g(2g-2)} (\mu)^{2-2g} \quad .$$

For $g = 0, 1$ a careful treatment of the above sum will in fact give the anticipated logarithmic corrections. If we now interpret μ as the *continuum* cosmological constant, we see that F_g is simply the generating function guessed at above and is given by equation (12). In other words the leading term of the singularity has precisely the structure needed to give back the Euler characteristic of moduli of Riemann surfaces *without punctures*, from the leading behaviour of Euler characteristic of moduli space of infinitely punctured surfaces! This behaviour we find rather remarkable and in need of a deep explanation. One could speculate about this as representing a kind of punctured world-sheet which is equivalent to the unpunctured one; something which might lie at the phase transition to $c > 1$ region.

The same result can be derived directly by considering the matrix integral which gives a quick derivation of (12). We wish to evaluate

$$e^F = \int d\phi \; e^{Nt \; tr \; (log(1-\phi)+\phi)} \tag{15}$$

$$= \int d\phi \; \det(1 - \phi)^{Nt} e^{Nt \; tr \; \phi} \quad .$$

Let $M = 1 - \phi$. The branch cut in $\det M^{Nt}$ would render an integral over all hermitian matrices M ill-defined. Instead, as noted in [10], we should integrate only over *positive definite* M. This does not alter the perturbative expansion of the integral discussed above, but does serve to make the matrix integral well-defined. After diagonalizing M, and integrating over the angular variables, we have to evaluate

$$e^F = e^{N^2 t} \int_0^\infty \prod_{i=1}^N d\lambda_i \; \Delta^2(\lambda) \prod_i (\lambda_i^{Nt} \; e^{-Nt\lambda_i}) \tag{16}$$

where the λ_i are the eigenvalues of M and the VanderMonde determinant, $\Delta(\lambda) = \prod_{i<j}(\lambda_i - \lambda_j)$, is the Jacobian resulting from this change of variables. To evaluate this, we introduce orthogonal "polynomials" $P_i(x)$ which satisfy

$$\int_0^\infty dx \; P_i(x) P_j(x) e^{-\alpha x} = h_i \delta_{i,j}$$

and $P_n(x) = x^{\alpha/2}(x^n + \ldots)$. In terms of these polynomials, we can write the integral (16) as

$$e^F = e^{\alpha^2/N} \int_0^\infty \prod_{i=1}^N d\lambda_i \; det(P_i(\lambda_j)) e^{-\alpha \Sigma \lambda_i} \tag{17}$$

where $\alpha = Nt$. The integral (17) is convergent for positive α, and can be analytically continued to the regime of interest, namely $\alpha = -(N + \mu)$. These polynomials satisfy a simple recursion relation

$$xP_n(x) = P_{n+1}(x) + \frac{2n + \alpha + 1}{\alpha} P_n(x) + \frac{n(n + \alpha)}{\alpha} P_{n-1}(x) \tag{18}$$

$$= P_{n+1}(x) + S_n P_n(x) + R_n P_{n-1}(x)$$

and

$$h_0 = \alpha^{-\alpha}\Gamma(\alpha), \qquad h_{n+1} = R_{n+1}h_n \quad .$$

The partition function is then given by

$$Z = e^F = N!\,h_0^N\,e^{\alpha^2/N}\prod_{k=1}^{N-1} R_{N-k}^k \quad .$$

To suppress the irrelevant (and divergent) constants of integration A, B, C in (9) we differentiate [1] the free energy three times with respect to μ.

$$\frac{\partial^3 F}{\partial\mu^3} = -\frac{\partial^3 F}{\partial\alpha^3} = -\left[N\left(\psi''(\alpha)+\frac{1}{\alpha^2}\right)+\sum_{k=1}^{N-1}\left(\frac{2k}{(N-k+\alpha)^3}-\frac{2k}{(\alpha)^3}\right)\right]$$

In the large-N limit, this simply becomes

$$\frac{\partial^3 F}{\partial\mu^3} \xrightarrow{N\to\infty} \sum_{k=1}^{\infty}\frac{2k}{(k+\mu)^3} = \frac{\partial^2}{\partial\mu^2}\,\mu\psi(\mu) \quad .$$

Integrating this with respect to μ, we obtain (12).

Clearly, the continuum limit we have just found for the Penner model is the first of a series of multicritical theories. More generally, we can consider matrix integrals of the form

$$Z \doteq \int dM \,(detM)^\alpha e^{-V(M)}$$

where M is a positive hermitian matrix. The model of Penner corresponds to the simple case of a linear potential $V(M) = \alpha\ trM$. It is natural to allow arbitrary polynomial potentials and classify the types of critical points we get.

As we have already seen, this is likely to be a little more subtle than the situation one usually encounters. Usually, the criterion for criticality is that the sphere contribution to the free energy scale homogeneously as a power of μ. Here we expect to have logarithmic corrections to the free energy on the sphere, and we must take some derivatives of the free energy in order to see the scaling behaviour.

A related problem is constructing the scaling operators in the Penner model. The simplest one is, of course, the puncture operator, $O_0 = P = \frac{\partial}{\partial\mu} = \sum_{n=2}^{\infty}\frac{1}{n}\phi^n$. The puncture operator simply adds a marked point to the surface. Another simple operator is the "dilaton" operator $D = \mu\frac{\partial}{\partial\mu}$, which shifts the string coupling constant.

Another interesting avenue to pursue is to construct the operators of "nonzero momentum". After all, if this really is a theory of $c = 1$, we ought to have such operators in our theory. To this end, introduce a complex $N \times N$ matrix ψ, and write (15) as

$$Z = e^F = e^{N^2 t}\int dM\,d\psi\,d\bar\psi\,e^{tr\left[\bar\psi M^{-t}\psi+Nt\,M\right]} \tag{19}$$

---

[1] It is routine to throw away regular terms (*i.e.* polynomials in μ) from F. Here, taking a few derivatives accomplishes this for free.

This theory has a U(1) symmetry under which $\psi \to e^{i\theta}\psi$. In fact, it has a SU(N) symmetry $\psi \to U\psi$, $M \to UMU^{-1}$, but this latter symmetry is fixed when we diagonalize M (more properly, we will, in the end, consider only $SU(N)$-singlet observables). We would like to tentatively identify states of nonzero U(1) charge as being the momentum states of a (compact) boson.

Writing $\psi = \psi_1 + i\psi_2$, where $\psi_{1,2}$ are hermitian matrices, we obtain a theory of two hermitian matrices coupled (nonpolynomially) to a positive hermitian matrix M. Unfortunately, we cannot simultaneously diagonalize these matrices in the action (19). The relative angular integrations do not decouple. This is analogous to the presence of $SU(N)/T$-nonsinglet states in the compact $d = 1$ string [2]. In fact, this is precisely the situation if we realize the $d = 1$ string as a closed matrix chain with nearest neighbour couplings [11][2]. One set of angular integrations does not decouple. Perhaps the correct path to follow here is to truncate to the singlet sector which, in the $d = 1$ string has the effect of suppressing vortices and leads to the usual continuum theory [2].

The form of (12) is very suggestive of the free energy one finds in other approaches to $c = 1$ [1][2][11], when one interprets μ is the difference of the fermi energy from its critical value, and $\psi(\mu)$ as the density of states. Such a description is roughly correct, but it seems that our natural variable μ which seems here to be identified with the cosmological constant is to be interpreted as the Legendre transformed variable dual to the comsological constant in the other approach. In fact Gross and Klebanov have pointed out [12] that the Legendre transform of their free energy (truncated to the singlet states) for strings propagating on a circle of radius R [2] *at the self dual radius* $R = 1$ is precisely (12) (after the wick rotation of $\mu \to i\mu$ discussed above).

We have not carried out the program of finding multicritical points of the positive hermitian matrix model. However, if this identification of the Penner Model as strings propagating on a circle of self-dual radius is correct, it is tempting to speculate that the m-th critical point is related to strings propagating on the circle whose radius is m times the self-dual radius. Remembering that the self-dual radius corresponds to conformal theory of $SU(2)$ at level one, these other radii are the orbifolds of it by cyclic subgroups of it. Indeed it is natural to expect that some kind of simplification should take place for all $c = 1$ models which are orbifolds of $SU(2)$ (see [13] for a discussion of $c = 1$ conformal theories). Maybe these are precisely the points where they can be described by a (finite) matrix models.

It is clear that more remains to be done on $c = 1$ theories coupled to gravity. I hope that I have convinced you that the Penner model is a useful vehicle for exploring some of these questions.

References

[1]   D. Gross and M. Miljković, Phys. Lett. **238B** (1990) 217;
      E. Brezin, V. Kazakov, and Al. B. Zamolodchikov, Nucl. Phys. **B338** (1990) 673;
      P. Ginsparg and J. Zinn-Justin, Phys. Lett. **240B** (1990) 333.
[2]   D. Gross and I. Klebanov, Princeton preprint PUPT-90-1172.

[3] R. C. Penner, *"Perturbative series and the Moduli Space of Riemann Surfaces"*, UCSD preprint (1986).

[4] J. Harer and D. Zagier, *"The Euler characteristic of the moduli space of curves"*, preprint (1985).

[5] K. Strebel, *"Quadratic Differentials"*, Erg. der Math. und ihrer Grenz. **5**, Springer-Verlag (1984).

[6] J. Harer, *"The Cohomology of the Moduli Space of Curves"*, preprint.

[7] V. Knizhnik, A. Polyakov and A. B. Zamolodchikov, Mod. Phys. Lett. **A3** (1988) 819;
J. Distler and H. Kawai, Nucl. Phys. **B321** (1989) 509;
F. David, Mod. Phys. Lett. **A3** (1988) 1651.

[8] E. Witten, *"On the structure of the topological phase of two dimensional gravity"*, IAS preprint IASSNS-HEP-89/66.

[9] E. Brezin, C. Itzykson, G. Parisi and J. Zuber, Comm. Math. Phys. **59** (1978) 35.

[10] C. Itzykson and J. Zuber, *"Matrix Integration and the Combinatorics of the Modular Group"*, Saclay preprint SPhT/90/004.

[11] G. Parisi, Roma Tor Vergata preprint, ROM2F-90/2.

[12] D. Gross, private communication.

[13] P. Ginsparg, Nucl. Phys. **B295** (1988) 153.

NON-PERTURBATIVE STRING THEORY

David J. Gross

Joseph Henry Laboratories
Princeton University,
Princeton, N.J. 08544

1. Introduction

The conventional approach to two dimensional gravity and to string theory is perturbative with respect to fluctuations of the topology. One sums over two dimensional geometries by first performing the functional integral for fixed topology (genus= number of handles) and then summing over genus. However, this sum is very badly behaved. The higher terms grow as factorials of the genus, and the positivity of these terms renders the series non Borel summable.[1]This situation is made worse by the lack of an adequate nonperturbative framework for the theory. Such a framework, (for example, a useful formulation of second quantized string theory) should be capable of reproducing the topological series as an asymptotic expansion, valid in the perturbative domain; but it should also provide a physical picture and a mathematical framework valid for strong coupling. It is essential that we develop nonperturbative methods if we are to relate unified string theories to the real world. At the perturbative level of string theory there are many too many possible worlds, *i.e.* classical vacua about which consistent perturbative expansions can be made. All of them have undesired features, such as unbroken supersymmetry and massless dilatons. One must hope that nonperturbative physics will lift the degeneracy and break the unwanted symmetries. This is strongly suggested by the divergence and non Borel summability of perturbation theory which can be taken as an indication of the nonperturbative instability of the classical vacua [1].

One of the main motivations for studying two dimensional gravity coupled to simple matter is that this provides a toy model for string theory, in which these nonperturbative issues might be explored with greater ease. There are other motivations. These theories are the simplest examples of quantum gravity, and might be used to probe issues of topology change in a finite and tame gravitational theory. They might also be used in the study to study of critical behavior in three dimensions, where the phase boundaries are two dimensional random surfaces. In addition, it has long been a goal to construct a string theory representation of QCD. For all of these purposes it is necessary to enlarge our understanding of string theories and to go beyond the special critical dimensions. In doing so we lose some of the simplicity of the critical string (the decoupling of the two dimensional metric from the matter) that gives rise to theenhanced symmetry of unified string theory. What we gain is the opportunity to consider simple models with a few degrees of freedom. The simplest of all toy models are those where we couple two dimensional gravity to the simplest two-dimensional field theories– the minimal models of matter with central charge less than one, which have only a finite number of degrees of freedom. Indeed the reduction in the number of degrees of freedom has enabled us recently to *find explicit analytic solutions to all orders in perturbation theory* for many such theories.

The study of these simple examples will hopefully teach us lessons that can be applied to critical string theory. Even more–it might be the case that there is no real distinction between critical and noncritical string theory. It is well known that the critical theory, say the 26-dimensional bosonic string, can be identified with a theory of $c = 25$ noncritical matter coupled to two dimensional gravity, wherein the Liouville field supplies the time coordinate[11]. The most radical point of view is that there is only one (or three including super and heterotic strings) string theory, the critical string representing being an expansion of the theory about a particularly symmetric classical solution. Thus in studying the toy models with central charge c we are exploring a $D = c + 1$ dimensional corner of critical string theory.

The recent progress is based on matrix model methods for generating cutoff, discrete representations of the perturbative expansion of string amplitudes. Here the geometry of the world sheet of the string (or two dimensional space in the case of pure gravity) is approximated by a dense Feynman graph, in the limit where the number of vertices becomes infinite. The topology is selected by means of the $\frac{1}{N}$ expansion of a SU_N invariant matrix model. The sum over all Feynman graphs of given genus and given number of vertices is a discrete version of the functional integral over metric tensors[2]. Remarkably, in many cases these discrete models can be handled with greater ease than their continuous analogs.

The great advance of last year was made by realizing that these matrix model representations simplify drastically in the *double scaling limit.* This is the limit where one adjusts the coupling constant, λ, to equal a critical value at which the loop expansion of the matrix model begins to diverge and sends N to infinity in just such a way that the string coupling, $g_s^2 \equiv \frac{1}{N^2(\lambda-\lambda_{cr})^{2+\gamma_s}}$, remains constant. This is precisely the limit which allows one to sum over all continuum surfaces of arbitrary topology. The double scaling limit of the matrix model representation of sums over discretized random surfaces has been used to calculate sums over all continuum surfaces to all orders in the topological expansion[3-6]. The remarkably rich structure that emerged has also illuminated the connection of two dimensional quantum gravity with topological field theories [7] and KdV hierarchies [4],[8], [9].

Matrix model methods have so far been employed mostly to study models where the matter content has central charge less or equal to one. In fact, $c = 1$, the model of one scalar matter field coupled to quantum gravity, is a transition point for this approach, where a phase transition probably takes place due to the vanishing mass of the tachyon degree of freedom, which becomes truly tachyonic for $c > 1$. $c = 1$ also represents a calculational barrier for matrix model technology. One of the main reasons that the matrix models are soluble for $c \leq 1$ is that, of the N^2 degrees of freedom of the matrices, only the N eigenvalues matter. The other $N(N - 1)$ angular degrees of freedom decouple[10]. This procedure breaks down when $c > 1$, where one finds that all N^2 degrees of freedom of the matrices become relevant. This is consistent with the emerging view of the string interpretation of these models as describing $D = c + 1$ string theory, in which the Liouville mode, which is identified with the space of eigenvalues of the matrix model, yields a (euclidean) time dimension[11-13]. The $D = 2$ string theory, described by the $c = 1$ matrix model, has only a limited number of degrees of freedom, since in two dimensions there are no transverse excitations and all we have are the center of mass of the string (the two-dimensional tachyon field), and a discrete infinity of quantum mechanical variables which occur at quantized momenta[14]. On the other hand, for $D > 2$ we find an infinite number of D-dimensional fields. Thus, it is not surprising that the number of relevant degrees of freedom suddenly increases as we pass the point $c = 1$.

The $c = 1$ theory is then, in many ways, the most complex of the soluble models to date, the one that is closest to realistic string theories in higher dimensions. In other ways it is it is the simplest of the soluble models where many of the results can be explicitly exhibited in terms of elementary functions. In this lecture I shall review the solution of this model, discuss the role of nonsinglet states and their relation to vortices and construct a two dimensionalfermionic field representation of the model.

2. The $c = 1$ Matrix Model

Let me first review the matrix model representation of the sum over random surfaces embedded in one dimension—$c = 1$ matter coupled to two dimensional quantum gravity[15]. The partition function can be generated by the Feynman diagrams of the large N limit of the euclidean quantum mechanics of a hermitean $N \times N$ matrix Φ[16],

$$Z_N(\beta) = \int D^{N^2} \Phi\, e^{-\beta \int dt\, \mathrm{Tr}\left[\frac{1}{2}\dot{\Phi}^2 + \mathrm{U}(\Phi)\right]} \tag{2.1}$$

where t runs from $-\infty$ to ∞ (from 0 to $2\pi R$) if the target space is the real line (a circle of radius R.)

The connection with $D = 1$ string theory is established by considering the perturbative expansion of the partition function. Each given order in $\frac{1}{N^2}$ corresponds to a sum over the surfaces of definite genus, generated by the Feynman graphs of that genus. Each graph is weighted by a factor $\left(\frac{N}{\beta}\right)^{\mathrm{Area}}$, and contains an integral over a product of propagators, $e^{-m|t_i - t_j|}$, connecting adjacent $\Phi(t)$'s on the graph. In the continuum limit, when we take $\frac{N}{\beta} \equiv g \to g_{critical}$ so that the perturbation series diverges and is dominated by infinite area terms, this should coincide with the continuum definition of a Gaussian variable coupled to two dimensional gravity.

The standard method for dealing with this problem as $N \to \infty$ is to reduce the number of degrees of freedom from the N^2 matrix elements of Φ to its N eigenvalues, λ_i, by writing the matrix as $\Phi(t) = \Omega(t)^\dagger \Lambda(t)\Omega(t)$, where Ω is unitary and Λ diagonal. In the case of the $c = 0$ model, in which Φ has no t dependence, the action does not depend on Ω, which can be integrated out. This is not the case for $c = 1$, due to the kinetic term $\mathrm{Tr}[\dot{\Phi}(t)^2]$, which can be written as $\mathrm{Tr}[\dot{\Phi}^2] = \mathrm{Tr}(\dot{\Lambda}^2 + [\Lambda, A][\Lambda, A])$, where we have introduced the pure gauge field $A(t) \equiv \dot{\Omega}\Omega^\dagger(t)$. Now, even though $A(t)$ does not decouple, the integral over it can be performed, following the classic work of Itzykson and Zuber and of Mehta [17]. The essential point is that the semi-classical evaluation of the integral over the Ω matrices is exact, at least if the Ω's are unconstrained. This is the content of the formula,

$$\int \mathcal{D}\Omega\, e^{\mathrm{Tr}\left[\Omega A \Omega^\dagger B\right]} = \sum_{p(i)} \frac{e^{\sum_i a_i b_{p(i)}}}{\Delta(a_i)\Delta(b_{p(i)})} = \frac{\mathrm{Det}[e^{a_i b_j}]}{\Delta(a_i)\Delta(b_i)} \tag{2.2}$$

where the a_i (b_i) are the eigenvalues of A (B), $p(i)$ is a permutation of the i's, and $\Delta(a_i)$ is the Vandermonde determinant $\Delta(a_i) = \prod_{i<j}(a_i - a_j)$. Using this one can then perform the Ω integral in (2.1) (most precisely by discretizing the t interval), and show that the effect is to precisely cancel the factors of $\Delta(\lambda)$ that come from the measure of integration, $\mathcal{D}\Phi = \mathcal{D}\Omega \prod_i d\lambda_i \Delta^2(\lambda)$. The result is that the partition function on a finite interval, $t_1 < t < t_2$, is given by

$$Z_N(\beta) = \int \prod_i d\lambda_i(t)\Delta(\lambda_i(t_2))e^{-\beta \int_{t_1}^{t_2} dt \sum_i [\frac{1}{2}\dot{\lambda}_i^2 + U(\lambda_i)]}\Delta(\lambda_i(t_1)) \tag{2.3}$$

The antisymmetric factors of $\Delta(\lambda_i)$, project out of any intermediate state the totally antisymmetric component. In other words, the λ_i are fermionic variables!

Alternatively, we may carry out canonical quantization of the $SU(N)$ symmetric matrix quantum mechanics[16,19]. The hamiltonian is

$$H = -\frac{1}{2\beta^2 \Delta(\lambda)}\sum_i \frac{d^2}{d\lambda_i^2}\Delta(\lambda) + \sum_i U(\lambda_i) + \sum_{i<j}\frac{\Pi_{ij}^2 + \tilde{\Pi}_{ij}^2}{(\lambda_i - \lambda_j)^2} \tag{2.4}$$

where the Π_{ij} and $\tilde{\Pi}_{ij}$ are generators of *left* rotations on Ω, $\Omega \to A\Omega$. When the target space is the infinite real line the only state that matters in the calculation of the matrix element of the time evolution operator e^{-HT}, as $T \to \infty$, is the ground state of the Hamiltonian, which is in the trivial representation of $SU(N)$, given by an $SU(N)$ singlet wave functions, $\psi(\lambda_i)$, which does not depend on the angular matrices Ω, and is a symmetric function of the eigenvalues λ_i. The hamiltonian of eqn (2.4), when acting on $\chi(\lambda_i)$, a fully antisymmetric wave function $\chi(\lambda_i) = \Delta(\lambda_i)\psi(\lambda_i)$, reduces to the sum of single particle hamiltonians,

$$h_i = -\frac{1}{2\beta^2}\frac{d^2}{d\lambda_i^2} + U(\lambda_i) \tag{2.5}$$

thus reducing the problem to the physics of N non-interacting fermions moving in the potential $U(\lambda)$, with the Planck constant $\hbar$ set equal to $\frac{1}{\beta} \sim \frac{1}{N}$. The free energy,

E, is given by $\beta \sum_{i=1}^{N} e_i$, where e_i are the lowest N eigenvalues of $\hat{h}$. In the limit $\beta \to \infty$, where β plays the role of $\hbar$, the problem becomes semiclassical. We would like to isolate the non-analytic behavior which occurs when the Fermi level of the N-fermion system μ_F approaches the top of the potential μ_c, ($\mu_c = U(x_c)$, $U'(x_c) = 0$). Introduce the density of eigenvalues $\rho(e) = \frac{1}{\beta} \sum_n \delta(e - e_n)$, in terms of which

$$g = \frac{N}{\beta} = \int_0^{\mu_F} \rho(e) de \; ; \quad \lim_{T \to \infty} -\frac{\ln Z_N(\beta)}{T} = E = \beta^2 \int_0^{\mu_F} \rho(e) e\, de \qquad (2.6)$$

We then tune the parameter g towards the critical value $g_c = 1$ where $\mu_F = \mu_c$, and the energy levels begin to spill over the top of the potential. $\Delta = 1 - g$ is a singular function of $\mu = \mu_c - \mu_F$ which leads to non-analyticity of $E(\Delta)$. It is this nonanalytic component of E, as a function of Δ, which can be identified as the cosmological constant, that controls the sum over finite surfaces in the continuum limit. Analytic pieces of $E(\Delta)$, when Laplace transformed, give contributions only for zero area. To calculate this function, we first use $\frac{\partial \Delta}{\partial \mu} = \rho(\mu_F)$, to find μ as a function of Δ. Subsequently, we integrate $\frac{\partial E}{\partial \Delta} = \beta^2(\mu - \mu_c)$, to find E. All that we need to know is the singular terms in the density of states near the top of the potential. In the WKB approximation [15] we have,

$$\rho(\mu_F) = \frac{1}{\pi} \int_{-x_c}^{x_c} \frac{dx}{\sqrt{2(\mu_F - U(x))}} \sim -\frac{1}{2\pi} \ln \mu + \mathcal{O}(1/\beta^2) \qquad (2.7)$$

from which it follows that $\Delta = -\frac{1}{2\pi} \mu \ln \mu + \ldots$, and $E = -N^2 \mu^2 \ln |\mu| over 4\pi + \ldots$. This result is in agreement with the prediction of the continuum methods $\gamma_{str} = 0$, where $E \sim N^2 \Delta^{2 - \gamma_{str}}$, up to logarithmic deviations. These logarithmic scaling violations are somewhat mysterious from the point of view of two dimensional gravity. Polchinski [13] argued that these scaling violations have a natural explanation if one regards the theory as a *two dimensional string theory*. One dimension arises directly as the spatial dimension of the target space of the matrix model (*i.e.*, the Euclidean time of the hermitean matrix quantum mechanics); the other dimension arises indirectly and is identified with the Liouville mode of quantum gravity. The unusual feature of the $D = 2$ string theory is the lack of translation invariance in the "Liouville dimension" ϕ, manifested by the ϕ-dependence of the background fields. This lack of translation invariance is actually necessary for maintaining the conformal invariance of the $D = 2$ string theory, and explains the apparent scaling violations of the matrix model solutions. The factor of $\ln \mu$ in the expression for the energy reflects the logarithmic divergence of the Liouville volume as $\mu \to 0$.

Since $\hbar = 1/\beta$, in order to find the sums over higher genus surfaces, it is sufficient to carry out a systematic higher-order WKB expansion for the density of states near the top of the potential. The calculation simplifies because one has to retain only the parabolic term near the maximum of the potential. This can be seen by examining the expression for the density of states $\rho(\mu_F) = \frac{1}{\pi\beta} \mathrm{Im\,Tr} \frac{1}{\hat{h} - \mu_F - i\epsilon}$, near the maximum of the potential. Expand the denominator about $y = x_c - x \sim 0$, and note that $\hat{h} - \mu_F \sim -\frac{1}{2\beta^2} \partial_y^2 + \mu - 2y^2 + O(y^3)$. In the double scaling limit we take $\beta \to \infty$ and

$\mu \sim \frac{1}{\beta}$. Thus if $y^2 \sim \frac{1}{\beta}$, then the denominator will scale as $\frac{1}{\beta}$, and the cubic and higher powers of y will be suppressed Even though the scaling will be spoiled by logarithms this does not affect this power suppression of all but the quadratic part of the potential! The hamiltonian that survives this scaling limit is simply an inverted harmonic oscillator. The calculation of $\rho(\mu_F)$ is then straightforward. We use the following trick. Consider the density of states of the normal harmonic oscillator of frequency ω, $\rho(E) = \frac{1}{\pi} \text{Im} \sum_n \frac{1}{(n+\frac{1}{2})\omega\hbar - E - i\epsilon}$, and continue ω to imaginary frequency. This yields, $\rho(\mu_F) = \frac{1}{\pi} \text{Re} \sum_n \frac{1}{2n+1+i\beta\mu}$. This expression is divergent, however the divergent part is μ independent and will not affect the critical behavior. We can fix it by demanding agreement with the WKB result in the limit of $\beta \to \infty$, and we obtain

$$\rho(\mu_F) = \frac{1}{2\pi} \text{Re}\left[\zeta(1, \frac{1+i\beta\mu}{2}) - \infty\right] = \frac{1}{2\pi}\left[\ln(\frac{\beta}{2}) - \text{Re}\psi(\frac{1+i\beta\mu}{2})\right] \tag{2.8}$$

The perturbative expansion of the density of states is

$$\rho(\mu_c - \mu) = \frac{1}{2\pi}\left\{ -\ln\mu + \sum_{m=1}^{\infty}(2^{2m-1} - 1)\frac{|B_{2m}|}{m}(\beta\mu)^{-2m} \right\} \tag{2.9}$$

which, since the Bernoulli numbers grow rapidly as $|B_{2m}| \sim (2m)!$, is a highly divergent asymptotic expansion of the kind that typically occur in string theories. In Ref. 6 an integral representation was constructed, whose asymptotic expansion coincides with eqn (2.9), $\frac{1}{\beta}\frac{\partial\rho}{\partial\mu} = \frac{1}{2\pi\beta\mu}\text{Im}\int_0^{\infty}dt e^{-it}\frac{t/\beta\mu}{\sinh(t/\beta\mu)}$, where we have differentiated to eliminate $\ln\mu$ and construct a function of $\beta\mu$. One can use this representation to explore the nonperturbative structure of the theory. One should remember that, in the absence of an alternate nonperturbative definition of the theory, the validity of this formula beyond its asymptotic expansion (2.9) is unclear. In terms of the Borel transform of the perturbation series (2.9), this expression corresponds to defining the integral over the Borel transform, which has an infinite number of poles on the real axis, by means of a principle value prescription.

3. Finite Radius and The Role of NonSinglet States

Both the path integral approach and the hamiltonian approach seem considerably more complicated when the target space is not simply connected, *eg*, a circle of finite radius. In the path integral approach, the Ω's are no longer unconstrained, $\Omega(t) = \Omega(t + 2\pi R)$, so that the integral over matrices can no longer be simply reduced to an integral over their eigenvalues. In the Hamiltonian approach, we now have to calculate the partition function at temperature equal to $\frac{1}{2\pi R}$, $Z = \text{Tre}^{-2\pi R\beta H}$, to which all energy levels, not just the ground state, contribute. If we neglect all but singlet states the calculation is no more difficult than before. Let us proceed then, ignoring for the moment the contribution of the non-singlet states.

Instead of working with a fixed number of fermions N, we will take the well-known route of introducing a chemical potential μ_F adjusted so that $g = \frac{N}{\beta} = \int_0^{\infty} \rho(e) \frac{1}{1+e^{2\pi R\beta(e-\mu_F)}}de$. In the thermodynamic limit, $N \to \infty$, the free energy satisfies $\frac{1}{\beta}\frac{\partial F}{\partial N} = \mu_F$. Defining $\mu = \mu_c - \mu_F$, $\lambda = \mu_c - e$, we can write these equations as

$$\frac{\partial \Delta}{\partial \mu} = \int d\lambda \rho(\mu_c - \lambda)\frac{\partial}{\partial \mu}\frac{1}{1 + e^{2\pi R\beta(\mu-\lambda)}} \quad ; \frac{\partial F}{\partial \Delta} = \beta^2(\mu - \mu_c) \qquad (3.1)$$

to emphasize resemblance with the equations for $R = \infty$. The change introduced by a finite R only affects the first of the equations. If we differentiate this equation, and use the integral representation for the density of states, we find, after performing the λ-integral, that

$$\frac{\partial \Delta}{\partial \mu} = \frac{1}{2\pi}\mathrm{Re}\int_\mu^\infty \frac{dt}{t}e^{-it}\frac{t/\beta\mu}{\sinh t/\beta\mu}\frac{t/2\beta\mu R}{\sinh(t/2\beta\mu R)} \qquad (3.2)$$

This relation has a remarkable duality symmetry under $2R \to \frac{1}{2R}$; $\beta \to 2R\beta$, the famous duality of string theory compactified on a circle.

The asymptotic expansion of eqn (3.2) is,

$$\begin{aligned}
\frac{\partial \Delta}{\partial \mu} &= \frac{1}{2\pi}\left[-\ln\mu + \sum_{m=1}^{\infty}\left(2R\beta^2\mu^2\right)^{-m}f_m(R)\right], \\
f_m(R) &= (2m-1)!\sum_{k=0}^{m}|2^{2k} - 2|\,|2^{2(m-k)} - 2|\frac{|B_{2k}|\,|B_{2(m-k)}|}{(2k)![2(m-k)]!}(2R)^{m-2k}
\end{aligned} \qquad (3.3)$$

where the dual functions $f_G(R)$ emerges at genus G. For instance, at genus 1, $f_1 \sim 2R + \frac{1}{2R}$; at genus 2, $f_2 \sim (2R)^2 + const + (\frac{1}{2R})^2$, and so forth. Using this one can evaluate the partition function to all orders in the genus expansion[10].

The fact that the above expressions are invariant under $R \to \alpha'/R$, certainly suggests that we correctly included the winding and momentum modes which distinguish the compactified theory from the theory on a real line. Furthermore, our answer agrees for genus one with a continuum Liouville calculation[19], which suggests that we have included all physical states. To justify this, we have estimated the non-singlet correction to the free energy and have shown that this correction is irrelevant for a sufficiently large radius, $R > R_c$. For $R < R_c$, the non-singlet contribution is more relevant than the singlet free energy, which indicates a phase transition at $R = R_c$. In fact, the non-singlet contribution implements the physical effects of vortices on the world sheet. Since the vortices are ignored in the standard continuum treatment of conformal field theory, it is not surprising that the continuum calculation[19]is in agreement with the singlet free energy.

In order to calculate the finite temperature partition function, we need to estimate all degeneracies and energy splittings. In studying non-trivial irreps of $SU(N)$, it is useful to remember that the ground state energy is $\mathcal{O}(N)$ while all splittings are $\mathcal{O}(1/N)$. Thus, the angular term in the hamiltonian can be treated as a perturbation[18]. For each singlet wave function, there is an infinite tower of angular excitations, labeled by n (the number of boxes in the Young tableaux for the representation). We have shown that the energy gaps between such different angular excitations grow as $|\ln\mu|$ in the limit $\mu \to 0$. On the other hand, splittings between different "radial" excitations simply carry over from the splittings in the singlet sector

$\sim 1/|\ln\mu|$. From this argument it follows that the total partition function can be factorized as $\mathrm{Tr}\,e^{-2\pi R\beta H} = \mathrm{Tr_{singlet}}\,e^{-2\pi R\beta H}\left(\sum_n D_n e^{-2\pi R\beta\delta E_n}\right)$, where δE_n is the energy gap between the ground state and the lowest state in the n th representation, and D_n is the degeneracy factor. The common factor $\mathrm{Tr_{singlet}}\,e^{-2\pi R\beta H}$, which we evaluated above, comes from summing over the "radial" excitations within each representation. The non-singlet correction to the free energy, $F_{ns} = -\frac{1}{2\pi R}\ln\sum_{n=0}^{\infty} D_n e^{-2\pi R\beta\delta E_n}$, can be estimated as follows. First, we assume that the gap δE_n between the lowest state of the n th representation and the absolute ground state is proportional to the quadratic Casimir invariant $C(n)$, as suggested by the form of the angular kinetic energy in the hamiltonian. (This assumption is not crucial to finding the leading non-singlet correction, but it simplifies some arguments.) The quadratic Casimir scales as $C(n) \approx Nn$. and the degeneracy factor D_n, which is the sum of dimensions of all representations with n white boxes, scales as $D_n \approx \frac{N^{2n}}{n!}$. Therefore, the free energy can be written as $F = F_s + F_{ns}$, where the non-singlet correction to the free energy has the simple form $F_{ns} = -\frac{1}{2\pi R}N^2 e^{-2\pi R\delta(\mu)}$.

The estimate of $\delta(\mu)$, which governs the size of gaps between different angular excitations, was derived following the work of Marchesini and Onofri[18], who studied this problem for a potential $U(\Phi) = \Phi^2 + g\Phi^4$, for *positive* values of g. For negative g, appropriately adjusted to get the double scaling limit, we have shown that $\delta(\mu) = \frac{1}{2\pi}|\ln\mu|$. Thus we find the scaling of the leading non-singlet correction to the free energy, $F_{ns} \sim N^2\mu^R$. Comparing this with the singlet free energy, $F_s \sim N^2\mu^2|\ln\mu|$, we see that the non-singlet contribution is irrelevant in the continuum limit $\mu\to 0$ for $R > 2$, but dominates the free energy for $R < 2$. This suggests a phase transition at $R = 2$ which is presumably of Kosterlitz-Thouless type. At this point the large degeneracy of the nonsinglet states (the entropy) overwhelms their suppression due to their large energy!

These matrix model results are in accord with the continuum approach to the Kosterlitz-Thouless phase transition. In ordinary conformal field theory the Kosterlitz-Thouless transition can be understood by considering the perturbation to the action of a scalar field that lives on a circle of radius R, of the form $O = \int d^2\sigma\sqrt{\hat{g}}\cos\left[\frac{R}{\alpha'}(X_L - X_R)\right]$, which is the sum of vertex operators which create states of winding number 1 and -1. The insertion of such an operator on a surface creates an endpoint of a cut in the values of X. The conformal weight of O is $h = \bar{h} = \frac{R^2}{4\alpha'}$; therefore, O becomes relevant for $R < 2\sqrt{\alpha'}$ thereby causing a phase transition at $R_c = 2\sqrt{\alpha'}$. The phase transition is connected with the instability towards creation of cuts in the values of X, *i.e.*, with vortex condensation. This picture of the Kosterlitz-Thouless phase transition can be easily adapted for coupling to two-dimensional quantum gravity. The gravitational dimension of the perturbation is then $\frac{R}{2\sqrt{\alpha'}} - 1$. For small values of the cut-off μ, and for $R > 2\sqrt{\alpha'}$, $\frac{Z}{N^2} = -\frac{\mu^2\ln\mu}{4\pi} + c\lambda^2\mu^{R/\sqrt{\alpha'}} + \mathcal{O}\left(\lambda^4\mu^{(2R/\sqrt{\alpha'})-2}\right)$, and the Kosterlitz-Thouless transition occurs at $R_c = \frac{2}{\sqrt{\alpha'}}$, the same value as in flat space. This leading correction has the same form as we found in the context of matrix quantum mechanics.

There is another exactly soluble example of a Kosterlitz-Thouless transition on a random surface, which arises in string theory defined on discretized real line with lattice spacing ϵ. This theory is related to string theory on a circle of radius $R \sim 1/\epsilon$ by a transformation to the dual lattice on the random surface[10]. This model was discussed by Parisi[20] and is interesting in its own right. The matrix model representation of the partition function is now in terms of an integral over a chain of M matrices with nearest neighbor couplings

$$Z(\epsilon) = \prod_{i=1}^{M} \int d\Phi_i \exp\left[-\beta \sum_i \left(\frac{1}{2\epsilon}\text{Tr}(\Phi_{i+1} - \Phi_i)^2 + \epsilon\text{Tr}W(\Phi_i)\right)\right] \qquad (3.4)$$

On the original lattice this model describes string theory on a discretized real line with lattice spacing $\epsilon = 1/R$. However, the dual lattice partition function defines string theory on a circle of radius R.

As before this integral can be expressed in terms of the eigenvalues of the matrices Φ_i. The only modification here is that, instead of quantum mechanics of N identical non-interacting fermions, we now find quantum mechanics with a discrete time step ϵ. Thus, $\lim_{M\to\infty} \frac{\ln Z(\epsilon)}{M} = \sum_{i=1}^{N} \ln \mu_i$, where μ_i are the N largest eigenvalues of the transfer matrix

$$\mu_i f_i(x) = \int_{-\infty}^{\infty} dy K(x,y) f_i(y), \quad K(x,y) = \sqrt{\frac{\beta}{2\pi\epsilon}} e^{-\frac{\beta}{2}\left[\frac{(x-y)^2}{\epsilon} + \epsilon\left(W(x)+W(y)\right)\right]} \qquad (3.5)$$

To find μ_i we construct a quantum mechanical hamiltonian $H(\epsilon)$ such that $K(x,y) = \langle x|e^{-\epsilon\beta H(\epsilon)}|y\rangle$. Then, $\mu_i = \exp(-\epsilon\beta e_i)$ where e_i are the N lowest eigenvalues of $H(\epsilon)$. Fortunately, it turns out that we do not need to know the exact form of $H(\epsilon)$ to find $Z(\epsilon)$ to all orders in $1/\beta^2$. In fact, as before, the only term in $W(x)$ that affects this expansion is the quadratic term about its maximum. Expanding $W(x) = x^2 - \lambda x^4$ about its maximum at $x = 1$ and rescaling, we have, in terms of $z = (x-1)\sqrt{\beta}$,

$$K(z,w) = \frac{1}{\sqrt{2\pi\epsilon}} \exp\left[-\frac{(z-w)^2}{2\epsilon} + \epsilon\left(z^2 + w^2 + \mathcal{O}\left(\frac{1}{\sqrt{\beta}}\right)(z^3 + w^3)\right)\right] \qquad (3.6)$$

Let us recall that, for an upside down harmonic oscillator with hamiltonian $H = \frac{p^2}{2m} - \frac{m\omega^2 x^2}{2}$, the propagator is

$$\langle x|e^{-\epsilon\beta H}|y\rangle = \sqrt{\frac{m\omega\beta}{2\pi \sin \omega\epsilon}} \exp\left[-\frac{m\omega\beta}{2}\left((x^2 + y^2)\cot \omega\epsilon - \frac{2xy}{\sin \omega\epsilon}\right)\right] \qquad (3.7)$$

Comparing eqs (3.6) and (3.7), we find $\cos \omega(\epsilon)\epsilon = 1 - 2\epsilon^2$, so that the energy levels of $H(\epsilon)$ are $i(n + 1/2)\omega(\epsilon)/\beta$. Thus, for small ϵ, changing the ϵ simply amounts to changing the energy scale of the quantum mechanics problem. As a result, the free energy, $E(\epsilon, \Delta) = \frac{\omega(\epsilon)}{2} E(\Delta)$, where $E(\Delta)$ is the N-fermion ground state energy of the infinite radius matrix quantum mechanics. As $\epsilon \to 0$, $\omega(\epsilon) \to 2$ and we recover matrix quantum mechanics from the infinite chain of matrices.

This model can be viewed as a representation of the string on a discretized real line where the identification with the Polyakov path integral is exact. We have shown that introducing a small lattice spacing into the target space does not change the critical properties of string theory. This result is quite astounding–it means that for all intensive purposes we can take the target space to be discrete. As long as its points are close enough together we will not notice the difference!

However this breaks down for large enough lattice spacing, which corresponds in the dual prescription to small enough radius. We find that, for $\epsilon > 1$, $\omega(\epsilon)$ and $H(\epsilon)$ become complex, which is a sign of instability of the $c = 1$ phase of string theory. This is the K-T transition. We have determined the precise location of the K-T transition on a random surface to lie at $R_c = \frac{1}{\epsilon_c} = 1$. What is the nature of the phase for $R < R_c$? Standard arguments suggest that the matter field acquires a mass and no longer affects the critical properties. Therefore, we expect the partition function to describe pure 2-d gravity, $c = 0$. This is in fact the case. In the limit $\epsilon \to \infty$ the matrix chain reduces to a collection of decoupled sites, each one described by the one-matrix model, well known to simulate $c = 0$ gravity.

4. Fermionic Field Theory Representation

The physics of the matrix model is, as we have seen, that of N non-interacting fermions, moving in the potential $U(\lambda)$, with Planck's constant equal to $\frac{1}{\beta} \sim \frac{1}{N}$. For a field theoretic description of second quantization, define a fermionic field $\Psi(\lambda, t) = \sum_i \alpha_i \psi_i(\lambda) e^{-i e_i t}$, where ψ_i are the single particle wave functions and α_i are the respective annihilation operators. The second quantized hamiltonian is then.

$$\hat{h} = \int d\lambda \left\{ \frac{1}{2\beta^2} \frac{\partial \Psi^\dagger}{\partial \lambda} \frac{\partial \Psi}{\partial \lambda} + U(\lambda)\Psi^\dagger \Psi - \mu_F(\Psi^\dagger \Psi - N) \right\} \tag{4.1}$$

where μ_F is the Lagrange multiplier necessary to fix the total number of fermions to equal N. An important feature of this field theory is that it is two-dimensional: in addition to the dependence on t, the field $\Psi(\lambda, t)$ depends on the eigenvalue coordinate λ. This is the simplest way to see how the hidden Liouville dimension emerges in the matrix model.

In the continuum limit of the matrix model representation of the sum of random surfaces, one adjusts its couplings so that the top of the potential, $U(\lambda_c)$, coincides with the Fermi energy. In this limit, the universal features of the continuum sum are due to the single particle eigenstates close to the Fermi level, which have an approximately linear energy spectrum, $e_n \approx \mu_F + n\omega$, $\omega \approx \frac{8\pi^2}{|\log \mu|}$ Given this linear spectrum, it is natural to express the non-relativistic hamiltonian of eqn (4.1) in terms of a relativistic hamiltonian for a Dirac particle[21]. Let us introduce new fermionic variables Ψ_L and Ψ_R,

$$\Psi(\lambda, t) = \frac{e^{-i\mu_F t}}{\sqrt{2v(\lambda)}} \left[e^{-i\beta \int^\lambda d\lambda' v(\lambda') + i\pi/4} \Psi_L(\lambda, t) + e^{i\beta \int^\lambda d\lambda' v(\lambda') - i\pi/4} \Psi_R(\lambda, t) \right] \tag{4.2}$$

where $v(\lambda)$ is the velocity of the classical trajectory of a particle in $U(\lambda)$ at the Fermi level, $v(\lambda) = \frac{d\lambda}{d\tau} = \sqrt{2(\mu_F - U(\lambda))}$. We substitute (4.2) into (4.1) and drop all terms which contain rapidly oscillating exponentials of the form $\exp\left[\pm 2i\beta \int^\lambda v(x)dx\right]$, since these give exponentially small terms as $\beta \sim N \to \infty$ and do not contribute to any order of perturbation theory. For the same reason we can restrict the coordinate

λ to lie between the two turning points of the classical motion, or equivalently restrict τ to lie between 0 and $\frac{T}{2}$, where T is the period of the classical motion. After some algebra we find

$$\mathcal{H} = 2\beta\hat{h} = \int_0^{\frac{T}{2}} d\tau \left[i\Psi_R^\dagger \partial\tau \Psi_R - i\Psi_L^\dagger \partial_\tau \Psi_L + \frac{1}{2\beta v^2}\left(\partial_\tau \Psi_L^\dagger \partial_\tau \Psi_L + \partial_\tau \Psi_R^\dagger \partial_\tau \Psi_R\right) \right.$$
$$\left. + \frac{1}{4\beta}\left(\Psi_L^\dagger \Psi_L + \Psi_R^\dagger \Psi_R\right)\left(\frac{v''}{v^3} - \frac{5(v')^2}{2v^4}\right) \right]$$

$$(4.3)$$

where $v' \equiv dv/d\tau$. Here we see that the natural spatial coordinate, in terms of which the fermion has a standard Dirac action to leading order in β, is τ – the classical time of motion at the Fermi level – rather than λ. As in the work of Das and Jevicki[12], we identify τ with the zero mode of the Liouville field.

The fermion fields are confined to lie in a box in the τ-direction and satisfy the boundary conditions, $\Psi_R(\tau = 0) = \Psi_L(\tau = 0)$, $\Psi_R(\tau = \frac{T}{2}) = \Psi_L(\tau = \frac{T}{2})$. These insure that the fermion number current not flow out of the finite interval, i.e., that $\bar{\Psi}(\tau)\gamma_1\Psi(\tau) = \Psi_R^\dagger \Psi_R - \Psi_L^\dagger \Psi_L$ vanish at the boundary. They also guarantee that Ψ_R and Ψ_L are not independent fields and that we are including the correct number of degrees of freedom.

Thus, we have succeeded in mapping the collection of N nonrelativisticfermions, which describe the eigenvalues of Φ, with Planck constant of order $\frac{1}{N}$, onto an action which, to leading order, is just the two-dimensional Dirac action with rather standard bag-like boundary conditions. However, the $\frac{1}{N}$ corrections in eqn (4.3) cannot be disregarded in the double scaling limit. A simple way to see this is to note that, as $\tau \to 0$, $v(\tau) = \sqrt{\mu}\sinh(2\tau)$. Then, it is easy to see that the $\frac{1}{\beta}$ terms in (4.3) are not negligible in the double scaling limit when we keep $\beta\mu$ fixed. In fact, they are given by

$$\mathcal{H}_{\frac{1}{N}} = \frac{1}{2\beta\mu} \int \frac{d\tau}{\sinh^2(2\tau)}\left[|\partial_\tau \Psi_i|^2 + 2|\Psi_i|^2\left(1 - \frac{5}{2}\coth^2(2\tau)\right)\right] \qquad (4.4)$$

where i runs over L and R. These $\frac{1}{N}$ corrections do not change the non-interacting nature of the fermions, but they do render the fermion propagator non-standard.

This fermionic field theoretic representation can be related to the bosonic formulation of Das and Jevicki [12], by simply bosonizing the fermion fields. A two dimensional free massless Dirac fermion is equivalent to a single free massless scalar boson. In our case, however, although the fermions are free, they are not truly relativistic beyond the semiclassical limit. This will give rise to interaction terms in the equivalent bosonic field theory. Following the standard bosonization rules for Dirac fermions we derive that

$$: \mathcal{H} := \frac{1}{2}\int_0^{T/2} d\tau : \left[P^2 + (X')^2 - \frac{\sqrt{\pi}}{\beta v^2}\left(PX'P + \frac{1}{3}(X')^3\right) - \frac{1}{2\beta\sqrt{\pi}}X'\left(\frac{v''}{3v^3} - \frac{(v')^2}{2v^4}\right)\right] :$$

$$(4.5)$$

where the field X is related to Ψ by $: \Psi_L^\dagger \Psi_L + \Psi_R^\dagger \Psi_R := -\frac{X'}{\sqrt{\pi}}$, and obeys Dirichlet boundary conditions. This field represents the deviation of the tachyon field, ϕ, from its background configuration, $\phi = \frac{1}{\pi}\left(v(\lambda) - \frac{\sqrt{\pi}}{\beta v}\partial_\tau X\right)$. It is not difficult to show that this is equivalent to the collective field formulation of Das and Jevicki[12]if their expression is properly normal ordered.

5. Conclusions

Much progress has been achieved in the study of simple theories of matter coupled to two dimensional gravity–noncritical string theory. However, many mysteries still remain and many directions remain to be explored. In the context of the c=1 theory work continues on the calculation of the correlation functions, on the field theoretic interpretation of the model and on the connection to the Liouville formulation. In addition one might ask:

1 Why is the space of eigenvalues of the the $N \times N$ matrices, used to generate random surfaces, identifiable as the Liouville mode of quantum gravity? Can one generalize ther matrix model methods to prove directly this identification .

2 What is the significance of the fact that the natural set of variables are fermionic fields? Is this simply an accident that occurs because the dimension of the string theory is two? Is there an equally simple representation of the $c < 1$ theories in terms of local fermion fields? What about $c > 1$?

3 The fact that the model is describable in terms of free fermion fields means that there are an infinite number of conservation laws. What are their geometric significance? Is the model a topolological field theory? Does this integrability extend to more complicated non-critical string theories?

4 Finally, there is the issue of the nonperturbative definition and construction of these theories. For example, it would be very instructive to understand the translation into the language of the string field theory of the instantons of the matrix models that are responsible for the large order behavior of the perturbation expansion and the possible singularities of the Borel transform. More generally it is important to extract all one can from these soluble models so that one can push forward towards the critical string.

References

[1] D. J. Gross and V. Periwal, *Phys. Rev. Lett.* **60**, 2105 (1988)

[2] V. Kazakov, *Phys. Lett.* **150**, 282 (1985); J. Ambjørn, B. Durhuus, and J. Fröhlich *Nucl. Phys.* **B257** , 433 (1985), F. David, *Nucl. Phys.* **B 257**, 45 (1985); V. Kazakov, I. Kostov and A. Migdal, *Phys. Lett.* **157**, 295 (1985)

[3] D. J. Gross and A. A. Migdal, *Phys. Rev. Lett.* **64** (1990) 717; M. Douglas and S. Shenker, *Nucl. Phys.* **B335** (1990) 635; E. Brezin and V. Kazakov, *Phys. Lett.* **236B** (1990) 144

[4] D. J. Gross and A. Migdal, *Nucl. Phys.* **340** (1990) 333; T. Banks, M. Douglas, N. Seiberg, and S. Shenker, *Phys. Lett.* **238B** (1990) 279

[5] D. J. Gross and A. Migdal, *Phys. Rev. Lett.* **64** (1990) 717; E. Brezin, M. Douglas, V. Kazakov and S. Shenker, *Phys. Lett.* **237B** (1990) 43; C. Crnkovic, P. Ginsparg and G. Moore, *Phys. Lett.* **237B** (1990) 196

[6] D. J. Gross and N. Miljković, *Phys. Lett.* **B238**(1990)217; E. Brezin, V. A. Kazakov and Al. B. Zamolodchikov, NP **B338**(1990) 673; P. Ginsparg and J. Zinn-Justin, *Phys. Lett.* **240B** (1990) 333

[7] E. Witten, *Nucl. Phys.* **B340** (1990) 281; J. Distler, *Nucl. Phys.* **B342** (1990) 523; R. Dijkgraaf and E. Witten, *Nucl. Phys.* **B342** (1990) 486; E. Verlinde and H. Verlinde, Princeton preprint PUPT-1176 (1990); R. Dijkgraaf, E. Verlinde and H. Verlinde, Princeton preprint (1990)

[8] M. Douglas, *Phys. Lett.* **238B** (1990) 176

[9] P. Di Francesco and D. Kutasov, *Nucl. Phys.* **B342** (1990) 589; M. Fukuma, H. Kawai and R. Nakayama, Tokyo preprint UT-562 May 1990

[10] D. J. Gross and I. R. Klebanov, *Nucl. Phys.* **B344** (1990) 475

[11] S. Das, S. Naik and S. Wadia, *Mod. Phys. Lett.* **A4** (1989) 1033; J. Polchinski, *Nucl. Phys.* **B324** (1989) 123; S. Das, A. Dhar and S. Wadia, *Mod. Phys. Lett.* **A5** (1990) 799; T. Banks and J. Lykken, *Nucl. Phys.* **B331** (1990) 173; A. Tseytlin, *Int. Jour. Mod. Phys.* **A5** (1990) 1833

[12] S. R. Das and A. Jevicki, Brown preprint BROWN-HET-750 (1990)

[13] J. Polchinski, *Nucl. Phys.* **346**,253 (1990)

[14] D. J. Gross, I. R. Klebanov and M. J. Newman, Princeton preprint PUPT-1192 (1990), to appear in *Nucl. Phys.* B

[15] V. Kazakov and A. Migdal, *Nucl. Phys.* **B311** (1989) 171

[16] E. Brezin, C. Itzykson, G. Parisi and J. Zuber, *Comm. Math. Phys.* **59** (1978) 35

[17] C. Itzykson and J.-B. Zuber, *J. Math. Phys.* **21** (1980) 411; M. L. Mehta, *Comm. Math. Phys.* **79** (1981) 327

[18] P. Marchesini and E. Onofri, *J. Math. Phys.* **21** (1980) 1103

[19] M. Bershadsky and I. R. Klebanov, Princeton preprint PUPT-1197; N. Sakai and Y. Tanii, Tokyo Inst. of Tech. preprint TIT/HEP-160

[20] G. Parisi, Roma Tor Vergata preprint, ROM2F-90/2

[21] D. J. Gross and I. R. Klebanov, Princeton preprint 1198; A. Sengupta and S. Wadia, Tata Preprint

BOSONIC STRINGS AND STRING FIELD THEORIES

IN ONE-DIMENSIONAL TARGET SPACE

Vladimir Kazakov

Lab. Physique Theorique
Ecole Normale Superieure
24 rue Lhomond, Paris 75005, France
(permanent address:
Academy of Sciences of USSR, Moscow)

Abstract: After a general review of the matrix model approach to the discretized two-dimensional quantum gravity a model of discretized bosonic strings in the $1D$ target space is considered in detail. A recently found non perturbative solution is discussed and investigated. Some aspects of this model, such as nonperturbative stability and renormalized physical obervables, are clarified. A compactified (finite temperature) version of this model is shown to be reducible to an integrable N-body problem of Calogero type.

1. Introduction

The most ambitious goal of the theory of quantum gravity is a possibility to describe all the richness of particle physics phenomena from the local properties of space-time, i.e. from the quantum fluctuations of the metric. For an effective description of quantum space-time one needs a geometrically natural and mathematically simple model of it. The traditional approach is based on the introduction of a global coordination system with the coordinates $\xi_1, \xi_2, ... \xi_d$, where d is a dimensionality of a curved manifold, representing a model of space-time. The properties of the curved space-time can be described by metric $g_{ab}(\xi)$, where $a, b = 1, 2, ...d$, defining an invariant interval dl between the close points marked by the coordinates $(\xi_1, ..., \xi_d)$ and $(\xi_1 + d\xi_1, ..., \xi_d + d\xi_d)$ on the manifold

$$(dl)^2 = g^{ab}(\xi)d\xi_a d\xi_b \tag{1.1}$$

The introduction of particular coordinates implies the existence of symmetry with respect to the diffeomorphisms, or general covariance:

Random Surfaces and Quantum Gravity
Edited by O. Alvarez *et al.*, *Plenum Press, New York, 1991*

$$\xi_a \rightarrow f_a(\xi_1, ..., \xi_d), \tag{1.2}$$

$$\tag{1.3}$$

$$a = 1, 2, ...d,$$

$$g_{ab} \rightarrow \frac{\partial f_c}{\partial \xi_a} g_{cd} \frac{\partial f_d}{\partial \xi_b} \tag{1.4}$$

All the lagrangians and measures of possible quantum gravity theories should obey this symmetry.

In fact, this symmetry looks very artificial: one introduces some coordinates and makes special attempts to verify that a theory does not depend of a particular choice of them. It is conceivable, that there should be another formulation of a theory of quantum gravity, which is not based on any coordinates, a coordinate-free formulation.

In order to do this, one has to define some local rules of construction of a curved manifold. The most natural way to do it is to imagine a collection of small pieces of flat space with the boundaries, having the topology of a sphere, and to glue them together along the boundaries in any possible way in order to create the closed manifold. Schematically, this procedure is shown in fig. 1 for the two-dimensional case. The curvature may appear along the lines of gluing.

Next step is to define the quantum version of this model. Let us note that we will always work with the Euclidean version of quantum gravity. Therefore we can use the statistical-mechanical interpretation of quantization. In this case one has to define the partition function of this Universe.

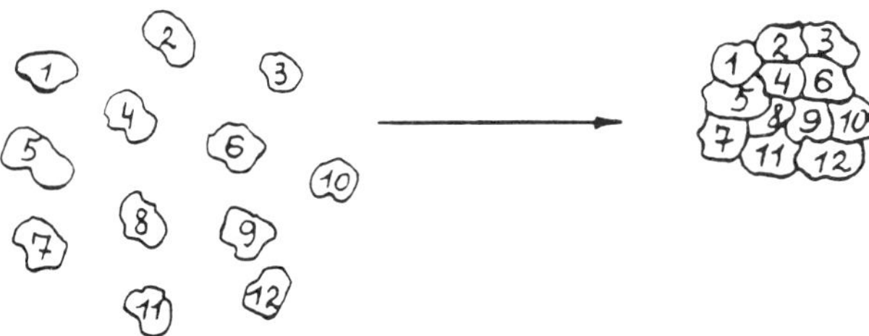

Figure 1. Building a curved manifold from a collection of flat pieces of two-dimensional space.

Again, let us do the most natural thing : let us define the partition function as the entropy of all gluings of a given number n of pieces (for certain forms of pieces and rules of gluings):

$$Z_n = \#gluings \tag{1.5}$$

2. Quantum Regge calculus in 2 dimensions

A convenient particular realization of these ideas in 2d-gravity is a construction, which can be called Quantum Regge Calculus (QRC). One uses in this case the equiliteral triangles of equal areas as abovementioned pieces of flat space and glues them together along the boundaries. By means of this operation, one obtains an abstract triangulation (see fig. 2) which may have or not have a boundary and obey any two-dimensional topology.

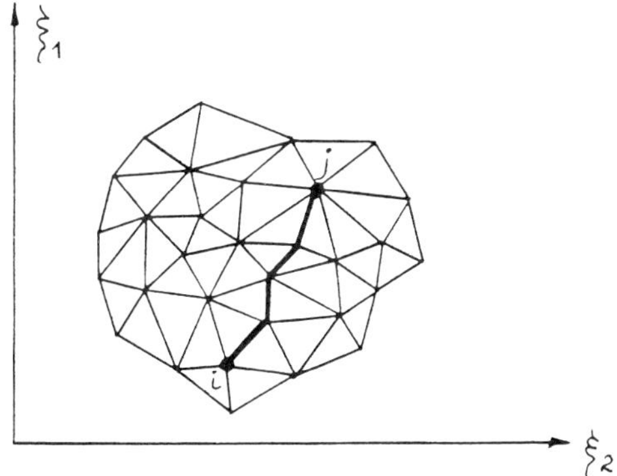

Figure 2. A model of two-dimensional curved manifold as an abstract triangulation, drown in a particular coordination system (ξ_1, ξ_2, with a "geodesics" (i,j) on it.

In this way we get (in general) a curved manifold. The curvature may appear in the vertices of triangulation: if the number of triangles, meeting at the i-th vertex, is not equal to 6, the deficit of the angle is equal to $\frac{\pi}{3}(q_i - 6)$. In principle, we may introduce some coordination system ξ_1, ξ_2 on this triangulation and define a metric $g_{ab}(\xi_1, \xi_2)$, which may be calculated in a particular gauge (see fig. 2). But there is no need in it since we can calculate any invariant quantities, like the curvatures or the invariant length of any path (drawn along the edges of triangulation, as it is shown in fig. 2), knowing only the adjacency matrix of the triangulation $G_{ij}^{(n)}$ (where n is the whole number of triangles, and i, j, = 1, 2, ..., n) defined as :

$$G_{ij}^{(n)} = \begin{cases} 1, & \text{if } i \text{ and } j \text{ are the neighbours} \\ 0, & \text{otherwise} \end{cases}$$

Let us note here, that the adjacency matrix describes unambiguously the triangulation, if we forbid for any two edges to connect the same two-vertices, and for any edge to have both ends at the same vertex (see [1]) for explanations).

One can define also an area element attached to an i-th vertex:

$$\sigma_i = \frac{1}{3} q_i = \frac{1}{3} \sum_j G_{ij}^{(n)} \tag{2.1}$$

The curvature per area unit, concentrated in the i-th vertex, is

$$R_i = \pi(\frac{6}{q_i} - 1) \tag{2.2}$$

One can see from (2.2), that a model for the flat 2-dimensional manifold is simply the regular triangular lattice with $q_i = 6$ for any i.

The well-known Euler theorem for graphs provides a discretized analogue of the Gauss-Bonet theorem (for closed manifold):

$$\int d^2\xi \sqrt{g} R(\xi) \to \sum_i R_i \sigma_i = 2\pi(2 - 2\Gamma) \tag{2.3}$$

where Γ is a genus of a triangulation.

Now, following the intuitive prescription (1.5), we can define the partition function $Z_\Gamma^{(n)}$ of the (pure) quantum 2d-gravity for the fixed number n of triangles (microcanonical ensemble) and for a given topology as:

$$Z_\Gamma^{(n)} = \#triangulations \ with \ fixed \ n \ and \ \Gamma \tag{2.4}$$

This model of quantum 2d-gravity and the corresponding Euclidean definitions of strings as discretized random surfaces were introduced in the papers [2, 3, 4]. It was inspired by Regge calculus [5, 6] for classical gravity and represents the most natural generalization for the quantum 2-dimensional case.

The next step is an introduction of matter fields on this discretized manifold, or, in other words, the definitions of various target spaces, where this manifold can be embedded - a natural way to the models of strings.

In order to do this we introduce the spins $\sigma_i, i = 1, ...n$, in every i-th vertex of a triangulation. They live in their own isotopic space (target space) and their dimensions, symmetries of interactions, quantum measures of integration may obey any desired properties, characteristic for that space.

The whole partition function for the 2d-gravity with matter fields σ_i can be defined as:

$$Z_\Gamma(\lambda, \beta, h) = \sum_{n=1}^{\infty} e^{-n\lambda} \sum_{G_\Gamma^{(n)}} \sum_{\{\sigma\}} \exp\left[-\beta \sum_{<ij>} E(\sigma_i, \sigma_j) - h \sum_i f(\sigma_i)\right] \tag{2.5}$$

were the first sum runs over all invariant areas n of manifolds with the weight $e^{-n\lambda}$, λ is a bare cosmological constant (canonical ensemble), the second sum runs over all possible triangulations with n triangles and a topology of a genus Γ, the third sum runs over all configurations of matter fields σ; $E(\sigma_i, \sigma_j)$ is the energy interaction of two spins and the sum $\sum_{<ij>}$ goes over all couples $< ij >$ of nearest neighbours on a graph $G_\Gamma^{(n)}$, β is the corresponding coupling constant ; the last term in the exponent represents an interaction of spins with the magnetic field h.

Various models of physical interest correspond to different choices of σ, E and f. Let us mention some of them.

The simplest example is the Ising model with $\sigma = \pm 1$:

$$E(\sigma_i, \sigma_j) = \sigma_i \sigma_j$$
$$f(\sigma) = \sigma \tag{2.6}$$

which was solved in [7, 8] in planar limit using the results of [9], and then in [10, 11, 12] for all topologies of lattices.

More general models, including Ising, are the Q-state Potts model [13, 15] with $\sigma = 1, 2, ...Q$:

$$E(\sigma_i, \sigma_j) = \delta_{\sigma_i \sigma_j}$$
$$f(\sigma) = \delta_{1,\sigma} \tag{2.7}$$

and a sigma model on dynamical lattices [16].

The model of Polyakov bosonic string, representing a D-dimensional embedding of 2d-gravity, will be of a particular interest for us in this paper. It is defined [17, 18, 4] for $x = (x_1, \cdots, x_D)$ - D-dimensional vectors, establishing the embedding of our lattice manifold in the D-dimensional Euclidean target space, having the gaussian interaction:

$$E(x_i, x_j) = (x_i - x_j)^2 \tag{2.8}$$

The action for this model may be represented in the form :

$$S = \sum_{<ij>} (x_i - x_j)^2 = \sum_{i,j} x_i \Delta_{ij} x_j \tag{2.9}$$

where

$$\Delta_{ij} = q_i \delta_{ij} - G_{ij} \tag{2.10}$$

is a discretized scalar Laplace operator.

In order to find the correspondence with the continuous Polyakov string, one can introduce explicitly the coordination system (ξ', ξ''), and place all the triangulation on the ξ-plane in such a way, that none of the links, connecting the vertices, are intersecting. This can always be done (see fig. 2), but for the price of making the triangles non-equilateral. The vertices will have the coordinates $\xi'_i, \xi''_i, i = 1, 2, \dots$. Of course, nothing depends on the choice of these coordinates and we have an enormous freedom of arbitrary shifts of the vertices on ξ-plane. One may associate these shifts with the diffeomorphisms. As we have noted above, the corresponding symmetry looks very artificial, and one can completely avoid mentioning it in our off-coordinate approach. It even seems, that it should exist a continuous analogue of this discrete off-coordinate construction, which would be very interesting to formulate.

One can define even the metric at a given vertex. Assuming that the points ξ_i and ξ_j for the neighbouring vertices i and j are not far from each other on the ξ-plane, we write

$$
\begin{aligned}
S &= \frac{1}{2} \sum_{<ij>} [x(\xi_i) - x(\xi_j)]^2 \\
&= \sum_i \partial_a x(\xi_i) \partial_b x(\xi_i) \sum_{j(neighb. \ of \ i)} (\xi_i - \xi_j)^{(a)} (\xi_i - \xi_j)^{(b)} = \\
&= \sum_i \sigma(\xi_i) g^{ab}(\xi_i) \partial_a x(\xi_i) \partial_b x(\xi_i)
\end{aligned}
\tag{2.11}
$$

where

$$
g^{ab}(\xi_i) = \frac{3}{q_i} \sum_{j(neighb. \ of \ i)} (\xi_i - \xi_j)^{(a)} (\xi_i - \xi_j)^{(b)}
$$

$$
\sigma(\xi_i) = \frac{q_i}{3}
\tag{2.12}
$$

In the continuous limit, when the number of triangles increases, we obtain the well-known Polyakov's action:

$$
Z_\Gamma(\tilde{\lambda}) = \int Dg(\xi) \int D^D x(\xi) \exp - \int d^2\xi \sqrt{g}(\tilde{\lambda} + g^{ab} \partial_a x^\mu \partial_b x^\mu)
\tag{2.13}
$$

The continuum limit for the measure of integration over metrics is more obscure, but there is a common belief now, that the partition function (2.5) in the case, say, of Polyakov's discretized bosonic string takes a well-known form:

$$
Z_\Gamma(\tilde{\lambda}) = \int Dg(\xi) \int D^D x(\xi) \exp - \int d^2\xi \sqrt{g}(\tilde{\lambda} + g^{ab} \partial_a x^\mu \partial_b x^\mu)
\tag{2.14}
$$

where $\tilde{\lambda}$ is a renormalized cosmological constant. The first two terms in (2.5) tend to the integral over metrics $\int Dg(\xi)$... in (2.14) (see [19] for the definition of continuous measure for the metrics), and $\sum_{\{\sigma\}}$ approaches the continuous integral over matter fields ($x(\xi)$ in this case).

Analogously, for the Ising model one can show [20] that it reduces to a theory of fermions, interacting with 2d-gravity, in the continuum limit, e.c.

3. Critical properties near the continuum limit

As is usual in the lattice formulations of field theories, a problem of a continuous limit is in fact a problem of universality properties of the thermodynamical (infinite volume) limit of the corresponding statistical-mechanical model.

In our case, consider the partition function of some model of the type (2.5) in the microcanonical ensemble with fixed area n of the triangulations, which is related to the canonical one (2.5) by:

$$Z_\Gamma(\lambda, \beta) = \sum_n e^{-\lambda n} Z_\Gamma^{(n)}(\beta) \tag{3.1}$$

In the thermodynamical limit, the asymptotics of $Z_\Gamma^{(n)}(\beta)$ is growing exponentially with n (the invariant area of a manifold):

$$Z_\Gamma^{(n)}(\beta) \sim C_\Gamma n^{-3+\gamma_{str}(\Gamma)} e^{n F_\Gamma(\beta)} \tag{3.2}$$

where $F_\Gamma(\beta)$ is a free energy per area unit, which should be independent of n (for $n \to \infty$) in the case of normal thermodynamical properties (extensivity of free energy). $F_\Gamma(\beta)$ is not a universal quantity (it depends on the pecularities of a microscopic definition of a model), but γ_{str} and C_Γ, as well as a character of singularities of $F_\Gamma(\beta)$ in β appear to be universal in many cases and depend only on macroscopical properties, such as symmetry, central charge of the matter, topology Γ, e.c. They define, as we shall see, the most important critical properties of the gravity itself.

Inserting eq.(3.2) in eq. (3.1) and substituting $\sum_n$ by $\int dn$ for large n (near the critical value $\lambda \sim \lambda_c$, i.e., in the continuous limit, only the terms with large n are essential), one obtains :

$$Z_\Gamma(\lambda, \beta) \sim C_\Gamma(\lambda - \lambda_c)^{-2+\gamma_{str}(\Gamma)} + regular\ terms \tag{3.3}$$

where

$$\lambda_c = F_\Gamma(\beta) \tag{3.4}$$

Hence, a simple recipe: if one wants to calculate the free energy, one has to find the radius of convergency of the series (3.1).

One can define a so-called string susceptibility:

$$\chi_\Gamma(\lambda) = \frac{\partial^2 Z_\Gamma(\lambda, \beta)}{\partial \lambda^2} \sim C_\Gamma(\lambda - \lambda_c)^{-\gamma_{str}(\Gamma)} + regular\ terms \tag{3.5}$$

with the universal properties defined by γ_{str} and C_Γ. $\chi_\Gamma(\lambda)$ represents a two-point function, in the sense that it can be defined as a sum over all triangulations, pinpointed in two vertices. This is clear from the fact that every marked vertex on a triangulation gives an extra factor n inside the sum in (3.1), and the same effect results from every differentiation in λ. One can define in this way, the m-point functions:

$$\chi_{\Gamma,m}(\lambda) = \frac{\partial^m Z_\Gamma(\lambda,\beta)}{\partial \lambda^m} \sim (\lambda - \lambda_c)^{-\gamma_{str}(\Gamma)-m+2} + regular\ terms \qquad (3.6)$$

The other interesting critical properties, of the matter itself, are defined by the quantities such as specific heat:

$$C(\beta) = \frac{\partial^2 F_\Gamma(\beta)}{\partial \beta^2} \sim (\beta - \beta_c)^{-\alpha} + regular\ terms \qquad (3.7)$$

magnetic susceptibility:

$$\kappa(\beta) = \frac{\partial^2 F_\Gamma(\beta,h)}{\partial h^2} \sim (\beta - \beta_c)^{-\gamma} + regular\ terms \qquad (3.8)$$

and others. The first and the simplest example of the calculation of all these quantities was given for the Ising model in [8].

An important step was made in the paper [21] and then in the papers [22, 23, 24], where by means of the continuum (Liouville) theory of 2d- gravity the scaling properties of all these models were calculated. It was found that the scaling dimensions Δ_{mn} of primary fields of unitary minimal models are defined by the central charge c of the matter itself by the formula

$$\Delta_{mn} = -\frac{1}{2r}[1 \pm [(1+r)m - n]] \qquad (3.9)$$

where r is related to c by

$$c = 1 - \frac{6}{r(r+1)} \qquad (3.10)$$

In the case of spherical topology γ_{str} corresponds to the operator with the most negative conformal weight Δ_N in the flat space:

$$\gamma_{str}(\Gamma = 2) = 2\Delta_{00} == \frac{c - 1 - 24\Delta_N - \sqrt{(1-c)(25 - c + 24\Delta_N)}}{12(1 - \Delta_N)} \qquad (3.11)$$

For the unitary minimal models $\Delta_N = 0$ but for the nonunitary models, as Li-Yang singularity, it can be nonzero [12].

Before these formulae were got in [22], a few particular models were solved directly from the dynamical triangulation formulation (2.5). Namely, these are bosonic string with c=D=0 [2, 3] and c=D=-2 [18, 1, 25] dimensional target spaces, and the Ising model on random triangulations [7, 8] with the central charge c=1/2. The results (3.9) and (3.11) appear to be in whole correspondence with them.

Some later results for various discrete models of 2d gravity with other central charges of matter fields [13, 14, 26, 27, 28, 29], and even with a continuously varying one [16, 30] exhibit a perfect correspondence as well.

The generalization of all these results to all genuses [23, 24] has shown a remarkable property - linearity of $\gamma_{str}(\Gamma)$ in the genus Γ:

$$\gamma_{str}(\Gamma) = -1/r + (2 + 1/r)\Gamma \tag{3.12}$$

the fact which was already found in [25] for a discrete model of bosonic string in d=-2 dimensions (r=1).

This enables us to formulate a specific double scaling limit of these models, which serves as a definition of string field theories [31, 32, 33]. As is known, the field theory of closed strings corresponds in the euclidean language to the sum over world sheets of strings, obeying all possible topologies, but with the factor $N^{-2\Gamma}$ for every given genus Γ of a world sheet. The new parameter 1/N corresponds to the vertex of interaction of three closed strings. Two such vertices create a handle on the sheet. This process corresponds to a creation of two strings from one, and then contraction into one again. Hence on the oriented closed surface every handle corresponds to the factor $1/N^2$.

The string susceptibility $\chi(\lambda, N)$ for such a string field theory is defined simply as

$$\chi(\lambda, N) = \sum_{\Gamma=0}^{\infty} N^{-2\Gamma}\chi_\Gamma(\lambda) \tag{3.13}$$

Now, inserting (3.5) and (3.12) into (3.13), we see that $\chi(\lambda, N)$ for $\lambda \to \lambda_c$ can be represented in the scaling form:

$$\chi(\lambda, N) = N^{\frac{-2}{2r+1}} f(x) + regular\ terms \tag{3.14}$$

where

$$x = (\lambda - \lambda_c)N^{\frac{r}{2r+1}} \tag{3.15}$$

is a scaling parameter and an universal scaling function f(x) is formally defined by the divergent expansion

$$f(x) = \sum_{\Gamma=0}^{\infty} C_\Gamma x^{-(2+1/r)\Gamma} \tag{3.16}$$

The double scaling prescription is the following: let $N \to \infty$ and $\lambda \to \lambda_c$ in such a way that x remains fixed. Then eqs. (3.14), (3.15) and (3.16) define universal scaling behaviour of the model.

One has to make the following remark. The coefficients C_Γ are usually growing as

(2Γ)! with Γ. Hence the function (3.16) can be represented as a series only up to a nonperturbative exponentially small terms of the order $\sim exp(-ax^{\frac{2r}{2r+1}})$, which is an inherent property of the string field theories [34]. The question of a nonperturbative definition of the expansion (3.16) seems still to be quite obscure, in spite of the very recent progress [35, 36, 37, 38, 39] (see also next chapters about $1D$ string). Many of the originally found nonperturbative solutions suffer from nonstabilities, leading to nonvanishing complex parts of physical quantities for finite x [23].

It would be useful to try to answer the general question: if we know only the coefficients of the divergent expansion of the scaling function (3.16), what is the space of all nonperturbative parameters of that function? At least, is it infinite or finite?

4. Dynamical triangulations and matrix models

A remarkable feature of the discretized models of 2d-gravity of the type (2.5) is their exact solvability in many interesting cases. Most of the methods are based on one-to-one correspondence of (2.5) to some integrals over matrices with specially chosen weights and measures.

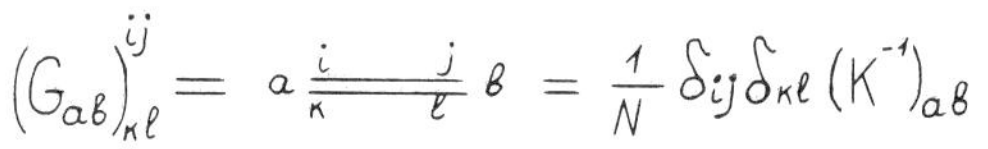

$$\left(G_{ab}\right)^{ij}_{\kappa\ell} = a \underset{\kappa}{\overset{i}{\rule{0pt}{0pt}}}\!\!\!\!\!\underset{\ell}{\overset{j}{\rule{3em}{0.4pt}}}\, b \;=\; \frac{1}{N}\delta_{ij}\delta_{\kappa\ell}\left(K^{-1}\right)_{ab}$$

Figure 3: A propagator of a zero dimensional Q-matrix ϕ^3-field theory.

To be more concrete, let us consider a rather big class of matrix models with the partition function

$$Z = \int \prod_{a=1}^{Q} d^{N^2}\phi_a \exp Ntr(\sum_{a>b} K_{ab}\phi_a\phi_b + \sum_{a} \lambda_a\phi_a^3) \tag{4.1}$$

where ϕ_a - are NxN hermitian matrices with matrix elements $(\phi_a)_{ij}$, $i, j = 1, 2, \cdots, N$, $a = 1, 2, \cdots, Q$. The gaussian part of the action is defined by the quadratic form K_{ab}, and λ_a serve as coupling constants of ϕ^3-type self-interaction.

To demonstrate the correspondence to (2.5), one has to expand (4.1) in powers of all λ_a's and formulate the rules of Feynman diagram technique.

It is useful to use the double line notation for the propagators G_{ab} of the diagrams, originally introduced by 't Hooft in the classical papers on 1/N-expansion [40]:

$$(G_{ab})^{ij}_{kl} = 1/N\delta_{ij}\delta_{kl}(K^{-1})_{ab} \tag{4.2}$$

Fig. 3 shows that every line carries (and conserves) a "colour" index (i=j, or k=l) and the whole propagator describes a propagation of an "isotopic" index a, which can be transformed to b with the "amplitude" given by $(K^{-1})_{ab}$.

The vertex $(\Lambda^{abc})_{ik,jm,ln}$ (see fig. 4) has a similar structure:

$$(\Lambda^{abc})_{ik,jm,ln} = N\lambda_a \delta_{ab}\delta_{bc}\delta_{ij}\delta_{kl}\delta_{mn} \tag{4.3}$$

Here the isotopic index is conserved, but one can have different types of ϕ^3-interactions, due to the abc-index structure.

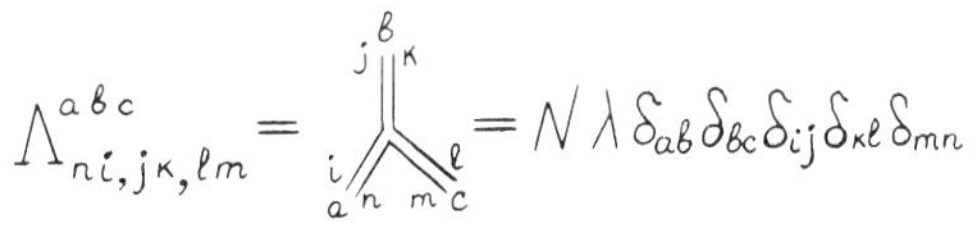

$$\Lambda^{abc}_{ni,jk,lm} = \; = N\lambda\,\delta_{ab}\delta_{bc}\delta_{ij}\delta_{kl}\delta_{mn}$$

Figure 4: A vertex of the interaction of a zero dimensional Q-matrix ϕ^3-field theory.

If one combines now all these elements into diagram technique, one gets ϕ^3 diagrams, drawn in the double line notation, as shown in fig. 5. Every index loop gives the factor $1/N$, according to (4.2) and (4.3). A given diagram gives a dependence on N as:

$$N^{\#vertices - \#propagators + \#loops} = N^{2-2\Gamma} \tag{4.4}$$

according to the Euler theorem for graphs.

This enables us to classify all the graphs by their genuses Γ , in terms of $1/N$ expansion [40].

The free energy $F(\lambda, N) = 1/N^2 \log Z$ of the model (4.1) can be written in the following way:

$$F(\lambda, N) - \sum_{N=0}^{\infty} N^{-2\Gamma} \sum_{\tilde{G}_\Gamma} \sum_{\{a\}} \prod_{<ij>\in\tilde{G}_\Gamma} (K^{-1})_{u_i u_j} \prod_{k\in\tilde{G}_\Gamma} \lambda_{a_k} \tag{4.5}$$

Here $\tilde{G}_\Gamma$ is a ϕ^3 graph of a genus Γ; $a_1, a_2, ...a_k...$ - is a collection of isotopic indices of matrices ϕ_a $(k \in \tilde{G}_\Gamma)$, appeared in a given term of expansion in λ_a's; $< ij >$ are the propagators, connecting the i-th and the j-th vertices of a graph; $\{a_i\}$ can be considered as a configuration of isotopic spins, taking the values $a_i = 1, 2, ..., Q$, as in the Potts model, defined by (2.7).

$(K^{-1})_{a_i a_j}$ is a Boltzmann weight of two neighbouring spins a_i and a_j. If one introduces the notation:

$$(K^{-1})_{a_i a_j} = \exp(\beta \tilde{E}(a_i, a_j)) \tag{4.6}$$

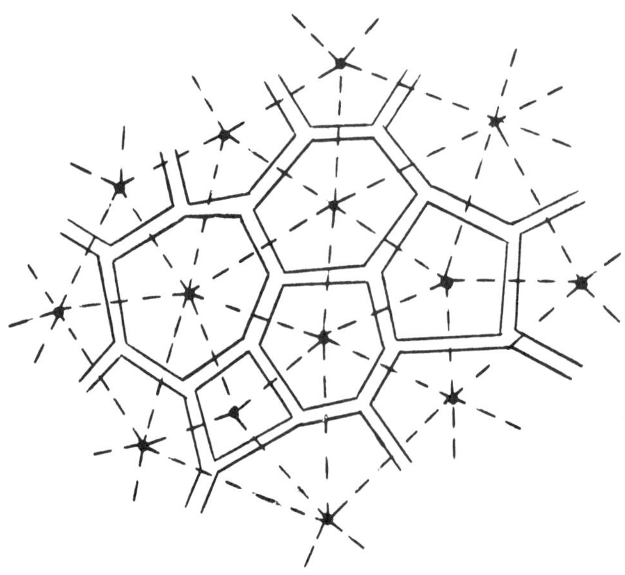

Figure 5. A piece of a ϕ^3-planar graph (double lines), and a dual triangulation (dotted lines).

and

$$\lambda_{a_i} = e^\lambda \exp(h\tilde{f}(a_i)) \tag{4.7}$$

one arrives to the discretized model of 2d gravity with matter fields, defined by eq.(2.5), with the only difference that it is formulated on ϕ^3 graphs, and not on the triangulations.

This last difference is not crucial at all. First of all, we hope that the universal properties of the models of physical interest do not depend on the microscopic properties of a lattice, and both, triangulated and ϕ^3, lattices, if they are macroscopically large, look similar for a "remote observer". The second important circumstance is the exact equivalence of many interesting models on both types of lattices. This can be established by duality transformation. For example, in the case of Potts model (2.7) a duality transformation brings the spins σ_i from the vertices of the original triangulated lattice G to the vertices of a dual one $\tilde{G}$, with the same type of a nearest neighbours interaction, but with a dual temperature $\tilde{\beta}$, related to the original one β by the formula ([13]):

$$\tilde{\beta} = \log tanh(1 + \frac{Q}{e^\beta - 1}) \tag{4.8}$$

Every triangulation corresponds to one and only one dual graph, as is clear from the fig. 5.

Hence, the equivalence is exact.

This matrix formulation of the models of 2d gravity has been proved to be a powerful tool for finding the exact solutions in many interesting cases. First used in the investigation of pure 2d gravity in [2, 3], and of the Ising matter, interacting with 2d gravity [7], it was applied later to a number of other important systems.

280

One has to underline, that not every model of the type (4.1) can be solved exactly by known methods. It is not surprising, since (4.1) includes nearly all interesting models of strings, and we could not hope in such a fortune. For example, one can easily formulate the models, obeying the central charge $c > 1$ of a matter, for which the KPZ formulae (3.9), (3.11) break down. Remarkably enough, this so-called c=1 barrier invisibly exists in our lattice models, preventing them from being exactly solvable.

To understand the arising difficulties on a purely technical level, let us recall, that the solution of any matrix model starts from the angular decomposition of hermitian matrices:

$$(\phi_a)_{ij} = \sum_k (\Omega_a^+)_{ik}(x_a)_k(\Omega_a)_{kj} \qquad (4.9)$$

where

$$x_a = diag((x_a)_1, (x_a)_2, ..., (x_a)_N) \qquad (4.10)$$

are the eigenvalues of ϕ_a, and $(\Omega_a)_{ij}$ are the U(N) group matrices (generalized "angles").

x_a and Ω_a should be understood as new variables. The corresponding (Dyson) measure is:

$$d^{N^2}\phi = (d\Omega)_{U(N)} \prod_{i=1}^{N} dx_i \Delta^2(x) \qquad (4.11)$$

where

$$\Delta(x) = \prod_{i<j}(x_i - x_j) \qquad (4.12)$$

is the Van-der-Monde determinant, and $(d\Omega)_{U(N)}$ is the standard Haar measure.

In the action of a matrix model (4.1) only the first, quadratic term is dependent on Ω's:

$$\sum_{a>b} K_{ab} tr(\phi_a \phi_b) = \sum_{a>b} K_{ab} tr[(\Omega_{ab}^+)(x_a)(\Omega_{ab})(x_b)] \qquad (4.13)$$

where

$$\Omega_{ab} = \Omega_a \Omega_b^+ \qquad (4.14)$$

In the case of two matrices ϕ_1 and ϕ_2 we have only one "angular" degree of freedom Ω_{12}, and the group integration can be performed by the well known formula [41]:

$$\int (d\Omega)_{U(N)} \exp[tr(x_{(1)}\Omega^+ x_{(2)}\Omega] = (\prod_{m=0}^{N-1} m!) \prod_{k=0}^{N-1} \frac{det_{ij}\exp[(x_1)_i(x_2)_j]}{\Delta(x_1)\Delta(x_2)} \qquad (4.15)$$

Then the problem reduces to the integration over 2N variables x_1, x_2, which is, in principle, a saddle point problem, since in the large N limit the action of a matrix model is of the order N^2, for only N variables of integration.

At first sight, it seems that all the models of the type (4.1) are reducible to the saddle point problem, by use of the substitution (4.9) and eq.(4.15). But it is wrong for the following reason.

Let us consider a graph K of a quadratic form K_{ab}. Q vertices of the graph correspond to the indices of the matrix K_{ab}, and the edges connect the vertices with the (nonequal) numbers a and b only for nonzero elements K_{ab}.

It is easy to see, that not all the matrices Ω_{ab}), corresponding to the edges $< ab >$ of the graph, are independent: in virtue of the definition (4.14) there exists a matrix identity

$$\prod_{<ab>\in L} \Omega_{(ab)} = I \tag{4.16}$$

for any independent loop L on K. These conditions should be included as δ-functional constraints to the integrals over "angular" variables $\Omega_{(ab)}$.

A rather general class of integrable (in the sense of reduction to the eigenvalue integrations only, even though the complete solution might be still nontrivial) models of the type (4.1) is represented by the tree-like quadratic forms, with the corresponding graph K having no loops.

Sometimes even the cases of non tree-like quadratic forms are integrable. For example, by the additional integration in ψ-matrix:

$$\exp[\frac{1}{2}\sum_{a,b=1}^{Q} tr(\phi_a\phi_b)] = \int d^{N^2}\psi \exp -tr(\frac{\psi^2}{2} + \psi \sum_{a=1}^{Q} \phi_a) \tag{4.17}$$

the quadratic form of the Potts model on dynamical graphs can be brought to the tree-like form [13] (see fig. 6).

Now we are going to demonstrate all these general principles on the example of D-dimensional bosonic string, mostly for D=1 case.

5. D-dimensional bosonic string as a matrix model

The discretized version of the bosonic Polyakov string can be formulated as a matrix field theory, where the isotopic space is the D-dimensional target space itself. Namely, we introduce a hermitian matrix field $\phi_{ij}(x)$ ($x = (x_1, x_2, ..., x_D), i, j = 1, 2, ...N$) is a D-dimensional vector), and define a matrix scalar field theory with the partition function:

$$\zeta(N, \lambda) = \int D^{N^2}\phi(x) \exp -\frac{N}{\lambda^2}tr \int d^D x[\frac{(\partial_\mu\phi)^2}{2} + V(\phi)] \tag{5.1}$$

where

$$V(\phi) = \phi^2/2 + \phi^3/3 \tag{5.2}$$

but might be a more general potential.

Let us compare the Feynman diagrammatic rules with the models of the type (2.5) for the free energy $f(N, \lambda) = \frac{1}{N^2}log\xi(N, \lambda)$. This will produce ϕ^3-diagrams, classified

282

by genuses Γ by means of $1/N$ expansion. The role of matter spines is played by the D-dimensional vectors x_i placed in the vertices of a ϕ^3-graph, labeled by $i = 1, 2, ..., n$. The spins establish an embedding of a graph in the target space. The Boltzmann weights correspond to the propagators

$$\exp E(x_i, x_j) = \int d^d p \, \frac{e^{ip(x_i - x_j)}}{p^2 + 1} \tag{5.3}$$

One can reformulate everything on the graphs dual to ϕ^3 graphs, i.e. triangulations, introducing the D-dimensional momenta p_a in the vertices $a = 1, 2, ..., \tilde{n}$ of a dual triangulation (see [1] for the details).

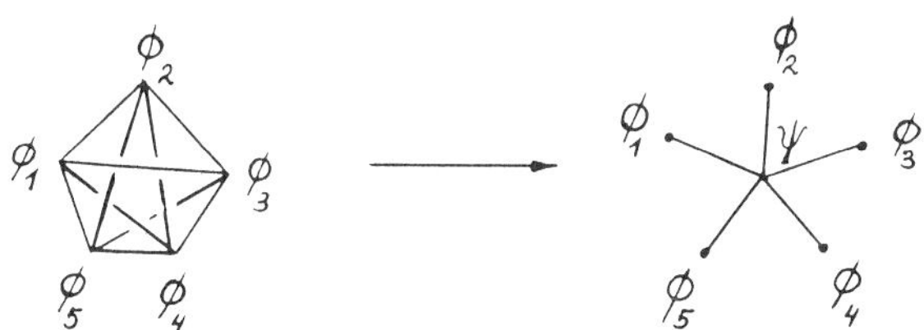

Figure 6. A transformation of the graph of the quadratic form of the Q-matrix integral, representing the Q=5 component Potts model on dynamical triangulations, to the tree-like type, by means of an additional gaussian ψ-integration.

The loops are in a one-to-one correspondence with the vertices of the triangulation. The corresponding Boltzmann weights are just the standard Feynman propagators in momentum space:

$$\exp E(p_a, p_b) = \frac{1}{(p_a - p_b)^2 + 1} \tag{5.4}$$

Now the momentum space plays the role of a target space, and both formulations are a priori suitable for the definition of bosonic strings.

The duality transformation is a discrete analog of a well known change of of variables to a dual scalar field $\partial^\mu p = i\varepsilon_{\mu\nu}\partial_\nu x$ in the continuous bosonic string theory.

Using Feynman prescriptions, one can now rewrite the free energy as a partition function of a D-dimensional string field theory:

$$\zeta(N,\lambda) = \frac{1}{N^2}log\zeta(N,\lambda) = \sum_{\Gamma=0}^{\infty} N^{-2\Gamma} \sum_{n=1}^{\infty} \lambda^n \sum_{G_\Gamma^{(n)}} \int d^D x_1...d_n^D \exp - \sum_{<ij>} E(x_i - x_j) \quad (5.5)$$

where *log* is necessary for keeping only connected graphs $G_\Gamma^{(n)}$, as it should be for the partition function of a string theory.

The expression (5.5) coincides with our definition (2.5) of a discretized Polyakov string up to the fact that $E(x_i - x_j)$ in the form (5.3) (or (5.4)) is different from the original gaussian expression (2.9). But one can hope, that both definitions give the same universality class. This should be true at least for $D < D_c$, where D_c is the critical dimension, below which the Feynman diagrams are ultraviolet superconvergent (as is well known, $D_c = 6$ for ϕ^3-interaction).

But even this limitation seems to be ruther technical, than physical, since the kinetic term $(\partial_\mu \phi)^2$ was chosen by us for the sake of simplicity, and the ultraviolete instability could be treated by adding higher derivatives.

As was pointed out in previous sections, the 1/N expansion in the expressions of the type (5.5) corresponds to a definition of a string field theory, where 1/N is a constant of a topological interaction of three closed free strings. To reach a continuum limit in this theory, one has, as usually, to take a double limit $\lambda \to \lambda_c$, where λ_c is a critical cosmological constant for every genus Γ, and $N \to \infty$, keeping some combination of these two parameters fixed.

In the case $D < 1$ (formal analytical continuation from the integer values of D) we expect from the model (5.5) the same power like scaling behaviour, as described by eqs. (3.9)-(3.12) (recall, that the central charge c=D).

For $D \geq 1$ one can expect the existence of a more complicated scaling parameter - some function $\mu(\lambda - \lambda_c, N)$, where the double scaling can be approached along a more complicated (then power-like) trajectory in the (λ, N) space.

As we shall demonstrate in the next sections, even in the limiting D=1 case the scaling parameter is nontrivial due to the logarithmic corrections.

6. Bosonic string field theory in one physical dimension

This is a model of 2d quantum gravity, "living"in the physical euclidean time t (D=1). The partition function (5.5) (with $x_i = t_i, d^D x = dt$) may be represented as a vacuum-vacuum matrix element of the quantum-mechanical evolution operator (during the time β) [29]:

$$\zeta(N,\lambda) = < 0|e^{-\hat{H}\beta}|0 > = \int D^{N^2}\phi(t) \exp - \frac{N}{\lambda^2} tr \int_0^\beta dt[\frac{(\dot{\phi})^2}{2} + V(\phi)] \quad (6.1)$$

In the corresponding ϕ^3 diagram technique we obtain the propagators

$$E(t_a - t_b) = e^{-|t_a - t_b|} \qquad (6.2)$$

instead of the original ones $E_P(t_a - t_b) = e^{-(t_a - t_b)^2}$. We again hope that the superconvergibility makes this difference irrelevant.

Next standard step is to diagonalize ϕ

$$\phi_{ij}(t) = \sum_{k=1}^{N} \Omega_{ik}^{+}(t) z_k(t) \Omega_{kj}(t) \qquad (6.3)$$

It is easy to see that

$$tr\dot{\phi}^2 = \sum_{i=1}^{N} \dot{z}_i^2 + \sum_{i \neq j}(z_i - z_j)^2 |A_{ij}|^2 \qquad (6.4)$$

where

$$A_{ij}(t) = (\Omega^{+}\dot{\Omega})_{ij} \qquad (6.5)$$

is a one dimensional "connectivity". The measure will be transformed into

$$D^{N^2}\phi(t) = \prod_{t \in [0,\beta]} \Delta^2(z(t)) \prod_{i=1}^{N} Dz_i(t) \prod_{t \in (0,\beta)} D^{N^2-N} A(t) \qquad (6.6)$$

Since there are no cycles in the quadratic form $\int dt(\dot{\phi}^2 + \phi^2)$ in (6.1) for free boundary conditions on $\phi(0)$ and $\phi(\beta)$ (see in the next section some comments about a more complicated periodic case $\phi(0) = \phi(\beta)$), the variables $A(t)$ are independent for any t, and the angular part of the integral becomes gaussian:

$$\int D^{N^2} A(t) \exp -\frac{N}{\lambda^2} \int_0^{\beta} dt \sum_{i \neq j}(z_i - z_j)^2 |A_{ij}|^2 \sim \prod_{t \in (0,\beta)} \Delta^{-2}(z(t)) \qquad (6.7)$$

If we take use (6.4), (6.6) and (6.7) in (6.1), we notice that the Van-der-Monde determinants of the measure (6.6) formally cancel with the corresponding determinants from (6.7). In fact, it is true only up to boundary effect, which gives extra factor $\Delta(z(0))\Delta(z(\beta))$ to the resulting integral over eigenvalues.

Finally, we arrive to the following representation

$$\zeta(N, \lambda, \beta) = \int \prod_i Dz_i(t) \exp -\frac{N}{\lambda^2} \int_0^{\beta} dt[\frac{\dot{z}_i^2}{2} + V(z_i)] \, \Delta(z(0))\Delta(z(\beta)) \qquad (6.8)$$

The existence of this extra factor follows from the fact, that every $\Delta^2(z(t))$ in (6.6) corresponds to a given point t, whereas the connectivity (6.5) is defined in the ε-vicinity of t, and one has to rewrite (6.4) in a more precise way:

$$\sum_{i\neq j}(z_i(t) - z_j(t))|A_{ij}(t)|^2 =$$

$$\sum_{i\neq j}(z_i(t) - z_j(t))(z_i(t + \varepsilon) - z_j(t + \varepsilon))A_{ij}(t, t + \varepsilon)A_{ij}^*(t, t + \varepsilon) \tag{6.9}$$

In this way it will be clear, that one Van-der-Monde determinant at every endpoint will be missing in (6.7).

A more direct way to prove this is to introduce the discrete time (lattice) version of this lattice model, with the action

$$S = tr[\sum_{a=1}^{\beta} \phi_{a-1}\phi_a + \sum_{a=0}^{\beta} V(\phi_a)] \tag{6.10}$$

A change of variables, similar to (4.9)-(4.14), and the use of the formula (4.15) will immediately convince us, that we deal with the discretized version of (6.8) (see ([42]) for details).

Eq.(6.8) defines a model of N noninteracting quantum-mechanical particles with the coordinates z_i, obeying the Fermi statistics, since the in- and out-states are anti-symmetrized due to the left Van-der-Monde determinants. The corresponding wave-eigenfunctions can be represented as Slater determinants:

$$\Psi_{n_1,n_2,...n_N} = \frac{1}{\sqrt{N!}}det_{km}\psi_k(\lambda_m) \tag{6.11}$$

where the one-particle eigenfunctions obey the Schrodinger equation:

$$[-\frac{1}{2N^2}\frac{d^2}{dz^2} + V(z) - \varepsilon_k]\psi_k(z) = 0 \tag{6.12}$$

Hence,the problem is completely integrable in the above-mentioned sense. It is no more a field theory problem, but rather a one-particle quantum mechanics. This nice result was found in the paper [43].

Now let us proceed with the calculation of the partition function $\zeta(N, \lambda, \beta)$. We shall do this directly in the double scaling limit.

For a sufficiently long evolution time β we have the asymptotics

$$\zeta \sim e^{-\beta E_0(N,\lambda)} \tag{6.13}$$

where $E_0(N, \lambda)$ is the partition function of the corresponding string field theory.

For a stable system (with a bounded below and growing for $|z| \to \infty$ potential $V(z)$) E_0 is just the sum of energies of N lowest states:

$$E_0 = \sum_{i=1}^{N} \varepsilon_i \tag{6.14}$$

The corresponding Fermi-level μ is simply

$$\mu = \varepsilon_N \qquad (6.15)$$

However, in the case of $\phi^2 + \phi^3$ potential we only have a metastable minimum (see fig.7), and the eigenvalues become resonances with finite widths. What rescues us in this situation is a fact, that the action of every fermion is of the order $\sim N$, and hence the amplitude of tunnelling through the barrier, defining the width of a resonance, $Im(\varepsilon)$, is exponentially small for large N:

$$Im(\varepsilon_i) \sim e^{-aN} \qquad (6.16)$$

Therefore this instability should be invisible in any order of the topological $(1/N)$ expansion, i.e. it can be seen only in the whole nonperturbative solution.

Later we shall see how to define the model nonperturbatively without this instability, but with the same topological expansion, as in the nonstable case, but so far we will ignore this problem.

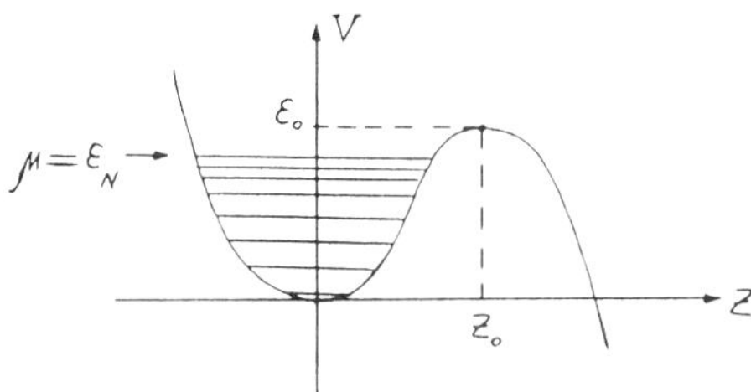

Figure 7. The spectrum of metastable states in the $\phi^2 + \phi^3$-potential; $\mu = \varepsilon_N$ represents the fermi level of the system of N fermions.

By changing the coupling λ in (6.8) we effectively change the form of the potential $V(\lambda)$, and at some critical value λ_c the N-th level, having the energy ε_N, may touch the top of the potential (with the coordinates z_0, ε_c in fig. 7). This λ_c corresponds to the critical point of divergency of a sum over diagrams for any fixed genus [43] and serves, as was pointed out in [29], as a critical cosmological constant, near which the continuous limit exists (see eq.(3.3)).

It is clear from all this, that the universal properties of the corresponding string field theory are defined by the vicinity of the top of the potential $V(z)$. Therefore it is useful to do the following change of variables in (6.12):

$$z = z_0 + a\frac{x}{\sqrt{N}} \tag{6.17}$$

$$\varepsilon_k = \varepsilon_0 - b\frac{\xi_k}{N} \tag{6.18}$$

where z_0, ε_0, a, b are the constants, chosen such that (in the case of the cubic potential) eq.(6.12) turns to

$$[-\frac{d^2}{dx^2} - \frac{x^2}{4} + \frac{\lambda}{3\sqrt{N}}x^3 + \xi_k]\psi_k(x) = 0 \tag{6.19}$$

For the potential V, more general than $\phi^2 + \phi^3$, one gets extra powers in (6.19), of a type $\sum_{k>3}\lambda_k(\frac{x}{\sqrt{N}})^k$.

If we now assume, that in the double scaling limit N tends to infinity, and λ to λ_c in such a way, that the essential ξ_k's and the coordinates x remain finite, we may neglect the last term in (6.19) (and all the next terms, if they exist, suppressed by extra powers $\frac{1}{\sqrt{N}}$), and we arrive finally to the parabolic cylindre equation:

$$[\frac{d^2}{dx^2} + \frac{x^2}{4} - \xi]\psi(x) = 0 \tag{6.20}$$

We expect, that (6.20) is a universal wave equation, describing the double scaling regime.

Of course, to have discrete levels in the inverted quadratic potential, we have to remember at some point, that there is a wall somewhere at a distance $\Delta x = \Lambda \sim \sqrt{N}$ from the position of the top, which prevents the particle from going to infinity. The characteristic depth of the potential is of the order $\Delta V \sim \Lambda^2 \sim N$.

It is natural to expect, that for such kind of potential the particle spends most of the time in the quasiclassical regime, i.e. it has almost always a big classical velocity $v(t) = \frac{1}{\sqrt{x^2/4-\xi}}$. Indeed, the classical time period of motion

$$\Delta T = 2\int_{2\sqrt{\xi}}^{\Lambda} dx \frac{1}{\sqrt{x^2/4 - \xi}} \sim log\Lambda \tag{6.21}$$

is very large, and the particle is almost always at distances $1 << |x| << \Lambda$ from the turning points of the potential.

All this enables us to use a quasiclassical approximation for the (exact) calculation of some basic physical quantities in the double scaling limit. Namely, we take quasiclassical asymptotics for the parabolic cylindre wave function [44]:

$$\psi_\xi(x) \approx \frac{1}{\sqrt{x}} \sin(x^2/4 + \xi \log x + \frac{\Phi(\xi)}{2} + \Phi_0)[a + b/x + c/x^2 + ...] \tag{6.22}$$

where Φ_0 is a constant, and

$$\Phi(\xi) = -\frac{i}{2}\log\frac{\Gamma(\frac{1}{2}+i\xi)}{\Gamma(\frac{1}{2}-i\xi)} \tag{6.23}$$

is a phase shift of the wave function due to the reflection of the inverted quadratic potential. In the Appendix we give the derivations of eqs.(6.22) and (6.23).

The Bohr-Sommerfeld quantization condition prescribes for the phase

$$\Phi_0 + \frac{1}{2}\Phi(\xi) + \xi\log\Lambda + \frac{\Lambda^2}{4} = -\pi n,$$
$$n = 0, 1, 2, ..., \tag{6.24}$$

which means in this case, that the wave function should be zero at the infinite wall, placed at a distance Λ from the top of the potential.

Note that such a simple treatment of this problem is possible due to the fact, that the whole phase factor is concentrated in the argument of sin, and the corrections contain only integer powers of x and hence do not contribute to the phase shift.

The one-particle density of states is

$$\rho(\xi) = \frac{1}{\pi}\frac{\partial n}{\partial\xi} = \frac{1}{\pi}[\frac{1}{2}\frac{\partial\Phi(\xi)}{\partial\xi} + \log\Lambda] =$$
$$= -\frac{1}{2\pi}Re\psi(\frac{1}{2}+i\xi) + \frac{1}{\pi}\log\Lambda \tag{6.25}$$

where $\psi = \frac{\Gamma\prime}{\Gamma}$ is the digamma function of Euler. It has a divergent $\frac{1}{\xi}$ expansion:

$$\rho(\xi) = \frac{1}{2\pi}\log\frac{\Lambda^2}{\xi} + \frac{1}{2\pi}\sum_{\Gamma=1}^{\infty}C_\Gamma\frac{1}{\xi^{2\Gamma}} + O(e^{-2\pi\xi}) \tag{6.26}$$

where

$$C_\Gamma = (-1)^{\Gamma+1}\frac{\frac{1}{2}-2^{-2\Gamma}}{\Gamma}B_{2\Gamma} \tag{6.27}$$

and $B_{2\Gamma}$ are the Bernoulli numbers:

$$B_0 = 1, B_2 = 1, B_4 = -1, B_6 = 1, B_8 = -1, B_{10} = 5, B_{12} = -691, \cdots \tag{6.28}$$

The result (6.26) is in fact exact, in the sense that it does not depend on the pecularities of the potential at the cut-off distances $\sim \Lambda$. Any change in the potential results in a constant extension of the whole spectrum for the energies $\xi \sim 1$.

How now to extract a physical information about the string field theory from all this?

Eq.(6.14) now reads as

$$\lambda^2 E_0(\lambda, N) = \int_{\mu_0}^{\mu}d\xi\rho(\xi)\xi \tag{6.29}$$

where μ_o is a cut-off dependent one-particle ground-state energy, which results only in an irrelevant shift in (6.29), and λ^2 appears in the left hand side due to the chosen normalization of the density of states:

$$\lambda^2 = \int_{\mu_0}^{\mu} d\xi \rho(\xi) \tag{6.30}$$

which follows from the presence of a $\frac{1}{\lambda^2}$ factor in front of the action in (6.8).

As we already mentioned, $E_0(\lambda, N)$ is equal to the partition function of the corresponding string field theory. Eqs. (6.29) and (6.30) give a representation of $E_0(\lambda, N)$ in a parametric way: excluding μ from both equations one gets, in principle, the needed result.

In practice, it is possible to exclude μ only order by order in the topological (1/N) expansion. To recover this expansion, let us recall that in the large N limit μ is of the order N (fermi level is far below the top of the potential). Introducing new notation

$$\mu = N\nu \tag{6.31}$$

and

$$\Delta = 2\pi(\lambda_c^2 - \lambda^2) \tag{6.32}$$

is the deviation of the cosmological constant from the critical value, depending on μ_0, we obtain from (6.26)

$$2\pi\rho(\nu) = -\log(\nu/\Lambda) + \frac{1}{24N^2}\frac{1}{\nu^2} + \frac{7}{2880N^4}\frac{1}{\nu^4} + \cdots \tag{6.33}$$

and from (6.26) and (6.30):

$$\Delta = -\nu\log(\nu/\Lambda) - \frac{1}{24N^2}\frac{1}{\nu} - \frac{7}{2880N^4}\frac{1}{\nu^3} - \cdots \tag{6.34}$$

It is useful, instead of the partition function, to consider $E_0(\lambda, N)$ a string susceptibility (two-point function) χ, defined by eq. (3.5). From (6.29) and (6.30) we obtain:

$$\chi(\lambda, N) = \frac{1}{\rho(\lambda, N)} + \quad regular\ part \tag{6.35}$$

which gives, for this physical quantity, the 1/N expansion in the form:

$$\chi(\lambda, N) = \chi_0(\lambda) + \frac{1}{N^2}\chi_1(\lambda) + \frac{1}{N^4}\chi_2(\lambda) + \cdots =$$
$$= \frac{1}{\log(\nu/\Lambda)} + \frac{1}{24N^2}\frac{1}{\nu^2\log^2\nu} + \frac{7}{2880N^4}\frac{1}{\nu^4\log^2\nu} + \cdots \tag{6.36}$$

with $\chi_0(\lambda)$, $\chi_1(\lambda)$, $\chi_2(\lambda)$, ..., being the two-point functions (susceptibilities) for the topologies $\Gamma = 0, 1, 2, \cdots$, respectively.

To exclude ν we have to solve eq.(6.34) with respect to ν order by order in $1/N$ expansion, and than insert in (6.33) and (6.35). The result may be expressed in terms of the function $\nu_0(\Delta)$, satisfying a transcendental equation

$$\nu_0(\Delta) \log \nu_0(\Delta) = \Delta \tag{6.37}$$

For small Δ it behaves as

$$\nu_0(\Delta) \approx \frac{\Delta}{\log \Delta} + \frac{\Delta \log(\log \Delta)}{\log^2 \Delta} + \cdots \tag{6.38}$$

By means of $\nu_0(\Delta)$ we obtain from eqs.(6.34) and (6.36) the string susceptibilities for different genera:

$$\chi_0(\lambda) = \frac{1}{\log \nu_0(\Delta)} \approx \frac{1}{\log \Delta} \tag{6.39}$$

$$\chi_1(\lambda) = \frac{1}{24} \frac{1}{\Delta^2} \tag{6.40}$$

$$\chi_2(\lambda) = \frac{7}{2880} \frac{1}{\nu_0^4 \log^2 \nu_0} \approx \frac{7}{2880} \frac{\log^2 \Delta}{\Delta^4} \tag{6.41}$$

e.c. The approximate expressions for χ_0 and χ_2 correspond to neglecting $\log(\log \Delta)$ corrections to $\nu_0(\Delta)$, whereas the result (6.40) is exact in that sense.

The result (6.39) was established in [29], and was confirmed recntly by the calculations on the bases of the Liouville theory of 2d gravity [45] [46] (see also [47, 48, 49, 50]. The result (6.40) appears to be also in the whole correspondence with Liouville theory [51] [52].

Unfortunately, the continuous Liouville theory only predicts by now the scaling behaviour in eq.(6.39)-(6.41), whereas the matrix model approach gives also the correct coefficients.

It was also shown in [29], that the characteristic mean square extent of a triangulated surface in the embedding t-space: $< t^2 >$, behaves near the critical cosmological constant as

$$< t^2 > \sim \log^2 \Delta \tag{6.42}$$

due to the fact, that the characteristic size of the surface is proportional to the inverse mass gap m_o^{-1} of our system of fermions, which is defined by the density of states in the vicinity of fermi level:

$$m_0 = \frac{1}{\rho(\mu)} \tag{6.43}$$

The result (6.42) follows from (6.43) and (6.33), and seems to be true for any genus. For the spherical topology it was confirmed by the calculations of the two-point propagator [53] (see also [54]).

7. Nonperturbatively stable definition of bosonic string field theory in $1D$ target space

As we mentioned in the previous section, all the definition of $1D$ bosonic string in terms of matrix quantum mechanics was justified only on the perturbative level - order by order in the topological $(1/N)$ expansion. The reason is simple: if $N \to \infty$, the fermi level $\mu = N\nu$, measured down from the top of the potential, is large, and the tunnelling of the particles through the barrier is downed by an exponential factor $\sim e^{-2\pi\mu}$.

If we consider a nonperturbative regime $\mu \sim 1$, we immediately realize that the only possibility to define the model is to find a stable quantum mechanical matrix system (as was pointed out in [36] and [37]), which has the same $1/N$ expansion.

A possible nonperturbative definition, which we are going to present here, is the following: let us put two infinite walls (instead of one, as before) symmetrically on the distances $\pm\Lambda$ from the top of the inverted quadratic potential. This inverted potential may be understood as a rescaled version of the potential $V = -x^2/2 + \frac{\lambda}{N}x^4$, for $N \to \infty$. Let us then allow to the particles (since we cannot forbid it) to be distributed over the whole $(-\Lambda, \Lambda)$ region.

It is clear, that if we would decrease the fermi level $(\mu \to +\infty)$ adiabatically, we would get two equivalent systems of $\sim \frac{N}{2}$ particles, on each side of the quadratic wall, each of them exhibiting the same critical behaviour, as a nonstable system of a previous section, order by order in $1/N$.

Note that unlike the one-matrix model of the paper [37], where the eigenvalues are distributed on two neighbouring supports, there is no interaction between two groups of eigenvalues in our $1D$ model, since the Van-der-Monde determinants of eigenvalues are cancelled, which makes them in the limit $\mu \to \infty$ completely noninteracting (except of the Pauli principle for the fermions). In one matrix-model with the two wells filled, the Coulomb interaction makes the system to be different even on the perturbative level: the topological expansion is generated now by Painleve II [55], and not by Painleve I equation.

Hence this redefinition of the model should be perfectly suitable, even though not unique, for the nonperturbative definition of the model [56]. But does it give the same results (6.26), (6.35) e.c., for the basic physical quantities on the nonperturbative level? Now we are going to show, that the answer is positive [57].

In the symmetric potential the states can be classified by their parity with respect to the reflection symmetry $\psi(-x) = \pm\psi(x)$ of the wave function. Even and odd solutions of eq.(6.19) can be represented through the degenerated hypergeometric function $M(a, b, z)$ as [44]:

$$\psi_{even} = e^{-ix^2/4} M\left(-i\frac{\xi}{2} + \frac{1}{4}, \frac{1}{2}, i\frac{x^2}{2}\right) \tag{7.1}$$

$$\psi_{odd} = x e^{-ix^2/4} M\left(-i\frac{\xi}{2} + \frac{3}{4}, \frac{3}{2}, i\frac{x^2}{2}\right) \tag{7.2}$$

The infinite walls imply

$$\psi_{even}(\pm\Lambda) = \psi_{odd}(\pm\Lambda) = 0 \tag{7.3}$$

Since $\Lambda \to \infty$, we can use a quasiclassical asymptotics for $M(a, b, z)$

$$M(a, b, z) \xrightarrow{z \to \pm\infty} \frac{\Gamma(b)e^{\pm i\pi a}}{\Gamma(b-a)} z^{-a}(1 + O(1/z)) \tag{7.4}$$

which gives

$$\psi_{even} \approx \frac{2}{\sqrt{\pi}} Re[\frac{e^{-ix^2/4 + \frac{i\pi}{4}(\frac{1}{2} - i\xi)}}{\Gamma(\frac{1}{4} + \frac{i\xi}{2})} (\frac{x^2}{2})^{-\frac{1}{4} + \frac{i\xi}{2}}] \tag{7.5}$$

$$\psi_{odd} \approx \sqrt{\pi} x Re[\frac{e^{-ix^2/4 + \frac{i\pi}{4}(\frac{3}{2} - i\xi)}}{\Gamma(\frac{3}{4} + \frac{i\xi}{2})} (\frac{x^2}{2})^{-\frac{3}{4} + \frac{i\xi}{2}}] \tag{7.6}$$

The conditions (7.3) imply the quantization of phases of the expressions (7.5) and (7.6), which are equal at the infinite walls to

$$\Phi_{even} = -\frac{\Lambda^2}{4} + \frac{\xi}{2}\log\frac{\Lambda^2}{2} - Arg\Gamma(\frac{1}{4} + \frac{i\xi}{2}) + \frac{\pi}{8} \tag{7.7}$$

$$\Phi_{odd} = -\frac{\Lambda^2}{4} + \frac{\xi}{2}\log\frac{\Lambda^2}{2} - Arg\Gamma(\frac{3}{4} + \frac{i\xi}{2}) + \frac{3\pi}{8} \tag{7.8}$$

The densities of even and odd states are, correspondingly

$$\rho_{even}(\xi) = -\frac{\partial\Phi_{even}}{\partial\xi} =$$
$$= \frac{1}{2}\log\frac{\Lambda^2}{2} - Re\psi(\frac{1}{4} + \frac{i\xi}{2}) \tag{7.9}$$

$$\rho_{odd}(\xi) = -\frac{\partial\Phi_{odd}}{\partial\xi} =$$
$$= \frac{1}{2}\log\frac{\Lambda^2}{2} - Re\psi(\frac{3}{4} + \frac{i\xi}{2}) \tag{7.10}$$

The whole density of states is

$$\rho(\xi) = \rho_{even}(\xi) + \rho_{odd}(\xi) = \log\Lambda - Repsi(\frac{1}{2} + i\xi) \tag{7.11}$$

which coincides with (6.26).

Hence, we checked that the nonperturbatively stable system, suggested here, generates correctly every order of the topological expansion, eventhough it is not true for the even and odd states separately. The results (6.26), (6.29) and (6.30) appear to be true even nonperturbatively.

8. Cut-off independant physical observables in the double scaling limit

The formula (6.26), as well as eqs. (6.29), (6.30) and (6.35) contain an explicit cut-off parameter $\Lambda \sim \sqrt{N}$, apart from the scaling parameter ξ. This makes the corresponding physical quantities ill-defined in the double scaling limit ($N \to \infty, \Delta \to 0, \xi(N, \Delta) = const$). For instance, $\chi_{sing} \sim \frac{1}{\rho} \sim \frac{1}{\log \Lambda} \to 0$, which wipes out all the dependence on ξ from the two-point function.

However, one can notice, that Λ-dependence enters purely additively to eq.(6.26), which gives the idea that one can define a new set of physical quantities, defined by polygamma functions, which do not have any dependence on Λ:

$$\rho_n(\xi) = (-\frac{\partial}{\partial \xi})^n \rho(\xi, \Lambda) = (-1)^n Re \psi^{(n)}(i\xi + \frac{1}{2}) \qquad (8.1)$$

which have the $\frac{1}{\xi}$ expansion

$$\rho_n(\xi) = \frac{n!}{2\pi} \frac{1}{\xi^n} + \frac{(n+1)!}{48\pi} \frac{1}{\xi^{n+2}} + \cdots \qquad (8.2)$$

These quantities obviously do not contain any Λ-dependence for n=1,2,... and are well defined in the double scaling limit.

But one has to clarify their physical meaning, relating them to the standard multi-point functions, which are a generalisation of those defined by the eq.(3.6):

$$\chi_m(\xi, \Lambda) = \frac{\partial^m E_0(\xi, \Lambda)}{\partial \Delta^m} \qquad (8.3)$$

From (6.30) and (6.32) we have a simple relation

$$\rho_n(\xi) = (-\frac{1}{\chi} \frac{\partial}{\partial \Delta})^n \frac{1}{\chi} \qquad (8.4)$$

where $\chi_2 = \chi$ is a (singular part of the) two-point function, or susceptibility.

Since $\chi_m = \frac{\partial^{m-2}\chi}{\partial \Delta^{m-2}}$, it is straightforward to express any ρ_n through a combination of χ_m's. First few relations are

$$\rho_1 = \frac{\chi_3}{\chi^3} \qquad (8.5)$$

$$\rho_2 = \frac{\chi_4}{\chi^4} - 3\frac{\chi_3^2}{\chi^5} \qquad (8.6)$$

$$\rho_3 = \frac{\chi_5}{\chi^5} - 10\frac{\chi_3\chi_4}{\chi^6} + 15\frac{\chi_3^3}{\chi^7} \qquad (8.7)$$

e.c.

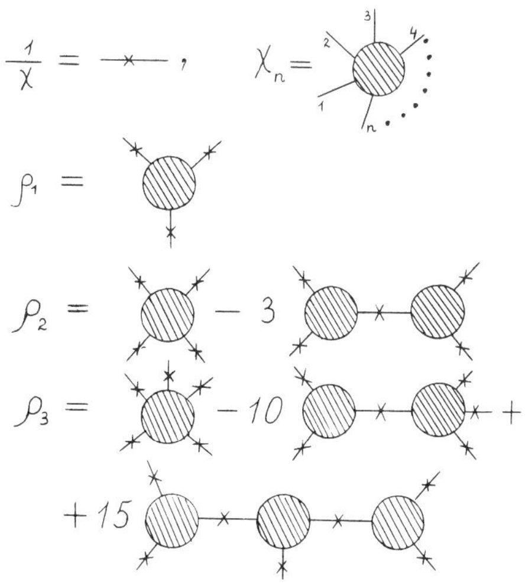

Figure 8. The relations between the nonrenormalized n-point functions χ_n and renormalized quantities ρ_n. The renormalization results in the consecutive removal of the two-point insertions on the tree level.

These equations have a nice graphical meaning [58]. Namely, let us interpret χ_n as n-leg amplitudes, and the division by χ - as an amputation of a leg. Than the formulae(8.5)-(8.7) can be interpreted in a graphical way, demonstrated in fig. 8. One can notice that ρ_n are expressible through all possible tree-like diagrams built from n-point amplitudes χ_n, with the external and internal legs removed. One can also show that all the combinatorial coefficients arising from the formula (8.4), are the correct coefficients of the corresponding diagram technique.

One can interpret (8.4)-(8.7) in terms of a Legendre transform:

$$E_0(\Delta) = \Delta\xi - F(\xi), \tag{8.8}$$

where one has to minimize E_0 with respect to ξ, and F plays the role of generating function of 1-particle irreducible amplitudes:

$$\frac{\partial^2 F(\mu)}{\partial\mu^2} = \frac{\partial\Delta}{\partial\mu} = \rho(\xi) = \frac{1}{\chi} \tag{8.9}$$

The procedure of the amputation of one-particle insertions, which appears to be necessery for removal of the cut-off Λ, brings to the mind the idea that there exists an instability in the sum over surfaces with respect to a creation of a thin tube. This might be related to the fact, that there exists a massless (tachyonic) excitation in the one-dimensional bosonic string, with divergent zero-momentum propagator, equal to the two-point function χ.

9. Bosonic strings in a periodic $1D$ target space and the role of angular degrees of freedom

One of the most interesting generalizations of bosonic strings in the infinite one-dimensional space is a compactified version of this model, when the embedding coordinate t is defined on a ring of radius β. It has lots of applications in such questions as finite temperature behaviour of strings, the analyses of free parameter dependence of c=1 models of conformal field theory, coupled to gravity, the construction of minimal models, e.c.

The periodic boundary conditions mean that we have the same one- dimensional matrix quantum mechanics, described by eq.(6.1), but with a restriction on every component of the matrix field:

$$\phi(t) = \phi(t + \beta) \tag{9.1}$$

The propagator of Feynman diagram technique, instead of (6.2), will now have a form

$$E(t_a - t_b) = \sum_{n=-\infty}^{\infty} e^{-|t_a-t_b+n\beta|} \tag{9.2}$$

For $\beta \to \infty$ we should recover in principle, the old model on the infinite t-line, since only n=0 term survives in (9.2).

For general β the situation is much more complicated, since the terms with $n \neq 0$ will contribute to the creation of the vortices, and their condensation may cause the Berezinski-Kosterlitz-Thouless phase transition at some β_c.

Here we shall point out few technical steps to the solution of this problem [59]. One has to admit, that in spite of the obvious solvability of this problem, the explicit results and the physical understanding are still missing.

Let us first note, that even for the boundary conditions (9.1) all the transformations, leading from (6.1) to (6.4), (6.5), are still valid, but the measure (6.6) should be modified by an additional constraint on the connectivity $A(t)$, due to the fact, that in (6.5) $\Omega(t)$ obeys also the periodicity condition

$$\Omega(t) = \Omega(t + \beta)) \tag{9.3}$$

Hence, solving (6.5) with (9.3), we get

$$\hat{T} \exp i \int_0^\beta A(t)dt = I \tag{9.4}$$

where $\hat{T}$ is the time ordering operator.

This constraint should be added to the measure (6.6), which gives instead of (6.7) an integral

$$\int D^{N^2} A(t) \exp[-\frac{N}{2\lambda^2} \int_0^\beta dt \sum_{i \neq j} (z_i - z_j)^2 |A_{ij}|^2] \, \delta[\hat{T} \exp i \int_0^\beta A(t)dt, I] \tag{9.5}$$

where δ-function is defined on the group space of U(N). The measure of integration over the hermitian matrices A(t) is now linear:

$$DA(t) = \prod_{i,j} DA_{ij}^*(t) DA_{ij}(t) \prod_k DA_{kk}(t) \tag{9.6}$$

One has to make a remark here. The whole expression (9.5) has a $[U(1)]^N$-gauge symmetry with respect to the transformations:

$$A_{km}(t) \rightarrow e^{i\alpha_k} A_{km} e^{-i\alpha_m} - \delta_{km}\dot{\alpha}_k \tag{9.7}$$

But since this symmetry is abelian and the group of symmetry is compact, it is possible to integrate in eq.(9.5) independently in all variables $A_{ij}(t)$, as we shall do in what follows.

Let us now use a character expansion for the δ-function:

$$\delta_{U(N)}(\Omega, I) = \sum_R d_R \chi_R(\Omega) \tag{9.8}$$

where d_R is the dimension of the R-th representation of U(N), $\chi_R(\Omega)$ is the corresponding character:

$$\chi_R(\hat{T} \exp i \int_0^\beta A dt) = tr_R(\hat{T} \exp i \int_0^\beta dt \sum_{i,j} \hat{\tau}_{ij}^R A_{ij}) \tag{9.9}$$

and $\hat{\tau}_{ij}^R = (\tau_{ij}^R)^{ab}$ are the generators of U(N) in the R-th representation (the indices a and b, the operations tr_R and $\hat{T}$ correspond to this representation). Now the integral in eq.(9.5) is ultralocal (up to the ordering, which we have to respect), so cutting the t-interval into small pieces of a size Δt and integrating independently on everyone in A_{ij} for $i \neq j$, we arrive to the following representation:

$$Tre^{-\beta\hat{H}} = \int \prod_i [Dz_i(t)e^{-S(z_i)}] \sum_R d_R tr_R \exp[-\frac{\lambda^2}{2N} \int_0^\beta dt \sum_{i \neq j} \frac{\tau_{ij}^R \tau_{ji}^R}{(z_i - z_j)^2}] \tag{9.10}$$

where

$$S(z) = \frac{N}{\lambda^2} \int_0^\beta dt[\frac{\dot{z}^2}{2} + V(z)] \tag{9.11}$$

Note that all the Van-der-Monde determinants have been cancelled here, but we still have here a fermionic statistics for the eigenvalues obeying now the periodicity condition:

$$z_i(0) = z_i(\beta), \quad for \, i = 1, 2, ..., N. \tag{9.12}$$

Unlike the case of open time interval the fermions are now interacting (except of the trivial representation).

The integration over the diagonal elements A_{ii} formally gives $\delta(\hat{\tau}_{ii}^R)$ which means, in fact, that we have to use, as intermediate states in (9.10), only the states $|\psi>$ with the zero weight of the representation R [60]:

$$\hat{\tau}_{ii}^R|\psi> = 0 \qquad for\ any\ i. \tag{9.13}$$

We can use as intermediate states in (9.10) the N-particle wave functions in the form of Slater determinants of various one-particle eigenfunctions $\psi_n(z)$ of the noninteracting fermions in the trivial representation (which is an obvious generalization of the ground state wave function (6.11)).

But for these intermediate states, due to the antisymmetry in all coordinates $z_1, ..., z_N$ one can substitute in (9.10):

$$\sum_{i \neq j} \frac{\hat{\tau}_{ij}^R \hat{\tau}_{ji}^R}{(z_i - z_j)^2} \to \sum_{i \neq j} \frac{\hat{\tau}_{ij}^R \hat{\tau}_{ji}^R}{N(N-1)} \sum_{i \neq j} \frac{1}{(z_i - z_j)^2} =$$

$$= \frac{C_R^{(2)}}{N(N-1)} \hat{P}_0^R \sum_{i \neq j} \frac{1}{(z_i - z_j)^2} \tag{9.14}$$

where $C_R^{(2)}$ is a value of a quadratic Casimir operator in the R-th representation, and $\hat{P}_0^R$ is a projector to the subspace of zero weight vectors in the space of an R-th representation. The condition (9.13) was essentially used here.

Eq.(9.10) transforms now into

$$Tre^{(-\beta\hat{H})} = \sum_R d_R d_R^{(0)} \int \prod_k Dz_k(t) e^{-S_R(z)} \tag{9.15}$$

where

$$S_R(z) = \frac{1}{\lambda^2} \int_0^\beta dt [N \sum_{i=1}^N (\frac{\dot{z}_i^2}{2} + V(z_i)) + \frac{C_R^{(2)}}{2N^3} \sum_{i \neq j} \frac{1}{(z_i - z_j)^2}] \tag{9.16}$$

and

$$d_R^{(0)} = tr\hat{P}_R^{(0)} \tag{9.17}$$

is the dimension of the subspace of vectors with zero weight in R.

In the hamiltonian language (9.15) looks like

$$Tre^{(-\beta\hat{H})} = \sum_R d_R d_R^{(0)} Tre^{(-\beta\hat{H}_R)} \tag{9.18}$$

with the hamiltonians

$$\hat{H}_R = N \sum_{i=1}^N (-\frac{1}{2N^2} \frac{\partial^2}{\partial z_i^2} + V(z_i)) + \frac{C_R^{(2)}}{2N^3(N-1)} \lambda^2 \sum_{i \neq j} \frac{1}{(z_i - z_j)^2} \tag{9.19}$$

298

If now we make the same change of variables as in (6.17)-(6.20) and drop the terms, irrelevant in the double scaling limit, we arrive to the hamiltonian:

$$\hat{H}_R = N \sum_{i=1}^{N} \left(\frac{-\partial^2}{\partial x_i^2} - \frac{1}{4} x_i^2 \right) + \frac{C_R^{(2)} \lambda^2}{2N^3} \sum_{i \neq j} \frac{1}{(x_i - x_j)^2} \tag{9.20}$$

All these terms are of the same order in N for $N \to \infty$, since $C_R^{(0)} \sim N^2$.

Formally, one can bring eq.(9.20) to the form

$$\hat{H}_R = N \sum_{i=1}^{(N+1)} \frac{-\partial^2}{\partial x_i^2} + \sum_{i \neq j} \left[\frac{1}{8} (x_i - x_j)^2 + \frac{C_R^{(2)}}{2N^3} \frac{1}{(x_i - x_j)^2} \right] \tag{9.21}$$

with an additional condition

$$\sum_{i=1}^{N+1} x_i = 0 \tag{9.22}$$

fixing the centre-of-mass coordinate in the otherwise translationary invariant system.

Eq. (9.21) looks like a (in principle, exactly solvable) Calogero chain [61], but with an opposite sign of the quadratic interaction. To stabilize the system, one can again put it in the box with the hard walls at the distances $x = \pm \Lambda$ from the origin. One has to understand, that these conditions will be different from the analogous conditions for the variables of translationary noninvariant eq.(9.20) and they will result in the change of the boundary conditions on the walls, but we hope again that both systems are in the same class of universality, since both boundary conditions obey the permutation symmetry with respect to the eigenvalues.

The next part of the problem - the sum of a type (9.15) over representations, also seems not to be hopeless, at least for the fixed genus of surface, where the sum over R's might be representable through the functional integration. On the other hand all we have to find performing this sum, is a function depending on N and only on one more parameter, namely, on the term

$$\sum_{i \neq j} \frac{1}{(x_i - x_j)^2} \tag{9.23}$$

itself. This shows that even after the sum over R we would deal with the same Calogero type interaction.

Let us also mention here an interesting hypothesis of the paper [54], where it was assumed that below the transition point of the condensation of vortices (which is a string version of the Berezinski-Kosterlitz-Thouless phase transiton), $\beta > \beta_c$, the representations higher than trivial are irrelevant. In that case one can again deal with the noninteracting fermions, obeying the wave functions, which satisfy equation (6.19), but with the periodic boundary conditions in time t. The calculation of the finite temperature partiton $Z(\beta, \lambda, N)$ function is now straghtforward, since we know the one-particle spectral density (6.26):

$$Z(\beta, \lambda, N) = \frac{1}{\beta} \int_0^\infty d\xi \rho(\xi) \log[1 + e^{-(\mu - \xi)\beta}] \tag{9.24}$$

where the fermi level μ is defined by the normalization condition:

$$N\lambda^2 = \frac{1}{\beta} \int_0^\infty d\xi \rho(\xi) \frac{1}{1 + e^{(\mu-\xi)\beta}} \tag{9.25}$$

Recall that we count the spectrum levels from the top of the potential down.

Introducing again $\Delta = \lambda_c^2 - \lambda^2$, we have for the string susceptibility:

$$\chi(\mu) = \frac{\partial^2 Z}{\partial \Delta^2} = -N \frac{\partial}{\partial \Delta} \frac{\partial Z}{\partial N\lambda^2} = \frac{N}{\lambda^2} \frac{\partial \mu}{\partial \Delta} \tag{9.26}$$

which enables us to represent it in the form:

$$\chi^{-1}(\mu) \sim \frac{\partial \lambda^2}{\partial \mu} = \frac{1}{\beta} \int_0^\infty d\xi \rho(\xi) \frac{1}{\cosh^2 \frac{(\mu-\xi)\beta}{2}} \tag{9.27}$$

If we take now for $\rho(\xi)$ the last expression from the Appendix and insert it in (9.27), we get after the integration in ξ [54]:

$$\chi^{-1}(\mu) \sim \int_{\Lambda^{-1}}^\infty dt \frac{t}{\mu} \frac{t}{\mu\beta} \frac{1}{\sinh \frac{t}{2\mu}} \frac{1}{\sinh \frac{\pi t}{\mu\beta}} \tag{9.28}$$

The most remarkable feature of this result is the dual symmetry with respect to the inversion of the radius of compactification (temperature):

$$\frac{\beta}{2\pi} \to \frac{2\pi}{\beta}, \quad \mu \to \frac{\beta}{2\pi}\mu \tag{9.29}$$

This remains to be true in all orders of the topological expansion, or μ-expansion ($\mu \sim N$):

$$\chi(\mu) = -\frac{1}{\log \mu} + \frac{1}{24}\mu^{-2}\left(\frac{\beta}{2\pi} + \frac{2\pi}{\beta}\right) + \frac{7}{2880}\mu^{-4}\left(\frac{\beta^2}{4\pi^2} + \frac{6}{7} + \frac{4\pi^2}{\beta^2}\right) + \dots \tag{9.30}$$

The dual symmetry of these formulae gives them a good chance to be correct, which is also confirmed by some recent Liouville theory calculations [51], [52]. But in any case, the contributions from higher representations are interesting by many physical reasons, especially, from the point of view of the phase transition of condensation of vortices, which are, most probably, missing in the approach of [54].

300

10. Conclusions

We did not consider here many interesting questions, concerning $1D$ dimensional bosonic strings. Let us mention some of them.

First, in order to obtain the maximum of physical information from this theory, one has to calculate the target space correlators in the double scaling limit, generalizing the tree-level results of papers [53] and [62].

Second, one has to understand the role of angular degrees of freedom of matrices for the existence of c=1 barrier. Especially, to try to develop sort of ε-expansion for $D = 1 + \varepsilon$, using the results for D=1, presented here, as a starting point.

Third, it is unclear whether all these results for $c = 1$ correspond to some topological field theory, in the sense of the papers [63, 64, 65]. Of course, even if it is so, it is a very special case of a topological theory, since the space of relevant operators is here certainly infinite dimensional, unlike the $c < 1$ theories. However, this makes the $1D$ field theory of closed strings even more interesting and resembling the physically realistic models.

Let us notice at the end that recently many new developments related to the questions considered here, have appeared, which we have not mentioned here. Therefore the list of references is far from being complete. We hope to fill this gap in [66].

Appendix: Calculation of the phase $\Phi(\xi)$ in eq.(6.23)

$\Phi(\xi)$ is an imaginary part of the logarithm of the scattering amplitude for the scattering from the inverted quadratic potential, or, in other words, of the logarithm of reflection coefficient. We can calculate the real part of reflection coefficient and relate the phase $\Phi(\xi)$ to it by means of a dispersion relation. As was shown by Kamble (1951), it is sufficient to use WKB-approximation for eq.(6.20).

Let us consider the reflection of a quantum-mechanical particle from the inverted quadratic potential. To the left of the top we have the wave function ψ_{left}, describing the falling and reflected waves:

$$\psi_{left}(x) \approx \exp(-\frac{i}{4}x^2)x^{i\xi-\frac{1}{2}} + A\exp(\frac{i}{4}x^2)x^{-i\xi-\frac{1}{2}} \tag{10.1}$$

and to the right of the top - the wave function ψ_{right} of the penetrated wave:

$$\psi_{right}(x) \approx B\exp(\frac{i}{4}x^2)x^{-i\xi-\frac{1}{2}} \tag{10.2}$$

Let us go around the top of the potential along some big circle in the complex plain of x. Unlike the vicinity of turning points of this potential, everywhere on this path the quasiclassical approximation is still available. We get a phase shift to x:

$$x \to (-x) = e^{i\pi}x \tag{10.3}$$

which results in the following relations:

$$A = e^{i\pi(-i\xi - \frac{1}{2})}B = -ie^{\pi\xi}B \tag{10.4}$$

Using an obvious consequence of the current conservation

$$|A|^2 + |B|^2 = 1 \tag{10.5}$$

we find from here

$$|A|^2 = \frac{1}{1 + e^{-2\pi\xi}} \tag{10.6}$$

The real part of the logarithm of reflection coeffitient is:

$$\log|A| = \log \cosh(\pi\xi) - \frac{\pi}{2}\xi + const \tag{10.7}$$

hence the phase of reflection follows from the following dispersion relation:

$$\Phi(\xi) = \frac{1}{2\pi} \oint \frac{d\eta}{\xi - \eta} \log|A| = -\frac{1}{2}\log \frac{\Gamma(\frac{1}{2} + i\xi)}{\Gamma(\frac{1}{2} - i\xi)} \tag{10.8}$$

coinciding with (6.23). The contour of integration encircles here all the poles.

The density of states is

$$\rho(\xi) = \frac{1}{\pi}\Phi'_\xi(\xi) = -\frac{1}{2\pi}Re\psi(i\xi + \frac{1}{2}) = Re \sum_{n=0}^{\infty} \frac{1}{i\xi - n - \frac{1}{2}} \tag{10.9}$$

Let us finally note that one can get $\rho(\xi)$ (in a simpler way, but not so rigorously) as an analytical continuation of the density of states of the usual harmonic oscillator:

$$\rho(\xi) = \sum_{n=0}^{\infty} \delta(\xi - \omega(n + \frac{1}{2})) == Im \sum_{n=0}^{\infty} \frac{1}{i\xi - \omega(n + \frac{1}{2}) + i\varepsilon} \tag{10.10}$$

Changing $\omega \to -i\omega$ we obtain the previous expression for $\rho(\xi)$.

The last formula may also be represented in the following way:

$$\rho(\xi) = \frac{1}{4\pi} \int_{\Lambda^{-1}}^{\infty} dt e^{-i\xi t} \frac{1}{\sinh \frac{\pi t}{2}} \tag{10.11}$$

which one can check expanding the integrand in $e^{-\pi t}$ and integrating term by term in t.

Acknowledgements

Many comments and clarifications, presented here, have appeared in the discussions with E.Brezin, D.Boulatov, M.Douglas, D.Gross, I.Klebanov, I.Kostov, A.A.Migdal, M.Olshanetski, N.Seiberg and Al.Zamolodchikov, to whom the author is very indebted.

Note Added

After the completion of this paper we received a paper of D.Gross and I.Klebanov, Princeton preprint PUPT-1210 (Oct.1990), where it is shown that the reduction of the one dimensional problem with periodic boundary conditions, considered in section 9 of our paper, to the (non-matrix) Calogero hamiltonian, can be true only in the large N limit.

References

[1] D. V. Boulatov, V. A. Kazakov, I. K. Kostov and A. A. Migdal, Nucl. Phys. B275 [FS17] (1986) 641.

[2] F. David, Nucl. Phys. B257 [FS14] (1985) 45.

[3] V. A. Kazakov, Phys. Lett. 150B (1985) 282.

[4] J. Ambjørn, B. Durhuus and J. Fröhlich, Nucl. Phys. B257 [FS14] (1985) 433.

[5] T. Regge, Nuovo Cimento 19 (1961) 558.

[6] M.Bander and C. Itzykson, Proc. Seminars on nonlinear equations in field theory (Paris VI and Meudon, September 1983 - May 1984) Springer, Berlin.

[7] V. A. Kazakov, Phys. Lett. 119A (1986) 140.

[8] D. V. Boulatov and V. A. Kazakov, Phys. Lett. B186 (1987) 379.

[9] M. L. Mehta, Comm. Math. Phys. 79 (1981) 327.

[10] C.Crnkovic, P. Ginsparg and G.Moore, Phys. Lett. 237B (1990) 196.

[11] D. Gross and A. Migdal, Phys. Rev. Lett. 64 (1990) 717.

[12] E. Brézin, M. Douglas, V. Kazakov and S. Shenker, Phys. Lett. 237B (1990) 43.

[13] V. A. Kazakov, Nucl. Phys.(Proc. Suppl.) B4 (1988) 93.

[14] V. A. Kazakov, Mod. Phys. Lett. A4 (1989) 1691.

[15] V. A. Kazakov and I.K.Kostov (unpublished), see some comments in I. K. Kostov, Nucl. Phys. B (Proc. Suppl.) 10A (1989) 295.

[16] I. K. Kostov, Mod. Phys. Lett. A4 (1989) 217.

[17] F. David, Nucl. Phys. B257 [FS14] (1985) 543.

[18] V. A. Kazakov, I. K. Kostov and A. A. Migdal, Phys. Lett. 157B (1985) 295.

[19] A. M. Polyakov, Phys. Lett. 103B (1981) 207.

[20] M.Bershadski and A. A. Migdal, Phys. Lett. 174B (1986) 393.

[21] A. M. Polyakov, Mod. Phys. Lett. A2(1987) 893.

[22] V. G. Knizhnik, A. M. Polyakov and A. B. Zamolodchikov, Mod. Phys. Lett. A3 (1988) 819.

[23] F. David, Mod. Phys. Lett. A3 (1988) 1651.

[24] J. Distler and H. Kawai, Nucl. Phys. B321 (1989) 509.

[25] I. K. Kostov and M. L. Mehta, Phys. Lett. 189B (1987) 118.

[26] B. Duplantier and I. K. Kostov, Phys. Rev. Lett. B (1989).

[27] M. Staudacher, Nucl. Phys. B336 (1990) 349.

[28] M. R. Douglas, Phys. Lett. 238B (1990) 176.

[29] V. A. Kazakov and A. Migdal, Nucl. Phys. B311 (1988) 171.

[30] I. K. Kostov, Nucl. Phys. B326 (1989) 583.

[31] M. Douglas and S. Shenker, Nucl. Phys. B335 (1990) 635.

[32] D. Gross and A. Migdal, Phys. Rev. Lett. 64 (1990) 127.

[33] E. Brézin and V. Kazakov, Phys. Lett. 236B (1990) 144.

[34] S. Shenker, proceedings of Cargese Workshop (1990).

[35] E. Marinari and G. Parisi, ROM2-90-05, January 1990.

[36] E. Brézin, E. Marinari and G. Parisi, Phys. Lett. B242 (1990) 35.

[37] M. Douglas, N. Seiberg and S. Shenker, Rutgers preprint, RU-90-19 (1990).

[38] M. Karliner and A. Migdal, Princeton preprint.

[39] F. David, Mod. Phys. Lett. A5 (1990) 1019.

[40] G.'t Hooft, Nucl. Phys. B72 (1974) 461.

[41] C. Itzykson and J.-B. Zuber, J. Math. Phys 21 (1980) 411.

[42] S. Chadha, G. Mahoux and M. L. Mehta, J. Phys. A: Math. Gen. 14 (1981) 579.

[43] E. Brézin, C. Itzykson, G. Parisi and J.-B. Zuber, Comm. Math. Phys 59 (1978) 35.

[44] M. Abramovitz and I. Stegun, Handbook of math. functions, National bureau of standards, Appl. Math series-55, Chap. 19 (1964).

[45] J. Polchinski, Texas preprint UTTG-15-90 (1990).

[46] S. R. Das and A. Jevicki, preprint Brown-HET-750 (1990).

[47] S. Das, A. Dhar and S. Wadia, Mod. Phys. Lett. A5.

[48] S. Das, A. Naik and S. Wadia, Mod. Phys. Lett. A4.

[49] T. Banks and J. Lykken, Nucl. Phys. B331 (1990) 173.

[50] A. Tseytlin,J. Mod. Phys. A5 (1990) 1883.

[51] M. Bershadski and I. Klebanov, Princeton preprint PUPT-1197 (September 1990).

[52] R. Sakai, Y. Tanii, preprint TIT/HEP-160, STUPP-90-114 (August 1990).

[53] I. K. Kostov, Phys. Lett. B215 (1988) 499.

[54] D. Gross and I. Klebanov, Princeton preprint PUPT-1172 (March 1990).

[55] V. Periwal and D. Shewitz, Phys. Rev. Lett. 64 (1990)1326.

[56] V. A. Kazakov, unpublished.

[57] E. Brézin, V. A. Kazakov and A. A. Migdal, unpublished.

[58] E. Brézin, V. A. Kazakov and N. Seiberg, unpublished.

[59] E. Brézin, V. A. Kazakov and I. K. Kostov, unpublished.

[60] D. V. Boulatov, private communication.

[61] F. Calogero, J. Math. Phys. 12 (1971) 419.

[62] D. Gross, I. Klebanov and M. Newman, Princeton preprint, PUPT- (1990). D. Gross and I. Klebanov, Princeton preprint PUPT-1198 (Sept. 1990).

[63] R. Dijkgraaf and E. Witten, Nucl. Phys. B342 (1990) 486.

[64] E. Verlinde and H. Verlinde, IAS preprint IASSNS-HEP-90/40, April 1990.

[65] R. Dijkgraaf, E. Verlinde and H. Verlinde, Princeton preprint PUPT-1184, May 1990.

[66] V. A. Kazakov and I. K. Kostov, review in Phys. Rep. (in preparation).

A FEW REMARKS ON LOW DIMENSIONAL STRING THEORY

Giorgio Parisi

Dipartimento di Fisica
Università di Roma *Tor Vergata*
Via E. Carnevale 00173 Roma (Italy)

Abstract: In this talk I will discuss some properties of low dimensional string theory. In particular I will concentrate my attention on the following items: the effects of the introduction of a lattice spacing, the definition of a pure Bosonic string beyond perturbation theory, the properties of a supersymmetric string and a possible strategy for obtaining results in two dimensions.

1. Introduction

The very exciting results that have been recently obtained on low dimensional string theory may lead to the hope of a better control of string theory in a physical relevant limit. Indeed it was shown that the discretized (in parameter space) string theory corresponds to a local field theory and the continuum limit may be reached at the critical point of the local field theory. Analytic computations have been done for some particular (low) values of the dimensions of the target space.

The main principles of the approach have been adequately reviewed in the other talks of this workshop. In this note I would like to present a few comments on the following problems, whose solution may be a crucial step for reaching our goals:

    a) The effects of introducing a lattice spacing in the discretized string.

    b) The meaning (if any) of the Bosonic string beyond perturbation theory.

    c) The definition a supersymmetric discretized string.

    d) How to extend the approach to two dimensions.

Random Surfaces and Quantum Gravity
Edited by O. Alvarez *et al.*, *Plenum Press*, New York, 1991

2. The Effects of a Lattice

The introduction of a lattice spacing in target space is likely to be compulsory at an intermediate stage, especially if we think to use numerical techniques to investigate the theory. We may fear that such an introduction can be dangerous because the theory (at least at the critical dimensions) contains gravity and the introduction of a lattice breaks general covariance.

In the one dimensional case using the perturbative expansion in the lattice spacing it has been shown that the introduction of the lattice has no effects at large distances. Indeed using the transfer matrix approach one finds [1] that the theory on the lattice may be described by a continuum Hamiltonian H_a, which can be computed by expanding it in powers of a^2. If we retain only a finite number of terms in this perturbative expansion, the long range behaviour of the two Hamiltonians at the critical point is the same.

It is quite likely that there is a critical value of a (a_c), such that for $a < a_c$, the lattice effects are negligible. This conjecture is partially reinforced by an explicit computation for the Hamiltonian $p^2 - x^2$ [2]. Unfortunately the results are somewhat doubtful as far as the computation was quite formal: the tunnel effects and the non essential selfadjointness of the Hamiltonian where both neglected.

The peculiarities of the model suggest that for large values of the lattice spacing (i.e. very small coupling between different points of the lattice) the behaviour of the theory is the same as in zero dimensions. This statement is not so trivial because in usual cases also a small value of the intersite coupling is able to influence strongly the transition and to change the value of the critical exponents. The models we consider here have however unusual properties, as far as a phase transition is present also in the zero dimensional case (on each lattice site there is an infinite number of degree of freedom). Numerical investigations have been done in order to verify if this picture is correct [3].

3. Beyond Perturbation Theory

The definition of the pure Bosonic string is a serious problem also in zero dimensions (two dimensional gravity without matter). Indeed the matrix model we have to start from contains a leading ϕ^3 interaction and the Euclidian action is not bounded from below and the functional integral is not defined. This result is hardly a surprise: is well known that the weight of surfaces of genus k is positive and increases as $k!$ for large k. The genus expansion is not Borel summable and a non perturbative definition of the theory is needed. Fermionic degrees of freedom (or non unitary Bosonic matter) may alleviate the problem, introducing the necessary cancellation to remove the unwanted singularities. This exactly what happens in the theories with odd m (in these theories the leading term in the potential is proportional to ϕ^{m+1}) or in the supersymmetric case, as we will see later. Indeed in the theories with odd m there is a simple prescription (the implementation of the correct boundary condition at infinity) which determines the theory without ambiguity [4] .

It is quite possible that the pure Bosonic string does not exist and it leads to a sick theory. However a non Borel summability of the perturbative expansion is not always a disaster: it may be present also when the theory is well defined but contains perturbative phenomena which cannot be controlled in perturbation theory. In known cases a benign non Borel summability of the perturbative expansion may be induced by instantons [5] or by infrared renormalons [6]. It is not clear if we are able to find an alternative way go beyond the perturbative expansion, saving to good properties of the theory, e.g causality and locality.

In my opinion it would be rather difficult to find a good definition of the theory (i.e. saving locality and positivity) without understanding better the physics of the problem. We can try however some general purpose techniques for defining Euclidians theories with unbounded from below actions.

The starting point may be an observation that Calogero made long time ago. He remarked the following interesting phenomenon: the Hamiltonian

$$H = p^2 + \left(\frac{dW}{dx}\right)^2 - \frac{d^2W}{dx^2} \tag{3.1}$$

has as formal eigenvector the function

$$\psi(x) = exp(-W(x)) \ . \tag{3.2}$$

The corresponding eigenvalue is zero: ψ is formally the ground state of the Hamiltonian H. When the function ψ is not normalizable, the true ground state energy is strictly positive.

In the case where the potential W has the form

$$W(x) = x - gx^2 \ , \tag{3.3}$$

g being a coupling constant, the energy of the true ground state of the Hamiltonian differs from zero by exponentially small terms, i.e. terms proportional to $exp(\frac{1}{g^2})$. These terms are invisible in perturbation theory. In other words the ground state of the Hamiltonian is a well defined probability distribution which differs from the unnormalizable probability distribution (ψ) by exponentially small terms.

This phenomenon was later related by Witten [7] to the existence of a supersymmetric extension of the Hamiltonian. The supersymmetry is broken spontaneously when ψ is not normalizable. The phenomenon was finally understood in simple physical terms [8] by noticing that the Hamiltonian in eq. (1) is is the Fokker- Plank Hamiltonian [1] associated to the following Langevin equation

$$\frac{dx}{dt} = -F(x) + \eta \ . \tag{3.4}$$

Supersymmetry is broken when there are trajectory which escape to infinity; more precisely the ground state energy is the probability to escape to infinity for unit time.

Later on, Greensite and Halpern [10] proposed to use the dimensionally increased

---

[1] This phenomenon is the parabolic counterpart of the supersymmetry and related dimensional reduction present in some elliptic stochastic differential equations [9].

supersymmetric theory (which is always well defined) as definition of a theory with an Hamiltonian unbounded from below. It was recently suggested that this approach may be useful for a non perturbative definition of quantum gravity [11].

In my opinion using the Fokker-Plank Hamiltonian (or similar definitions like summing over all the solution of the Langevin equation which do not escape to infinity) is a dangerous procedure: there are no theoretical arguments which suggests that the equation of motions of the original theory should be preserved (or only slightly modified). Moreover the terms that one is adding are essentially non local; also the Markov properties and the analyticity properties of the theory may be ruined.

A typical example of this disaster may be obtained by applying this construction to the D-dimensional field theory with the following Euclidean action:

$$S = \int dx \left(\frac{d\phi}{dx}\right)^2 - m^2 \phi^2 \ . \tag{3.5}$$

Here the mass has the wrong sign. The Euclidean action is not bounded from below and we cannot associate to it any probability distribution. On the other end, if we use the the Fokker-Plank formalism to compute the probability distribution of the field ϕ, we find that it is has formally a Gaussian distribution. A simple computation shows that the propagator is given by

$$\frac{1}{|p^2 - m^2|} \ . \tag{3.6}$$

The analytic properties of this propagator cannot be accepted in a local quantum field theory. This is not a surprise because there is no reason for which locality should be saved in this approach.

The recent suggestion [12] of using the Langevin formalism, compactifying the space at infinity may, have not these drawbacks. Indeed the equations of motion are only slightly modified and one may hope that this fact is sufficient to enforce the locality of the theory. Unfortunately at the present moment it is not clear if and how this proposal, which is formulated in 0 dimensions, can be extended to higher dimensions. A better understanding of this suggestion should be very interesting because at the present moment is seems to be the only possible strategy for defining pure gravity.

4. Supersymmetry

The introduction of supersymmetry would be very interesting because we expect that the problems connected with the divergence of the perturbative expansion should not be present, i.e. the perturbative expansion may be non Borel summable, but there should be a simple unambiguous prescription to define the the theory. As usual it is rather difficult to introduce supersymmetry in parameter space: a lattice breaks translational invariance. In principle here the task is not so desperate because translational invariance in parameter space is restored in the end; unfortunately things are not so simple.

Just for a moment let us consider a theory in D dimensions invariant under the $O(D)$ group. Our aim is to find the supersymmetric extension to the group $O(D/2)$. A possible form for the partition function on a random lattice is the following

$$\int dx_i d\phi(x_i) exp\left(-\sum_{i,j} exp(-(x_i - x_j)^2)(\phi(x_i) - \phi(x_j))^2 \right) , \qquad (4.1)$$

where we have introduced N points labeled by the index i and we have only written the kinetic term in the action.

The supersymmetric extension can be very easily done by substituting the small case with the upper case in (4.1) . The final formulae are not simple at all, moreover this formulation is not reparametrization invariant from the starting point. A reparametrization invariant formulation should be obtained by starting with supertriangulations. Unfortunately the very definition of a supertriangulation of a supersurface (in the sense of a manifold with two commuting and two anticommuting dimensions embedded in a commuting space) is unclear to me.

The simplest possibility for introducing a supersymmetry consists in considering only supersymmetry in target space, not in parameter space. In this case the supersurface is a much more simple object: it is a mapping from a two dimensional space to a superspace. The usual matrix approach can be easily generalized to this supersymmetric case by substituting the small case with the upper case in the appropriate formulae.

Let us see how this works in details [13] .

The superstring theory may defined by integrating overall the supersurfaces of arbitrary genus, using a weight that is reparametrization invariant in parameter space and supersymmetric invariant in target space. The simplest way to discretize the theory consists in introducing a triangulation of the surface in the parameter space; we also associate to each triangle of the surface a point in superspace. Each supertriangulation is thus fixed by the topology of the two dimensional triangulation and by the position of the centers of the triangles in superspace.

We can assign to each triangulation a weight which is the product of a function of the relative superdistances of all contiguous triangles. The weights get also an extra multiplicative factor proportional to v^g, where g is the genus of the two dimensional surface and v is the parameter which controls the perturbative genus expansion. We must sum over all possible topologies of the triangulations and integrate in the super space the corresponding weights. A very convenient form for the weights is the free propagator of the superfield .

If we use this prescription, a triangulation with k vertices corresponds to the k'th order of the usual perturbative expansion for a supersymmetric theory, of a Wess–Zumino kind, with a Φ^3 interaction in superspace. The supersymmetric string should be described by an appropriate Wess–Zumino model with a $N \times N$ superfield, in the scaling limit $N \to \infty$.

This proposal may work in any dimensions, explicit computations have been done in one dimensions. Following ref./MP/ the Action is given

$$S = \int dt d\theta d\bar{\theta} \ \text{Tr}\left(-\Phi \bar{D} D \Phi + \mathbf{W}(\Phi)\right) , \qquad (4.2)$$

where the fields Φ are $N \times N$ Hermitian matrices, and D is the usual differential operator in superspace, and

$$\mathbf{W}(\Phi) \equiv \tfrac{1}{2}\Phi^2 - \frac{\lambda}{3\sqrt{N}}\Phi^3 . \qquad (4.3)$$

The content of the superfield in ordinary space is

$$\Phi \equiv \mathbf{X} + \theta \mathbf{\Psi} + \bar{\theta}\bar{\mathbf{\Psi}} + \mathbf{A}\theta\bar{\theta} . \qquad (4.4)$$

It is well known that in this case fermions may be integrated out explicitly, recovering in this way Witten supersymmetric quantum mechanics.

In the simplest case ($N = 1$) the Hamiltonian in the Bosonic sector is given by

$$H = p^2 + F^2(x) - \frac{dF}{dx} , \qquad (4.5)$$

Similar results are obtained for generic N. An explicit computation shows that in our case we can compute the spectrum of H by solving a Schrödinger equation with potential $N \ \text{Tr} \ \mathbf{V}(\mathbf{X}/\sqrt{N})$, where the function V is defined by

$$V(x) \equiv (x(1 - \lambda x))^2 - 1 + 2\lambda x . \qquad (4.6)$$

The properties of the Hamiltonian can be computed in the limit $N \to \infty,$. In this way we are able to identify the value of λ_c. It is well known [14] that the Hamiltonians like those in eq. (4.5) describe an equivalent 1 dimensional system composed of N non interacting Fermions with a one fermion Hamiltonian given by

$$\mathcal{H} = p^2 + NV(x/\sqrt{N}) . \qquad (4.7)$$

We start to analyze the system in the WKB regime. For small λ the potential has two minima and a maximum while for λ large enough (i.e. greater than λ_c) there is only one minimum. At the critical point the Hamiltonian coincides with that of the $m = 3$ Bosonic tricritical theory.

In the WKB approach we have to find the Fermi energy ϵ_F, which is fixed by the condition

$$\frac{1}{\pi}\int dx \ \sqrt{\epsilon_F - V(x)} = 1 . \qquad (4.8)$$

The ground state energy of $H^{(B)}$, $\mathcal{E}_0$ is given by

$$\mathcal{E}_0 = \sum_{i=1,N} E_i , \qquad (4.9)$$

where N also coincides with the number of states with $E < \epsilon_F$. In the WKB approach $\mathcal{E}_0$ can be explicitly computed. When λ is smaller that λ_c, $\mathcal{E}_0$ turns out to be identically zero, as it must be when supersymmetry is not spontaneously broken. In this region the tunneling between the two wells is exponentially suppressed, and it cannot be seen in the WKB expansion.

On the contrary in the large λ region the vacuum energy is different from zero in the WKB approach and for $\lambda \to \lambda_c$ the vacuum energy vanishes as $N^2(\lambda - \lambda_c)^{5/2}$. Here supersymmetry is spontaneously broken.

In the case of unbroken supersymmetry the wave function of the vacuum would be given by

$$\psi \simeq e^{-Tr\mathbf{W}(\mathbf{X})} \, , \tag{4.10}$$

The Hamiltonian H is essentially the usual forward Fokker–Plank operator corresponding to the Langevin equation:

$$\frac{d\mathbf{X}}{dt} = -\mathbf{F}(\mathbf{X}) + \eta \, . \tag{4.11}$$

If the wave function ψ is not normalizable there are trajectories which escape to ∞, and the lowest eigenvalue of the Fokker–Plank operator is slightly larger than zero.

In our case the wave function ψ is not normalizable, supersymmetry is broken and the vacuum energy is exponentially small when $(\lambda - \lambda_c) \to 0$.

We expect therefore that in the planar limit the vacuum energy is identically zero and supersymmetry breaking is invisible. This conclusion holds also for all the corrections of order N^{-2g}.

A careful analysis may be done to compute the scaling relations needed to extrapolate the finite N results in the limit $N \to \infty$. In the scaling limit the one dimensional potential is simply given by $ax^3 + bx$, a and b being appropriate constants. The corresponding Hamiltonian has many selfadjoint extensions corresponding to different boundary conditions at infinity as happens in the usual non supersymmetric case [15]. Each of the extension has a discrete spectrum (I disagree on this point with [16], where a continuous spectrum is claimed). The correct supersymmetric selfadjoint extension can be found by comparison with the WKB expression for the high levels. It is also possible to do numerical computations of the vacuum energy for not too large N by evaluating the first N energy levels of the one dimensional Schrödinger equation.

It turns out that supersymmetry is spontaneously broken for any value of λ also in the scaling limit where N goes to infinity. Indeed we have seen that the ground state energy is related the probability for unit time that the solution of the associated Langevin equation escape to infinity. It is convenient to write the associated Langevin equation for the eigenvalues of $\mathbf{X}$ (let us call then x_i). The final equation is

$$\frac{dx_i}{dt} = -\frac{\partial U}{\partial x_i} + \eta_i \, , \tag{4.12}$$

where

$$U(x) = \sum_i V(x_i) - \sum_{i,j} ln(x_i - x_j) \, . \tag{4.13}$$

In the limit of very large N the effective potential U may be evaluated and coincide with the real part of the F function introduced in the original paper [14] . A barrier is present for λ is smaller that λ_c. By computing explicitly the tunneling one finds [17] that the vacuum energy is proportional to $\exp(-N(\lambda_c - \lambda)^{\frac{5}{4}})$.

The supersymmetric theory is well defined and no ambiguity are present. There are no serious objections to the extension of this approach to dimensions larger than 1.

As it should be clear form the previous discussion the equal time expectation values of this supersymmetric theory coincide, neglecting exponentially small terms (like $\exp(-A\, s^{\frac{5}{4}})$), with those of the ill defined Bosonic theory in $d = 0$ for $m = 2$. This coincidence is likely to be present only in one dimension and it should disappear in high dimensions.

5. Toward Two Dimensions

The brute force computations we have discussed here do not seem to allow us to perform computations in two dimensions. The most simple questions have unknown answers. At a preliminary stage we would like to compute the critical exponents of the theory, i.e. how the mass gap goes to zero at the critical coupling constant (a power law is expected). A new approach is needed.

Fortunately conformal invariance may help us. In the scaling limit, where the distance is much smaller of the inverse of the mass gap and much larger of the scale set by the lattice in parameter space, the theory should be conformal invariant and symmetric under the $SU(\infty)$ group. It seem reasonable that the theory in this limit is invariant under the Kac–Moody extension of the $SU(\infty)$ group. It may be possible that the set of unitary representation of this group is sufficiently constrained that useful results may be obtained with sufficient physical insight.

We notice that the transitions we are speaking about disappear for finite N so that the representation we are looking for are likely not the limit of unitary representation for Kac–Moody extension of the group $SU(\infty)$ in the limit where N goes to infinity. This proposal is not easy to implement, but it is at the present moment the only open possibility for doing analytic computations in $D = 2$.

Acknowledgements

A great part of the work presented here has been done in collaboration with E. Marinari. It is a pleasure for me to thank him for many interesting discussions on the subject of this talk.

References

[1]   G. Parisi, Phys. Lett. B238 (1990) 213.
[2]   D. Gross and I. Klebanov, Princeton preprint, March 1990, PUPT-1172.
[3]   G. Parisi, G. Salina and S. Vladikas, *Tor Vergata* preprint ROM2F/90/52.
[4]   E. Brézin, E. Marinari and G. Parisi, Phys. Lett. B242 (1990) 35.
[5]   E. Brézin, G. Parisi and J. Zinn Justin, Phys. Rev. D16, 408 (1977).

[6] G. Parisi, Phys. Rep. 49, 215 (1979).

[7] E. Witten, Nucl. Phys. **B185** (1981) 513; Nucl. Phys. **B202** (1983) 253.

[8] G. Parisi and S. Sourlas, Nucl. Phys, 217B, 203, 1983.

[9] G. Parisi and S. Sourlas, Phys. Rev. Lett. 43, 744 (1979).

[10] J. Greensite and M. Halpern, Nucl. Phys. B242 (1984) 167.

[11] J. Ambjorn, J. Greensite and S. Varsted, Niels Bohr Institute preprint NBI-HE-90-3.

[12] E. Marinari and G. Parisi, Phys. Lett. B247 (1990)247.

[13] E. Marinari and G. Parisi, Phys. Lett. B240 (1990) 375.

[14] E. Brézin, C. Itzykson, G. Parisi and J. Zuber, Comm. Math. Phys. 59 (1978) 35.

[15] G. Parisi, Phys. Lett. B238 (1990) 209; Euro. Phys. Lett. 15, 595 (1990)

[16] M. Karliner and A. Migdal, Princeton preprint, July 1990.

[17] E. Marinari, G. Parisi, unpublished.

EXCITATIONS AND INTERACTIONS IN A ONE DIMENSIONAL STRING THEORY

Spenta R. Wadia

Tata Institute of Fundamenta Research
Homi Bhabha Road
Bombay 400005, India

Abstract:We discuss the singlet sector of the $d = 1$ matrix model in terms of a Dirac fermion formalism. The leading order two- and three- point functions of the density fluctuations are obtained by this method. This allows us to construct the effective action to that order and hence provide the equation of motion.

Introduction: In recent works[1] [2] [3] we have studied the problem of two-dimensional quantum field theories coupled to gravity. Our original motivation to do this was to arrive at a natural setting for the theory space formulation[4] [5] of string theory, where, (1) there is no restriction on the central charge of the matter sector, and, (2) the theory has, within it, the ingredients to describe trajectories which join special points in the theory space, namely the classical vacua which correspond to conformally invariant theories. One of our main results has been that the matter + gravity system can be regarded as a field theory of the Liouville mode and matter fields in the background of the fiducial metric. Generic couplings or backgrounds now depend both on the Liouville mode and on the matter degrees of freedom and satisfy equations of motion in $d + 1$ variables (d = matter, 1 = Liouville). Other related works are due to J. Polchinski[6], and, T. Banks and J. Lykken[7].

These ideas were illustrated in various situations:

(a) For d-scalar fields interacting with 2-dim. gravity, we proved that this system quantized in the light cone gauge is exactly mapped into the conformally invariant field theory of $d + 1$-scalar fields, with background charge. At $d = 25$ we obtained the exact tree level S-matrix and spectrum of the "$d = 26$ critical string".

(b) In the case of $d < 1$, we considered the $(m, m + 1)$ minimal models coupled to gravity, and could effectively describe the interpolation between two minimal models, for m very large, by means of a 'string field' that depends only on the Liouville mode, a function $\kappa(\eta)$ which satisfies the field equation[3]

$$\left(\partial_\eta{}^2 + Q\partial_\eta + h\right)\kappa(\eta) = b\kappa^2(\eta) + o(\kappa^3) \tag{1}$$

(c) In the case of d-scalar fields coupled to gravity, perturbed by a 'tachyon' background, the tachyon coupling T,which depends on d coordinates ϕ_i and the Liouville mode η satisfies the $d+1$ dimensional field equation[1][3]

$$\left(\partial_\eta^{\;2} + Q\partial_\eta + \partial_i\partial_i + 2\right) T(\phi,\eta) + T^2(\phi,\eta) + \cdots = 0 \tag{2}$$

where

$$Q = \sqrt{\frac{25-d}{3}} \tag{3}$$

The coupling of other backgrounds like the metric, antisymmetric tensor and dilaton can be discussed similarly. To see the spectrum from equation (2), one eliminates the linear derivative piece by defining $\tilde{T} = e^{-\frac{1}{2}Q\eta}T$

$$\left(\partial_\eta^{\;2} + \partial_i\partial_i + \frac{1}{4}(8 - Q^2))\right)\tilde{T}(\phi,\eta) + e^{-\frac{1}{2}Q\eta}\tilde{T}^2(\phi,\eta) + \cdots = 0 \tag{4}$$

This equation tells us that the spectrum at $d=1$ (i.e., $Q^2 = 8$) is that of a massless particle. For $d > 1$, there is a tachyon in the spectrum and hence for much the same reasons as in 26-dimensional critical string theories, where it ruins the perturbation expansion, these theories may not exist. It is likely that the tachyon perturbation drives $d > 1$ theories to a stable point which is $d = 1$. It would also be interesting to understand how one can reach models with $d < 1$ by appropriate perturbations of the $d = 1$ model.

Our main purpose here is to discuss the cutoff string field theory at $d = 1$[8][9][10], formulated as the quantum mechanics of the matrix hamiltonian which was originally discussed by Brézin, Itzykson, Parisi and Zuber[11].

$$H = -\frac{1}{2N}\nabla_M^2 + N\ tr\ V(M) \tag{5}$$

where ∇_H^2 is the laplacian in the space of hermetian matrices and $V(M)$ is a polynomial. We can expect the results of the continuum theory and that from the matrix model approach to agree in the low momentum region only.

Since this hamiltonian is invariant under $U(N)$ transformations, $M \to UMU^\dagger$, there would be wavefunctions transforming according to various different representations of $U(N)$. (To be more precise, these consist of the trivial representation and the representations that can be generated by taking products of the adjoint.) It is likely that states which transform nontrivially under $U(N)$ are not related to the string degrees of freedom. Presently we will analyse the singlet sector of model. We use the fermionic representation of this sector as explained below. This representation has two major advantages.

(a) The model is well defined even for finite N and for noncritical values of the coupling. Hence the nature of the various regularizations are most clearly recognized in this picture.

(b) It is easier to see various approximate and exact symmetries of the system from this point of view. The string theory discussed here is an exactly integrable system by virtue of the fact that it is a theory of free non-relativistic fermions.

As in well known, ∇^2_M acting on the singlet sector wave function $\psi(\Lambda)$ has the form

$$\nabla^2_M \psi(\Lambda) = \frac{1}{\Delta(\Lambda)} \left(\sum_i \frac{\partial^2}{\partial \lambda_i^2} \right) \Delta(\Lambda)\psi(\Lambda) \tag{6}$$

where $\Lambda = (\lambda_1, \cdots, \lambda_N)$, λ_i being the eigenvalues of M. $\Delta(\Lambda)$ is the Vandermonde determinant

$$\Delta(\Lambda) = \prod_{i<j}(\lambda_i - \lambda_j) \tag{7}$$

If we use $\chi(\Lambda) = \Delta(\Lambda)\psi(\Lambda)$ instead of $\psi(\Lambda)$ as the wave function, the effective hamiltonian becomes

$$H_f = -\frac{1}{2N} \sum_i \frac{\partial^2}{\partial \lambda_i^2} + N \sum_i V(\lambda_i) \tag{8}$$

This is the hamiltonian for non-interacting particles. However, since $\psi(\Lambda)$ is symmetric, $\chi(\Lambda)$ is antisymmetric. Hence the problem reduces to that of noninteracting fermions moving in an external potential.

The ground state is obtained by filling the N states which are lowermost in energy. The corresponding total energy becomes singular when the Fermi energy of the system approaches the value of the potential at a stationary point. This is related to the fact that the time period of the classical orbits corresponding to the states near the fermi surface starts diverging when they can approach the stationary point. In the semiclassical analysis one can obtain the nature of the singularities. Also one can take the double-scaling limit by keeping fixed the energy difference between the value of the potential at the stationary point and that of Fermi energy as N goes to infinity. Inverse of this energy difference can be identified as the string coupling constant, g_{str}[9].

We have developed a Dirac fermion formalism for the states near the Fermi surface which works well for the leading contributions and lends valuable insight into many results.

To obtain the equations of motion we write this theory as a second quantized theory with the action

$$S = \int dt\, d\lambda\, \chi^\dagger(\lambda,t)(i\partial_t + \frac{1}{2N}\partial^2_\lambda - NV(\lambda))\chi(\lambda,t) + \int dt\, d\lambda\, N a_0(\lambda,t)\chi^\dagger(\lambda,t)\chi(\lambda,t) \tag{9}$$

where $\chi(\lambda,t)$ is a second-quantized fermion field in two dimensions and $a_0(\lambda,t)$ is the source function conjugate to density. It corresponds a coupling of the form $\sum_n tr\, M^n(t)a_n(t)$ in the original matrix model, where $a_n(t) = \int d\lambda\, \lambda^n\, a_0^n(\lambda,t)$. The corresponding vacuum to vacuum amplitude $Z[a_0]$ contains the information of all correlation function of density. Let

$$F[a_0] = \ln Z[a_0] \tag{10}$$

By taking the Legendre transformation of $F[a_0]$ we obtain the effective action $\Gamma[\rho]$

where

$$\rho(\lambda, t) = \frac{\delta F[a_0]}{\delta a_0(\lambda, t)} \tag{11}$$

and

$$\Gamma[\rho] = \int d\lambda \ dt \ \rho(\lambda, t) a_0(\lambda, t) - F[a_0] \tag{12}$$

When $a_0 = 0$ we have

$$\frac{\delta \Gamma[\rho]}{\delta \rho(\lambda, t)} = 0 \tag{13}$$

which is the quantum equation of motion.

Since we are looking at an effective bosonic theory we define the field variable that is used for bosonizing relativistic fermion theories,

$$\phi(\lambda, t) = \int^{\lambda} (\rho(\lambda, t) - \ <\rho(\lambda, t)>) \tag{14}$$

We will find, perturbatively, the leading order terms in the equation of motion in terms of this variable.

Dirac fermion representation

In the following we briefly indicate how this result was obtained using a Dirac fermion representation. This is essentially by a change of variables from $\lambda \to \tau = \int^{\lambda} \rho_o(\lambda')d\lambda'$, and scaling the hamiltonian by the fermi level wave function,

$$\begin{aligned}
\hat{H} &= \left(e^{\mp iN\Theta_f} \rho_f^{-1/2} \right) H \left(\rho_f^{1/2} e^{\pm iN\Theta_f} \right) \\
&= \mp i \frac{d}{d\tau} - \frac{1}{2N} \frac{d}{d\tau} \rho_o^2 \frac{d}{d\tau}.
\end{aligned} \tag{15}$$

where $H(\rho_f^{1/2} e^{\pm iN\Theta_f}) = E_f(\rho_f^{1/2} e^{\pm iN\Theta_f})$.

To get the scales right, let us make some estimates. The leading order large N solution of the equation

$$H\phi = E_f \phi \tag{16}$$

is

$$\phi = \text{constant} \times \frac{1}{\sqrt[4]{2(\epsilon_f - V(\lambda))}} e^{\pm iN \int^{\lambda} d\lambda' \sqrt{2(\epsilon_f - V(\lambda'))}} \tag{17}$$

where

$$\epsilon_f = \frac{E_f}{N} \tag{18}$$

If we choose the constant to be 1, we have

$$\rho_f(\lambda) = \frac{1}{\sqrt{2(\epsilon_f - V(\lambda))}}$$
$$\Theta_f(\lambda) = \int^{\lambda} d\lambda' \sqrt{2(\epsilon_0 - V(\lambda))} \tag{19}$$

and

$$\rho_f \Theta'_f = 1 \tag{20}$$

Let the potential have a maximum at λ_0 with $V''(\lambda_0) \neq 0$. If we take a solution for ϵ_0 very near $V(\lambda_0)$ then most of the probability is concentrated near that tip. Classically this is manifested by the particle spending a lot of time near the turning point, which is very close to the rather flat region around the potential maximum.

In this region $V(\lambda) \approx V(\lambda_0) - \frac{1}{2}|V''(\lambda_0)|(\lambda - \lambda_0)^2$

$$\tau - a \sim \int_{\lambda_1}^{\lambda} d\lambda' \frac{1}{\sqrt{|V''(\lambda_0)|(\lambda' - \lambda_0)^2 - 2(V(\lambda_0) - \epsilon_0)}}$$
$$\sim \frac{1}{(|V''(\lambda_0)|)^{1/2}} \int_{\lambda_0 + \sqrt{2\mu}}^{\lambda} d\lambda' \frac{1}{((\lambda' - \lambda_0)^2 - 2\mu)^{1/2}} \tag{21}$$

By convention we make $V''(\lambda_0) = -1$ and define μ to be $V(\lambda_0) - \epsilon_0$.

Upon integration we get

$$\tau - a = \cosh^{-1}\left(\frac{\lambda - \lambda_0}{\sqrt{2\mu}}\right) \tag{22}$$

or

$$\lambda = \lambda_0 + \sqrt{2\mu} \cosh(\tau - a) \tag{23}$$

where a is the value of τ at the turning point.

Now

$$\rho_f^2 \sim \frac{1}{4\mu \sinh^2(\tau - a)} \tag{24}$$

$$\hat{H} \sim \mp i \frac{\partial}{\partial \tau} - \frac{1}{8N\mu} \frac{\partial}{\partial \tau} \frac{1}{\sinh^2(\tau - a)} \frac{\partial}{\partial \tau} \tag{25}$$

This estimate can be trusted, when τ is not too near a.

To recover an approximate relativistic fermion picture from a nonrelativistic one, the most natural thing to do is to take the reference energy level E_0 to be the Fermi level E_f. If we now want the expression of $\hat{H}$ in terms of τ to be a scaled expression, that is, if we want to keep $\tau - a$ as a scaled variable, we have to have $N\mu = \text{fixed}$. (This is true irrespective of the semiclassical approximation that we made to reach this expression of $\hat{H}$.)

Strictly speaking, for this problem, the wave functions are not exactly like $\rho^{1/2}e^{+iN\Theta}$ and $\rho^{1/2}e^{-iN\Theta}$, but a specific linear combination which depends upon the energy and the the boundary conditions. In terms of τ variables, $\rho^{1/2}e^{\pm iN\Theta}$, after the relevant transformation looks like a plane wave in the leading order. The extent of classically allowed $\lambda - \lambda_0$ is roughly from 0 to say 1. Corresponding range of $\tau - a$ is from 0 to $\ln\frac{1}{\sqrt{\mu}}$. The level spacing goes as inverse of this range. Hence the boundary condition can give rise to mixing of left moving and right moving plane waves which can change the energy atmost by $\frac{1}{\ln\frac{1}{\sqrt{\mu}}}$. This vanishes in the scaling limit.

Thus we are allowed, in the scaling limit to deal with chiral states which are almost exact eigenstates. The hamiltonian which makes the right moving states near the Fermi surface look like plane waves is

$$\hat{H}_R = -i\partial_\tau - \frac{1}{2N}\partial_\tau \rho_f^2 \partial_\tau \qquad (26)$$

The hamiltonian which does the same for the left moving states is

$$\hat{H}_L = i\partial_\tau - \frac{1}{2N}\partial_\tau \rho_f^2 \partial_\tau \qquad (27)$$

Both the hamiltonians have the information about all the states. However, for the left moving states the second term in $\hat{H}_R$ cannot be considered as a small perturbation. Similar problem arise for right moving states and $\hat{H}_L$.

Thus, for the calculations where only states near the Fermi surface matter, one can describe the left moving states by $\hat{H}_L$ and the right moving by $\hat{H}_R$. This gives a Dirac hamiltonian. In the second-quantized notation the hamiltonian is

$$H = \int d\tau \left(\psi_+^\dagger \hat{H}_L \psi_+ + \psi_-^\dagger \hat{H}_R \psi_- \right) \qquad (28)$$

To be honest one should discard half the solutions of each of the hamiltonians, not to over count the states. This would be some ultraviolet cut off in the theory. This cutoff would refer to the value of the momenta where the second term starts dominating over the first. For calculations involving processes near the Fermi surface, this cutoff is not important.

In many of the leading order calculations, this problem does not show up. This effective ultraviolet cutoff parameter, in a certain region of τ, is finite in the scaled picture (as opposed to the semiclassical case). Hence one has to be careful about it.

We now briefly indicate the results of the calculation of the 2 and 3 point functions of the density which corroborate our guess of the effective action to leading order.

The two-point function of density fluctuations

$$G^{(2)}(1,2) = <0|T\rho(\lambda_1,t_1)\rho(\lambda_2,t_2)|0>_c \qquad (29)$$

$\rho(\lambda,t)$ is the eigenvalue/fermion density. If we look only at the connected part, we would see the correlation of density fluctuation, $\rho(\lambda,t) - <\rho(\lambda,t)>$, only. This density fluctuation can be represented also by $\chi^\dagger\chi$ normal ordered with respect to the Fermi sea.

If we change over to τ variables we have

$$\tilde{G}^{(2)}(1,2) = < 0|T\tilde{\rho}(\tau_1, t_1)\tilde{\rho}(\tau_2, t_2)|0 >_c$$
$$= \frac{d\lambda_1}{d\tau_1}\frac{d\lambda_2}{d\tau_2} < 0|T\rho(\lambda_1, t_1)\rho(\lambda_2, t_2)|0 > \tag{30}$$

We call $\tilde{\chi}_L = \psi_+$ and $\tilde{\chi}_R = \psi_-$

$$: \chi^\dagger \chi : \rightarrow : \psi_+^\dagger \psi_+ : + : \psi_-^\dagger \psi_- : \tag{31}$$

$$\tilde{G}^{(2)}(1,2) = < 0|T : \psi_+^\dagger(1)\psi_+(1) :: \psi_+^\dagger(2)\psi_+(2) : |0 > +(+ \rightarrow -) \tag{32}$$

Take $t_1 > t_2$ and consider

$$< 0| : \psi_-^\dagger(1)\psi_-(1) :: \psi_-^\dagger(2)\psi_-(2) :: |0 >$$
$$= S_p^{(-)}(1,2)S_h^{(-)}(1,2) \tag{33}$$

In the leading order the particle and hole propagators are identical and charge conjugation symmetry is explicit,

$$S_p^{(-)} = S_h^{(-)} = \frac{\omega}{2\pi} \sum_{n=0}^{\infty} e^{-i(n+\frac{1}{2})\omega t_{12}^-} \tag{34}$$

Using these formulae the 2-point function is calculated to be

$$\tilde{G}^{(2)}(1,2) = \frac{\partial}{\partial \tau_1}\frac{\partial}{\partial \tau_2}\left[\sum_{j=-\infty}^{\infty} \frac{i\omega}{4\pi^3} \int_{-\infty}^{\infty} dE \frac{e^{-i(Et_{12}-j\omega\tau_{12})}}{E^2 - (j\omega)^2 + i\epsilon}\right] \tag{35}$$

The expression inside the square bracket is the correlator of a free bose field. This is not surprising since what we have done is to bosonize the noninteracting fermions in a finite volume. We identify the free Bose field through the well known relation

$$: \psi^\dagger\psi := \partial_\tau \phi \tag{36}$$

One can then see equation (37) coming out immediately from the Bose field correlator.

The three-point function of density fluctuations

For fermions satisfying the Dirac equation, the three point function of density is zero. This is a consequence of the charge conjugation symmetry of the Dirac hamiltonian. In other words, it is a consequence of the symmetry of the problem under reflection about the Fermi level. However, we know that this symmetry is broken in the nonrelativistic model and this is caused by the second term in the hamiltonian. This term, treated as a perturbation, should provide systematic order by order contributions to the three point function.

The lowest order contribution to the three-point function

$$\tilde{G}^{(3)}(1,2,3) = <0| : \psi_+^\dagger(1)\psi_+(1) :: \psi_+^\dagger(2)\psi_+(2) :: \psi_+^\dagger(3)\psi_+(3) : |0> +(+ \rightarrow -) \quad (37)$$

$(t_1 > t_2 > t_3)$ turns out to be the following lengthy expression after a long calculation.

$$\tilde{G}^{(3)}(1,2,3) = \sum_{j_1,j_2=1}^{\infty} \frac{\omega^4}{8\pi^3} e^{-ij_1\omega t_1^- -i(j_2-j_1)t_2^- +ij_2 t_3^-}$$

$$\left[\omega^{-1}\frac{\partial\omega}{\partial\epsilon}\left\{ j_2(j_1-j_2)\Theta(j_1-j_2)(1-ij_1\omega t_1^-) \right.\right.$$

$$+ j_2 j_1(1-i(j_2-j_1)\omega t_2^-)$$

$$\left. + j_1(j_2-j_1)\Theta(j_2-j_1)(1+ij_2\omega t_3^-)\right\}$$

$$- j_2(j_1-j_2)\Theta(j_1-j_2)\left(\frac{1}{2(\epsilon-V(1))} + ij_1\omega \int^{\tau_1}\frac{d\tau}{2(\epsilon-V)}\right) \quad (38)$$

$$- j_2 j_1 \left(\frac{1}{2(\epsilon-V(2))} + i(j_2-j_1)\omega \int^{\tau_2}\frac{d\tau}{2(\epsilon-V)}\right)$$

$$\left. - j_1(j_2-j_1)\Theta(j_2-j_1)\left(\frac{1}{2(\epsilon-V(3))} - ij_2\omega \int^{\tau_3}\frac{d\tau}{2(\epsilon-V)}\right)\right]$$

$$+ \left(t^- \rightarrow t^+ \text{ and } \int^{\tau}\frac{d\tau'}{2(\epsilon-V)} \rightarrow -\int^{\tau}\frac{d\tau'}{2(\epsilon-V)}\right)$$

$\Theta(x)$ being the Heaviside function.

The structure of the effective action

We want to keep only terms upto the order of $\frac{1}{N}$ in the equation of motion. It can be easily seen from N counting that if we normalize the two-point connected Green's function to be order 1, then the order of the n-point connected function is N^{2-n}. Hence we need to consider only the two-point and the three-point function. The leading contribution to the three-point function is of the order $1/N$. The $1/N$ contribution to the two-point function cancells off. This is because the two-point function of the density is $S_h(1,2)S_p(1,2)$ which is $S(1,2)^2$ in the lowest order. The next order is $\delta(S_h(1,2)S_p(1,2))$. Since, in the lowest order $\Delta S_h = -\Delta S_p$, the first correction to the two-point funcion is zero. The correction is therefore $\sim O(\frac{1}{N^2})$.

Hence, from what we have done till now, we can reconstruct in the lowest order quadratic and cubic pieces of the effective action. The quadratic piece is going to be that of a free boson field which is $2\pi \int dt d\tau \partial_+\phi\partial_-\phi$. We need to choose a three-vertex which gives the correct three point function. This three point function has two pieces. One is proportional to $\omega^{-1}\frac{\partial\omega}{\partial\epsilon}$, the other is not. This first term is the dominant one for fixed λ_i, if we calculate $< \prod_i \rho(\lambda_i, t_i) >$. However if we change over to scaled variable like $\tau - a$ then since

$$\omega \sim |\ln \Delta\epsilon|^{-1}, \quad \Delta\epsilon = V(\lambda_0) - \epsilon$$
$$\frac{1}{N\omega}\frac{\partial\omega}{\partial\epsilon} \sim \frac{1}{|\ln\Delta\epsilon|}\frac{1}{N\Delta\epsilon} \to 0 \tag{39}$$

if $N\Delta\epsilon$ is held fixed when $N \to \infty$. On the other hand quantities like

$$\frac{1}{N}\frac{1}{2(\epsilon - V)} \sim \frac{1}{N\Delta\epsilon}\frac{1}{\sinh^2(\tau - a)} \tag{40}$$

remain finite. Hence we pay less attention to the piece proportional to $\frac{\partial\omega}{\partial\epsilon}$. The other piece is a sum of two chiral contributions. This indicates that the vertex is made of $\partial_+\phi$ and $\partial_-\phi$. $\partial_\pm = \frac{\partial}{\partial t^\mp}$. In fact one can show that the required interaction piece of the effective action is of the form

$$\Gamma_{\mathrm{int}} = \frac{-2\pi^2}{3N} \int dt d\tau \rho_f^2(\tau)\{(\partial_+\phi)^3 - (\partial_-\phi)^3\}. \tag{41}$$

It is remarkable that some very similar action can be obtained if one tries to bosonize the fermion theory naively by using Mandelstam formulae[12],

$$\psi_\pm^\dagger(\tau_1)\psi_\pm(\tau_2) = \mp\frac{i}{2\pi(\tau_1 - \tau_2)} \exp\left\{-\pi i \int_{\tau_1}^{\tau_2} d\tau(\dot\phi \pm \phi') + O(\tau_1 - \tau_2)^2\right\} \tag{42}$$

(Note that our normalization of ϕ is different from Mandelstam's.) Now, one can separately differentiate in τ_1 and τ_2 and then take the limit $\tau_1 \to \tau_2$ and use the result in equation (42) to obtain the bosonic expression for the perturbation.

We know that Mandelstam formulae depend crucially on the short distance properties of the Green's function, which can be modified if the perturbation is singular. This is precisely the case here. Yet this procedure gives the same leading order effective action,except for a $\frac{1}{N}\int dt d\tau \rho_f^2 \partial_\tau^3 \phi$ term (which, if genuinely present, should shift the background ϕ from zero to a value $\sim 0(\frac{1}{N})$ and in that process give $0(\frac{1}{N^2})$ correction to the two point function which no longer remain translation-invariant). It is possible that there is a generalization of the Mandelstam formulae in our case, where terms more singular than $\frac{1}{\tau_1 - \tau_2}$ appear, but they are always multiplied by higher powers of $1/N$ (or g_{str}).

The equation of motion in the lowest order looks like

$$\partial_+\partial_-\phi = \frac{\pi}{2N}\left[\partial_+\{\rho_f^2(\partial_+\phi)^2\} - \partial_-\{\rho_f^2(\partial_-\phi)^2\}\right] \tag{43}$$

since

$$\rho_f^2(\tau) \sim \frac{1}{4\mu\sinh^2(\tau - a)} \tag{44}$$

for large $\tau - a$, i.e. for points far away from the turning point,

$$\rho_f^2(\tau) \sim \frac{e^{-2(\tau-a)}}{\mu} \tag{45}$$

Then

$$\partial_+\partial_-\phi = \frac{\pi}{N\mu} e^{-2(\tau-a)} \left[-(\partial_+\phi)^2 + (\partial_-\phi)^2 + \partial_+\phi\partial_+^2\phi - \partial_-\phi\partial_-^2\phi \right] \tag{46}$$

This is very similar to the tachyon equation. Note, however, that the interaction terms consist solely of derivatives of ϕ and not ϕ itself. Also it can be written entirely in terms of the currents $j_\pm = \partial_\pm\phi +$ higher order terms.

Note added

While this work was in progress we became aware of similar works by S.R. Das and A. Jevicki[13] and[14]J. Polchinski.

Acknowledgements

The work presented here was done with Anirvan Sengupta[15]. I thank Orlando Alvarez, Enzo Marinari and Paul Windey for organizing a most exciting meeting and also for the opportunity to present a preliminary form of this work. I thank E. Brezin, S.R. Das, A. Dhar, D.J. Gross, V. Kazakov, G. Mandal and S. Shenker for useful discussions.

REFERENCES

1. S. R. Das, S. Naik and S. R. Wadia, Mod. Phys. Lett. A4 (1989) 1033.

2. A. Dhar, T. Jayaraman, K. S. Narain and S. R. Wadia, Mod. Phys. Lett. A5 (1990) 863.

3. S. R. Das, A. Dhar and S. R. Wadia, Mod. Phys. Lett. A5 (1990) 799.

4. A. B. Zamolodchikov, JETP Lett. 43 (1986) 731.

5. S. R. Das, G. Mandal and S. R. Wadia, Mod. Phys. Let. A4 (1989) 745.

6. J. Polchinski, Texas preprint UTTG-02-89(1989).

7. T. Banks and J. Lykken, Univ. of California, Santa Cruz preprint (May, 1989).

8. V. A. Kazakov and A. A. Migdal, Nucl. Phys. B320 (1989) 654.

9. E. Brézin, V.A. Kazakov and Al.B. Zamolodchikov, Ecole Normale preprint LPS-ENS 99-19R; G. Parisi, Phys. Lett. B238 (1990) 209; D. Gross and N. Miljković, Phys. Lett. B238 (1990) 217; P. Ginsparg and J. Zinn-Justin, Harvard preprint, HUTP-90/A004 (1990).

10. S. R. Das, A. Dhar, A. M. Sengupta and S. R. Wadia, Mod. Phys. Lett. A5 (1990) 891.

11. E. Brézin, C. Itzykson, G. Parisi and J.B. Zuber, Comm. Math. Phys. 59, 35 (1978).

12. S. Mandelstam, Phys. Rev. D11 (1975) 3026.

13. S.R. Das and A. Jevicki, Brown Preprint, 1990.

14. J. Polchinski, Texas Preprint, 1990.

15. A. Sengupta and S.R.Wadia, Tata Preprint 90-33, July 1990 (to appear in Int.J.Mod.Phys. A).

RANDOM SURFACES IN DIMENSIONS LARGER THAN ONE

J. Ambjørn

The Niels Bohr Institute
Blegdamsvej 17
Copenhagen Ø, Denmark

INTRODUCTION

Whenever we want to address the question of a random surface representation of the bosonic string for $d > 1$, we have to ask ourselves what goal we want to achieve. All evidence hint that the bosonic string has a tachyonic mode all the way down to $d = 1$. Although we formally describe the bosonic string by a path integral over random surfaces, it cannot be strictly true. If we have a statistical theory of random surfaces embedded in R^d, $d \geq 2$, where all surfaces are counted with positive weight, we expect it to be formally reflection positive[1]. This being the case, we know that the spectrum of excitations, determined from the two-point function, must be positive semidefinite. We can have no tachyons. When the random surface models were first considered 5-6 years ago, the hope was that one could derive a scaling limit for a "healthy" tachyon-free bosonic string. The hope was tied to the fact, that one in these models automatically included the Liouville mode in the correct way. Although the question of inclusion of the Liouville field is still not yet solved in $d > 1$ from a field theoretical point of view, we have seen no evidence of the disappearance of the tachyon. (See for instance N. Seibergs contribution to this conference). I will describe how each class of the discretized random surface models suggested so far seems to avoid scaling to a genuine string theory. The

---

[1]The principle of reflection positivity can be extended from Euclidean field theory to random surfaces[1]. It can be shown to be valid for the hypercubic random surface theory. In Euclidean field theory it will strictly speaking not be satisfied for any regularization, which is Euclidean invariant and gives a finite short distance two-point function. This follows from the Lehmann spectral representation of the propagator. As two examples one can take the hypercubic random walk representation of free field theory (which satisfy the principle) and the piecewise linear random walk representation in R^d, which is Euclidean invariant and finite at short distances. Of course their scaling limits agree. We expect the same situation for random surfaces, where we can define hypercubic models and Euclidean invariant piecewise linear surfaces. This is why I use the phase "formally reflection positive"

Random Surfaces and Quantum Gravity
Edited by O. Alvarez *et al.*, *Plenum Press, New York, 1991*

problems can in a natural way be called a "tachyonic disease", although these models have no tachyons, as already mentioned. Instead the surfaces will consist of spikes, or collapse to one-dimensional objects called branched polymers, while the string tension will not scale. A simple argument by Cates shows that even the Liouville theory predicts a (KT) transition to a spiky phase for $d > 1$.

The disease of the pure bosonic random surfaces also contains a hint of a possible cure of the problem: we need smoother surfaces. This is most easily accomplished by adding terms to the string Lagrangian, which depend explicitly on the extrinsic geometry of the string in target space. It is reassuring that such terms are present if we consider the way the continuum string theory cures the tachyonic disease: It becomes supersymmetric. If we integrate out the woldsheet fermions of the superstring we get an effective bosonic action which depends on the normal bundle of the surface spanned by the string and contains both a WZ-like term and an extrinsic curvature term. I describe how such terms might influence the critical behaviour of bosonic random surfaces.

DISCRETIZED MODELS OF THE BOSONIC STRING

The oldest attempt to formulate the bosonic string as an unrestricted geometrical sum of random surfaces with weight $\exp(-\beta \cdot Area)$ goes back to Durhuus, Jonsson and Fröhlich [1]. In many ways it is the most appealing approach. The surfaces are all hypercubic surfaces of a given topology (say spherical) and the weight given to each surface is the one dictated by geometry. By using surfaces attached to a hypercubic lattice one breaks explicitly Euclidean invariance, but *a priori* the situation seems no different from the use of a random walk representation on a hypercubic lattice for the free field propagator. In the scaling limit Euclidean invariance is restored. The lowest mass excitation is determined by the exponential fall off of the two-point function, which can be defined as follows:

$$G(\Box_x, \Box_y) = \sum_{S(\Box_x, \Box_y)} e^{-\beta A(S)}. \tag{1}$$

In eq. (1) x, y denotes two lattice points, $\Box_x, \Box_y$ the associated plaquettes, S the lattice surfaces of fixed topology having the two plaquettes as boundary, $A(S)$ the area of the surface S, which is simply number of plaquettes, and β finally the bare string tension. It can be shown that the theory has a critical point $\beta_c > 0$. Since the lattice action is reflection positive the mass gap $m(\beta)$ defined by

$$G(\Box_x, \Box_y) \sim e^{-m(\beta)|x-y|} \quad \text{for} \quad |x - y| \to \infty \tag{2}$$

must be positive. Under mild assumptions it can be shown that the mass $m(\beta)$ scales to zero at the critical point β_c. This uniquely defines the scaling of the lattice spacing in terms of the a physical mass m_{ph}:

$$m(\beta) = m_{ph}a(\beta).\qquad(3)$$

Since $m(\beta)$ scales to zero, the lattice spacing $a(\beta)$ goes to zero too. The whole problem arises because we have an independent variable: the string tension. It can be shown that the string tension does *not* scale[2]. The implication of this is as follows: since the scaling is already fixed by the lowest mass excitation (3) we are forced to write for the string tension τ

$$0 < \tau(\beta_c) \leq \tau(\beta) = \tau_{ph}a^2(\beta).\qquad(4)$$

Here $\tau(\beta)$ is the bare, dimensionless string tension, while τ_{ph} is the physical string tension. Since $a(\beta) \to 0$ for $\beta \to \beta_c$ this relation does not allow a finite physical string tension for $\beta \to \beta_c$. In the scaling limit no surfaces except the one with minimal area (which depends on the boundary conditions) will survived. The only excitations allowed are thin tubes of "no" area, called branched polymers.

At this point one could try to argue that the scaling of the string tension is the most important feature and one should insist on that in the model. From eq. (4) we see that it is not possible unless we go beyond the critical point β_c. Formally the two-point function will become infinite, indicating some kind of tachyonic behaviour. It is intriguing to attempt an analytic continuation of the partition function beyond β_c, such that the string tension scales to zero and $m^2(\beta)$ at the same time scales to zero through negative values. This would give the continuum bosonic string, but clearly the analytic continuation needed will make an interpretation in terms of random surfaces with positive weight impossible. Unfortunately it has not been possible to substantiate this scenario.

An attempt to change the situation to the better was suggested by D. Gross [2]. Instead of using hypercubic surfaces he attempted to discretize the Nambu-Goto action by use of a fixed triangulation. The vertices of the triangulation is mapped in R^d and the weight of a particular configuration taken as $\exp(-\beta \cdot Area)$, where the area is that of the surface in embedding space. This prescription was analyzed carefully in [3], where it was shown that it in general is ill defined. For a generic triangulation the partition function will simply be infinite. Local spikes, where vertices wander out to infinity without any increase in the area are not sufficient suppressed by entropy. In exceptional cases of very regular triangulations, the partition function might be finite, but the spiky nature of the surfaces persists even in these cases.

The spikes of the Nambu-Goto action motivated the authors of [3] to used instead a discretized version of the Polyakov formulation of string theory[3], although it is much less

---

²The proof is slightly non-trivial and I have to refer to the original article [1]. I will present a simple proof in the case of dynamical triangulated surfaces.

³The discretized model was independently suggested in ref. [4, 5]

geometrical in nature than the Nambu-Goto version. As is wellknown we first have to integrate over the coordinates x_μ in target space and afterwards over all internal metrics. For each fixed internal metric the action is Gaussian and a discretization of this Gaussian action will not lead to any spikes. The summation over metrics is identified with a summation over different triangulations. One can partly justify this by appealing to Regge calculus and viewing the internal length of the triangles as fixed. In this way the dynamics is entirely associated with the connectivity of the triangulation and each choice of triangulation corresponds to a choice of metric by Regge calculus. With the hindsight of history[4] one should maybe have suspected that the trouble with the bosonic string would only be shuffled into the integration of metrics, which in the discretized model means the summation over triangulations. This is indeed what seems to happen. The dynamical triangulated model agree in the scaling limit with the continuum bosonic string for $d < 1$, as is now general accepted. For $d > 1$ we have only numerical simulations to rely on. There is evidence that at least for $d > 4$ the surfaces degenerate into what can be called branched polymers [8, 9, 10, 11]. The Hausdorff dimension in target space has been measured to go to 4 or infinity when $d \to \infty$, depending on the weight factor attached to the triangulations. This behaviour is a characteristic feature of generalized branched polymers with positive weight: either the Hausdorff dimension is 4 or it is infinite. In fact it is possible to prove that in the $d \to \infty$ limit the dynamical triangulated random surface model degenerates into a generalized branched polymer model [8, 9]. It is less clear whether this phase really extends all the way down to $d = 1$.

Irrespectively of the validity of the branched polymer approximation one can prove[12] that the string tension does not scale for any integer dimension $d \geq 2$. The proof is surprisingly simple.

The string tension $\tau(\beta)$ is defined by the exponential decay of the one-loop Green function for a large square loop $\gamma_{L,L}$ of area L^2 :

$$G_\beta(\gamma_{L,L}) \sim e^{-\tau(\beta)L^2} \quad \text{for} \quad L \to \infty. \tag{5}$$

where the definitions of the Green function is as follows:

$$G_\beta(\gamma_{L,L}) = \sum_{T \in \mathcal{T}_{\gamma_{L,L}}} \int \prod_{i \in T/\partial T} dx_i e^{-\beta A(S)} \tag{6}$$

$$A_T(S) \equiv \sum_{<i,j>} (x_i - x_j)^2 \tag{7}$$

In these formulae T denotes a triangulation, S the corresponding surface in R^d, that is the coordinates x_i, $i = 1, 2, \ldots, |T|$, whose connectivity is defined by means of T. γ is the boundary $S(\partial T)$ in R^d. We assume that $|\partial T| \propto L$. For the given (abstract) triangulation

---

[4]There are now several proofs of the equivalence between the Nambu-Goto formulation and the formulation of Polyakov [6, 7]

330

T we let $A_{min}(T, \gamma_{L,L})$ denote the minimum of $A(S)$ for all such surfaces $S : T \to R^d$ satisfying

$$S(\partial T) = \gamma_{L,L} \qquad (8)$$

We let $S' : T \to R^d$ denote a surface where

$$S'(\partial T) = 0_L \qquad (9)$$

We can imagine S' as coming from S by contracting the boundary loop $\gamma_{L,L}$ to a single point 0_L of order $|\partial T| \propto L$. We now have the following decomposition:

$$A_T(S) = A_{min}(T, \gamma_{L,L}) + A(S') \qquad (10)$$

The decomposition (10) follows simply from the quadratic nature (7) of $A_T(S)$. Inserting (10) in (6) we get

$$G_\beta(\gamma_{L,L}) = \sum_{T \in \mathcal{T}_{0_L}} e^{-\beta A_{min}(T, \gamma_{L,L})} \int \prod_{i \in T/\partial T} dx_i e^{-\beta A(S')} \qquad (11)$$

Next we note that the sum of squares of the length of any two sides of a triangle is ≥ 2 times its area. It follows that

$$A_{min}(T, \gamma_{L,L}) \geq 2L^2 \qquad (12)$$

and from (11) we can therefore write

$$G_\beta(\gamma_{L,L}) \leq e^{-2\beta L^2} G_\beta(0_L) \qquad (13)$$

In eq. (13) $G_\beta(0_L)$ denotes the loop Green function where the loop $\gamma_{L,L}$ is contracted to one point of order $|\gamma_{L,L}| \propto L$. The following can now be proven for this Green function [3] : it has the same critical point β_c as ordinary Green functions like $G_\beta(\gamma_{L,L})$ and it can be bounded by

$$G_\beta(0_L) \leq e^{c(\beta)L} \qquad (14)$$

where $c(\beta)$ is finite for $\beta > \beta_c$. From the definition of the string tension (5) it follows finally that

$$\tau(\beta) \geq 2\beta \qquad (15)$$

and since $\beta_c > 0$ $\tau(\beta)$ does not vanish at β_c.

People doing strong coupling expansions will recognize estimates like (13) as typical strong coupling estimates. What usually happens is that the function $G_\beta(0_L)$, which is based on a strong coupling approximation, becomes dominant before one reaches β_c. However, in this case we can control it all the way down to β_c.

I should emphasize that the above derivation is valid for all topologies. It seems unlikely that any summation over genus will help us to get scaling of the string tension.

Does there exist any hint from the continuum formulation of the Polyakov string theory which support the picture of too spiky surfaces? The answer is yes. The following simple argument by Cates [13] shows a clash between the quantum fluctuations of the Liouville mode and the requirement of reparametrization invariance. Let us define a "spiky" configuration as one where the numerical value of the internal curvature R is large:

$$R(\xi) = 2\pi\mu\delta^2(\xi - \xi_0) \tag{16}$$

Since the Liouville mode is connected to R by $R = \partial^2\phi$ we have

$$\phi(\xi) = -\mu\log|\xi - \xi_0| \tag{17}$$

However, the invariant area is given by

$$\int d^2\xi \, e^\phi \sim \int d^2\xi \, |\xi - \xi_0|^{-\mu} \tag{18}$$

and it follows that *the strength μ of the spike has to be smaller than 2 in order to be compatible with reparametrization invariance*. If spiky configurations with strength μ larger than 2 are important in the functional integral for the Liouville mode we have a breakdown of the fundamental assumption of the theory. Since the "spiky" configurations (17) are similar to static charged configurations in electrostatics, Cates showed that the importance of these configurations could be evaluated in the same way as the importance of vortex configurations in the X-Y model, and one finds that for a given dimension d only spikes of strength

$$|\mu| < \sqrt{\frac{96}{25 - d}} \tag{19}$$

are important. (19) is in agreement with the intuition that for $d \to -\infty$ all spikes will be suppressed, since we can use the semiclassical saddlepoint expansion around smooth surfaces. When d grows towards 1 the strength of the important spikes will grow to the critical value 2, and for $d > 1$ the strength of important spikes is incompatible with a reparametrization invariant cut off of the type $\epsilon^2 = e^\phi(d\xi)^2$.

SMOOTH STRINGS

From the discussion in the previous section it seems as if a possible cure of the (tachyonic) disease of the bosonic string would be to introduce some penalty for the outgrow of spikes or branches. While it is not very natural to put in such terms by hand, one can ask whether world sheet supersymmetry is able of providing such a penalty.

One can turn to the random walk as a convenient laboratory, since all questions can there be answered by analytical means. Will the random walk be smoother at short distances if we introduce local world line supersymmetry? It is well known how to do that: Let $e(s)$ denote the "einbein" of the world line and $\chi(s)$ the "gravitino". The standard continuum action for the relativistic bosonic particle is

$$S = \int d\tau \, (\frac{\dot{x}_\mu^2}{2e} + me) \tag{20}$$

and the corresponding action with world line supersymmetry (with ψ_5 being a Lagrange multiplier field)

$$S = \int d\tau \, \frac{1}{2} \left\{ \frac{\dot{x}_\mu^2}{e} - \psi_\mu \dot{\psi}_\mu + +\chi \psi_\mu \dot{x}_\mu \right\} + m \left\{ e - \psi_5 \dot{\psi}_5 + \psi_5 \chi \right\}. \tag{21}$$

It is also wellknown how to integrate over ψ_5, χ and ψ_μ in the gauge $\dot{e} = 0$ [14]. After these integrations we end with an expression for the propagator involving only the integration over the bosonic variable $x_\mu(\tau)$:

$$G(0, x) = \int_{X(\tau):0 \to x} \mathcal{D} X_\mu(\tau) \, e^{-\int d\tau e(\tau)} \, \mathrm{P} \exp \int d\tau \, \omega_{\mu\nu}(\tau) s_{\mu\nu} \tag{22}$$

where P stands for path ordering, and

$$s_{\mu\nu} = \frac{1}{2} \sigma_{\mu\nu} = \frac{i}{4} [\gamma_\mu, \gamma_\nu] \tag{23}$$

$$\omega_{\mu\nu}(\tau) = \frac{1}{2} (\dot{x}_\mu \ddot{x}_\nu - \dot{x}_\mu \ddot{x}_\mu). \tag{24}$$

This is a remarkable formula, but quite formal because of the path ordered phase factor. It can however be given a precise mathematical meaning by summing over discretized random walks [15]. Let me only mention here that the phase factor $\mathrm{P} \exp \int d\tau \, \omega_{\mu\nu}(\tau) s_{\mu\nu}$ is responsible for the cancellation of contributions from many paths. Only "smooth" paths will survive in the scaling limit, representing a class of random walks of Hausdorff dimension one. This is in contrast to the ordinary bosonic walk where the Hausdorff dimension is two. In fact when the scaling limit is taken we get the Dirac propagator [15], which is known to have a short distance Hausdorff dimension of one.

One could hope for a similar phenomena in the case of the string if we impose local world sheet supersymmetry. To address this question, it is useful again to integrate out the worldsheet fermions [16]. After this integration two types of terms are produced, which both depend on the *extrinsic* geometry of the world sheet :

$$S_{eff} = S_{bosonic} + \frac{\tau}{8} \int d^2\xi \, \left\{ (e_\alpha^\mu \partial_\beta e_\gamma^\mu)^2 + (D_\alpha n_i^\mu)^2 \right\} + W_k(A^{(n)}). \tag{25}$$

Here n_i^μ , $i = 1, \ldots, d-2$ are normals to the surface, e_α^μ $\alpha = 1, 2$ are the tangents and D_α denotes the covariant derivative with respect to the connection $A^{(n)}$ in the normal bundle :

$$A_{ij}^{(n)\alpha} = n_i^\mu \partial_\alpha n_j^\mu - n_j^\mu \partial_\alpha n_i^\mu. \tag{26}$$

τ is a Dynkin factor coming from the fermionic representation and finally $W_k(A^{(n)})$ denotes the Wess-Zumino action :

$$W_k(A^{(n)}) = \frac{ik}{8}\mathrm{Tr}\ \left(\frac{1}{2}\int d^2\xi\ A\wedge A + \frac{1}{3\pi}\int_D d^3x\ A\wedge A\wedge A\right) \tag{27}$$

where we have used the notation $A^\alpha = A^{(n)\alpha}_{ij} M_{ij}$, M_{ij} being the generators of SO$(d-2)$ and D a three-dimensional disc bounded by the world sheet.

The Wess-Zumino term is in some way the analog of the phase factor term we encountered for the world line of the fermionic particle (this can be made more precise), but in addition we have terms which depend very explicitly on the extrinsic geometry. If $K(\xi)$ denotes the extrinsic curvature of the world sheet we have

$$K(\xi) = \frac{1}{r_1(\xi)} + \frac{1}{r_2(\xi)} \tag{28}$$

where $r_1(\xi)$ and $r_2(\xi)$ are the principal curvatures of the surface. Since

$$\int d^2\xi\ (D_\alpha n_i)^2 = \int d^2\xi\ K^2(\xi) \tag{29}$$

this term clearly favours smooth surfaces.

From a numerical point of view it is not clear how to incorporate the WZ-term since it is a phase factor. One could instead try to be more modest and (appealing to universality) hope that the extrinsic curvature term alone would do the job of smoothing the surfaces. This was first suggested in [17]. It is now easy to define a triangulated model having only the extrinsic curvature term. For a given triangulation T we define the action by

$$S = \beta \sum_{\langle i,j\rangle} (x_i - x_j)^2 + \lambda \sum_l \sin^2(\frac{\theta_{\Delta.\Delta'}}{2}) \tag{30}$$

where β and λ are non-negative coupling constants. The last summation is over all links l and Δ,Δ' denote the triangles having the common link l. $\theta_{\Delta,\Delta'} \in [0,\pi]$ is the angle between the embedded neighbour triangles Δ and Δ' in R^d. The extrinsic curvature term in (30) shares with (29) the property of being scale invariant.

In the (β,λ) coupling constant plane we will now have a critical line $\lambda = \lambda_c(\beta)$ [17]. There is numerical evidence [18, 19, 20] indicating the existence of a critical point $\tilde\beta_c$ with the property that the approach to $\tilde\lambda = \lambda_c(\tilde\beta_c)$ will result in *smooth* surfaces. This *crumpling transition* is well documented from numerical works if the triangulation of the surface is kept fixed [21, 22, 23, 24], but is harder to understand when the triangulation (the metric) is allowed to fluctuate. In fact the current belief in solid physics is that there should be no transition. Nevertheless the numerical simulations strongly suggest that it remains essentially unchanged even when the fluctuation of triangulations are taken into account. The phase transition seems to be of second order and this allows us to dream of a new continuum limit dominated by smoother string configurations.

The Monte Carlo simulations suggest that the Hausdorff dimension of the surface ensemble jumps to 2 above the transition. We get indeed smoother surfaces. One can try to to analyze (numerically) the behaviour of the mass gap and the string tension when approaching the second order transition at $(\tilde{\lambda}_c, \tilde{\beta}_c)$ along the critical line $\lambda = \lambda_c(\beta)$, coming from $\lambda < \tilde{\lambda}_c$ [25]. At this time the numerical results can only be considered as preliminary, but they indicate that both the mass gap and the string tension scales in the right way for an interesting continuum limit to exist. This opens the possibility of studying a new class of strings by numerical methods. But I should stress once more that the interpretation of the results is still preliminary and so far the simulations for the string tension have only been carried out in $d = 3$.

If a continuum limit exists at $(\tilde{\lambda}_c, \tilde{\beta}_c)$, it immediately raises a number of questions. It cannot be the superstring, since it only exists in $d = 10$. It might be impossible to get the correct theory for the superstring unless we include also the WZ-like term in the effective bosonic theory. It is possible that we by adding only extrinsic curvature have obtained a theory, which can serve as an effective string theory for QCD, as originally suggested by Polyakov. Although he reached the conclusion that one could not get such a theory by adding extrinsic curvature, it was only based on a perturbative 1-loop calculation. A true nonperturbative analysis could lead to a different answer. At least one can say that the surprising results obtained by Monte Carlo simulations illustrate how poorly we understand the universality classes of random surfaces.

References

[1] B. Durhuus, J. Fröhlich and T. Jonsson, Nucl.Phys. B240[FS12] (1984) 453.

[2] D.J. Gross, Phys.Lett. 138B (1984) 185.

[3] J. Ambjørn, B. Durhuus and J. Fröhlich, Nucl.Phys. B257[FS14] (1985) 433.

[4] F. David, Nucl.Phys. B257[FS14] (1985) 543.

[5] V.A, Kazakov, I.K. Kostov and A.A. Migdal, Phys.Lett. 157B (1985) 295.

[6] T. Morris, Nucl.Phys.B341 (1990) 443.

[7] J. Govaerts, Int.Jour. Mod. Phys. A4 (1989) 173.

[8] J. Ambjørn, B. Durhuus, J. Fröhlich and P. Orland, Nucl.Phys. B270[FS16] (1986) 457.

[9] J. Ambjørn, B. Durhuus, J. Fröhlich, Nucl.Phys. B275[FS17] (1986) 161.

[10] D.V. Boulatov, V.A. Kazakov, I.K. Kostov and A.A. Migdal, Nucl.Phys. B275[FS17] (1986) 641.

[11] F. David, J. Jurkiewiecz, A. Krzywicki and B. Petersson, Nucl.Phys. B290[FS20] (1987) 218.

[12] J. Ambjørn and B. Durhuus, Phys.Lett. 188B (1987) 253.

[13] Cates, Europhys.Lett 8 (1988) 719.

[14] A.M. Polyakov, "Gauge fields and strings", Harwell Academic Press (1987).

[15] J. Ambjørn, B. Durhuus and T. Jonsson, Nucl.Phys.B330 (1990) 509.

[16] P. Wiegmann, Nucl.Phys. B323 (1989) 330.

[17] J. Ambjørn, B. Durhuus, J. Fröhlich and T. Jonsson, Nucl.Phys. B290[FS20] (1987) 480.

[18] S. Catterall, Phys.Lett. 220B (1989) 207.

[19] C.F Baillie, D.A. Johnston and R.D. Williams, Nucl.Phys.B335 (1990) 469.

[20] R. Renken and J. Kogut, *The Crumpling Transition in the Presence of Quantum Gravity*, Preprint ILL-TH-90-30.

[21] Y. Kantor and D.R Nelson, Phys.Rev.Lett 58 (1987) 2774.

[22] J. Ambjørn, B. Durhuus and T. Jonsson, Nucl.Phys. B316 (1989) 526.

[23] M. Baig, D. Espriu and J.F. Wheater Nucl.Phys. B316 (1989).

[24] R. G. Harnish and J.F. Wheater, *The Crumpling Transition of Crystalline Random Surfaces*, Oxford preprint OUTP-90-13P.

[25] S. Catterall, *Scaling in Dynamical Random Surfaces*, Cambridge Preprint DAMTP-90-14.

ON GAUGE INVARIANCES IN STOCHASTIC QUANTIZATION

Laurent Baulieu

Laboratoire de Physique Théorique et Hautes Energies
Université de Paris VI
4 Place Jussieu, F-75252 Paris Cedex 05, France

Abstract: We show that the symmetry of stochastically quantized gauge theories
is governed by a single differential operator. The latter combines supersymmetry
and ordinary gauge transformations. Quantum field theory can be defined on the
basis of a parabolic differential operator $\frac{\partial}{\partial t} - (\frac{\partial}{\partial x^i})^2$, with a Hamiltonian of the type
$H = \frac{1}{2}[Q, \bar{Q}]$. The nilpotent operator Q has deep relationship with the conserved
charge of a topological gauge theory. We display the example of the Yang-Mills
theory. We also show the relevant equations for gravity. The 2-dimensional case has
interesting particularities studied with more details in a separate publication. For a
first order action, the stochastic quantization leads to a quantum field theory based
on path integral, with a second order supersymmetric action which maintains gauge
invariance. As an example, we relate the quantum theory of the two dimensional
Chern Simons action $\int_{M_2} Tr\phi F$ to a topological four dimensional theory, obtained by
gauge fixing the second Chern class $\int_{M_4} TrFF$.

1. Introduction

Stochastic quantization is an alternative to Feynman path integral for quantizing
a theory. In [1], Parisi and Wu have suggested to apply stochastic quantization to
gauge theories. Numerous works have followed [2]. One of the motivations of Parisi
and Wu was that no gauge-fixing is necessary to compute gauge invariants quantities
in stochastic quantization, since the stochastic evolution can be consistently defined
from a drift force equal to minus the gradient of the classical action with respect to
the gauge field, with no reference to the ghosts which occur in the ordinary path
integral formalism. However, it has been realized that it is usefull to introduce a kind

Random Surfaces and Quantum Gravity
Edited by O. Alvarez *et al., Plenum Press, New York, 1991*

of gauge fixing in stochastic quantization: a drift force can be defined along gauge orbits [3]. This permits a consistent renormalizability of the stochastically quantized gauge theory. Moreover, with a particular choice of this drift force, it seems that the gauge field is confined within the first Gribov horizon, and so one escapes naturally the Gribov problem [3,4]. The freedom in the Langevin equation of a gauge theory, which permits the introduction of the gauge dependent drift force, follows in fact from the simple geometrical principle that stochastic evolution be compatible with the gauge symmetry [5].

There is a supersymmetry which is inherent to any stochastically quantized theory, wether or not it has a gauge invariance [2]. This is a very general result, linked to the possibility of interpreting the Langevin equation as a constraint between the noises and the fields: this constraint can be exponentiated in the Boltzman weight involving the noise, provided a relevant Jacobian multiplies the measure; in turn, this Jacobian can be exponentiated, and one ends up with a supersymmetric action. It has been observed that this supersymmetry has deep relationship with the notion of a topological gauge symmetry [6,7]. The stochastic supersymmetry is technically usefull, since it implies Ward identities for correlation functions. The latter permit one to control and consistently renormalize the divergences which generally occur in stochastic quantization [8,9].

If one considers a theory with a gauge invariance, the stochastic supersymmetry must be supplemented by some other symmetry acting as a reminder of the original gauge invariance. Previous attempts for expressing the gauge invariance in stochastic quantization to control the divergences of a stochastically quantized Yang-Mills theory can be found in [8]. In [9], two separate Ward identities were written, one for the stochastic supersymmetry and one for the gauge symmetry. A basic tool in [10] is the definition of the ghost through its own Langevin equation, first introduced in [5].

Here we show that there is in fact a single symmetry which combines both stochastic supersymmetry and gauge invariance. We work out in details the case of theories with a Yang-Mills invariance, and briefly sketch the case of the invariance under changes of coordinates, for which interesting phenomena seem to occur in two dimensions [10]. We find that the underlying invariance is of the topological type and has thus a geometrical meaning.

We mainly consider the case where the supersymmetric path integral representation is based on the parabolic differential operator $\frac{\partial}{\partial t} - \Delta$.

The other case where the supersymmetric path integral representation is based on elliptic or hyperbolic operators $\frac{\partial^2}{\partial t^2} \pm \Delta$ necessitates another interpretation of the drift force along gauge orbits. This case has interesting applications for quantizating first order systems. We briefly display the method in the case of the 2-D Chern Simons action $\int_{M_2} Tr F\phi$, and show how it provides a theory in three dimensions (the third dimension is stochastic time), which is related to the topological action for the second Chern class $\int_{M_4} Tr FF$. The quantization of the three-dimensional Chern-Simon theory is presented elsewhere [10], with a conclusion which supports Atiyah's conjecture of a link between Donaldson polynomials and Jones polynomials.

2 . Langevin equations for theories with a Yang-Mills symmetry

Following earlier results [5], we first construct the Langevin equations for the quantum theory by postulating that the ordinary Yang-Mills BRST transformations (which include as a subset the ordinary infinitesimal gauge transformations) must commute with the evolution along stochastic time. We start from the expression of the generator s of the ordinary Yang-Mills BRST symmetry. s is a graded differential operator, defined by its action on the G-valued Yang-Mills field one-form $A = A_i dx^i$, with $1 \leq i \leq n$ and n the dimension of the physical space of the theory, and on the G-valued anticommuting zero-form ghost c:

$$sA = -Dc \qquad\qquad sc = -\frac{1}{2}[c, c]. \qquad\qquad (1-a)$$

d is the ordinary exterior derivative $d = dx^i \partial_i$ and $D = d + [A, \ ]$ is the covariant derivative. We define the covariant BRST differential operator $S = s + [c, \ ]$. By definition s and d anticommute. One has $s^2 = 0$, $S^2 = 0$ and $SD + DS = 0$. The BRST equations (1-a) can be rewritten as:

$$(d + s)(A + c) + \frac{1}{2}[A + c, A + c] = dA + [A, A]. \qquad\qquad (1-b)$$

In what follows, $\partial_o = \frac{\partial}{\partial t}$ denotes the derivative with respect to the stochastic time t and all fields are assumed to depend on x^i and t. Our postulate is that s and ∂_o commute:

$$s\partial_o = \partial_o s. \qquad\qquad (2)$$

One has also $s\partial_i = \partial_i s$. Applying ∂_o to both sides of (1 b), and using (2), we obtain:

$$(D + S)(\partial_o(A + c)) = D(\partial_o A). \qquad\qquad (3)$$

This equation, once expanded in ghost number, gives:

$$S(\partial_o c) = 0 \qquad\qquad (4-a)$$

$$S(\partial_o A) + D(\partial_o c) = 0. \qquad\qquad (4-b)$$

Equations (4) can be easily solved, simply by using power counting (the dimensions and ghost numbers of A and c are as usual) and the properties $S^2 = 0$ and $SD + DS = 0$. One obtains:

$$\partial_o c = Sv - \gamma = sv + [c, v] - \gamma \qquad\qquad (5-a)$$

$$\partial_o A_i = D_i v + \frac{\delta I_{cl}[A]}{\delta A_i} + b_i, \qquad\qquad (5-b)$$

where γ and b_i, which can be identified as noises for c and A_i respectively, are submitted to the BRST constraints:

$$S\gamma = 0 \qquad\qquad Sb_i = D_i\gamma. \qquad\qquad (6)$$

I_{cl} can be any given functional of the A_i's, provided it is s-invariant, so that the equation of motion is s-covariant, $S\frac{\delta I_{cl}[A]}{\delta A_i} = 0$. This means that I_{cl} is a gauge-invariant classical action.

v is an arbitrary G-valued function, left undetermined when solving (3). The freedom in the choice of v, and thus in part of the drift force in the Langevin equations (5), can be understood as the manifestation of gauge invariance.

Translating the Langevin equation (5-b) into a Fokker-Planck equation, it is possible to prove, in the limit $t \to \infty$, the independence on the choice of v of correlation functions of gauge-independent functionals of the A_i's, computed from the Langevin equations (5-b).

In the next section we shall take v as a functional of the A_i's, $v = v[A_i]$. In this case, (5-b) is the modified Langevin equation introduced in [3] for inducing a drift force along the gauge orbits, without affecting the values of gauge invariant quantities (a natural choice is $v = \partial_i A^i$). The stochastic ghost equation (5-a) means now $\partial_o c = \frac{\delta v}{\delta A_i} D_i c + [c, v] - \gamma$. The convergence toward the usual Faddeev distribution of the Fokker-Planck equation associated to the Langevin equation (5-b) has been demonstrated in [11], by using a certain functional $v[A]$.

If we compare (5-a) and (5-b), we see that the evolution of A is not correlated to that of c, while that of c is correlated to that of A. (After introducing antighosts, one could imagine a more general situation where v is ghost and antighost dependent, which implies a spurious ghost dependence in the evolution of A.)

The Langevin equation (5-b) for the ghost has been introduced in [5]. It can be generalized for any given gauge theory. It was rederived by other means in [9], with $\gamma = 0$ and the choice $v = \partial_i A^i$, and used to investigate the Ward identities and the renormalization of the stochastically quantized Yang-Mills theory. (Choosing $\gamma = 0$, and thus $sb_i = -[c, b_i]$, reproduces the usual convention that the noise of the gauge field transforms covariantly and not as a gauge field.) Here, we will consider the general situation $\gamma \neq 0$ in order to construct a symmetry operator which unifies the stochastic supersymmetry and the ordinary gauge symmetry.

3 . The partition function

Up to the obvious requirement that the stochastic process induced by the Langevin equations (5-b) is meaningful, the correlation functions of gauge-independent functionals of the A_i's do not depend on the choice of v. Let $J_i(x, t)$ be the sources of the A_i's. The stochastic partition function is:

$$Z[J] = \int [Db_i] \exp \int dt dx \, \mathbf{Tr}(-\frac{1}{2}b_i^2 + J_i(x, t)A_i(x, t)), \tag{7}$$

where the $b(x, t)$'s are submitted to the constraints (5-b).

We wish to construct a supersymmetric functional representation of the Langevin equation which involves the G-valued Faddev-Popov ghost c. We insert in the generating functional $Z[J]$ the identity $1 = \int [d\bar{c}][d\gamma] \exp - \int dt dx \bar{c}\gamma$ ($\bar{c}$ and γ are G-valued anticommuting fields). In this way, we have:

$$Z[J] = \int [Db_i][dc][d\gamma] \exp \int dt dx \, \mathbf{Tr}(-\frac{1}{2}b_i^2 - \bar{c}\gamma + J_i(x, t)A_i(x, t)). \tag{8}$$

It is convenient to define $F_{oi} = \partial_o A_i - D_i v$ and $D_o = \partial_o + [v, \ ]$. Assuming that the superJacobian of the transformation $(A_i, c) \to (F_{oi} - \frac{\delta I_{cl}}{\delta A_i}, D_o c - \frac{\delta v}{\delta A_i} D_i c)$ is not singular, we can insert in (8) the formal identity:

$$1 = \int [DA_i][Dc] \delta(b_i - F_{oi} + \frac{\delta I_{cl}}{\delta A_i}) \delta(\gamma - D_o c + \frac{\delta v}{\delta A_i} D_i c) \det \frac{\delta(b_i, \gamma)}{\delta(A_i, c)}. \tag{9}$$

The integration over b_i and γ is trivial. We get:

$$Z[J] = \int [DA_i][D\bar{c}][Dc] \det \frac{\delta(b_i, \gamma)}{\delta(A_i, c)}$$

$$\exp \int dt dx\, \mathbf{Tr}(-\frac{1}{2}(F_{oi} - \frac{\delta I_{cl}}{\delta A_i})^2$$

$$-\bar{c}(D_o c - D_i c \frac{\delta v}{\delta A_i}) + J_i(x, t) A_i(x, t)). \tag{10}$$

To exponentiate the superdeterminant in (10), we introduce G-valued anticommuting ghosts Ψ_i and $\bar{\Psi}_i$ and G-valued commuting ghosts for ghosts Φ and $\bar{\Phi}$. One has:

$$\det \frac{\delta(b_i, \gamma)}{\delta(A_i, c)} = \int [D\Psi_i][D\bar{\Psi}_i][D\Phi][D\bar{\Phi}]$$

$$\exp \int dt dx\, \mathbf{Tr}\left(\bar{\Psi}_i, \ \bar{\Phi} \right) \begin{pmatrix} D_o \delta_{ij} - \frac{\delta^2 I_{cl}}{\delta A_i \delta A_j} - D_i \frac{\delta v}{\delta A_j} & 0 \\ -\frac{\delta^2 v}{\delta A_j \delta A_k} D_k c^a + [\frac{\delta v}{\delta A_j}, c] & D_o - \frac{\delta v}{\delta A_k} D_k \end{pmatrix} \begin{pmatrix} \Psi_j \\ \Phi \end{pmatrix}. \tag{11}$$

We can therefore express the partition function as follows:

$$Z[J] = \int [DA_i][D\Psi_i][D\bar{\Psi}_i][Db_i][D\bar{c}][Dc][D\Phi][D\bar{\Phi}] \exp(I_{GF} + \int dt dx\, \mathbf{Tr} J_i A_i)$$

with

$$I_{GF} = \int dt dx\, \mathbf{Tr}(\frac{1}{2} b_i^2 + b_i(F_{oi} - \frac{\delta I_{cl}}{\delta A_i})$$

$$-\bar{c}(D_o c - D_i c \frac{\delta v}{\delta A_i}) - \bar{\Psi}_i(D_o \delta_{ij} - D_i \frac{\delta v}{\delta A_j} - \frac{\delta^2 I_{cl}}{\delta A_i \delta A_j}) \Psi_j$$

$$-\bar{\Phi}(D_o - \frac{\delta v}{\delta A_k} D_k) \Phi$$

$$-\bar{\Phi} \frac{\delta^2 v}{\delta A_j \delta A_k} D_k c \Psi_j + \bar{\Phi}[[\frac{\delta v}{\delta A_j}, c], \Psi_j]). \tag{12}$$

(We have reintroduced Lagrange multiplier fields b in (12).) A non trivial feature of our action (12) is the trilinearity of the last term in the ghosts $c, \Psi, \bar{\Phi}$.

We can verify that I_{GF} is invariant under the action of the graded differential operator s_{top} defined as follows:

$$s_{top} A_i = \Psi_i$$

$$s_{top} c = \Phi$$

$$s_{top} \Psi_i = 0$$

$$s_{top}\Phi = 0 \tag{13-a}$$

$$s_{top}\bar{\Psi}_i = b_i \qquad\qquad s_{top}b_i = 0$$

$$s_{top}\bar{\Phi} = \bar{c} \qquad\qquad s_{top}\bar{c} = 0. \tag{13-b}$$

(All gradings are summarized by attributing ghost numbers to all fields: 0 for A_i and b_i; 1 for c and Ψ_i; -1 for $\bar{c}$ and $\bar{\Psi}_i$; 2 for Φ and -2 for $\bar{\Phi}$. Moreover the gradings of all forms and operators are defined as the sum modulo two of the ghost numbers and form degrees.)

One has $s_{top}^2 = 0$. The s_{top}-invariance of the action is obvious, since one can write I_{GF} as an s_{top}-exact term:

$$I_{GF} = \int dt dx\, s_{top}(\bar{\Psi}_i(F_{oi} - \frac{\delta I_{cl}}{\delta A_i} + \frac{1}{2}b_i) - \bar{\Phi}(D_o c - \frac{\delta v}{\delta A_i}D_i c)). \tag{14}$$

The last equation shows that $F_{oi} - \frac{\delta I_{cl}}{\delta A_i}$ and $D_o c - D_i c \frac{\delta v}{\delta A_i}$ can be interpreted as gauge functions: functional integration is concentrated around the the domains where these functions vanish.

If one eliminates the fields b_i by their algebraic equation of motion, the action of s_{top} on $\bar{\Psi}$, $s_{top}\bar{\Psi}_i = b_i$, is changed into $s_{top}\bar{\Psi}_i = F_{oi} - \frac{\delta I_{cl}}{\delta A_i}$. If one further Legendre transforms I_{GF} into a Hamiltonian, one gets a supersymmetric Hamiltonian $H_{GF} = \frac{1}{2}[Q,\bar{Q}]$, where Q is the charge associated to s_{top}, with $Q^2 = 0$, and $\bar{Q}$ is the adjoint of Q. Thus H_{GF} can be interpretated à la Witten.

s_{top} can be identified with the topological BRST Yang-Mills operator constructed in [12]: if one performs the change of field-variables $\Psi_i \rightarrow \Psi_i + D_i c$, $\Phi \rightarrow \Phi - \frac{1}{2}[c,c]$, one gets indeed:

$$s_{top}A_i = \Psi_i + D_i c$$

$$s_{top}c = \Phi - \frac{1}{2}[c,c]$$

$$s_{top}\Psi_i = D_i\Phi - [c,\Psi_i]$$

$$s_{top}\Phi = -[c,\Phi]. \tag{15}$$

Under the form (15), one sees that the stochastic supersymmetry is combined with the ordinary BRST symmetry. This phenomenon is made possible thanks to the existence of the ghost of ghost Φ, a 0-form with ghost number two. The geometrical interpretation follows from the possibility of expressing (15) as a generalization of (1-b) [12]:

$$(d+s)(A+c) + \frac{1}{2}[A+c, A+c] = F + \Psi_i dx^i + \Phi. \tag{16}$$

If we consider the ordinary Yang-Mills action, $I_{cl} = \int dx F_{ij}^2$, the case of interest is $v = \partial_i A^i$ (Zwanziger gauge). The action becomes:

$$I_{GF} = \int dt dx\, \mathbf{Tr}(\frac{1}{2}(\partial_o A_i - D_j \partial^j A_i - [F_{ij}, A_j])^2$$

$$-\bar{c}(\partial_o c - D_i \partial_i c)$$

$$-\bar{\Psi}^i((\partial_o - D_j D^j)\Psi_i - 2[F_{ij}, \Psi_j] + D_i[A^j, \Psi_j]$$

$$-\bar{\Phi}((\partial_o + D^j \partial_j)\Phi - [\Psi^i, \partial_i c)). \qquad (17)$$

We see that the theory is based on a parabolic differential operator of the type $\frac{\partial}{\partial t} - (\frac{\partial}{\partial x^i})^2$. For a space dimension smaller or equal than four one has renormalizability by power counting, and the stability of the theory is due to the symmetry of (17), $s_{top} I_{GF} = 0$, with:

$$s_{top} A_i = \Psi_i$$

$$s_{top} c = \Phi$$

$$s_{top} \Psi_i = 0$$

$$s_{top} \Phi = 0 \qquad (18-a)$$

$$s_{top} \bar{\Psi}_i = \partial o A_i - D_j \partial^j A_i - [F_{ij}, A^j]$$

$$s_{top} \bar{\Phi} = \bar{c} \qquad\qquad s_{top} \bar{c} = 0. \qquad (18-b)$$

4 . The case of diffeomorphism invariance

The approach followed to obtain the Langevin equations (5) can be applied to other gauge invariances [5]. If one considers for instance diffeomorphism invariance, equations (15) are replaced by the following ones:

$$s_{top} g_{ij} = \Psi_{ij} + g_{ik}\partial_j \xi^k + g_{jk}\partial_i \xi^k + \xi^k \partial_k g_{ij}$$

$$s_{top} \xi^i = \Phi^i + \xi^j \partial_j \xi^i$$

$$s_{top} \Psi_{ij} = g_{ik}\partial_j \Phi^k + g_{jk}\partial_i \Phi^k + \Phi^k \partial_k g_{ij} + \Psi_{ik}\partial_j \xi^k + \Psi_{jk}\partial_i \xi^k + \xi^k \partial_k \Psi_{ij}$$

$$s_{top} \Phi = \Phi^j \partial_j \xi^i - \xi^j \partial_j \Phi^i. \qquad (19)$$

In this case the drift force v is a vector field v^i. Again, there is an interpretation in terms of a topological gauge symmetry. If we consider the case of $2-D$-gravity, one should take as conformally invariant variables the Beltrami differentials, so that $g_{ij}dx^i dx^j = exp(\phi)(dz + \mu_{\bar{z}}^z d\bar{z})(d\bar{z} + \mu_z^{\bar{z}} dz)$. The stochastic equations are separated into holomorphic and antiholomorphic sectors. The holomorphic sector is:

$$s_{top}\mu_{\bar{z}}^z = \Psi_{\bar{z}}^z + \partial_{\bar{z}} c^z + c^z \partial_z \mu_{\bar{z}}^z - \mu_{\bar{z}}^z \partial_z c^z$$

$$s_{top} c^z = \Phi^z + c^z \partial_z c^z$$

$$s_{top}\Psi_{\bar{z}}^z = \partial_{\bar{z}}\Phi^z + \Phi^z \partial_z \mu_{\bar{z}}^z - \mu_{\bar{z}}^z \partial_z \Phi^z + c^z \partial_z \Psi_{\bar{z}}^z - \Psi_{\bar{z}}^z \partial_z c^z$$

$$s_{top} \Phi^z = \Phi^z \partial_z c^z - c^z \partial_z \Phi^z. \qquad (20)$$

It is convenient to rename the analog of v as $\mu_o^{\bar{z}}$. The Langevin equations (5) become:

$$\partial_o \mu_{\bar{z}}^z = T^{zz} + \partial_{\bar{z}}\mu_o^z + \mu_o^z \partial_z \mu_{\bar{z}}^z - \mu_{\bar{z}}^z \partial_z \mu_o^z + b_{o\bar{z}}^z$$

$$\partial_o c^z = s\mu_o^z + c^z \partial_z \mu_o^z - \mu_o^z \partial_z c^z - \Psi_o^z, \tag{21}$$

where T is the energy momentum tensor. A most interesting fact is the possibility of expressing all relevant equations for the stochastic quantization of a worldsheet as follows:

$$(d + s_{top})\tilde{A} + \frac{1}{2}[\tilde{A}, \tilde{A}] = \tilde{B}$$

$$(d + s_{top})\tilde{B} + [\tilde{A}, \tilde{B}] = 0. \tag{22}$$

We have defined:

$$d = dt\partial_o + dz\partial_z + d\bar{z}\partial_{\bar{z}}$$

$$\tilde{A} = (dz + d\bar{z}\mu_{\bar{z}}^z + dt\mu_o^z + c^z)\partial_z$$

$$\tilde{B} = (dtd\bar{z}(T^{zz} + b_{o\bar{z}}^z) + d\bar{z}\Psi_{\bar{z}}^z + dt\Psi_o^z + \Phi^z)\partial_z. \tag{23}$$

The analogy between (15) and (20-23) is quite striking, and will be used in a separate publication [15]. There, we show a link between the stochastic quantization of 2-D gravity and the path integral quantization of 4-D topological gravity.

5 . The case of first order systems

First order actions, i.e actions only linear in the velocities, have vanishing Hamiltonians. Their quadratic aproximations are not definite positive, and therefore their quantization through ordinary Feynman path integral formalism is conceptually difficult to understand. In [10], in the case of the three-dimensional pure Chern Simons action, stochastic quantization has been shown to go around this difficulty, by giving a four-dimensional supersymmetric action whose bosonic part is of the ordinary Yang-Mills type, and thus second order.

To show the generality of this transmutation of a first order action into a second order one, we present here an other case, that of the "two dimensional Chern Simon action":

$$I_{cl} = \int_{M_2} Tr(F\phi). \tag{24}$$

This action has a physical interest since, for particular choices of the gauge group, there are arguments for its relation to 2-D gravity. $F = dA + AA$ is the curvature of a connection $A = A_z dz + A_{\bar{z}} d\bar{z}$. ϕ is a scalar field, valued in the same fundamental representation as A.

I_{cl} is first order, and the Hamiltonian vanishes modulo the classical constraint on A. The equations of motion are:

$$\frac{\delta I_{cl}}{\delta A_i} = D_i \phi = 0$$

$$\frac{\delta I_{cl}}{\delta \phi} = \epsilon^{ij} F_{ij} = 0. \tag{25}$$

344

(i,j refer to the indices in M_2.) The Langevin equations which describe stochastic quantization for the action (24) are therefore:

$$F_{oi} = D_i\phi + b_i$$

$$D_o\phi = \epsilon^{ij}F_{ij} + b. \tag{26}$$

As in the previous sections, the index o refer to stochatic time and we have defined $F_{oi} = \partial_o A_i - \partial_i A_o + [A_o, A_i]$ and $D_o = \partial_o + [A_o, \ ]$. (We have renamed the arbitrary function v as A_o). Since gauge invariant quantities do not depend on the choice of $v = A_o$, we can functionally integrate over all possible choices of A_o, provided we define a stochastic evolution for A_o, for instance:

$$\partial_o A_o = \partial_i A_i + b_o, \tag{27}$$

where b_o is a Gaussian noise for A_o.

If we write the stochastic partition $\int [db] \exp - \int_{M_3} d^2z\, dt(\frac{1}{2}b_i^2 + \frac{1}{2}b_o^2)$ by expressing the noises in function of the fields as in the previous sections, (see reference [10] for the details in the case of the the-three dimensional Chern Simon action), we end up with a supersymmetric functional integral representation of the Langevin equations (26-27) defined from the following action:

$$I_{GF} = \int_{M_3} d^2z\, dt((F_{oi} - D_i\phi)^2 + (D_o\phi - \epsilon^{ij}F_{ij})^2 + (\partial_o A_o - \partial_i A_i)^2$$

$$+\text{supersymmetric terms})$$

$$= \int_{M_3} d^2z\, dt((F_{\alpha\beta}^2 + (D_\alpha\phi)^2 - 2\epsilon^{\alpha\beta\gamma}F_{\alpha\beta}D_\gamma\phi + (\partial_\alpha A^\alpha)^2$$

$$+\text{supersymmetric terms}). \tag{28}$$

The greek indices $\alpha, \beta, \gamma, \ldots$ stand for three-dimensional indices for $M_2 \times S$, where S is the one-dimensional manifold in which the stochastic time runs. (The term $\epsilon^{\alpha\beta\gamma}F_{\alpha\beta}D_\gamma\phi$ is a pure derivative and can be omitted.)

The right hand side of (28) shows that the field A_o can be truly interpreted as a gauge field component along the the stochsstic direction. This interpretation of A_o was already quite clear by contemplating equation (25) †.

Equation (28) shows also that stochastic quantization provides us an action which is second order: its quadratic field approximation is based on the elliptic operator $\sum_{i=1}^{3}(\frac{\partial}{\partial x^i})^2$. The method which has yield the second order action (28) is quite general. Presumably, it can be used for any given theory with a classical first order action. Notice that gauge invariance (under a BRST form) has been maintained for the supersymmetric stochastic action by summing over all possibilities on the freedom of the Langevin equation.

---

† The idea of interpreting v as an additional gauge field component has first appeared in the work of Chern and Halpern [13], in a different context.

As far as the above specific example is concerned, it is interesting to observe that the action (28) is the same as the one constructed in [14] for defining a quantum field theory from the magnetic monopole topological charge $\int_{M_3} TrFD\phi$. Moreover, by a trivial dimensional reduction, the latter is itself linked to the quantum field theory associated to the four dimensional topological invariant $\int_{M_4} TrFF$ (the scalar field ϕ can be seen as the fourth component of a connection over M_4). We have thus an example of a bidimensional quantum theory which has deep relationship with a theory in four dimensions. What does stochastic quantization is a jump of one dimension, which could be called a generalization of Stokes theorem at the quantum level.

References

1. G. Parisi and Y.S. Wu, *Sci. Sinica* **24** (1981) 484.

2. For reviews, see D. Zwanziger, Stochastic Quantization Of Gauge Fields, Proceedings of the 1985 Erice School on Fundamental Problems of Gauge Field Theory, Eds. G. Velo and A. Wightman, (Plenum, New-York, 1986); E. Seiler, *Acta Physica Austriaca* **26** (1984) 259; P.H. Damgaard and H. Huffel ,*Phys. Reports* **152** (1987) 227, M.B. Halpern, these proceedings.

3. D. Zwanziger, *Nucl. Phys.* **B 192** (1981) 259.

4. D. Zwanziger, *Nucl. Phys.* **B 209** (1982) 336; E. Seiler, I.O. Stamatescu and D. Zwanziger, *Nucl. Phys.* **B 239** (1984) 204.

5. L. Baulieu, *Phys. Lett.* **B 167** (1986) 421; L. Baulieu, *Nucl. Phys.* **B 270** (1986) 507.

6. L. Baulieu and B. Grossman, *Phys. Lett.* **B 212** (1988) 351.

7. D. Birmingham, M. Rakowski and G. Thompson, *Phys. Lett.* **B 214** (1988) 381.

8. E. Gozzi, *Phys. Rev.* **D28** (1983) 1922; J. Zinn-Justin, *Nucl. Phys.* **B 275** (1986) 135;30 (1984) 1218; R.F. Alvarez-Estrada and A. Munoz Sudupe, *Phys. Lett.* **B 164** (1985) 102; **166B** (1986) 58; K. Okono, *Nucl. Phys.* **B 289** (1987) 109.

9. D. Zwanziger and J. Zinn-Justin, *Nucl. Phys.* **B 295** (1988) 297.

10. L. Baulieu, Stochastic And Topological Gauge Theories, *Phys. Lett.* **B 232**(1989) 479, Yue-Yu, Beijing preprint BIHEP-Th-893.

11. L. Baulieu and D. Zwanziger, *Nucl. Phys.* **B 193** (1981) 163.

12. L. Baulieu and I.M.Singer, *Nucl. Phys. Proc. Suppl.* **B 5** (1988) 12.

13. H.S. Chan and M.B. Halpern, *Phys. Rev.D33* (1986) 540.

14. L. Baulieu and B. Grossman, *Phys. Lett.* **B 214** (1988) 223.

15. L. Baulieu, A. Bilal and M. Picco, CERN preprint, to be published in *Nucl. Phys. B* .

THE QUANTUM GROUP STRUCTURE OF

QUANTUM GRAVITY IN TWO DIMENSIONS

Jean-Loup Gervais

Physique Théorique, École Normale Supérieure
24, rue Lhomond
75231 Paris Cédex 05 - France

1. INTRODUCTION

These lectures summarize recent progress of the algebraic approach to quantum gravity in the conformal gauge (that is to the Liouville field theory) [1,2,3,4,5,6]. The basic point[1,2] is that there exist decompositions of inverse powers of the metric into operators that precisely transform under irreducible representations of the quantum group $U_q(sl(2))$ in the standard form. Their non-commutativity as quantum-field operators coincides with the non-commutativity that is induced by the "quantum" deformation of this group in the mathematical sense. Their braiding and fusion properties are known explicitly, since they are given by the universal R matrix and q-Clebsch-Gordan coefficients respectively[2,5]. In addition, some background material scattered in the early papers[7,8,9,10,11,12] are included for completeness.

2. THE CLASSICAL STRUCTURE

First recall some basic points about the weak coupling regime. In the conformal gauge, the classical dynamics is governed by the action:

$$S = \frac{1}{4\pi\gamma} \int d_2 x \sqrt{\widehat{g}} \left\{ \frac{1}{2} \widehat{g}^{ab} \partial_a \Phi \partial_b \Phi + e^{2\Phi} - \frac{1}{4} R_0 \Phi \right\} \tag{2.1}$$

$\widehat{g}_{ab}$ is the fixed background metric. We work for fixed genus, and do not integrate over the moduli. As is well known, one can choose a local coordinate system such that $\widehat{g}_{ab} = \delta_{ab}$. Thus we are reduced to the action

$$S = \frac{1}{4\pi\gamma} \int d\sigma d\tau \left(\frac{1}{2}(\frac{\partial\Phi}{\partial\sigma})^2 + \frac{1}{2}(\frac{\partial\Phi}{\partial\tau})^2 + e^{2\Phi} \right) \tag{2.2}$$

where σ and τ are the local coordinates. The complex structure is assumed to be such that the curves with constant σ and τ are everywhere tangent to the local imaginary and real axis respectively.

This last action corresponds to a conformal theory such that $\exp(2\Phi)$ is conformal with weights (1,1). The chiral modes may be separated very simply using the fact that

Random Surfaces and Quantum Gravity
Edited by O. Alvarez *et al., Plenum Press, New York, 1991*

the field $\Phi(\sigma, \tau)$ satisfies the equation

$$\frac{\partial^2 \Phi}{\partial \sigma^2} + \frac{\partial^2 \Phi}{\partial \tau^2} = 2e^{2\Phi} \tag{2.3}$$

if and only if†

$$e^{-\Phi} = \frac{i}{\sqrt{2}} \sum_{j=1,2} f_j(x_+)g_j(x_-); \quad x_\pm = \sigma \pm i\tau \tag{2.4}$$

where f_j (resp.(g_j)), which are functions of a single variable, are solutions of the same Schrödinger equation

$$-f_j'' + T(x_+)f_j = 0, \quad (\text{ resp. } - g_j'' + \overline{T}(x_-)g_j). \tag{2.5}$$

The solutions are normalized such that their Wronskians $f_1'f_2 - f_1f_2'$ and $g_1'g_2 - g_1g_2'$ are equal to one. The proof goes as follows.

1) First check that (2.4) is indeed solution. Taking the Laplacian of the logarithm of the right-hand side gives

$$\frac{\partial^2 \Phi}{\partial \sigma^2} + \frac{\partial^2 \Phi}{\partial \tau^2} \equiv 4\partial_+\partial_-\Phi = -4 \Big/ \Big(\sum_{i=1,2} f_i g_i \Big)^2$$

where $\partial_\pm = (\partial/\partial\sigma \mp i\partial/\partial\tau)/2$. The numerator has been simplified by means of the Wronskian condition. This is equivalent to (2.3).

2) Conversely check that any solution of (2.3) may be put under the form (2.4). If (2.3) holds one deduces

$$\partial_\mp T^{(\pm)} = 0; \quad \text{with } T^{(\pm)} := e^\Phi \partial_\pm^2 e^{-\Phi} \tag{2.6}$$

$T^{(\pm)}$ are thus functions of a single variable. Next the equation involving $T^{(+)}$ may be rewritten as

$$(-\partial_+^2 + T^{(+)})e^{-\Phi} = 0 \tag{2.7}$$

with solution

$$e^{-\Phi} = \frac{i}{\sqrt{2}} \sum_{j=1,2} f_j(x_+)g_j(x_-); \quad \text{with } - f_j'' + T^{(+)}f_j = 0$$

where the g_j are arbitrary functions of x_-. Using the equation (2.5) that involves $T^{(-)}$, one finally derives the Schrödinger equation $-g_j'' + T^{(-)}g_j = 0$. Thus the theorem holds with $T = T^{(+)}$ and $\overline{T} = T^{(-)}$ □

One may deduce from (2.6) that the potentials of the two Schrödinger equations coincide with the two chiral components of the stress-energy tensor. **Thus these equations are the classical equivalent of the Ward identities that ensure the decoupling of Virasoro null vectors.** From the canonical Poisson brackets (P.B.) one finds two P.B. realizations of the Virasoro algebra such that the f_j's (resp. g_j's) are primary fields with weights $(-1/2, 0)$ (resp. $(0, -1/2)$. At the classical level it is trivial to compute powers of $e^{-\Phi}$:

$$e^{-N\Phi} = \Big(\frac{i}{\sqrt{2}}\Big)^N \sum_{p=0}^N \frac{N!}{p!(N-p)!}(f_1g_1)^p(f_2g_2)^{N-p} \tag{2.8}$$

---

† The factor i means that these solutions should be considered in Minkowsky space-time

which is primary with weight $(-N/2, -N/2)$. $e^{-N\Phi}$ is thus built up from powers of the solutions of the basic fields f_i and g_i. For positive N one has a finite number of term but the weights are negative. Operators with positive weights have N negative so that (2.8) involves an infinite number of terms. Setting $N = -2$ gives weights $(1,1)$ in agreement with the fact that the potential term of (2.2) is equal to $e^{2\Phi}$ which must be a marginal operator. An important point to keep in mind is that, a priori, any two pairs f_j and g_j of linearly independent solutions of (2.5) are suitable. In this connection one easily sees that (2.4) is left unchanged if f_j and g_j are replaced by $\sum_k M_{jk} f_k$ and $\sum_k (M^{-1})_{jk} g_k$, respectively, where M_{jk} is an arbitrary constant matrix. It is natural to choose the determinant of M to be equal to one in order to preserve the normalisation of the wronskians. Eq. (2.4) is a $sl(2, C)$-invariant with f_j transforming as a representation of spin $1/2$. The higher powers (2.8) may be regarded in a similar way. The set of functions of x_+ wich appear, that is $(f_1)^p (f_2)^{N-p}$, $0 \le p \le N$, transform as a representation of spin $N/2$ under the above transformation of the f_j. This group structure will be replaced by a quantum group one when we turn to the quantum mechanical problem. This is natural, since the f's and the g's become non-commutative objects. At the classical level, the present viewpoint moreover shows that the definition of positive powers of the metric is connected with the problem of representation with negative spins (More about this below).

For the time being we shall concentrate on one of the two chiral components. Consider for instance the $-$ chiral components which are analytic functions of $z = \tau + i\sigma$. In a typical situation, σ and τ may be taken as coordinates of a cylinder obtained by conformal mapping from a particular handle of the Riemann surface considered. τ plays the role of imaginary time and σ is a space variable. One may work at $\tau = 0$ without loss of generality. The potential $T(\sigma)$ is periodic with period say 2π and we are working on the unit circle. Any two independent solutions of the Schrödinger equation is suitable. It seems natural at first sight to diagonalize the monodromy matrix, that is to choose two solutions noted ψ_j, $j = 1, 2$, that are periodic up to a multiplicative constant†. It is convenient to introduce

$$\phi_j(\sigma) := \ln(\psi_j)/\sqrt{\gamma} - \ln d_j,$$

d_j are suitable normalization constants. The fields ϕ_j are periodic up to additive constants and have the expansion

$$\phi_j(\sigma) = q_0^{(j)} + p_0^{(j)}\sigma + i \sum_{n\neq 0} e^{-in\sigma}\, p_n^{(j)}/n, \quad j - 1, 2, \tag{2.9}$$

As shown in Ref. 8,9, the canonical P.B. structure of the action (2.2) is such that the chiral fields ϕ_j satisfy

$$\{\phi_1'(\sigma_1), \phi_1'(\sigma_2)\}_{\text{P.B.}} = \{\phi_2'(\sigma_1), \phi_2'(\sigma_2)\}_{\text{P.B.}} - 2\pi\,\delta'(v_1 - v_2), \tag{2.10}$$

$$\{q_0^{(j)}, p_0^{(j)}\}_{\text{P.B.}} = 1 \tag{2.11}$$

$$T/\gamma = (\phi_1')^2 + \phi_1''/\sqrt{\gamma} = (\phi_2')^2 + \phi_2''/\sqrt{\gamma}. \tag{2.12}$$

$$p_0^{(1)} = -p_0^{(2)}, \tag{2.13}$$

Equations (2.12) are trivial to derive from the Schrödinger equation (2.5). They are the associated Riccati equation. Eq. (2.13) follows from the fact that the product of the two eigenvalues of the monodromy matrix is equal to one, as a standard Wronskian arguement shows.

---

† *We assume that the monodromy matrix is diagonalizable*

At this point a parenthesical remark is in order. Equations (2.6) give

$$T^{(-)} = (\partial_- \Phi)^2 - \partial_-^2 \Phi \tag{2.14}$$

that is very similar to (2.12). As a result, there is often a confusion, in the current literature, between ϕ and Φ. It should be stressed that the quadratic expressions (2.14) cannot be directly used to set up the canonical Hamiltonian formalism since it involves the second derivative of Φ with respect to τ, while a point in phase space is entirely determined by Φ and $\partial\Phi/\partial\tau$ at a given time. Indeed, in the canonical formalism, $\partial^2\Phi/\partial\tau^2$ is a dependent variable to be eliminated by using the field equations, before writing Hamilton's equations. When one does this, the potential term $\exp(2\Phi)$ reappears in (2.14) which is thus not trivially equivalent to a free-field expression. The equivalence is more involved and was just summarized. For the ϕ_j fields, the field equation is simply $\partial\phi_j/\partial\tau = i\partial\phi_j/\partial\sigma$ and (2.10) do remain quadratic when $\partial^2\phi_j/\partial\tau^2$ is eliminated.

In the language of field theory, the ϕ's are two equivalent free fields such that (2.12) takes the form of a U_1–Sugawara stress–tensor with a linear term. The latter is responsible for the classical Virasoro central charge $C_{class.} = 3/\gamma$. Clearly the two free fields play a symmetric role and one could as well build $e^{-\Phi}$ from different sets of Schrödinger solutions. Such a possibility is at the origin of the quantum group action, as recalled above. This is discussed in Ref. 2,4,5 and below.

Before leaving the classical problem, it is worth pointing out another interpretation of the existence of the two free fields we just recalled. This is related with the Drinfeld-Sokolov[13] Hamiltonian reduction from the affine Kac-Moody algebra $sl(2)^A$ to the Virasoro algebra. In standard notations, one imposes the constraint $J_- = 1$. After reduction, the currents have different conformal weights, that is, 0 for J_-, 1 for J_0, and 2 for J_+. This type of operators may be found in the present scheme as follows: Eq. (2.12) trivially gives:

$$\frac{T}{\gamma} = \frac{1}{2}\Big[(\phi_1')^2 + \frac{\phi_1''}{\sqrt{\gamma}} + (\phi_2')^2 + \frac{\phi_2''}{\sqrt{\gamma}}\Big] \equiv \frac{1}{2}\Big[(J_0)^2 + J_+ J_- + \frac{J_0'}{\sqrt{\gamma}}\Big], \tag{2.15}$$

where we have let

$$J_- = 1, \quad J_0 \equiv \phi_1' + \phi_2', \quad J_+ = -2\phi_1'' \phi_2'' \tag{2.16}$$

One arrives at a deformed $SU(2)$-Sugawara expression for T, and the relation between the ϕ_j fields and the current does give the correct spectrum of conformal weights, up to central terms. In this way, one should be able to regard the present discussion as coming from a WZW model by Hamiltonian reduction with a special choice of gauge.

3. THE QUANTUM (GROUP) STRUCTURE

Let us now come to the quantum case. The basic point of the method[9,10,12] is to quantize the above classical structure in such a way that the conformal structure is maintained. In particular powers of the metric tensor must be primary fields. This is ensured by the following result of Ref. 9:

On the unit circle, $z = e^{i\sigma}$, and for generic γ, there exist two equivalent free fields:

$$\phi_j(\sigma) = q_0^{(j)} + p_0^{(j)}\sigma + i\sum_{n\neq 0} e^{-in\sigma}\, p_n^{(j)}/n, \quad j = 1, 2, \tag{3.1}$$

such that

$$\Big[\phi_1'(\sigma_1), \phi_1'(\sigma_2)\Big] = \Big[\phi_2'(\sigma_1), \phi_2'(\sigma_2)\Big] = 2\pi i\, \delta'(\sigma_1 - \sigma_2), \qquad p_0^{(1)} = -p_0^{(2)}, \tag{3.2}$$

$$N^{(1)}(\phi_1')^2 + \phi_1''/\sqrt{\gamma} = N^{(2)}(\phi_2')^2 + \phi_2''/\sqrt{\gamma}. \tag{3.3}$$

$N^{(1)}$ (resp. $N^{(2)}$) denote normal orderings with respect to the modes of ϕ_1 (resp. ϕ_2). Eq. (3.3) defines the stress-energy tensor and the coupling constant γ of the quantum theory. The former generates a representation of the Virasoro algebra with central charge $C_{Liou} = 3 + 1/\gamma$.

The chiral family is built up[10,14,11,2] from the following operators

$$\psi_j = d_j\, N^{(j)}(e^{\sqrt{h/2\pi}\,\phi_j}), \quad \widehat{\psi}_j = \widehat{d}_j\, N^{(j)}(e^{\sqrt{h/2\pi}\,\phi_j}), \quad j = 1,\,2, \tag{3.4}$$

$$h = \frac{\pi}{12}\Big(C_{Liou} - 13 - \sqrt{(C_{Liou} - 25)(C_{Liou} - 1)}\Big),$$

$$\widehat{h} = \frac{\pi}{12}\Big(C_{Liou} - 13 + \sqrt{(C_{Liou} - 25)(C_{Liou} - 1)}\Big), \tag{3.5}$$

where d_j and $\widehat{d}_j$ are normalization constants. They are determined as solutions of the equations

$$-\psi_j'' + (\frac{h}{\pi})(\sum_{n<0} L_n\, e^{-in\sigma} + \frac{L_0}{2})\psi_j + (\frac{h}{\pi})\psi_j(\sum_{n>0} L_n\, e^{-in\sigma} + \frac{L_0}{2}) = 0 \tag{3.6}$$

$$-\widehat{\psi}_j'' + (\frac{\widehat{h}}{\pi})(\sum_{n<0} L_n\, e^{-in\sigma} + \frac{L_0}{2})\widehat{\psi}_j + (\frac{\widehat{h}}{\pi})\widehat{\psi}_j(\sum_{n>0} L_n\, e^{-in\sigma} + \frac{L_0}{2}) = 0 \tag{3.7}$$

These are operator Schrödinger equations equivalent to the decoupling of Virasoro null-vectors[10,14,11]. They are the quantum versions of Eq. (2.5). Since there are two possible quantum modifications h and $\widehat{h}$, there are four solutions. By operator product ψ_j, $j = 1$, 2, and $\widehat{\psi}_j$, $j = 1$, 2, generate two infinite families of chiral fields which are denoted $\psi_m^{(J)}$, $-J \leq m \leq J$, and $\widehat{\psi}_{\widehat{m}}^{(\widehat{J})}$, $-\widehat{J} \leq \widehat{m} \leq \widehat{J}$; respectively, with $\psi_{-1/2}^{(1/2)} = \psi_1$, $\psi_{1/2}^{(1/2)} = \psi_2$, and $\widehat{\psi}_{-1/2}^{(1/2)} = \widehat{\psi}_1$, $\widehat{\psi}_{1/2}^{(1/2)} = \widehat{\psi}_2$. An easy computation shows that the standard screening charges $\alpha_\pm$ are given by

$$\alpha_- = -\sqrt{\frac{2h}{\pi}}, \quad \alpha_+ = -\sqrt{\frac{2\widehat{h}}{\pi}} \tag{3.8}$$

$\psi_m^{(J)}$, $\widehat{\psi}_{\widehat{m}}^{(\widehat{J})}$, are of the type $(0, 2J)$ and $(2\widehat{J}, 0)$, respectively, in the BPZ classification. For the zero-modes, it is simpler[2] to define the rescaled variables

$$\varpi = ip_0^{(1)}\sqrt{\frac{2\pi}{h}}; \quad \widehat{\varpi} = ip_0^{(1)}\sqrt{\frac{2\pi}{\widehat{h}}}; \quad \widehat{\varpi} = \varpi\,\frac{h}{\pi}; \quad \varpi = \widehat{\varpi}\,\frac{\widehat{h}}{\pi}. \tag{3.9}$$

At this point a pedagogical parenthesis may be in order: the hatted and unhatted ψ fields have the same chirality; if we go to $\tau \neq 0$ they are both functions of x_-; there are two counterparts $\overline{\psi}_m^{(J)}(x_+)$ and $\widehat{\overline{\psi}}_m^{(J)}(x_+)$ which may be discussed in essentially the same way. Returning to our main line we recall that the Hilbert space in which the operators ψ and $\widehat{\psi}$ live, is a direct sum[2,3,4,5,12] of Fock spaces $\mathcal{F}(\varpi)$ spanned by the harmonic excitations of highest-weight Virasoro states noted $|\varpi, 0>$. They are eigenstates of the quasi momentum ϖ, and satisfy $L_n |\varpi, 0 > = 0$, $n > 0$; $(L_0 - \Delta(\varpi))|\varpi, 0 > = 0$. The corresponding highest weights $\Delta(\varpi)$ may be rewritten as

$$\Delta(\varpi) \equiv \frac{1}{8\gamma} + \frac{(p_0^{(1)})^2}{2} = \frac{h}{4\pi}(1 + \frac{\pi}{h})^2 - \frac{h}{4\pi}\varpi^2. \tag{3.10}$$

The commutation relations (3.2) are to be supplemented by the zero mode ones:

$$[q_0^{(1)}, p_0^{(1)}] = [q_0^{(2)}, p_0^{(2)}] = i.$$

The fields ψ and $\widehat{\psi}$ shift the quasi momentum $p_0^{(1)} = -p_0^{(2)}$ by a fixed amount. For an arbitrary c-number function f one has

$$\psi_m^{(J)} f(\varpi) = f(\varpi + 2m) \psi_m^{(J)}, \quad \widehat{\psi}_{\widehat{m}}^{(\widehat{J})} f(\varpi) = f(\varpi + 2\widehat{m}\,\pi/h)\, \widehat{\psi}_{\widehat{m}}^{(\widehat{J})}. \tag{3.11}$$

The fields ψ and $\widehat{\psi}$ together with their products may be naturally restricted to discrete values of ϖ. They thus live in Hilbert spaces† of the form

$$\mathcal{H}(\varpi_0) \equiv \bigoplus_{n,\widehat{n}=-\infty}^{+\infty} \mathcal{F}(\varpi_0 + n + \widehat{n}\,\pi/h). \tag{3.12}$$

ϖ^0 is a constant which is arbitrary so far. The $sl(2,C)$–invariant vacuum corresponds to $\varpi_0 = 1 + \pi/h,^2$ but other choices are also appropriate, as we shall see.

At the quantum level, one makes use of the above chiral conformal family, since the quantum field equation is likely to imply that the quantum Schrödinger equation holds for each chiral component. Associated with each quantum modification one finds a quantum version of (2.4). Since (3.4) involes h or $\widehat{h}$ instead of γ, it should be considered as defining different powers of the metric than in the classical case. In order to agree with standard notations, we write them as $\exp(-\alpha_\pm \Phi/2)$. For $\gamma \to 0$, $\alpha_- \sim 2\sqrt{\gamma}$, and $\exp(-\alpha_-\Phi/2) \sim \exp(-\sqrt{\gamma}\Phi)$. This does give back the classical metric field of section 2 since we have chosen a quantum definition that agrees with the the conventional one, but corresponds to rescaling our classical field by $\sqrt{\gamma}$. By short-distance operator-product expansion, one generates the set of fields $\exp[-(J\alpha_- + \widehat{J}\alpha_-)\Phi]$. From the quantum-group viewpoint, J and $\widehat{J}$ are the spins of the representations. Thus one sees that the cosmological term, that is $\exp[\alpha_-\Phi]$ corresponds to $J = -1$! We shall come back to this below. The basic point of introducing the two normal orderings $N^{(i)}$, $i = 1$, 2, was to obtain a conformal regularization of the metric tensor operators $\exp(-\alpha_\pm \Phi/2)$. In terms of the Liouville field Φ, it is rather involved and field dependent. For γ going to zero, α_+ blows up. Thus if one wants to keep a smooth classical limit, only α_- should appear. This is possible with open boundary condition.[10,6] With closed boundary conditions, both α's should be kept in order to couple rational theories with gravity.[12]

In any case, the quantum modifications are real only if $C_{Liou} > 25$ or $C_{Liou} < 1$ (The latter case describes 2D matter by analytic continuation of 2D gravity). Thus the construction of the metric tensor operator just recalled fails for $1 < C_{Liou} < 25$, which is the region of strongly coupled gravity. The chiral families may be continued, however, and this is taken to be the the way to deal with 2D gravity in the strong coupling regime, if a consistent truncation may be found as shown in Ref. 4,5.

---

† *Mathematically they are not really Hilbert spaces since their metrics are not positive definite*

352

Next we display the quantum-group structure of the chiral fields. The operators ψ and $\widehat{\psi}$ are closed under O.P.E. and braiding. Each family obeys a quantum group symmetry of the $U_q(sl(2))$ type. However, the fusion coefficients and R–matrix elements depend upon ϖ and thus do not commute with the ψ's and $\widehat{\psi}$'s. Their explicit form is unusual, therefore. One may exhibit the standard $U_q(sl(2))$-quantum-group structure by changing basis to new families. Following my recent work,[2] let us introduce

$$\xi_M^{(J)}(\sigma) := \sum_{-J \leq m \leq J} |J, \varpi)_M^m \, \psi_m^{(J)}(\sigma), \quad -J \leq M \leq J; \tag{3.13}$$

$$|J, \varpi)_M^m = \sqrt{\binom{2J}{J+M}} \, e^{ihm/2} \times$$
$$\sum_{(\frac{J-M+m-t}{2}) \text{ integer}} e^{iht(\varpi+m)} \binom{J-M}{(J-M+m-t)/2} \binom{J+M}{(J+M+m+t)/2}; \tag{3.14}$$

$$\binom{P}{Q} \equiv \frac{\lfloor P \rfloor!}{\lfloor Q \rfloor! \lfloor P-Q \rfloor!} \quad \lfloor n \rfloor! \equiv \prod_{r=1}^{n} \lfloor r \rfloor \quad \lfloor r \rfloor \equiv \frac{\sin(hr)}{\sin h}. \tag{3.15}$$

The last equation introduces q–deformed factorials and binomial coefficients. The other fields $\widehat{\xi}_{\widehat{M}}^{(\widehat{J})}$ are defined in exactly the same way replacing h by $\widehat{h}$ everywhere. The symbols are the same with hats, e.g.

$$\lfloor \widehat{n} \rfloor! \equiv \prod_{r=1}^{n} \lfloor \widehat{r} \rfloor, \qquad \lfloor \widehat{r} \rfloor \equiv \frac{\sin(\widehat{h}r)}{\sin \widehat{h}}, \qquad \text{and so on.} \tag{3.16}$$

The above transformation may be explicitly inverted [5]. One has

$$\psi_m^{(J)}(\sigma) - \sum_{M=-J}^{J} \xi_M^{(J)}(\upsilon) \, (J, \varpi|_m^M, \tag{3.17}$$

$$(J, \varpi|_m^M := (-1)^{J-M} \, e^{ih(J-M)} |J, \varpi)_{-M}^{-m} \Big/ C_m^{(J)}(\varpi), \tag{3.18}$$

$$C_m^{(J)}(\varpi) := (-1)^{J-m} (2i \sin h)^{2J} e^{ihJ} \frac{\binom{2J}{J-m} \lfloor \varpi - J + m \rfloor_{2J+1}}{\lfloor \varpi + 2m \rfloor}. \tag{3.19}$$

The quantum group structure of the exchange algebra is exhibited by introducing group theoretic states $|J, M>, \ -J \leq M \leq J$ and operators $J_\pm$, J_3 such that

$$J_\pm|J, M> = \sqrt{\lfloor J \mp M \rfloor \lfloor J \pm M + 1 \rfloor}|J, M \pm 1>, \quad J_3|J, M> = M |J, M>. \tag{3.20}$$

These operators satisfy the $U_q(sl(2))$-commutation-relations

$$\left[J_+, J_- \right] = \lfloor 2J_3 \rfloor, \quad \left[J_3, J_\pm \right] = \pm J_\pm. \tag{3.21}$$

In [2,5] the operator algebra of the ξ fields was completely determined.

1) For $\pi > \sigma > \sigma' > 0$, these operators obey the exchange algebra

$$\xi_M^{(J)}(\sigma)\,\xi_{M'}^{(J')}(\sigma') = \sum_{-J\leq N\leq J;\,-J'\leq N'\leq J'} (J,J')_{M\,M'}^{N'\,N}\,\xi_{N'}^{(J')}(\sigma')\,\xi_N^{(J)}(\sigma), \qquad (3.22)$$

$$(J,J')_{M\,M'}^{N'\,N} = \Big(< J,M|\otimes < J',M'|\Big)\,\mathbf{R}\,\Big(|J,N>\otimes|J',N'>\Big), \qquad (3.23)$$

$$\mathbf{R} = e^{(-2ihJ_3\otimes J_3)}\Big(1 + \sum_{n=1}^{\infty} \frac{(1-e^{2ih})^n\,e^{ihn(n-1)/2}}{\lfloor n\rfloor!}e^{-ihnJ_3}(J_+)^n \otimes e^{ihnJ_3}(J_-)^n\Big). \qquad (3.24)$$

$\mathbf{R}$ coincides with the universal R matrix of $U_q(sl(2))$.

2) For $0 < \sigma < \sigma' < \pi$, the ξ fields obey the exchange algebra

$$\xi_M^{(J)}(\sigma)\,\xi_{M'}^{(J')}(\sigma') = \sum_{-J\leq N\leq J;\,-J'\leq N'\leq J'} \overline{(J,J')}_{M\,M'}^{N'\,N}\,\xi_{N'}^{(J')}(\sigma')\,\xi_N^{(J)}(\sigma), \qquad (3.25)$$

$$\overline{(J,J')}_{M\,M'}^{N'\,N} = \Big(< J,M|\otimes < J',M'|\Big)\,\overline{\mathbf{R}}\,\Big(|J,N>\otimes|J',N'>\Big), \qquad (3.26)$$

$$\overline{\mathbf{R}} = e^{(2ihJ_3\otimes J_3)}\Big(1 + \sum_{n=1}^{\infty} \frac{(1-e^{-2ih})^n\,e^{-ihn(n-1)/2}}{\lfloor n\rfloor!}e^{-ihnJ_3}(J_-)^n \otimes e^{ihnJ_3}(J_+)^n\Big). \qquad (3.27)$$

3) The two exchange formulae are related by the inverse relation

$$\sum_{-J\leq N\leq J;\,-J'\leq N'\leq J'} (J,J')_{M\,M'}^{N'\,N}\,\overline{(J',J)}_{N'\,N}^{P\,P'} = \delta_{M,P}\,\delta_{M',P'}. \qquad (3.28)$$

4) The short-distance operator-product expansion of the ξ fields is of the form:

$$\xi_{M_1}^{(J_1)}(\sigma)\,\xi_{M_2}^{(J_2)}(\sigma') = \sum_{J=|J_1-J_2|}^{J_1+J_2} \Big\{(d(\sigma-\sigma'))^{\Delta(J)-\Delta(J_1)-\Delta(J_2)}$$

$$(J_1,M_1;J_2,M_2|J_1,J_2;J,M_1+M_2)\,\Big(\xi_{M_1+M_2}^{(J)}(\sigma) + \text{descendants}\Big)\Big\}, \qquad (3.29)$$

where $d(\sigma-\sigma') \equiv 1 - e^{-i(\sigma-\sigma')}$, $(J_1,M_1;J_2,M_2|J_1,J_2;J,M_1+M_2)$ denotes the Clebsch-Gordan coefficients of $U_q(sl(2))$, and $\Delta(J) := -hJ(J+1)/\pi - J$ is the Virasoro-weight of $\xi_M^{(J)}(\sigma)$.

5) Define the quantum group action on the ξ fields by

$$J_3\,\xi_M^{(J)} = M\xi_M^{(J)}, \qquad J_\pm\,\xi_M^{(J)} = \sqrt{\lfloor J\mp M\rfloor\lfloor J\pm M+1\rfloor}\,\xi_{M\pm 1}^{(J)}. \qquad (3.30)$$

Then the operator-product $\xi_{M_1}^{(J_1)}(\sigma)\,\xi_{M_2}^{(J_2)}(\sigma')$ gives a representation of the quantum group algebra (2.6) with the co-product generators

$$\mathbf{J}_\pm := J_\pm \otimes e^{ihJ_3} + e^{-ihJ_3} \otimes J_\pm, \qquad \mathbf{J}_3 := J_3 \otimes 1 + 1 \otimes J_3, \qquad (3.31)$$

where the tensor product is defined so that

$$(A \otimes B)\left(\xi_{M_1}^{(J_1)}(\sigma)\,\xi_{M_2}^{(J_2)}(\sigma')\right) := (A\xi_{M_1}^{(J_1)}(\sigma))\,(B\xi_{M_2}^{(J_2)}(\sigma')), \qquad (3.32)$$

and where each term in the expansion over J transforms according to a representation of spin J.

Similar formulae hold in the other half circle. Eq. (3.31) coincides with the standard co-product, and is thus non-symmetric, the two definitions being related by the universal R matrix. The exchange properties of the ξ fields (Eqs (3.22)-(3.27)) show that their quantum mechanical structure precisely matches this asymmetry, so that the transformation law (3.30) is fully consistent with the operator-algebra.

Obviously the same structure holds for the hatted fields. One replaces h by $\widehat{h}$ everywhere. Moreover the hatted and unhatted fields have simple braiding and fusions.[2] The most general $(2\widehat{J}, 2J)$ field $\xi_{M\,\widehat{M}}^{(J\,\widehat{J})} \sim \xi_M^{(J)}\,\widehat{\xi}_{\widehat{M}}^{(\widehat{J})}$ has weight

$$\Delta_{Kac}(J,\widehat{J}; C_{Liou}) = \frac{C_{Liou}-1}{24} - \frac{1}{24}\left((J+\widehat{J}+1)\sqrt{C_{Liou}-1} - (J-\widehat{J})\sqrt{C_{Liou}-25}\right)^2,$$
$$(3.33)$$

in agreement with Kac's formula.

4. THE CASE OF REAL SCREENING CHARGES

If $C_{Liou} > 25$ the screning charges $\alpha_{\pm}$ are real. This is the weak coupling regime which is connected with the classical limit ($\gamma \to 0$). In this region h and $\widehat{h}$ are real and the structure recalled above is directly handy. Let us briefly discuss how the powers of the metric are reconstructed. Consider for instance $\exp(-J\alpha_-\Phi)$. There are two types of cases one may distinguish.

1) One may consider, as is most usual, closed surfaces without boundary. Then the natural region is the whole circle $0 \leq \sigma \leq 2\pi$. We are aiming at the quantum version of Eq. (2.8). It will involve the fields $\xi_M^{(J)}$, together with their counterparts $\overline{\xi}_M^{(J)}(x_+)$ whose exchange properties are similar. Concerning the latter one should remember that they are functions of z^*, that is, are anti-analytic functions. so that the orientation of the complex plane is reversed. This may be taken into account simply by replacing i by $-i$ in the above formulae for the ξ-fields, that is by taking the complex conjugate of all the c-numbers **without taking the Hermitian conjugate of the operators**. The appropriate definition of $\overline{\xi}_M^{(J)}(\sigma)$ is

$$\overline{\xi}_M^{(J)}(\sigma) := \sum_{-J \leq m \leq J} c^{ih(J+M)}\,(|J,\overline{\omega})_M^m)^*\,\overline{\psi}_m^{(J)}(\sigma), \quad -J \leq M \leq J; \qquad (4.1)$$

where $(|J,\overline{\omega})_M^m)^*$ is the complex conjugate of $(|J,\overline{\omega})_M^m)$. In addition to taking the complex conjugate of $|J,\overline{\omega})_M^m$, we have introduced an additional phase factor for later convenience. One may see that it does not change the braiding and fusion properties, up to overall normalisations. Concerning braiding, for instance, this is true because $(J_1,J_2)_{M_1\,M_2}^{P_2\,P_1}$ is non-zero only if $P_1 + P_2 = M_1 + M_2$. As is usual in conformally invariant field theory we assume that the right- and left- movers commute. Thus we take the ξ-fields to commute with the $\overline{\xi}$-fields.

There are two basic requirements that determine $\exp(-J\alpha_-\Phi)$. The first one is locality, that is, that it commutes with any other power of the metric at equal τ. The second one concerns the Hilbert space of states where the physical operator algebra is realized. The point is that, since we took the fields $\xi_M^{(J)}$ and $\bar\xi_M^{(J)}$ to commute, the quasi momenta ϖ and $\bar\varpi$ of the left and right movers are unrelated, while periodicity in σ requires that they be equal. This last condition is replaced by the requirement that $\exp(-J\alpha_-\Phi)$ leave the subspace of states with $\varpi = \bar\varpi$ invariant. The latter condition defines the physical Hilbert space $\mathcal{H}_{phys}$ where it must be possible to restrict the operator-algebra consistently. At $\tau = 0$, the appropriate definition is:

$$e^{-J\alpha_-\Phi(\sigma)} = c_J \sum_{M=-J}^{J} (-1)^{J-M}\, \xi_M^{(J)}(\sigma)\bar\xi_{-M}^{(J)}(\sigma) \tag{4.2}$$

where c_J is a normalisation constant. It is invariant under the quantum group action (3.30) if the $\bar\xi$ fields transform in the same way as the ξ fields:

$$J_3\,\bar\xi_M^{(J)} = M\bar\xi_M^{(J)}, \quad J_\pm\,\bar\xi_M^{(J)} = \sqrt{\lfloor J \mp M\rfloor\lfloor J \pm M + 1\rfloor}\,\bar\xi_{M\pm1}^{(J)}. \tag{4.3}$$

Locality is checked by making use of Eqs (3.22)-(3.24). Choose $\pi > \sigma > \sigma' > 0$. In agreement with the above discussion we have

$$\bar\xi_M^{(J)}(\sigma)\bar\xi_{M'}^{(J')}(\sigma') = \sum_{-J\le N\le J;\, -J'\le N'\le J'} ((J, J')_{M\,M'}^{N'\,N})^*\, \bar\xi_{N'}^{(J')}(\sigma')\bar\xi_{N}^{(J)}(\sigma), \tag{4.4}$$

In checking locality, one encounters the product of two R matrices. It is handled by means of the identities

$$((J_1, J_2)_{-M_1\,-M_2}^{-N_2\,-N_1})^* = ((J_2, J_1)_{-N_2\,-N_1}^{-M_1\,-M_2})^* = \overline{(J_2, J_1)}_{N_2\,N_1}^{M_1\,M_2} \tag{4.5}$$

that follow from the explicit expressions (3.24), (3.27). In this way one deduces the equation

$$\sum_{M_1 M_2} (J_1, J_2)_{M_1\,M_2}^{P_2\,P_1} ((J_1, J_2)_{-M_1\,-M_2}^{-N_2\,-N_1})^* = \delta_{P_1,N_1}\,\delta_{P_2,N_2} \tag{4.6}$$

from the inverse relation (3.28), and the desired locality relation follows:

$$e^{-J_1\alpha_-\Phi(\sigma_1)}\,e^{-J_2\alpha_-\Phi(\sigma_2)} = e^{-J_2\alpha_-\Phi(\sigma_2)}\,e^{-J_1\alpha_-\Phi(\sigma_1)} \tag{4.7}$$

On the other hand, the precise form of the factor $e^{ih(J+M)}$ in (4.1) is dictated by the requirement that $\mathcal{H}_{phys}$ be left invariant, as we show next. This is seen by re-expressing (4.2) in terms of ψ fields. One gets, at first

$$e^{-J\alpha_-\Phi(\sigma,\tau)} = c_J \sum_{M=-J}^{J} (-1)^{J-M} e^{ih(J-M)}\, |J,\varpi)_M^m\, (|J,\bar\varpi)_{-M}^p)^*\, \psi_m^{(J)}(\sigma)\bar\psi_p^{(J)}(-\sigma) \tag{4.8}$$

Using Eq. (3.22) of Ref. 2 one writes

$$\sum_{M=-J}^{J} (-1)^{J-M} e^{ih(J-M)}\, |J,\varpi)_M^m\, (|J,\bar\varpi)_{-M}^p)^* =$$

$$\sum_{M=-J}^{J} (-1)^{J-M} e^{ih(J-M)}\, |J,\varpi)_M^m\, |J,\bar\varpi + 2p)_{-M}^{-p} \tag{4.9}$$

If $\varpi = \overline{\varpi}$, this becomes, according to (3.17,3.18),

$$\sum_{M=-J}^{J} (-1)^{J-M} e^{ih(J-M)} |J, \varpi\rangle_M^m |J, \varpi + 2p\rangle_{-M}^{-p} = \delta_{m,p}\, C_m^{(J)}(\varpi). \tag{4.10}$$

As a consequence, and when it is restricted to $\mathcal{H}_{phys}$, Eq. (4.2) is equivalent to

$$e^{-J\alpha_-\Phi(\sigma)} = c_J \sum_{m=-J}^{J} C_m^{(J)}(\varpi)\, \psi_m^{(J)}(\sigma)\, \overline{\psi}_m^{(J)}(\sigma) \tag{4.11}$$

and the condition $\varpi = \overline{\varpi}$ is indeed left invariant, according to (3.11).

2) One may also consider gravity with boundary, following Ref. 7,8,6. A typical situation is the half circle $0 \leq \sigma \leq \pi$. One may set up boundary conditions such that the system remains conformal, albeit with one type of Virasoro generators only. The left- and right-movers become related as is the case for open strings. The appropriate definition of the metric becomes[6]:

$$e^{-J\alpha_-\Phi(\sigma)} = c_J \sum_{M,N} A_{M,N}^{(J)}\, \xi_M^{(J)}(\sigma)\, \xi_N^{(J)}(2\pi - \sigma) \tag{4.12}$$

where

$$A_{M,N}^{(J)} = \langle J, M| \left\{ e^{-ihJ_3^2} \sum_{r,s=0}^{\infty} e^{ih(r+s)J_3} \frac{(J_+)^{r+s}}{\lfloor r\rfloor!\lfloor s\rfloor!} q^{a(r-s)} q^{rs/2} \right\} |J, N\rangle \tag{4.13}$$

where a depends upon the boundary condition chosen. These operators are mutually local and closed by fusion[6].

So far the present discussion assumes that q is not a root of unity, that is, deals with irrational theories. The problem of specializing q to a root of unity has, however, been essentially reduced to the equivalent limit in the representation theory of $U_q(sl(2))$ which is a much studied problem. Clearly, the present discussion applies whenever the screening charges are real, so that it also describes the $C < 1$ models, as a continuation of the Liouville theory.

There still remain the difficulty already pointed out in the classical case, that positive powers of the metric are difficult to handle. This is a major difference between the proper region of the Liouville theory ($C_{Liou} > 1$) and the region of statistical models ($\widehat{C} < 1$), since (3.33) gives negative or complex weights for positive J and $\widehat{J}$ in the former case, so that one must deal with negative J or $\widehat{J}$. The above discussion must be continued to negative spins[4,5]. One may show that equation (3.14) is equivalent to

$$|J, \varpi\rangle_m^M = \sqrt{\binom{2J}{J+M}}\, e^{ih(m/2 + (\varpi+m)(J-M+m))} F_q(a, b;\, c;\, e^{-2ih(\varpi+m)}), \tag{4.14}$$

where $a = M - J$ (resp. $a = -M - J$), $b = -m - J$ (resp. $b = m - J$), $c = 1 + M - m$ (resp. $c = 1 - M + m$) for $M > m$ (resp. $M < m$) and $F_q(a, b;\, c;\, z)$ is a q-deformed—so called basic— hypergeometric function. The continuation to negative J is a direct consequence of Rodgers identity[15]

$$F_q(a, b;\, c;\, e^{-2ihu}) = (2i\sin h)^{c-a-b}\, e^{ihu(a+b-c)}\, \frac{\Gamma_q(u - (a+b-c-1)/2)}{\Gamma_q(u + (a+b-c+1)/2)} \times$$
$$F_q(c-a, c-b;\, c;\, e^{-2ihu}). \tag{4.15}$$

where Γ_q denotes the q-deformed gamma-function. Equations (4.14,4.15) give

$$|J,\varpi)_m^M = (2i\sin(h))^{1+2J} \binom{2J}{J+M} \frac{\Gamma_q(\varpi+m+J+1)}{\Gamma_q(\varpi+m-J)} |-J-1,\varpi)_m^M. \qquad (4.16)$$

This exhibits a symmetry between J and $-J-1$ which is also shared by the R matrices and Clebsch-Gordan coefficients[4,5]. It is the basis for defining operators with negative J. The crucial point of (4.16) is to show that $\psi_m^{(-J-1)}$ and $\xi_M^{(-J-1)}$ are to be considered for $-J \leq m \leq J$, and $-J \leq M \leq J$, respectively.

5. SOLVING THE REALITY PROBLEM OF STRONGLY COUPLED GRAVITY: THE UNITARY TRUNCATION THEOREM

Next, we consider the region $1 < C_{Liou} < 25$, which is relevant to the strong coupling regime of 2D gravity. In this case, h and $\widehat{h}$ are complex and $\widehat{h} = h^*$. We choose the imaginary part of h to be negative for definiteness. In the weak coupling regime of gravity, the solution of the conformal bootstrap we just outlined arised in a natural way from the chiral decomposition of the 2D metric tensor in the conformal gauge, that is by solving Liouville's equation. It is thus legitimate to study the strong coupling regime by continuing this chiral structure below $C_{Liou} = 25$. Complex numbers appear all over the place. However—in a way that is reminiscent of the truncations that give the minimal unitary models— for $C_{Liou} = 7$, 13, 19; there is a consistent truncation of the above general family down to a unitary theory involving operators with real Virasoro conformal weights only. The main points of this theorem[4,5] are briefly summarized next. The truncated family is as follows:

<u>a) The physical Hilbert space.</u> It is given by[16,3,4,5]:

$$\mathcal{H}_{phys} \equiv \bigoplus_{r=0}^{1-s} \mathcal{H}_-(\varpi_0^r) \equiv \bigoplus_{r=0}^{1-s} \bigoplus_{n=-\infty}^{\infty} \mathcal{F}(\varpi_{r,n}), \qquad (5.1)$$

$$\varpi_{r,n} \equiv \varpi_0^r + n(1 - \frac{\pi}{h}) \equiv \left(\frac{r}{2-s} + n\right)\left(1 - \frac{\pi}{h}\right). \qquad (5.2)$$

The integer s is such that the special values correspond to

$$C_{Liou} = 1 + 6(s+2), \quad s = 0, \pm 1; \quad h + \widehat{h} = s\pi. \qquad (5.3)$$

$\Delta(\varpi_{r,n})$ is positive and in $\mathcal{H}_{phys}$ the representation of the Virasoro algebra is unitary. The torus partition function corresponds to compactification on a circle with radius $R = \sqrt{2(2-s)}$ (see Ref. 3).

<u>b) The restricted set of conformal weights.</u> The truncated family only involves operators of the type $(2J,2J)$ noted $\chi_-^{(J)}$ and $(2(-J-1),2J)$ noted $\chi_+^{(J)}$. Their Virasoro conformal weights[12,3,4,5] which are respectively given by

$$\Delta^-(J) = -\frac{C_{Liou}-1}{6}J(J+1), \quad \Delta^+(J) = 1 + \frac{25-C_{Liou}}{6}J(J+1), \qquad (5.4)$$

are real. $\Delta^-(J)$ in negative for all J (except for $J = -1/2$ where it becomes equal to $\Delta^+(-1/2) = (s+2)/4$). $\Delta^+(J)$ is always positive, and is larger than one if $J \neq -1/2$.

<u>c) The truncated families</u>: $\mathcal{A}^{\pm}_{phys}$ is the set of operators $\chi^{(J)}_{\pm}$, $J \geq 0$ of the form[4,5]

$$\chi^{(J)}_{-} = \sum_{M=-J}^{J} \kappa^{J-M} (-1)^{s(J-M)(J-M-1)/2} \, \xi^{(J,J)}_{M,-M}, \tag{5.5a}$$

$$\chi^{(J)}_{+} = \sum_{M=-J}^{J} \kappa^{J-M} (-1)^{s(J-M)(J-M-1)/2} \, \xi^{(-J-1,J)}_{M,-M}. \tag{5.5b}$$

κ will be defined below.

THE UNITARY TRUNCATION THEOREM:

For $C_{Liou} = 1 + 6(s+2)$, $s = 0, \pm 1$, and when it acts on $\mathcal{H}_{phys}$; the set $\mathcal{A}^{+}_{phys}$ (resp. $\mathcal{A}^{-}_{phys}$) of operators $\chi^{(J)}_{+}$ (resp. $\chi^{(J)}_{-}$) is closed by fusion and braiding, and only gives states that belong to $\mathcal{H}_{phys}$.

<u>PROOF</u>

Conditions (5.2) and (5.3) are instrumental since they allow us to relate hatted and unhatted quantities. In particular, for N integer, one has

$$\lfloor \widehat{N} \rfloor = e^{-i(N-1)s\pi} \lfloor N \rfloor, \qquad \widehat{h}\widehat{\varpi}_{r,n} - h\varpi_{r,n} = \pi(r + n(2-s)). \tag{5.6}$$

The truncation was originally observed[16,3] in terms of ψ and $\widehat{\psi}$ fields, for the braiding of $\chi^{(1/2)}_{-}$ with itself. The expression of $\chi^{(1/2)}_{-}$ using ξ fields was written in Ref. 3. From this one may derive the general formula for $\chi^{(J)}_{-}$ recursively: one deduces $\chi^{(J+1/2)}_{-}$ from $\chi^{(J)}_{-}$ by fusion with $\chi^{(1/2)}_{-}$ to leading order in the singularity, using (3.29). The form of $\chi^{(1/2)}_{+}$ may be infered from the symmetry between J and $-J-1$ recalled above. The proof of the unitary truncation theorem relies on the following three special properties:

Theorem (1). *If (5.3) holds the hatted and unhatted Clebsch-Gordan coefficients are related by*

$$\left(J_1, M_1; J_2, M_2 \widehat{|} J_1, J_2; J, M\right) = (J_1, -M_1; J_2, -M_2 | J_1, J_2; J, -M) \times$$

$$(-1)^{s\{(J_1-M_1)(J_2+M_2)+(J_1+J_2-J)(J_1+M_1+J_2-M_2)\}}(-1)^{J_1+J_2-J}. \tag{5.7}$$

Thus they satisfy the orthogonality relation

$$\sum_{M_1, M_2} \left(J_1, -M_1; J_2, -M_2 \widehat{|} J_1, J_2, \widehat{J}, -M\right)(J_1, M_1; J_2, M_2 | J_1, J_2; J, M) \times$$

$$(-1)^{s\{(J_1+M_1)(J_2-M_2)+(J_1+J_2-J)(J_1-M_1+J_2+M_2)\}} = (-1)^{J_1+J_2-J} \, \delta_{J\widehat{J}}. \tag{5.8}$$

Theorem (2). *If (5.3) holds, the hatted and unhatted R–matrices are related by the relation*

$$\left(\widehat{J_1, J_2}\right)^{N_2 N_1}_{M_1 M_2} = \overline{(J_2, J_1)}^{-M_1 -M_2}_{-N_2 -N_1} \, e^{is\pi\{(J_1-M_1)(J_2+M_2)-J_1(J_2+N_2)+N_1(J_2-N_2)\}}. \tag{5.9}$$

Theorem (3). *Introduce the operator*

$$\kappa := -e^{i(\widehat{h}\widehat{\varpi}-h\varpi)} \, e^{i(h-\widehat{h})/2}. \tag{5.10}$$

In $\mathcal{H}_{phys}$ one has

$$\kappa^{J+M}\, \hat{|J,\,\hat{\varpi}})_M^m = (-1)^{s(J+M)(J+M-1)/2+J+M} \times$$

$$e^{ih(J+M-m)}e^{is\pi(-J^2+m/2)}e^{-im(\widehat{h\varpi}-h\varpi)}|J,\,\varpi-2m)_M^m. \qquad (5.11)$$

Closure by fusion and braiding are consequences of Theorems 1 and 2 respectively. Theorem 3 combined with relation (4.10) shows that the χ fields may be rewritten in terms of the ψ fields with only terms with $m+\hat{m}=0$ appearing, so that $\mathcal{H}_{phys}$ is indeed left invariant. Details are given in Ref. 5.

6. PHYSICAL ASPECTS OF THE STRONGLY COUPLED GRAVITY THEORIES

1) String theories

First, taking D free fields as worldsheet matter,[16,12,17,18] one sees that one may construct consistent string emission vertices if $D = 26 - C_{grav} = 19, 13, 7$. The mass squared of the emitted string ground state is $m^2 = 2(\Delta - 1)$, where Δ is the conformal weight of the 2D-gravity dressing-operator. Since an infinite number of tachyons is unacceptable, this selects the $\mathcal{A}^+_{phys}$ family with positive weights Δ^+. Bilal and I have already unravelled striking properties of the associated Liouville strings.[17,18]. Remarkably, one finds that the spectrum of conformal weights, which is selected by the truncation theorem, automatically gives modular partition functions on the torus so that the associated string theories are consistent, at least up to one loop. The extension of the present discussion to $N = 1$ super-Liouville theory is in progress[19]. The main features of the corresponding Liouville superstrings may be predicted[17,18]. The possible dimensions are $D = 3, 5, 7$. The first two models have striking features: First their total number of degree of freedom, that is 4 and 6 coincide with two choices of space-time dimensions where classical Green-Schwarz actions may be written, but could not be quantized consistently. These two theories may be regarded as the correct quantum theories associated with these classical actions where Lorentz invariance is broken from $D + 1$ to D dimensions. Indeed, the physical number of degrees of freedom— that is $D - 1$— are equal to 2 and 4 which coincide with the real dimensions of the division algebras of complex numbers and quaternion, so that light-cone Green-Schwarz formalisms exist. The $O(8)$ triality of the ten-dimensional model is replaced by those of $U(1) \otimes U(1)$ and $SU(2) \otimes SU(2) \otimes SU(2)$ respectively. Both theories are space-time supersymmetric once the appropriate GSO projections are performed.

The existence of a three-dimensional model raises the hope of verifying the long-standing conjecture of Polyakov that the 3D Ising model is equivalent to a string theory. Indeed under plausible assumptions, it has been possible[18] to obtain a relation between critical exponents that is exactly satisfied by the numerical studies of the 3D Ising model.

2) Two-dimensional critical systems

Clearly, $\mathcal{A}^+_{phys}$ is also selected if we consider the associated conformal theories by themselves, in order to avoid correlation functions that grow at very large distance. One may play the game of fractal gravity, since Eq. (5.4) shows that $\Delta^-(J,C) + \Delta^+(J, 26 - C) = 1$, and since the set of values 7, 13, 19 is left invariant by $C \to 26 - C$. One will have two copies of the theories discussed above. One describes gravity, and, following what happens for non-critical strings, one would make use of the family $\mathcal{A}^+_{phys}$. Then the other copy, would only involve $\mathcal{A}^-_{phys}$ and correspond to a non-unitary matter theory. It is a challenge to derive these models from the matrix approach to 2D gravity.

Finally, the truncation theorem holds for any integer s so that it applies to $C_{Liou} = 1$ ($s = -2$), and $C_{Liou} = 25$ ($s = 2$), as well as for $C_{Liou} < 1$ ($s < -2$) and $C_{Liou} > 25$ ($s > 2$).

REFERENCES

1. O. Babelon, *Phys. Lett.* **B215**, 523 (1988).

2. J.-L. Gervais, *Comm. Math. Phys.* **130**, 257 (1990).

3. J.-L. Gervais, B. Rostand, *Nucl. Phys.* **B346**, 473 (1990).

4. J.-L. Gervais, *Phys. Lett.* **B243**, 85 (1990).

5. J.-L. Gervais, "Solving the strongly coupled 2D gravity: unitary trucation and quantum group structure" LPTENS preprint 90/13 submitted to *Comm. in Math. Phys.*

6. E. Cremmer, J.-L. Gervais, to be published.

7. J.-L. Gervais, A. Neveu, *Nucl. Phys.* **B199**, 59 (1982).

8. J.-L. Gervais, A. Neveu, *Nucl. Phys.* **B202**, 125 (1982).

9. J.-L. Gervais, A. Neveu, *Nucl. Phys.* **B224**, 329 (1983).

10. J.-L. Gervais, A. Neveu *Nucl. Phys.* **B238**, 125 (1984); *Nucl. Phys.* **B238**, 396 (1984).

11. J.-L. Gervais, A. Neveu, *Nucl. Phys.* **B264**, 557 (1986).

12. For reviews see, J.-L. Gervais, "Liouville Superstrings" in "Perspectives in string the proceedings of the Niels Bohr/Nordita Meeting (1987) World Scientific; DST workshop on particle physics-Superstring theory, proceedings of the I.I.T. Kanpur meeting (1987) World Scientific; J.-L. Gervais, " Systematic approach to conformal theories", *Nucl. Phys. B (Proc. Supp.)* **5B**, 119-136 (1988) 119; A. Bilal, J.-L. Gervais, " Conformal theories with non-linearly-extended Virasoro symmetries and Lie-algebra classification", Conference Proceedings: "Infinite dimensional Lie algebras and Lie groups", edited by V. Kac, Marseille 1988, World-Scientific.

13. see, e.g. B. Feigin, E. Frenkel, "Quantization of the Drinfeld-Sokolov reduction " Harvard preprint submitted to Phys. Lett. B.

14. J.-L. Gervais, A. Neveu, *Nucl. Phys.* **B257[FS14]**, 59 (1985).

15. See, e.g., G. Andrews, Conference board of the math. sciences, Regional conference series in math. # 66 A.M.S. ed.

16. J.-L. Gervais, A. Neveu, *Phys. Lett.* **151B**, 271 (1985).

17. A. Bilal, J.-L. Gervais, *Nucl. Phys.* **B284**, 397 (1987); *Phys. Lett.* **B187**, 39 (1987). *Nucl. Phys.* **B293**, 1 (1987); for reviews see Ref. 12.

18. A. Bilal, J.-L. Gervais, *Nucl. Phys.* **B295[FS21]**, 277 (1988).

19. J.-L. Gervais, B. Rostand, in preparation.

NOTES ON QUANTUM LIOUVILLE THEORY
AND QUANTUM GRAVITY

Nathan Seiberg

1. Introduction

There are two motivations to study two-dimensional quantum gravity. First, this theory is a toy model for four-dimensional gravity. Second, two-dimensional quantum gravity is the theory on the world-sheet of both "critical" as well as "non-critical" string theories. Some of our conclusions apply to these two more general situations. Since we would like to draw lessons from this simple theory, which are generic and valid in more complicated systems, we should study it in continuum field theory language. In these notes we review the status of the continuum Liouville approach to quantum gravity. Our understanding of the subject is far from complete. We will point out what we think are the most important open problems, and will speculate about their solutions.

After understanding some of the results presented here I learned that they had been known to Polyakov and Zamolodchikov [1]. I would like to thank A.B. Zamolodchikov for discussing their results with me and for sharing his insights. This has helped me to develop these ideas further.

The problem of quantum gravity is the problem of integrating over all metrics modulo diffeomorphisms. We will use the conformal gauge

$$g_{ab} = e^{\gamma\phi}\hat{g}_{ab}^{(\tau)} \tag{1.1}$$

where γ is a parameter and will refer to $\hat{g}$ as the fiducial metric. τ are the moduli and ϕ is known as the Liouville mode.

It is known [2] that we need to study the Liouville field theory based on the action

$$S_L = \frac{1}{4\pi} \int \sqrt{\hat{g}}(\frac{1}{2}\hat{g}^{ab}\partial_a\phi\partial_b\phi + \frac{1}{\gamma}\phi R(\hat{g}) + \frac{\mu}{2\gamma^2}e^{\gamma\phi}) \tag{1.2}$$

Random Surfaces and Quantum Gravity
Edited by O. Alvarez *et al.*, *Plenum Press, New York, 1991*

with the coupling constant γ ($\hbar = \gamma^2$). μ is the cosmological constant. We ignore in (1.2) possible total derivatives which do not affect the local analysis.

Locally, we can choose $\hat{g}_{ab} = e^{\rho}\delta_{ab}$, shift ϕ to set $\rho = 0$ and use complex coordinates z and $\bar{z}$. At least classically (1.2) describes a conformal field theory; invariant under

$$z \to \tilde{z} = f(z)$$

$$\phi(z) \to \tilde{\phi}(\tilde{z}) = \phi(z) - \frac{1}{\gamma}\log|\partial f|^2 \tag{1.3}$$

such that $e^{\gamma\phi}dzd\bar{z}$ is invariant.

Classical Liouville field theory has been well understood since the 19'th century [3]. It is the theory of negatively curved Riemann surfaces. We will review some of its relevant properties in section 2. Despite a lot of very interesting work [1] [4] [5] [6] [7] [8], the quantum theory is still much less understood. As we will see, some of the difficulties are related to the subtleties of quantum gravity.

In section 3 we will analyze Liouville theory in the semi-classical approximation and in section 4 we will study the exact quantum theory. We will show that although the action (1.2) describes a conformal field theory, the standard identification of states and local operators is not valid. An insertion of a state is obtained by cutting a little hole in the manifold and evaluating the functional integral with boundary conditions ϕ at the hole and later integrating over the boundary values ϕ with a weight given by the wave function of the state $\psi(\phi)$. The hole can be arbitrarily small in the fiducial metric $\hat{g}$. But if we want the insertion to be local in the physical metric $g = e^{\gamma\phi}\hat{g}$, the wave function $\psi(\phi)$ should be (infinitely) peaked on small holes – it should grow as the circumference of the hole $L = \oint e^{\frac{1}{2}\phi} \to 0$. Such a wave function is not normalizable. Normalizable wave functions cut macroscopic holes in the manifold and are not local disturbances.

In section 5 we will couple the system to a matter conformal field theory and study quantum gravity. When the minimal models of [9] are coupled to gravity all the physical states are microscopic (non-normalizable wave functions) and the models are finite. However, in the generic conformal field theory, even for $c < 1$, there are macroscopic physical states corresponding to non-local operators. These states are similar to the tachyons of the critical string and they lead to an instability of the two dimensional surface. This is the problem of the so called $c = 1$ barrier (which, can happen both for c larger and smaller than 1). This problem can perhaps be solved by fine tuning the coefficients of all the macroscopic operators to zero. Another possibility is to relax some of the axioms of field theory in the two-dimensional matter system.

2. Classical Liouville

The action for $\hat{g}_{ab} = \delta_{ab}$ is

$$S = \int d^2x \frac{1}{8\pi}\partial_{\alpha}\phi\partial^{\alpha}\phi + \frac{\mu}{8\pi\gamma^2}e^{\gamma\phi} + \frac{1}{4\pi\gamma}\phi\sqrt{\hat{g}}R(\hat{g}) \; . \tag{2.1}$$

We have included the coupling to R which is necessary when we couple the theory to curved $\hat{g}$ for later use. S is bounded from below for $\mu > 0$.

The equation of motion $\frac{\delta S}{\delta \phi} = 0$ is

$$-e^{-\gamma\phi}\triangle\gamma\phi = -\frac{\mu}{2} \ , \tag{2.2}$$

i.e. the metric $g = e^{\gamma\phi}\hat{g}$ has constant negative curvature.

The stress tensor is found by $T_{ab} = 2\pi\frac{\delta S}{\delta\hat{g}^{ab}}$. Notice that we have to differentiate the coupling to $R(\hat{g})$ in (2.1) before setting $\hat{g}_{ab} = \delta_{ab}$. The contribution from this term is known as "an improvement term in T." Using the equation of motion (2.2), we find

$$T_{z\bar{z}} = 0$$
$$T_{zz} = -\frac{1}{2}(\partial\phi)^2 + \frac{1}{\gamma}\partial^2\phi \tag{2.3}$$

(∂ denotes ∂_z). $T_{z\bar{z}} = 0$ is a signal of conformal invariance. Under the transformation law (1.3) T does not transform like a tensor but is shifted by the Schwarzian derivative with $c = \frac{12}{\gamma^2}$

$$T(\phi(z)) = T(\tilde{\phi}(\tilde{z}))(\partial f)^2 + T(\frac{1}{\gamma}\log\partial f) \ . \tag{2.4}$$

2.1. Canonical formalism

It is useful to derive the same result about the central extension c by canonical formalism in Minkowski space. We consider the system on a flat cylinder where $\sigma \in [0, 2\pi)$ parametrizes space and the non-compact coordinate t parametrizes time.

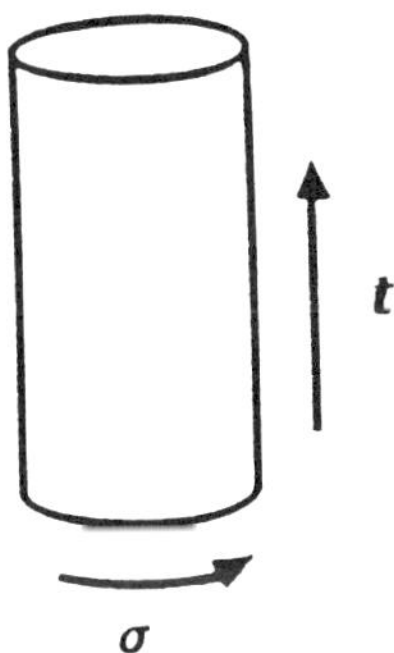

Cylinder used for canonical formalism

The momentum conjugate to ϕ is $\Pi = \frac{\delta S}{\delta\dot{\phi}} = \frac{1}{4\pi}\dot{\phi}$. It satisfies the Poisson bracket

$$\{\phi(\sigma, t), \Pi(\sigma', t)\}_{PB} = \delta(\sigma - \sigma') \ . \tag{2.5}$$

In the light-cone coordinates $x^{\pm} = \sigma \pm t$ the stress tensor is

$$T_{+-} = 0$$
$$T_{++} = \frac{1}{2}(\partial_+\phi)^2 - \frac{1}{\gamma}\partial_+^2\phi + \frac{1}{2\gamma^2} \tag{2.6}$$
$$= \frac{1}{8}(4\pi\Pi + \phi')^2 - \frac{1}{2\gamma}(4\pi\Pi + \phi')' + \frac{\mu}{8\gamma^2}e^{\gamma\phi} + \frac{1}{2\gamma^2} \ .$$

The shift term $\frac{1}{2\gamma^2}$ arises from changing variables from the sphere to the cylinder $z = e^{t+i\sigma}$ using the Schwarzian derivative.

The Fourier components

$$L_n = \int_0^{2\pi} \frac{d\sigma}{2\pi} e^{in\sigma} T_{++}(\sigma, t) \tag{2.7}$$

satisfy the Virasoro algebra

$$i\{L_n(t), L_m(t)\}_{PB} = (n - m)L_{n+m}(t) + \frac{c}{12}(n^3 - n)\delta_{n+m,0} \tag{2.8}$$

with $c = \frac{12}{\gamma^2}$. The Poisson bracket of L_n and $e^{\alpha\phi}$ shows that $e^{\alpha\phi}$ is a primary field with $\Delta = \frac{\alpha}{\gamma}$. Of course, the same result can be found by using the transformation law of ϕ.

2.2. Classical solutions in Minkowski space

We will consider for simplicity $\phi = \phi_0$ which is independent of σ. Restricting attention to such configurations is known as the "mini-superspace approximation." The Minkowski space equation of motion of ϕ_0 is that of a particle moving in a potential $V = \frac{\mu}{\gamma^2} e^{\gamma\phi_0} + \frac{1}{2\gamma^2}$.

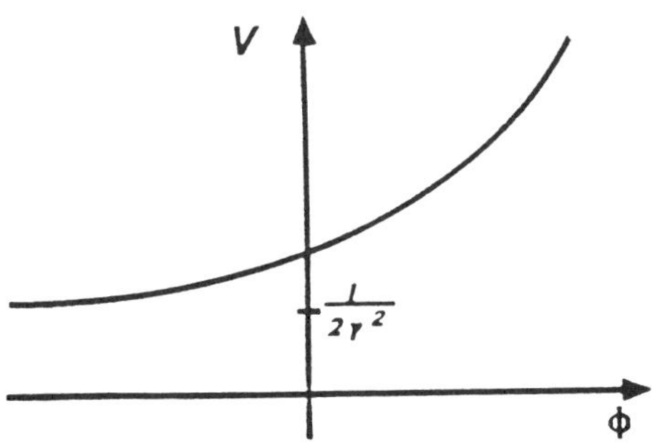

The potential

All classical solutions satisfy $\lim_{t\to\pm\infty} \phi_0 \sim \mp 2pt$ where the particle is essentially free and $p > 0$ is its momentum. The general solution (up to time translation) is

$$e^{\gamma\phi_0(t)} = \frac{16\gamma p^2}{\mu} \frac{e^{2p\gamma t}}{(1 - e^{2p\gamma t})^2} \tag{2.9}$$

and it represents reflection off the potential. The surface with such a metric looks like

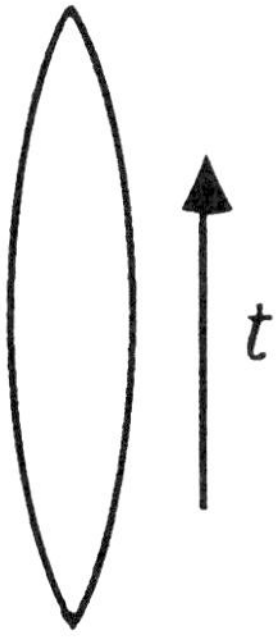

A σ independent solution in Minkowski space

Notice that there is no solution with $p = 0$ and the solution with p is the same as with $-p$. The energy of the solution is

$$\Delta(p) = \frac{1}{2}p^2 + \frac{1}{2\gamma^2} > \frac{1}{2\gamma^2} \ . \tag{2.10}$$

More general solutions which depend on σ are given locally by

$$e^{\gamma\phi} = \frac{16}{\mu} \frac{A'(x^+)B'(x^-)}{(1 - AB)^2} \tag{2.11}$$

with A (B) a function of x^+ (x^-). A and B need not be single valued on the cylinder – only ϕ has to be single valued.

2.3. Classical solutions in Euclidean space

In Euclidean space there are more interesting solutions. The most general local solution is similar to (2.11)

$$e^{\gamma\psi}dzd\bar{z} = \frac{16}{\mu} \frac{\partial A(z)\bar{\partial}B(\bar{z})}{(1 - AB)^2}dzd\bar{z} \tag{2.12}$$

with A (B) a function of z $(\bar{z})$. As $z \to e^{2\pi i}z$, A and B transform by an $SL(2, R)$ transformation. The solution (2.12) is obtained by taking the quotient of a disk with constant negative curvature metric by a subgroup of $SL(2, R)$. The stress tensor T of the classical solution (2.12)

$$T(z) = -\frac{1}{2\gamma^2}[\partial(\log \partial A)]^2 + \frac{1}{\gamma^2}\partial^2(\log \partial A) \tag{2.13}$$

is the Schwarzian derivative of A. It is manifestly holomorphic.

Depending on the conjugacy class of the monodromy of A (and of B) there are three classes of local solutions: elliptic, parabolic and hyperbolic[1].

1. Elliptic solutions. For ϕ independent of σ ($z = e^{t+i\sigma}$) these are (up to dilation)

$$A = z^a$$

$$B = \bar{z}^a$$

$$e^{\gamma\phi}dzd\bar{z} = \frac{16}{\mu}\frac{a^2}{(z\bar{z})^{1-a}[1-(z\bar{z})^a]^2}dzd\bar{z} \qquad (2.14)$$

$$T(z) = \frac{1}{z^2}\left(-\frac{a^2}{2\gamma^2} + \frac{1}{2\gamma^2}\right)$$

with a real and hence the monodromy is elliptic, $A \to e^{2\pi i a}A$. The solution with a is the same as that with $-a$, so we will let $a > 0$.

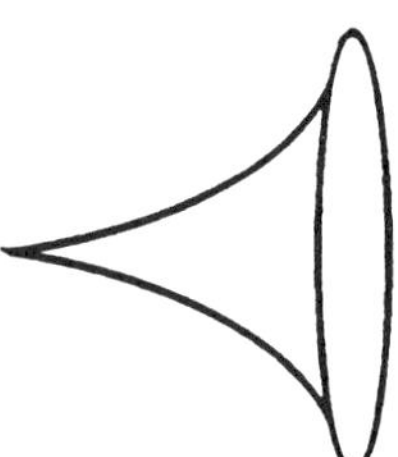

A σ independent elliptic solution

This solution has a curvature singularity at $z = 0$ and it satisfies

$$\frac{1}{4\pi}\triangle\phi - \frac{\mu}{8\pi\gamma}e^{\gamma\phi} + \frac{1-a}{\gamma}\delta^{(2)}(z) = 0 \ . \qquad (2.15)$$

As $z \to 0$ the metric approaches a flat metric with a conical singularity. In this region the momentum of ϕ is $p = \lim_{t\to-\infty}\frac{1}{2}\dot{\phi} = \frac{a}{\gamma}$ (after changing variables to the cylinder). It is real in Euclidean space and hence imaginary in Minkowski space.

2. Parabolic solutions. These are obtained in the limit $a \to 0$ of the elliptic solutions. For ϕ independent of σ they are (up to dilation)

$$A = i\log z$$

$$B = \frac{i}{\log\bar{z}}$$

$$e^{\gamma\phi}dzd\bar{z} = \frac{16}{\mu}\frac{1}{z\bar{z}[\log z\bar{z}]^2}dzd\bar{z} \qquad (2.16)$$

$$T(z) = \frac{1}{z^2}\frac{1}{2\gamma^2}$$

---

[1] We thank A.B. Zamolodchikov for a very useful discussion on this subject.

and the monodromy is parabolic, $A \to A - 2\pi$.

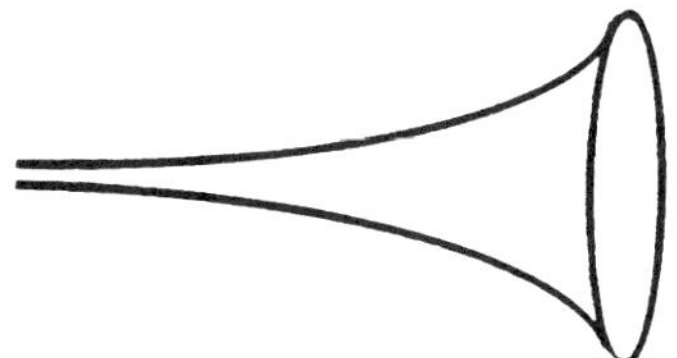

A σ independent parabolic solution

This solution has a curvature singularity at $z = 0$ which is half the curvature of the sphere, i.e. it corresponds to a puncture

$$\frac{1}{4\pi}\triangle\phi - \frac{\mu}{8\pi\gamma}e^{\gamma\phi} + \frac{1}{\gamma}\delta^{(2)}(z) = 0 \quad . \tag{2.17}$$

As $z \to 0$ the momentum vanishes. In this region the metric does not approach a flat metric because of the logarithmic correction.

3. Hyperbolic solutions. These correspond to a imaginary. For ϕ independent of σ they are (up to dilation)

$$A = z^{im}$$
$$B = \bar{z}^{im}$$
$$e^{\gamma\phi}dzd\bar{z} = \frac{4}{\mu}\frac{m^2}{z\bar{z}[\sin(\frac{m}{2}\log z\bar{z})]^2}dzd\bar{z} \tag{2.18}$$
$$T(z) = \frac{1}{z^2}(\frac{m^2}{2\gamma^2} + \frac{1}{2\gamma^2})$$

the monodromy is hyperbolic, $A \to e^{-2\pi m}A$, and the metric is the constant negative curvature metric on the annulus – the plumbing fixture metric. Again, the solution depends only on $|m|$ and not on its sign. As $m \to 0$ we find the metric of the parabolic solution.

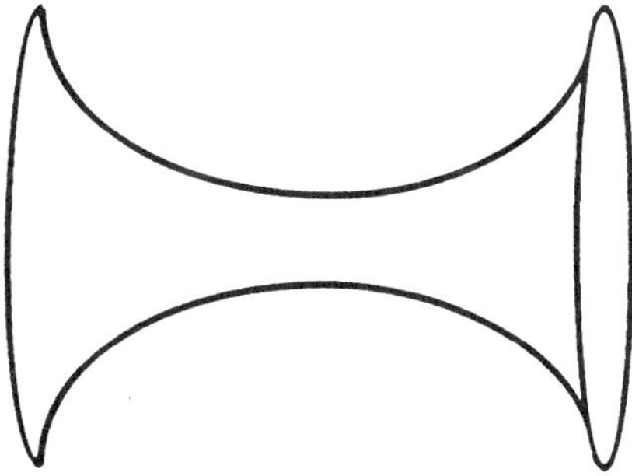

A σ independent hyperbolic solution

The value of T for such solutions is larger than $\frac{1}{2\gamma^2}\frac{1}{z^2}$ and hence is the same as in the Minkowski space solutions. There are no Minkowski space solutions with energies (values of T) equal to those of the elliptic and parabolic solutions.

2.4. Relation to $SL(2,R)$

The form of the classical solution

$$e^{\gamma\phi}\,dz\,d\bar{z} = \frac{16}{\mu}\frac{\partial A(z)\bar{\partial}B(\bar{z})}{(1-AB)^2}\,dz\,d\bar{z} \qquad (2.19)$$

suggests considering

$$e^{-j\gamma\phi} = \left(\frac{16}{\mu}\right)^{-j}\left(\frac{1}{\sqrt{\partial A}}\frac{1}{\sqrt{\bar{\partial}B}} - \frac{A}{\sqrt{\partial A}}\frac{B}{\sqrt{\bar{\partial}B}}\right)^{2j} \ . \qquad (2.20)$$

For $2j$ a positive integer

$$e^{-j\gamma\phi} = \left(\frac{16}{\mu}\right)^{-j}\sum_{m=-j}^{j}\psi_m^j(z)\psi^{jm}(\bar{z}) \qquad (2.21)$$

is a finite sum of holomorphic times antiholomorphic fields. It is easy to check that under $SL(2,R)$ transformations ψ_m^j transform like the spin j $(2j+1$ dimensional) representation of $SL(2,R)$. For $2j$ a negative integer, the decomposition into holomorphic fields is infinite. These fields also form an $SL(2,R)$ representation[6][7].

The two $j=\frac{1}{2}$ fields $\psi_{-\frac{1}{2}} = \frac{1}{\sqrt{\partial A}}$, $\psi_{\frac{1}{2}} = \frac{A}{\sqrt{\partial A}}$ satisfy

$$\left[\partial^2 + \frac{\gamma^2}{2}T(z)\right]\psi_{\pm\frac{1}{2}}(z) = 0 \ . \qquad (2.22)$$

They can be "bosonized"

$$\psi_{\pm\frac{1}{2}} = \exp\left(-\frac{\gamma}{2}\Phi_\pm\right) \ . \qquad (2.23)$$

In terms of $\Phi_\pm$

$$T = -\frac{1}{2}(\partial\Phi_\pm)^2 + \frac{1}{\gamma}\partial^2\Phi_\pm \ . \qquad (2.24)$$

Gervais and Neveu [6] have shown that the two fields $\Phi_\pm$ and their momenta satisfy free field Poisson brackets. However the bracket of Φ_+ and Φ_- is complicated[6]. The reason for this is that Φ_+ and Φ_- are not independent. One of them, say Φ_+ can be solved in terms of the other. Then, the theory can be described by a single free field. This is known as the Backlund transformation.

3. Semi-classical Liouville

In the quantum theory we are interested in correlation functions of the form

$$\left\langle \prod_i e^{\alpha_i \phi(z_i)} \right\rangle = \int D\phi \, e^{-S(\phi)} \prod_i e^{\alpha_i \phi(z_i)} \quad . \tag{3.1}$$

In this section we will study them in the semi-classical approximation which is valid for $\gamma \ll 1$. We should stress that while the expansion in γ makes sense, one cannot expand in the cosmological constant μ. By shifting ϕ its value can be changed. Therefore, there is no sense in which it is small. More explicitly, the correlation functions are given by a power of μ which is in general fractional times a function of the moduli (z_i and the moduli of the surface) which is independent of μ. This function cannot be found[2] by studying the theory perturbatively in μ.

3.1. Semi-classical correlation functions

We first analyze (3.1) for α_i large of order $\frac{1}{\gamma}$. Then the functional integral (3.1) is dominated by configurations ϕ_{cl} satisfying the classical equation of motion

$$-\frac{\delta S(\phi_{\mathrm{cl}})}{\delta \phi} + \sum_i \alpha_i \delta^{(2)}(z_i) = \frac{1}{4\pi} \triangle \phi_{\mathrm{cl}} - \frac{\mu}{8\pi\gamma} e^{\gamma \phi_{\mathrm{cl}}} + \sum_i \alpha_i \delta^{(2)}(z_i) = 0 \quad . \tag{3.2}$$

By integrating (3.2) over the surface we find

$$\gamma \sum_i \alpha_i + (2h - 2) - \frac{\mu}{8\pi} A = 0 \tag{3.3}$$

where A is the area of the surface and h is the number of handles. We see that a classical solution exists only when

$$\gamma \sum_i \alpha_i + (2h - 2) > 0 \quad . \tag{3.4}$$

We will first assume that this condition is satisfied and will return to the other case below.

The solutions of (3.2) are constant negative curvature solutions. The operators $e^{\alpha \phi}$ of the quantum theory appear as sources of curvature in the classical equation of motion and lead to solutions with local elliptic monodromy with $a = 1 - \gamma\alpha$. An operator with $\alpha = \frac{1}{\gamma}$ creates a puncture in the surface and will be called the puncture operator[3]. For $\alpha < \frac{1}{\gamma}$ the disturbance to the surface is milder than a puncture – only

---

[2] There is one exception to this fact, some correlation functions are analytic in μ and can be analyzed perturbatively in it. We will discuss this case below.

[3] This terminology should not be confused with the term puncture operator in topological gravity in two dimensions [10].

a spike with a curvature singularity is formed. For $\alpha > \frac{1}{\gamma}$ there is no classical solution
– we cannot localize so much curvature at a point.

After finding the classical solution, (3.1) is approximated by

$$\int D\phi e^{-S(\phi)} \prod_i e^{\alpha_i \phi(z_i)} = e^{-S(\phi_{\rm cl}) + \sum_i \alpha_i \phi_{\rm cl}(z_i)} \det\left(\frac{\delta^2 S(\phi_{\rm cl})}{\delta \phi^2}\right)(1 + O(\gamma^2)) \ . \qquad (3.5)$$

The exponent $-S(\phi_{\rm cl}) + \sum_i \alpha_i \phi_{\rm cl}(z_i)$ is infinite. The divergences arise from the vicinity of the sources. These are most easily calculated by writing the first term in S as $-\int \frac{1}{8\pi} \phi_{\rm cl} \triangle \phi_{\rm cl}$ and using (3.2) to eliminate $\triangle \phi_{\rm cl}$. Then, the divergences appear as terms of the form $\log |z_i - z_i|^2$ which are regulated as $-\log \Lambda^2$ with a short distance cutoff Λ. The divergence from the vicinity of a source $e^{\alpha\phi}$ with $\alpha \le \frac{1}{\gamma}$ contributes $(\Lambda^2)^{\frac{\alpha^2}{2}}$ to (3.5). Therefore, finite answers are obtained for correlation functions of the renormalized operators $(\Lambda^2)^{-\frac{\alpha^2}{2}} e^{\alpha\phi}$. This divergence has a standard interpretation. It represents the fact that the quantum dimension of the operator $e^{\alpha\phi}$ is not $\frac{\alpha}{\gamma}$ as in the classical theory but

$$\Delta(e^{\alpha\phi}) = \frac{\alpha}{\gamma} - \frac{1}{2}\alpha^2 + O(1) \ . \qquad (3.6)$$

This interpretation is consistent with the expressions for T in this background (2.14)(2.16). The coefficient of $\frac{1}{z^2}$ is the dimension of the operator. The shift in the dimension occurs because of the need to regularize the quantum fluctuations of ϕ. With $\alpha \sim \frac{1}{\gamma}$ this appears as a divergence in the classical action. We see that despite the interaction term $e^{\gamma\phi}$, the divergence is as if ϕ is a free field.

An important lesson from this analysis is that classical solutions with hyperbolic monodromies do not correspond to local operators. The circumference of the surface does not become arbitrarily small at any point.

3.2. Semi-classical correlation functions with fixed area

If $X = \sum_i \alpha_i + \frac{1}{\gamma}(2h - 2) \le 0$, there is no classical solution. In this case, it is convenient to insert $1 = \int dA\delta(\int e^{\gamma\phi} - A)$ into the functional integral (3.1) and to define the functional integral for fixed area A as

$$\langle \prod_i e^{\alpha_i \phi(z_i)} \rangle_A = \int D\phi e^{-S(\phi) + \frac{\mu}{8\pi\gamma^2}A} \prod_i e^{\alpha_i \phi(z_i)} \delta\left(\int e^{\gamma\phi} - A\right)$$

$$\langle \prod_i e^{\alpha_i \phi(z_i)} \rangle = \int dA \langle \prod_i e^{\alpha_i \phi(z_i)} \rangle_A e^{-\frac{\mu}{8\pi\gamma^2}A} \ . \qquad (3.7)$$

$\langle \prod_i e^{\alpha_i \phi(z_i)} \rangle_A$ can be evaluated in the semi-classical approximation. The constraint on the area is represented by a Lagrange multiplier and the classical equation of motion states that the surface has constant curvature (except at the sources). If $X > 0$ the solution has negative curvature and it is as in the previous subsection. If $X = 0$ it is flat and if $X < 0$, it has constant positive curvature.

By shifting

$$\phi \to \phi + \frac{1}{\gamma} \log A \tag{3.8}$$

we learn that

$$\langle \prod_i e^{\alpha_i \phi(z_i)} \rangle_A = A^{\frac{X}{\gamma} - 1} \langle \prod_i e^{\alpha_i \phi(z_i)} \rangle_{A=1} \ . \tag{3.9}$$

For $X > 0$ the integral over A in (3.7) is convergent and reproduces the answer we found earlier. For $X = 0$ it is logarithmically divergent and for $X < 0$ it diverges like a power.

These divergences arise from the region of small A ($\phi \to -\infty$). In order to regularize them one needs a cutoff on field space. From the point of view of Liouville theory with a background metric $\hat{g}$ this is not a UV divergence. It is not associated with a small distance measured in the $\hat{g}$ metric. However, remembering that the metric on the surface is g, this is a short distance problem. Therefore, every coordinate invariant regulator in the integral over g (like in the matrix models) cuts them off. With such a regulator

$$\langle \prod_i e^{\alpha_i \phi(z_i)} \rangle = P(\mu) + \mu^{-\frac{X}{\gamma}} C \langle \prod_i e^{\alpha_i \phi(z_i)} \rangle_{A=1} \tag{3.10}$$

where C is a constant independent of the moduli and μ, and $P(\mu)$ is a polynomial of degree $n = [-\frac{X}{\gamma}]$. For $-\frac{X}{\gamma}$ a non-negative integer the factor $\mu^{-\frac{X}{\gamma}}$ is multiplied by $\log \mu$. The analytic part $P(\mu)$ depends on the UV cutoff. It does not describe macroscopic surfaces and is not expected to be universal. The non-analytic contribution, proportional to $\langle \prod_i e^{\alpha_i \phi(z_i)} \rangle_{A=1}$ is universal and can be studied in the semi-classical approximation. If $\langle \prod_i e^{\alpha_i \phi(z_i)} \rangle_{A=1} = 0$, the correlation function comes entirely from the region of small (of the order of the UV cutoff) area and is analytic in μ.

We now turn to a few interesting examples:

1. $h = 1$ without insertions. Here $X = 0$. With the flat metric $\hat{g} = \delta_{z\bar{z}}$ on the torus the classical solution is

$$e^{\gamma \phi_{\rm cl}} = \frac{A}{\tau_2} \ . \tag{3.11}$$

Expanding $\phi = \phi_{\rm cl} + \phi_{\rm q}$ the measure and the delta function constraints become

$$\| \delta\phi \|^2 = \int e^{\gamma\phi} (\delta\phi)^2 \approx e^{\gamma\phi_{\rm cl}} \int (\delta\phi_{\rm q})^2$$
$$\delta(\int e^{\gamma\phi} - A) \approx \frac{1}{A\gamma} \delta(\frac{1}{\tau_2} \int \phi_{\rm q}) \ . \tag{3.12}$$

The functional integral over $\phi_{\rm q}$ is the same as in free field theory

$$\langle 1 \rangle_A \approx \frac{1}{2\pi A\gamma \sqrt{2\tau_2} |\eta(q)|^2} = \frac{1}{A\gamma} \int_0^\infty \frac{dp}{\pi} \frac{(q\bar{q})^{\frac{1}{2}p^2}}{|\eta(q)|^2} \tag{3.13}$$

and as expected, the integral over A is logarithmically divergent

$$\int dA \langle 1 \rangle_A e^{-\frac{\mu}{8\pi\gamma^2}} \approx \frac{1}{2\pi\gamma\sqrt{2\tau_2}|\eta(q)|^2} \log \frac{\Lambda^2}{\mu} \ . \tag{3.14}$$

2. n point function on the sphere $\langle \prod_i e^{\alpha_i(z_i)} \rangle$ with $\alpha_i \sim 1$. Here $X = -\frac{2}{\gamma} + \sum_i \alpha_i \approx -\frac{2}{\gamma} < 0$ and we have to constrain the area. This situation is similar to the one studied in [11]. Here, the classical solution is not affected by the existence of the operators. It is the round metric on the sphere and is labeled by three real parameters associated with the action of $PSL(2,C)$ on the classical solution (there are only three real continuous parameters and not six because of the $SO(3)$ isometry)

$$e^{\gamma\phi_{cl}} \sim \frac{A}{(|az+b|^2 + |cz+d|^2)^2} \tag{3.15}$$

with $ad - bc = 1$. The correlation function is given by an integral over these collective coordinates with the invariant measure $\int d^2a \, d^2b \, d^2c \, d^2d \, \delta^{(2)}(ad - bc - 1)$

$$\langle \prod_i e^{\alpha_i(z_i)} \rangle_A \sim A^{\frac{1}{\gamma}\sum_i \alpha_i - \frac{2}{\gamma^2} - 1} \int_{PSL(2,C)} \prod_i e^{\alpha_i \phi_{cl}(z_i)} \tag{3.16}$$

up to a factor independent of the z_i's arising from the determinant of the non-zero modes. Although the answer (3.16) does not look translational invariant (more generally Mobius invariant), the integral over the collective coordinates restores this symmetry. Because of the non-compact integration region, for some α_i's the integral diverges . This divergence can be handled by analytic continuation in α_i or in a fashion similar to [12], i.e. by regulating the integration region and dropping the power divergences.

For the two point function $\langle e^{\alpha\phi(\infty)} e^{\beta\phi(0)} \rangle$ we find a divergent integral over the subgroup of the Mobius group leaving 0 and ∞ invariant – the dilation group. We would like to interpret it as a delta function

$$\int_0^\infty \frac{d\lambda}{\lambda} \lambda^{2\frac{\beta-\alpha}{\gamma}} = \int_{-\infty}^\infty ds \, e^{2s\frac{\beta-\alpha}{\gamma}} \sim \delta(\alpha - \beta) \ . \tag{3.17}$$

This prescription leads to an $SL(2,C)$ invariant answer

$$\langle e^{\alpha\phi(0)} e^{\beta\phi(x)} \rangle_A \sim A^{2\frac{\alpha}{\gamma} - \frac{2}{\gamma^2} - 1} \frac{\delta(\alpha - \beta)}{|x|^{4\Delta_\alpha}} \tag{3.18}$$

where for $\alpha \sim 1$, $\Delta_\alpha = \frac{\alpha}{\gamma}$. For $\alpha = \beta$ (3.18) is infinite and proportional to the volume of the dilation group $\mathrm{Vol(dil.)} = \int_0^\infty \frac{d\lambda}{\lambda} = \infty$. Another special case, studied in [11], is the zero point function on the sphere where the integral leads to an infinite answer proportional to the volume of $SL(2,C)$. We should stress that these divergence is unrelated to the power divergence in the integral over A. These infinities are very subtle and can lead to surprises. For example, The one point function of $e^{\gamma\phi}$ can be thought of as the two point function of $e^{\gamma\phi}$ and the identity. By (3.18) we should expect to find zero. However, the expectation value of $\int e^{\gamma\phi}$ is clearly equal to A – we integrate a vanishing quantity over an infinite range and find a finite answer.

3. Two point function on the sphere $\langle e^{\alpha\phi(0)}e^{\beta\phi(\infty)}\rangle$ with $\alpha,\beta \sim \frac{1}{\gamma}$. Since both α and β are smaller than $\frac{1}{\gamma}$, $X < 0$ and we need to constrain the area. For $\alpha = \beta$ there is a one parameter family of solutions associated with dilation

$$e^{\gamma\phi_{\rm cl}} \sim \frac{A\lambda^{2-2\alpha\gamma}}{(z\bar{z})^{\alpha\gamma}(1+(\frac{z\bar{z}}{\lambda^2})^{1-\alpha\gamma})^2} \quad . \tag{3.19}$$

As for $\alpha \sim 1$, integrating over λ we find the infinite volume of the dilation group. For $\alpha \neq \beta$ there is no classical solution, i.e. the minimum action configuration is at the boundary of the space of ϕ's. More explicitly, dilating a configuration $\phi(z) \to \phi(\lambda z) - \frac{1}{\gamma}\log|\lambda|^2$, $-S + \alpha\phi(0) + \beta\phi(\infty)$ is shifted by $\frac{\beta-\alpha}{\gamma}\log|\lambda|^2$ (remember to change coordinates at infinity). For $\alpha > \beta$ the minimum action configuration occurs for $\lambda = 0$. Separating the mode λ from the functional integral, we encounter again $\int_0^\infty \frac{d\lambda}{\lambda}\lambda^{2\frac{\beta-\alpha}{\gamma}}$ which we interpret as in (3.17) as proportional to $\delta(\alpha - \beta)$.

The last two examples demonstrate the difference between Liouville field theory and free field theory. As in free field theory the spectrum is continuous and the two point function on the sphere is divergent. However, higher n point functions are finite. The divergence is of a geometrical origin and is not associated with an integration over the zero mode of the field. Related to this fact is a lack of conservation law in correlation functions. $\langle \prod_i e^{\alpha_i(z_i)}\rangle$ does not vanish for generic α_i's.

4. Quantum Liouville

We now turn to the full quantum theory and should compute higher order corrections in γ. A diagrammatic formalism for these corrections was developed by D. Friedan[5]. It is clear that to every order in γ, the interaction of $\phi_{\rm q} = \phi - \phi_{\rm cl}$ is polynomial and hence super-renormalizable. Therefore, all the divergences are as if ϕ is a free field. We have already seen that in the leading order in the semi-classical expansion. From an operatorial point of view, this means that free field normal ordering should remove all divergences in $e^{\alpha\phi}$ to all orders in perturbation theory in γ. It is amusing that quantum Liouville theory is both conformally invariant for every γ and is super-renormalizable.

As is standard in quantum field theory, the parameters might be renormalized. This renormalization depends on the precise way the quantum theory is regularized and renormalized. Therefore, the value of the coupling constant γ does not give an unambiguous definition of the theory. We define the coupling constant γ in terms of the cosmological constant $e^{\gamma\phi}$ where ϕ is normalized such that its kinetic term is $\frac{1}{8\pi}\partial_a\phi\partial^a\phi$. Physical quantities like the value of the conformal anomaly c and the dimensions of the operators Δ are independent of such ambiguities.

4.1. Canonical quantization

The easiest way to quantize the theory is to follow Curtright and Thorn [4] and to use canonical quantization. They constructed the operators using free field normal ordering and checked the commutation relations as follows. The Fourier decomposition of the fields on the cylinder is

$$\phi(\sigma, t) = \phi_0(t) + \sum_{n \neq 0} \frac{i}{n} (a_n(t) e^{-in\sigma} + b_n(t) e^{in\sigma})$$

$$\Pi(\sigma, t) = p_0(t) + \sum_{n \neq 0} \frac{1}{4\pi} (a_n(t) e^{-in\sigma} + b_n(t) e^{in\sigma}) \tag{4.1}$$

with $a_n^\dagger = a_{-n}$, $b_n^\dagger = b_{-n}$. Since ϕ is not a free field, the time dependence of the components is complicated. Quantization amounts to imposing the equal time commutator

$$[\phi(\sigma, t), \Pi(\sigma', t)] = i\delta(\sigma - \sigma') \ . \tag{4.2}$$

With the Fourier decomposition (4.1), the commutator (4.2) turns into simple commutators of the modes: a_n, b_n for $n < 0$ are creation operators and for $n > 0$ are annihilation operators. Every classical observable is a function $f(\phi, \phi', ..., \Pi, \Pi'...)$ and can be written in terms of the modes. The quantum mechanical operator f is defined by normal ordering, i.e. all the creation operators are to the left of all the annihilation operators. As explained above, this definition of the operators is expected to remove all the divergences (at least for simple operators like those in the action).

Upon quantization, we should allow for renormalization of the parameters which depend on our regularization scheme. Therefore, we introduce a new parameter $Q = \frac{2}{\gamma} + O(1)$ which will be determined later and modify the stress tensor (2.6)

$$T_{+-} = 0$$

$$T_{++} = \frac{1}{8}(4\pi\Pi + \phi')^2 - \frac{Q}{4}(4\pi\Pi + \phi')' + \frac{\mu}{8\gamma^2} e^{\gamma\phi} + \frac{Q^2}{8} \ . \tag{4.3}$$

From the coefficient of the improvement term we learn that the quantum action is

$$S = \int d^2x \frac{1}{8\pi} \partial_\alpha \phi \partial^\alpha \phi + \frac{\mu}{8\pi\gamma^2} e^{\gamma\phi} + \frac{Q}{8\pi} \phi \sqrt{\hat{g}} R(\hat{g}) \ . \tag{4.4}$$

Since we have set $T_{+-} = 0$, we expect conformal invariance. If the quantization is consistent, at equal time T_{++} should satisfy the Virasoro algebra and commute with T_{--}. These two conditions are satisfied [4], if $Q = \frac{2}{\gamma} + \gamma$ and the value of the central charge is found to be

$$c = 1 + 3Q^2 \ . \tag{4.5}$$

The equal time commutator $[T_{++}(\sigma, t), e^{\alpha\phi(\sigma', t)}]$ shows that $e^{\alpha\phi}$ is a primary field with conformal dimension

$$\Delta(e^{\alpha\phi}) = -\frac{1}{2}\alpha^2 + \frac{1}{2}\alpha Q = -\frac{1}{2}(\alpha - \frac{Q}{2})^2 + \frac{c-1}{24} \ . \tag{4.6}$$

These results for c and Δ have been confirmed [5] using a diagrammatic expansion and agree with the semi-classical analysis valid for small γ. It is interesting that despite the interaction, the values of c and $\Delta(e^{\alpha\phi})$ are as if ϕ is a free field. However, unlike free field theory, the operators $e^{\alpha\phi}$ with $\alpha > \frac{Q}{2}$ do not exist and the correlation functions are not subject to selection rules on the sum of the exponents.

Two comments are in order here:

1. Since γ is real ($e^{\gamma\phi}$ is the metric), the central charge is bounded from below $c \geq 25$. This fact will be important in the next section. For $\gamma = \sqrt{2}$, $c = 25$ and the cosmological constant is the puncture operator.

2. In a generic conformal field theory the spectrum of conformal dimensions Δ is bounded from below. Here, on the other hand, $\Delta \leq \frac{Q^2}{8}$ and is not bounded from below.

As in the semi-classical approximation, we can constrain the area $\int e^{\gamma\phi} = A$ of the surface in evaluating correlation functions. The scaling argument used there shows that the exact scaling of the correlation functions is $A^{\frac{X}{\gamma}-1}$ for $X = \sum_i \alpha_i + \frac{Q}{2}(2h - 2)$. If $X > 0$, the integral over A is convergent and the correlation function scales like $\mu^{-\frac{X}{\gamma}}$. If $X \leq 0$ the correlation function is divergent at small area. After regularization, a non-universal analytic dependence on μ appears. The non-analytic term is proportional to $\mu^{-\frac{X}{\gamma}}$ and is multiplied by $\log \mu$ if $-\frac{X}{\gamma}$ is a non-negative integer.

There exists an alternative quantization procedure which is based on the free fields $\Phi_\pm$ (and their antiholomorphic counterparts) of section 2.4. Since these fields are free, their quantization is straightforward. All the complications of the interacting theory are in the Backlund transformation – expressing the Liouville field ϕ as a complicated non-local function of one of the free fields, say Φ_-. The authors of [4] and [6] have carried out the Backlund transformation at the quantum level and have proved its consistency. Braaten, Curtright and Thorn [4] used only Φ_-. They have constructed the quantum operators which are functions of the Liouville field ϕ as functions of Φ_-. This has enabled them to compute exactly some correlation functions[4].

Gervais and Neveu[6], (see also [7]) preferred not to solve for Φ_+ in terms of Φ_-. Keeping the two fields, the underlying $SL(2, R)$ symmetry is manifest. Expressing the results for the central charge and the conformal dimensions in terms of the natural quantities in $SL(2, R)$,

$$k + 2 = -(\tilde{k} + 2) = -\frac{2}{\gamma^2}$$

$$\alpha = -j\gamma$$

(4.7)

we find

$$c = \frac{3k}{k + 2} - 6k - 2 = -\frac{3\tilde{k}}{\tilde{k} + 2} + 6\tilde{k} + 28$$

$$\Delta(e^{-j\gamma\phi}) = \frac{j(j + 1)}{k + 2} - j = -\frac{j(j + 1)}{\tilde{k} + 2} - j \ .$$

(4.8)

The classical dimension $-j$ is renormalized by a familiar expression for the dimension in $SL(2,R)$ Kac-Moody. Furthermore, splitting holomorphically the exponentials $e^{-j\gamma\phi}$ where $2j$ is an integer (positive or negative), [6] and [7] have shown that the holomorphic components ψ_m^j satisfy an exchange algebra with braiding and fusion matrices of the quantum group $SL(2)$ with $q = \exp\frac{\pi i}{k+2}$ indicating further relation to an underlying $SL(2,R)$ Kac-Moody symmetry.

4.2. The spectrum; states vs. operators

We now turn to the important issue of the spectrum of the theory. The standard rules of quantization are easily implemented in the Liouville theory [4]. Since the subtleties of the theory are associated with the zero mode of ϕ, we simplify our discussion by considering the mini-superspace approximation, i.e. σ independent ϕ's denoted by ϕ_0. The quantum mechanics problem of ϕ_0 was studied in [13]. The Schrodinger problem for the zero mode is

$$H\psi = (\frac{1}{2}p_0^2 + \frac{\mu}{8\gamma^2}e^{\gamma\phi_0} + \frac{Q^2}{8})\psi = \Delta\psi \ . \tag{4.9}$$

As is standard in quantum mechanics, the momentum conjugate to ϕ_0, $p = -i\frac{\partial}{\partial\phi_0}$ is hermitian. There are two reasons for that. First, classically the Minkowski space momentum $\dot{\phi}_0$ is real. Second, with real p the wave functions are normalizable (more precisely delta function normalizable).

More explicitly, the wave functions are labeled by a continuous parameter p. For $\phi_0 \sim -\infty$ the first term in the potential $V = e^{\gamma\phi_0} + \frac{Q^2}{8}$ is small and the wave function $\psi_p(\phi_0)$ is a linear combination of $e^{\pm ip\phi_0}$. Because of the complete reflection off the potential, the wave function in this region satisfies $\psi_p \sim \sin(p\phi_0)$ with energy $\Delta = \frac{p^2}{2} + \frac{Q^2}{8}$ and the linearly independent states have $p > 0$. In particular, there is no $p = 0$ state.

Including the oscillators in ϕ and remembering that the theory is conformally invariant Curtright and Thorn [4] suggested that the spectrum is

$$\mathcal{H} = \oplus_p H_{\Delta(p)} \otimes H_{\bar{\Delta}=\Delta(p)} \tag{4.10}$$

where $\oplus_p$ denotes a direct integral over $p > 0$ and H_Δ is the irreducible representation of Virasoro with $\Delta = \frac{1}{2}p^2 + \frac{Q^2}{8}$. Since $\Delta - \frac{c}{24} = \frac{1}{2}p^2 - \frac{1}{24}$, this spectrum is consistent with the semi-classical evaluation of the torus partition function in section 3.1.

In a standard conformal field theory there is a one to one correspondence between states and local operators. This is not the case in Liouville field theory. Here the primary operators are $e^{\alpha\phi}$ with $\Delta = -\frac{1}{2}(\alpha - \frac{Q}{2})^2 + \frac{Q^2}{8}$ and the set of operators and the set of states are distinct.

The standard map from an operator $\mathcal{O}$ to the state $\mathcal{O}(z = 0)|0\rangle$ cannot be used here because the $SL(2,C)$ invariant state $|0\rangle$ is not in $\mathcal{H}$. Alternatively, one could have constructed the state corresponding to $\mathcal{O}$ by performing the functional integral over a disk with an insertion of $\mathcal{O}$ in the center.

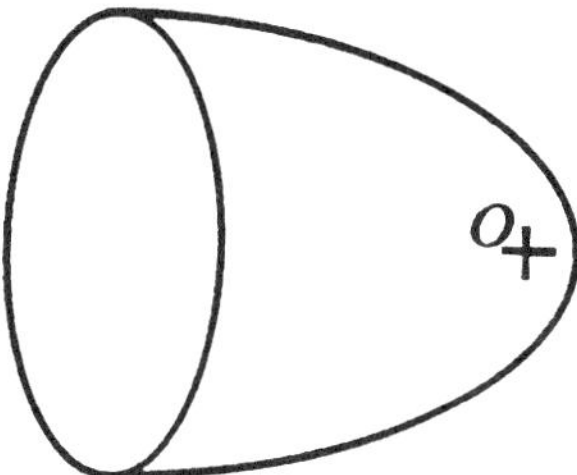

A definition of a state corresponding to $\mathcal{O}$ by a functional integral

This construction is close in spirit to the suggestion of Hartle and Hawking [14] about the wave function of the universe. In the mini-superspace approximation the wave function corresponding to $\mathcal{O} = e^{\alpha\phi}$ behaves for $\phi_0 \to -\infty$ like $\psi_{\mathcal{O}}(\phi) = e^{-(\frac{Q}{2}-\alpha)\phi_0}$. It diverges as $\phi_0 \to -\infty$ and is not normalizable. The norm of the state can be evaluated by gluing two such disks along their boundaries. As we discussed, the functional integral over the resulting sphere is divergent at small area. Therefore, the wave function defined by the disk with an insertion is not normalizable. Such non-normalizable wave functions can be regularized (as in the matrix model – see below) by cutting them off at some ϕ_0. As the regulator is removed, we can either keep the norm finite and have $\psi(\phi_0) \to 0$ at any finite ϕ_0 or keep $\psi(\phi_0)$ fixed and let the norm diverge.

Because of the similarity to the Hartle-Hawking construction (their construction corresponds to $\mathcal{O}$ being the identity and the resulting state is the $SL(2,C)$ invariant state) we will say that such states are in $\mathcal{H}_{HH}$ even though they are not in any standard Hilbert space.

The functional integral over a manifold with a boundary leads to a state. For a manifold with h handles, insertions of operators $e^{\alpha_i\phi}$ and one boundary the state is normalizable (in $\mathcal{H}$) if $X = \sum_i \alpha_i + \frac{Q}{2}(2h - 1) > 0$ and non-normalizable (with a component in $\mathcal{H}_{HH}$) if $X \le 0$.

After understanding why the standard map from operators to states cannot be used here, we turn to discuss the map from states to local operators. To insert a state in a standard conformal field theory, we cut a little hole in the surface and perform the functional integral with boundary conditions $\phi(\sigma)$ at the hole. Then we integrate over the boundary values $\phi(\sigma)$ with the weight given by the wave function of the state we want to insert $\psi[\phi(\sigma)]$. This procedure can be implemented for arbitrarily small holes, and therefore, it corresponds to a local operator. Liouville theory is different because $g_{ab} = e^{\gamma\phi}\hat{g}_{ab}$ is the physical metric on the surface. A local operator should be local with respect to g_{ab} and not only with respect to $\hat{g}_{ab}$. Therefore, the wave function $\psi(\phi)$ should be peaked on small circumference, i.e. at $\phi_0 \to -\infty$, and hence $\psi(\phi)$ is not normalizable.

The distinction between the two kinds of states (those in $\mathcal{H}$ and those in $\mathcal{H}_{HH}$) can be made clearer by examining the quantum mechanics problem (4.9). The local operators lead to eigenfunctions of the Hamiltonian which diverge as $\phi_0 \to -\infty$. Their momenta p are imaginary and therefore $\Delta < \frac{Q^2}{8}$. The normalizable states have real momenta and hence $\Delta > \frac{Q^2}{8}$. The fact that the local operators $e^{\alpha\phi}$ have $\alpha < \frac{Q}{2}$ has a simple interpretation from this point of view. The wave function associated with $e^{\alpha\phi}$ behaves like $e^{(-\frac{Q}{2}+\alpha)\phi_0}$ for $\phi_0 \to -\infty$. It grows in this region only if $\alpha < \frac{Q}{2}$. The puncture operator with $\Delta = \frac{Q^2}{8}$ has zero momentum. The two linearly independent eigenfunctions of the Hamiltonian (4.9) are constant and linear. The linear one grows at infinity and therefore leads to a local operator. Hence, the puncture operator is $\phi e^{\frac{Q}{2}\phi}$. It is easy to check that this operator is a Virasoro primary with the correct dimension. Unlike the other tensors, the puncture operator is not an exponential of ϕ. Therefore, correlation functions involving $\phi e^{\frac{Q}{2}\phi}$ are not given by powers of μ. This fact is particularly important for $\gamma = \sqrt{2}$ where the puncture operator is the cosmological constant and it appears in the action. In this case the dependence of correlation functions on μ is more complicated [15].

The semi-classical calculation of the genus one partition function describes the trace over the Hilbert space in this approximation. It seems like the trace is over the states in $\mathcal{H}$. Since the spectrum in $\mathcal{H}$ is continuous, the states have delta function normalization and therefore the integral over A diverges logarithmically (like in a free particle in a box of size $\log(\Lambda^2/\mu)$).

The semi-classical calculations of the two point function in section 3.2 are an attempt to study what might be called "inner products" of states in $\mathcal{H}_{HH}$. The ill-defined divergent expressions that we found reflect the fact that $\mathcal{H}_{HH}$ is not an ordinary Hilbert space. The small area divergence is associated with the bad behavior of these states as $\phi \to -\infty$. The definition of the divergent integral (3.17) can be interpreted as an attempt to guarantee that states with different imaginary momenta are orthogonal (with a delta function norm). The infinite norm of a state arises from the integral over the dilation group and not from the integral over A (which leads to a power divergence). Our treatment of these divergent quantities in the inner products, is very suspicious. Even if it turns out to be correct, it should certainly be made more rigorous.

It is natural to name the two kinds of states microscopic and macroscopic. The macroscopic states have normalizable wave functions with hermitian momenta. They correspond to non-local operators. The microscopic states are associated with local operators. Their wave functions are not normalizable and the momenta are antihermitian.

This description of the wave functions is consistent with the matrix model [16] [17] [18] [19] [20] [21] results. The exact scaling operators on the lattice are [19]

$$\mathcal{O}_k = Tr(1-M)^{k+\frac{1}{2}} = \sum_{n=0}^{\infty} a_n Tr M^n \ . \tag{4.11}$$

As $n \to \infty$

$$a_n \sim (-1)^{k+1} \frac{(2k+1)!}{k!\sqrt{\pi}2^{2k+1}} n^{-(k+\frac{3}{2})} \ . \tag{4.12}$$

The operator $Tr M^n$ creates a hole of size na where a is the lattice spacing. Therefore, we can interpret the insertion of the operator $\mathcal{O}_k$ as an insertion of a state $|k\rangle$ and the coefficients a_n in (4.11) as the amplitude that the hole has size na in this state. a_n can be thought of as a regularized version of the wave function. Changing variables to the Liouville mode $na = e^{\frac{\gamma}{2}\phi_0}$ and accounting for the measure in the integral, we find that the wave function for macroscopic circumference ($n \to \infty$) behaves like

$$\psi_k(\phi_0) \sim e^{-\frac{\gamma}{2}\phi_0(k+\frac{1}{2})} \ . \tag{4.13}$$

Keeping $\psi(\phi_0)$ finite and removing the cutoff, the wave function is not normalizable and grows exponentially as $\phi_0 \to -\infty$. The corresponding operator is

$$V_k \sim e^{(-\frac{\gamma}{2}(k+\frac{1}{2})+\frac{Q}{2})\phi} \ . \tag{4.14}$$

The other kind of states appears in the matrix model in the form of macroscopic loops [19][20] $\lim_{a\to 0} Tr M^{\frac{L}{a}}$ with $L = e^{\frac{\gamma}{2}\phi_0}$ held fixed. They correspond to wave functions in $\mathcal{H}$ of the form $\psi(\phi_0) = \delta(\phi_0 - \frac{2}{\gamma}\log L)$.

The distinction between microscopic states corresponding to local operators and macroscopic states follows from the fact that the metric is a dynamical variable and hence, it is an argument of the wave function. Therefore, we expect it to be more general and to be present in any theory of quantum gravity.

Since the structure of the space of states of the theory is unusual, the issue of factorization of correlation functions is subtle (see the discussion in [8]). Correlation functions of local operators correspond to insertions of microscopic states in the external lines. Which states propagate in the intermediate channels? Since the Riemann surface described by the fluctuating metric g has non-zero circumference, we expect to find the macroscopic states in internal lines. Furthermore, the $SL(2,R)$ monodromy of the classical solution can be either hyperbolic, parabolic or elliptic around various non-contractible cycles. The monodromy determines a state which propagates through the cycle (we found only one state and not a sum over states because semi-classically the sum is dominated by one term) which can be either microscopic or macroscopic. Therefore, it seems that both kinds of states can propagate in intermediate channels.

The issue of factorization in some channel is best studied by cutting the Riemann surface Σ along a closed loop. For simplicity, let as assume that the surface is separated into two disconnected surfaces with boundaries (the conclusion is easily generalized to the other case). The functional integral over the two manifolds with boundaries Σ_1 and Σ_2 defines two states $|1\rangle$ and $|2\rangle$. The original functional integral can be described as an inner product of the two states $\langle 1|2\rangle$ by gluing back the two boundaries. The question of factorization is to find the energy eigenstates which overlap both with $|1\rangle$ and with $|2\rangle$.

Let Σ_1 have insertions α_i and h_1 handles and Σ_2 have insertions β_i and h_2 handles. Consider first the case where both $X_1 = \sum_i \alpha_i + \frac{Q}{2}(2h_1 - 1)$ and $X_2 = \sum_i \beta_i + \frac{Q}{2}(2h_2 - 1)$ are positive. Then, both $|1\rangle$ and $|2\rangle$ are normalizable and can be expanded in momentum eigenstates $|p\rangle$ with real momenta. Hence, all the intermediate states are macroscopic. This fact has important consequences. Consider the original surface Σ obtained by gluing back Σ_1 and Σ_2 along a closed loop. Let q be a Fenchel-Nielsen modular parameter of Σ associated with the length and twist of the region of that loop. As $q \to 0$, the functional integral is dominated by the propagation of the lowest energy states in this channel. These are momentum eigenstates $|p\rangle$ with real p near zero and hence with $\Delta > \frac{Q^2}{8}$. Since this dimension is positive (and typically large), the behavior of the functional integral in this limit is less singular than in an ordinary conformal field theory.

For example, consider the four point function on the sphere

$$\langle e^{\alpha_1 \phi(0)} e^{\alpha_2 \phi(z)} e^{\alpha_3 \phi(1)} e^{\alpha_4 \phi(\infty)}\rangle \tag{4.15}$$

with $X_1 = \alpha_1 + \alpha_2 - \frac{Q}{2} > 0$ and $X_2 = \alpha_3 + \alpha_4 - \frac{Q}{2} > 0$ and factorize in the limit $z \to 0$. The correlation function behaves there like $(z\bar{z})^{\frac{Q^2}{8} - \Delta_{\alpha_1} - \Delta_{\alpha_2}} = (z\bar{z})^{\frac{1}{2}(\alpha_1 - \frac{Q}{2})^2 + \frac{1}{2}(\alpha_2 - \frac{Q}{2})^2 - \frac{Q^2}{8}}$ (perhaps times a logarithmic factor from the integral over p) which is less singular than in free field theory.

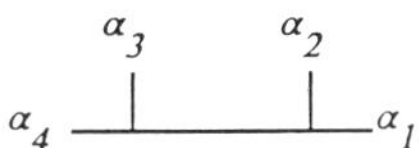

Factorization of a four point function

Another simple example is any high genus surface without insertions. As we pinch any non-contractible cycle with a modular parameter $q \to 0$, the functional integral behaves either as $(q\bar{q})^\Delta$ or as $(q\bar{q})^{\Delta - \frac{c}{24}}$. With $\Delta > \frac{c-1}{24}$ both are less singular than in an ordinary conformal field theory with that value of c. The semi-classical calculation of the torus amplitude in section 3.2 is an example of this phenomenon. This softening of the singularities near the boundary of moduli space has also been noticed and discussed in [8].

If $X = X_1 + X_2 > 0$ but $X_1 \leq 0$, $|2\rangle$ is normalizable and can be expanded in macroscopic states but $|1\rangle$ is not normalizable. Continuing formally, one might expect that again, the functional integral can be computed by summing only over the macroscopic states. In order to establish this result, we need a better definition and understanding of the inner products of non-normalizable states.

Clearly, the issue of factorization is not yet completely understood. Related to that is the need to compute correlation functions in the full quantum theory. Without explicit expressions for the correlation functions, it will be difficult to interpret them and to understand the structure of the Hilbert space. We view this problem as the most important open problem in Liouville theory.

4.3. A generalization of Liouville field theory

If a two dimensional field theory which is not conformally invariant is coupled to quantum gravity, a generalization of the Liouville action has to be considered. Let L_0 be the Lagrangian of a conformal field theory and $\mathcal{O}_i$ are scalar operators of dimensions Δ_i. The massive theory $L = L_0 + \sum_i m_i \mathcal{O}_i$ becomes when coupled to gravity

$$\int L_0 + \frac{1}{8\pi}\partial_\alpha\phi\partial^\alpha\phi + \frac{\mu}{8\pi\gamma^2}e^{\gamma\phi} + \sum_i m_i e^{\alpha_i\phi}\mathcal{O}_i + \frac{Q}{8\pi}\phi\sqrt{\hat{g}}R(\hat{g}) \ . \tag{4.16}$$

From this, more general, point of view the coupling of the cosmological constant is a coupling to a particular operator – the identity – in the conformal field theory and is in no sense different than the other couplings. As in the previous case $(m_i = 0)$, here one of the coefficients μ and m_i can be rescaled and we cannot expand in it. The other coefficients can be assumed small and used as expansion parameters. Because of the underlying general covariance, (4.16) should be a conformal field theory. Fixing μ and expanding in m_i we find in the leading order that the exponents α_i should satisfy

$$-\frac{1}{2}\alpha_i^2 + \frac{1}{2}\alpha_i Q + \Delta_i = 1 \ . \tag{4.17}$$

Higher orders in m_i are more complicated. If, however, we set μ to zero, at least one of the m_i's, should be non-zero. In this case we should study the action

$$\int L_0 + \frac{1}{8\pi}\partial_\alpha\phi\partial^\alpha\phi + m e^{\alpha\phi}\mathcal{O} + \frac{Q}{8\pi}\phi\sqrt{\hat{g}}R(\hat{g}) \tag{4.18}$$

(we have dropped the subscript i). It is easy to see after using the ϕ equations of motion that

$$T_{+-} = 0$$

$$T_{++} = \frac{1}{2}(\partial_+\phi)^2 - \frac{Q}{2}\partial_+^2\phi + \frac{Q^2}{8} + T_{++}^{(0)} \tag{4.19}$$

$$= \frac{1}{8}(4\pi\Pi + \phi')^2 - \frac{Q}{4}(4\pi\Pi + \phi')' + m\pi e^{\alpha\phi}\mathcal{O} + \frac{Q^2}{8} + T_{++}^{(0)}$$

where $T_{++}^{(0)}$ is the stress tensor of the conformal field theory L_0. Repeating the analysis of [4], we find that if $\mathcal{O}$ is a primary field of $T^{(0)}$ with dimension Δ and the equal time commutator of $\mathcal{O}$ with itself vanishes, T_{++} satisfies a Virasoro algebra provided $-\frac{1}{2}\alpha^2 + \frac{1}{2}\alpha Q + \Delta = 1$. The central charge of the algebra is $c = 1 + 3Q^2 + c^{(0)}$ ($c^{(0)}$ is the central charge of $T^{(0)}$). A scaling argument shows that the correlation functions scale like $m^{-\frac{X}{\alpha}}$ with $X = \sum \alpha_i + \frac{Q}{2}(2h - 2)$.

Two simple examples of such conformal field theories are a massive Majorana fermion

$$\int \bar{\psi}\partial\bar{\psi} + \psi\bar{\partial}\psi + \frac{1}{8\pi}\partial_\alpha\phi\partial^\alpha\phi + me^{\alpha\phi}\bar{\psi}\psi + \frac{Q}{8\pi}\phi\sqrt{\hat{g}}R(\hat{g}) \tag{4.20}$$

with $c = \frac{3}{2} + 3Q^2$ and a Sine-Gordon model

$$\int \frac{1}{8\pi}\partial_\alpha X\partial^\alpha X + \frac{1}{8\pi}\partial_\alpha\phi\partial^\alpha\phi + me^{\alpha\phi}\cos px + \frac{Q}{8\pi}\phi\sqrt{\hat{g}}R(\hat{g}) \tag{4.21}$$

with $c = 2 + 3Q^2$. We should note that we have assumed in this analysis that the models exist and did not prove this fact. Such an assumption might not be satisfied in cases where the Euclidean space Lagrangian is not bounded from below (as in (4.21)).

By setting $\mu = 0$, we are no longer limited by $Q^2 \geq 8$ as in ordinary Liouville theory. Here, the condition that $me^{\alpha\phi}\mathcal{O}$ be local, constrains $Q^2 \geq 8(1 - \Delta)$. If $8 > Q^2 \geq 8(1 - \Delta)$, there is no local operator of dimension 1 which is made purely of the Liouville field. The state with these properties is a macroscopic state.

5. Quantum Gravity

5.1. The general situation

We now return to our original problem of quantum gravity coupled to matter. We consider a conformal field theory with central charge c coupled to gravity. Repeating the steps in critical string theory, David, Distler and Kawai [22] have set the total central charge of the system which is the sum of the matter, Liouville and the ghosts central charges to zero

$$c^{total} = c + 1 + 3Q^2 - 26 = 0 \tag{5.1}$$

and hence

$$Q = \sqrt{\frac{25 - c}{3}} \ . \tag{5.2}$$

The value of γ is then fixed to

$$\gamma = \frac{\sqrt{25 - c} - \sqrt{1 - c}}{\sqrt{12}} \ . \tag{5.3}$$

We choose the branch of the solution corresponding to $\gamma \leq \frac{Q}{2}$ because otherwise the cosmological constant operator $e^{\gamma\phi}$ does not exist. The physical states of the theory are products of a matter primary state $\mathcal{O}_i$ times a Liouville primary state such that the total dimension is one. Notice that $\mathcal{O}_i$ has to be scalar, i.e. $\Delta_i = \bar{\Delta}_i$. If the

Liouville state is microscopic, it corresponds to the local operator $e^{\alpha_i \phi}$. The value of α_i is determined

$$\alpha_i = \frac{\sqrt{25-c} - \sqrt{24\Delta_i + 1 - c}}{\sqrt{12}} = \frac{Q}{2} - \sqrt{2(\Delta_i + \frac{1-c}{24})} \qquad (5.4)$$

where again, the branch is picked by $\alpha_i \leq \frac{Q}{2}$.

The momentum of the Liouville state is $p = i\sqrt{2(\Delta_i + \frac{1-c}{24})}$. Depending on whether it is imaginary or real the state is microscopic or macroscopic. We should distinguish between three situations

1. $\Delta_i + \frac{1-c}{24} > 0$. The Liouville momentum is imaginary and the state is microscopic. The corresponding operator $\mathcal{O}_i e^{\alpha_i \phi}$ is local. We refer to such operators as massive. The term puncture operator in topological gravity [10] refers to an operator of this kind.

2. $\Delta_i + \frac{1-c}{24} = 0$. The Liouville momentum vanishes and the corresponding operator is the puncture operator. We refer to such operators as massless.

3. $\Delta_i + \frac{1-c}{24} < 0$. The Liouville momentum is real and the state is macroscopic. Its wave function is normalizable and it is in $\mathcal{H}$. Even though the operator $\mathcal{O}_i$ is local, the coupling to gravity makes it non-local. An insertion of such an operator cuts a macroscopic hole in the surface. We refer to such operators as tachyonic.

The origin of the terminology massive, massless and tachyon comes from critical string theory. Interpreting the Liouville field as Euclidean time, its on shell momentum $p = i\sqrt{2(\Delta_i + \frac{1-c}{24})}$ is real for tachyons, imaginary for massive states and vanishes for massless.

5.2. Comments on correlation functions

Correlation functions in quantum gravity should be computable as in critical string theory by multiplying the correlation functions of the matter theory and the Liouville sector and integrating them over moduli space. Unlike the matrix model techniques, the continuum Liouville approach is not yet powerful enough to find the correlation functions. As stressed above, the main problem is to compute the Liouville part of the answer.

Let us first assume that the integral of the correlation function over moduli space converges. In this case, the answer should involve the power of the cosmological constant we found in Liouville theory. This scaling with respect to the cosmological constant is known as KPZ scaling [23] and this derivation of it is due to [22]. An insertion of an operator $\mathcal{O}_i e^{\alpha_i \phi}$ multiplies the correlation function by $\mu^{-\frac{\alpha_i}{\gamma}}$. The obvious interpretation of this factor is that the flat space dimension of the operator, Δ_i, is renormalized by the gravitational corrections to

$$\Delta_i^{\text{ren}} = 1 - \frac{\alpha_i}{\gamma} = \frac{\sqrt{24\Delta_i + 1 - c} - \sqrt{1-c}}{\sqrt{25-c} - \sqrt{1-c}} \ . \qquad (5.5)$$

The analysis of Knizhnik, Polyakov and Zamolodchikov used the light cone gauge [24]. It is based on an $SL(2,R)$ Kac-Moody algebra which exists in this gauge. Its level k is as in equation (4.7). The renormalized dimension Δ^{ren} is given in terms of the $SL(2,R)$ weight of the field $\Delta^{\text{ren}} = 1 + j$ which agrees with (4.8)(5.5).

When a non-unitary theory is coupled to gravity, the dimension of the lowest dimension operator in the spectrum is typically negative. A generic deformation of the conformal field theory couples to this operator with coefficient m. In this case Liouville field theory should be replaced by a theory of the kind discussed in section 4.3. Simple scaling occurs when m is the only non-zero coupling and in particular, $\mu = 0$ [25]. Alternatively, the coupling to the negative dimension operator, m, can be set to zero and be replaced by the coupling to the identity. This can be achieved only by fine tuning. If this fine tuning is done, KPZ scaling [23] is recovered.

If the integral over moduli space of the correlation function diverges, there is a source of scaling violation. This divergence has to be regularized and this may introduce new dependence on μ. In critical string theory the divergences are associated with on-shell intermediate states. If we assume that the Liouville intermediate states are only the macroscopic states, such divergences can occur only when tachyon or massless states propagate (a massless state is not on shell because the $p = 0$ state is not macroscopic but the contribution of states with $p \approx 0$ can lead to a divergence). As an example, consider the functional integral over the torus of quantum gravity coupled to a conformal field theory with c large and negative. In this case we can use the semi-classical answer (3.14) for the Liouville part of the integrand. With the proper normalization of the torus integral [26] we find an integral over a fundamental domain of the modular group

$$\int \frac{d^2\tau}{4\pi\tau_2^2}(\sqrt{2\pi\tau_2}|\eta|^2)^2 \left(\frac{\log \frac{\Lambda^2}{\mu}}{2\pi\gamma\sqrt{2\tau_2}|\eta|^2} \right) Z_{\text{CFT}} \qquad (5.6)$$

where the first factor is from the ghosts, the second is from Liouville and Z_{CFT} is the conformal field theory partition function. Let Δ be the dimension of the lowest dimension spin zero field in the conformal field theory. In the region near $\tau = i\infty$, the integrand in (5.6) behaves like $\tau_2^{-\frac{3}{2}}e^{-4\pi\tau_2(\Delta - \frac{c-1}{24})}$ and (5.6) diverges, if there are tachyons. This fact together with the assumption that only macroscopic states in Liouville can propagate in intermediate lines is consistent with the interpretation of Liouville as Euclidean time. The amplitudes are Euclidean space correlation functions. In the Euclidean regime massive states cannot be on shell – only tachyons and massless fields can lead to divergences.

We have seen that the zero point function of Liouville field theory on the sphere with fixed area is divergent and proportional to the volume of $SL(2,C)$. Similarly, the two point function is proportional to the volume of the dilation group. When we couple Liouville to matter and consider quantum gravity, these two divergences disappear. As in critical string theory we have to divide by the volume of the conformal

Killing vectors which precisely cancels the divergence from Liouville. More rigorously, in the process of fixing the diffeomorphism invariance of the functional integral, we fix the collective coordinates of the Liouville classical solution by Mobius invariance and we never have to integrate over them[11]. A similar argument applies to the one point function of the cosmological constant. Hence, in quantum gravity, both the zero [11], the one and the two point functions do not vanish. This is unlike the situation in critical string theory where these amplitudes vanish because the matter functional integral does not exhibit these geometrical divergences.

Since the amplitudes in quantum gravity are obtained as integrated correlation functions, we should be careful not to miss important contributions from the boundary of the integration region. Two kinds of boundaries are important:

1. In correlation functions on the sphere, we need to integrate over the locations of the operators. There might be delta function contributions from the region where two operators are one on top of the other. Such delta function correlations are common in field theory and are known as contact terms. They typically arise when one uses the equation of motion to simplify the operators. This is allowed at generic points but not at the boundary. Some operators vanish when the equation of motion is used. They are known as redundant operators. Their correlation functions vanish at separated points but are not zero at the boundary. A simple example of such an operator is $\partial\bar{\partial}X$ where X is a free field. Such operators do not correspond to physical states. This is easily seen by examining the mapping between states and operators. However, they do have non-trivial integrated correlation functions. This is the reason that matrix models and therefore also quantum gravity have an infinite number of operators while the continuum approach predicts only a finite number of states – all the other operators are redundant. Furthermore, such contact terms might explain why correlation functions in the matrix model do not satisfy the conformal field theory fusion rules and exhibit strange positivity properties [27]. Similar delta function contributions can appear in other boundaries of moduli space, for example, when a handle degenerates. It is difficult to find these contact terms unambiguously in the continuum Liouville approach. They cannot be determined without imposing some physical requirements on the correlation functions. These contact terms arise naturally in the context of the topological interpretation of these theories where they have been calculated explicitly [10][28].

2. Another boundary of field space is zero area. As remarked above, contributions from this boundary appear as analytic dependence on the cosmological constant μ. As in the previous case, it is not easy to find such a contribution in the continuum Liouville approach. It is not clear whether it is universal and if so, what physical principle determines it.

5.3. Minimal models

The minimal models of BPZ [9] are special conformal field theories because they have only a finite number of Virasoro primaries. Therefore, once they are coupled to gravity, there are only a finite number of physical operators. One expects that a theory with a finite number of operators would be simple and solvable.

The theories are labeled by two relatively prime integers p and p' where one can assume without loss of generality that $p > p'$. The central charge is

$$c_{pp'} = 1 - 6\frac{(p - p')^2}{pp'} \ . \tag{5.7}$$

The operators are labeled by $1 \leq n \leq p - 1$ and $1 \leq n' \leq p' - 1$ and the (n, n') operator is identified with $(p - n, p' - n')$. It is convenient to pick a fundamental domain of the identification by $n'p - np' > 0$. The dimensions of the operators are

$$\Delta_{nn'} = \frac{(n'p - np')^2 - (p - p')^2}{4pp'} \ . \tag{5.8}$$

The coupling of these theories to gravity was first studied by KPZ [23]. Here, we use the conformal gauge and follow [22]. First, we notice that

$$\Delta_{nn'} - \frac{c - 1}{24} = \frac{(n'p - np')^2}{4pp'} > 0 \tag{5.9}$$

and therefore all the operators are massive. Using (5.2)(5.3)(5.4) we find the values of Q and the dressing $e^{\alpha_{nn'}\phi}$ of the (n, n') operator

$$\begin{aligned}
\gamma &= \alpha_{11} = \sqrt{\frac{2p'}{p}} \\
Q &= \frac{2(p + p')}{\sqrt{2pp'}} = \gamma(1 + \frac{p}{p'}) \\
\alpha_{nn'} &= \frac{p + p' - n'p + np'}{\sqrt{2pp'}} = \frac{Q}{2} - \gamma\frac{n'p - np'}{2p'} = \gamma(\frac{1}{2} + \frac{p}{2p'} - \frac{n'p - np'}{2p'}) \ .
\end{aligned} \tag{5.10}$$

Since we do not expect divergences in the integration over the moduli, the scaling in μ of the correlation functions should be simple. These expectations have been confirmed and the numerical coefficients have been found in the matrix models [16][17][18][19][21] and in the topological field theory approach [10][28].

A simple calculable example in the continuum is the following[4]. Consider the series of models $(p = 2l - 1, p' = 2)$ corresponding[25] to the multi-critical points of the one matrix model. As $l \to \infty$, $c \to -\infty$ and the semi-classical analysis of Liouville is applicable. The conformal field theory partition function Z_{CFT} becomes in this limit

$$Z_{\mathrm{CFT}}(l) \approx \frac{\sqrt{l}}{2\sqrt{\tau_2}|\eta(q)|^2} \tag{5.11}$$

---

[4] I understand that V. Fateev and A.M. Polyakov have performed a similar calculation.

Using the approximate value of $\gamma \approx \sqrt{\frac{2}{l}}$ and (5.6) the functional integral over the torus in this limit is

$$\frac{l}{16\pi}\log\frac{\Lambda^2}{\mu}\int\frac{d^2\tau}{\tau_2^2} = \frac{l}{48}\log\frac{\Lambda^2}{\mu} \ . \tag{5.12}$$

If instead of μ we set the scale by the coupling m to the lowest dimension operator in these theories [25], we should replace μ by $m^{2/l}$ and find

$$\frac{l}{48}\log\Lambda^2 - \frac{1}{24}\log m \tag{5.13}$$

The coefficient of $\log m$ is universal and agrees with the matrix model prediction in this limit[5].

Since all the operators in the minimal models are massive, the Liouville states are microscopic. Hence, the wave functions of all the physical states are not normalizable. Had we limited ourselves to normalizable wave functions in quantum gravity, we would not have found any physical state.

It is interesting to find the Liouville wave functions of the physical states in the mini-superspace approximation. It is convenient to replace the labels (n, n') by $k = [\frac{pn'}{p'}] - n$ $(k = 0, ..., [\frac{pn'}{p'}] - 1)$ and $r = n'p - p'[\frac{n'p}{p'}]$ $(r = 1, ..., p' - 1)$ where $[\frac{pn'}{p'}]$ denotes the integer part. Then the (r, k) matter state is dressed by the Liouville wave function

$$\psi_{rk}(\phi_0) = e^{-\frac{1}{2}\gamma(k+\frac{r}{p'})\phi_0} = L^{-(k+\frac{r}{p'})} \tag{5.14}$$

which is independent of p. Notice that for $p' = 2$, (5.14) agrees with the expression derived from the matrix model (4.13). Since these wave functions are independent of p, they must also be the correct wave functions for every massive model interpolating between conformal field theories with different p's but the same value of p'. Therefore, these are also the wave functions of the redundant operators in the p theory which are physical for a larger value of p, i.e. for $k \geq [\frac{pn'}{p'}]$. It is amusing to note that the form of the wave functions $L^{-(k+\frac{r}{p'})}$ is similar to the form of the operators as suggested by Douglas [21].

In the topological field theory interpretation [10] the states with $k = 0$ are called primary states. They are labeled by r. States with larger values of k are called descendants. The $SL(2, R)$ spin of these states is

$$j - -\frac{1}{2} - \frac{p}{2p'} + \frac{r}{2p'} + \frac{k}{2} \tag{5.15}$$

It is intriguing that the descendants differ from each other by spin $\frac{1}{2}$ which is the fundamental field in the underlying $SL(2, R)$.

---

[5] As pointed out in [29] the answer in a matrix model based on an even potential is twice that of a matrix model based on a general potential. The reason for this is that with an even potential the eigenvalue distribution has two critical end points and with a general potential only one such end point. This leaves open the question of which of the two answers corresponds to the conformal field theory coupled to gravity. Our semi-classical result agrees with that of a general potential and is half the prediction of the even potential models.

Most conformal field theories are not minimal. Therefore, it is important to study non-minimal theories in order to learn which of the results found for the minimal models are generic. We start our discussion by mentioning some relevant results in conformal field theory (without gravity).

For values of c not in the BPZ list there is at most one null vector in the Verma module and its Verma module has no null vectors[6]. Therefore, all the characters are either

$$\chi_\Delta(\tau) = \frac{q^{\Delta - \frac{c-1}{24}}}{\eta(q)} \ , \tag{5.16}$$

if there is no null vector, or

$$\chi_\Delta = \frac{q^{\Delta - \frac{c-1}{24}}}{\eta(q)}(1 - q^N) \ , \tag{5.17}$$

if there is a null vector at the N'th grading.

We now generalize an argument due to Cardy [30]. The full partition function of the conformal field theory is

$$Z(\tau) = \sum_{\Delta \bar{\Delta}} N_{\Delta \bar{\Delta}} \chi_\Delta(\tau) \bar{\chi}_{\bar{\Delta}}(\bar{\tau}) \tag{5.18}$$

with non-negative integer coefficients $N_{\Delta \bar{\Delta}}$. We study the limit $\tau \to i\infty$ with $q = e^{2\pi i \tau}$ in two different ways. If the spectrum has a gap,

$$\lim_{\tau \to i\infty} Z = \lim M(q\bar{q})^{d - \frac{c}{24}} \tag{5.19}$$

where $d = \min(\Delta + \bar{\Delta})/2$ and M is the number of states satisfying the minimum (if there is no gap, (5.18) has an integral over Δ and the right hand side of (5.19) is multiplied by a power of τ_2). Because of (5.16)(5.17), for q real

$$\frac{q^{\Delta - \frac{c-1}{24}}}{\eta} \ge \chi_\Delta \ge \frac{q^{\Delta - \frac{c-1}{24}}}{\eta}(1 - q) \tag{5.20}$$

for every Δ. Using modular invariance, $Z(\tau) = Z(-\frac{1}{\tau})$

$$\lim_{\tau \to i\infty} Z(\tau) = \lim Z(-\frac{1}{\tau}) \ge \lim \frac{|1 - q'|^2}{|\eta(q')|^2} \sum_{\Delta \bar{\Delta}} N_{\Delta \bar{\Delta}} q'^{\Delta - \frac{c-1}{24}} \bar{q}'^{\bar{\Delta} - \frac{c-1}{24}} = \lim \frac{4\pi^2}{\tau_2^3} N(\tau)(q\bar{q})^{-\frac{1}{24}}$$

$$\lim_{\tau \to i\infty} Z(\tau) = \lim Z(-\frac{1}{\tau}) \le \lim \frac{1}{|\eta(q')|^2} \sum_{\Delta \bar{\Delta}} N_{\Delta \bar{\Delta}} q'^{\Delta - \frac{c-1}{24}} \bar{q}'^{\bar{\Delta} - \frac{c-1}{24}} = \lim \frac{1}{\tau_2} N(\tau)(q\bar{q})^{-\frac{1}{24}}$$

$$\tag{5.21}$$

---

[6] We thank D. Friedan for a useful discussion on this point.

where $q' = e^{(-\frac{2\pi i}{\tau})}$ and $N(\tau) = \sum_{\Delta\bar{\Delta}} N_{\Delta\bar{\Delta}} q'^{\Delta-\frac{c-1}{24}} \bar{q}'^{\bar{\Delta}-\frac{c-1}{24}}$. Clearly, $N = \lim_{\tau\to i\infty} N(\tau)$ is the number of Virasoro primaries in the theory. We learn that

$$\lim \frac{1}{\tau_2} N(\tau) \geq \lim_{\tau\to i\infty} M(q\bar{q})^{d-\frac{c-1}{24}} \geq \lim \frac{4\pi^2}{\tau_2^3} N(\tau) \ . \tag{5.22}$$

If we assume that N is finite, $d - \frac{c-1}{24} > 0$ and then $N = 0$. Hence, N must be infinite. We conclude that if c is not in the BPZ list, there must be an infinite number of Virasoro primaries (generalizing [30] to non-unitary theories). In this case we learn from (5.22) that $N(\tau) \sim \tau_2^a e^{-4\pi\tau_2(d-\frac{c-1}{24})}$ (up to a possible power of $\log \tau_2$) with $1 \leq a \leq 3$. Hence, $N = \infty$, implies $d - \frac{c-1}{24} \leq 0$ and there are operators with

$$\Delta - \frac{c-1}{24} \leq 0 \tag{5.23}$$

in the spectrum.

We now return to the study of quantum gravity. The physical states are spinless ($\Delta = \bar{\Delta}$) Virasoro primaries dressed with a Liouville wave function. We have shown that the lowest dimension operators in the spectrum have $\Delta - \frac{c-1}{24} \leq 0$. However, they are not necessarily spinless. A stronger result would be that the lowest scalar operator has $\Delta - \frac{c-1}{24} \leq 0$. We did not succeed to prove it. Nevertheless, we believe that such a result is reasonable because of the following heuristic argument.

In the critical string, divergences in amplitudes are associated with on-shell intermediate states. By duality and modular invariance every amplitude can also be analyzed from the "cross-channel" point of view. There, the divergences do not correspond to on-shell states but to a sum over an infinite number of off-shell states. Our result about Δ in non-minimal models is in this spirit. We examined the behavior of the amplitude $Z(\tau)$ near the boundary of moduli space and found that the divergence resulting from $\Delta - \frac{c-1}{24} \leq 0$ can be understood as a consequence of $N = \infty$ from the cross channel point of view.

A theory with massless and tachyon states suffers from divergences in amplitudes arising from the integral over the moduli. The divergences are correlated with the existence of an infinite number of physical states in the theory. Since the generic conformal field theory has an infinite number of Virasoro primaries, we expect it to have tachyons.

5.5. Theories with tachyons

The simplest examples of theories with tachyons are conformal field theories with $c > 1$ where the identity is a tachyon. This is the origin of the so-called $c = 1$ barrier. However, as we have just seen, the problem has nothing to do with the value of c – this is not a problem of strongly coupled gravity but a problem in the spectrum of the matter theory. Therefore, by considering a non-minimal conformal field theory with c large and negative, we can study the problem in the semi-classical approximation.

As we have seen in the previous sections, a tachyon operator is dressed by a macroscopic state in Liouville. An insertion of such an operator is not a local disturbance to the surface; it creates a macroscopic hole and tears the surface apart. Therefore, it is not easy to study the correlation functions of tachyon operators. The problem is even more serious if we try to deform the theory by adding a tachyon operator to the action. Studying such a deformation in perturbation theory, we find that the surface of the perturbed theory has holes in it. Every order in perturbation theory adds a hole to the surface. Summing up such a perturbation expansion, the surface is ruined completely. The perturbed theory is not a theory of continuous surfaces – all but a microscopic (of the order of the cutoff) fraction of the world-sheet is holes.

A continuum two-dimensional field theory cannot be obtained if a tachyon operator is in the action. Therefore, we should set its coefficient to zero. In the flat space theory (without gravity), the tachyon operators are the most relevant operators in the spectrum and hence a generic deformation of the conformal field theory couples to them. The coupling to gravity does not ruin the surface only if these operators are fine tuned to zero. When c is larger than 1, the cosmological constant is tachyonic and must be fine tuned to zero. We have stressed above that one cannot simply set $\mu = 0$ because the value of μ sets the scale. However, we can have $\mu = 0$, if the scale is set by the coupling to another operator as in the theories in section 4.3. The world-sheet theory is then a theory with massive matter and vanishing cosmological constant[7].

Even if all tachyon operators are fine tuned to zero, it is not clear that the resulting theory makes sense. In particular, we should check that no pathologies exist when a tachyon appears as an intermediate state in some channel. Let z be the Fenchel-Nielsen coordinate associated with the channel and study the $z \to 0$ region. Motivated by the semi-classical expression, and the experience in critical string theory, we expect the integral for the amplitude to behave near $z = 0$ like

$$\int d^2 z \int dp\, f(p)(z\bar{z})^{\Delta + \frac{1}{2}p^2 - \frac{c-1}{24} - 1} \tag{5.24}$$

where Δ is the dimension of the matter operator in the channel and $f(p)$ is some weight function depending on the operator product coefficients. As in critical string theory, we interchange the order of the integrals over z and over p and perform the integral over z by analytic continuation in p from a region where the integral converges to find that (5.24) is proportional to

$$\int dp\, f(p) \frac{1}{p^2 + 2(\Delta - \frac{c-1}{24})} \; . \tag{5.25}$$

---

[7] It is tempting to speculate that perhaps a similar mechanism can be relevant to the problem of the cosmological constant in four dimensions.

Interpreting Liouville as Euclidean time, this expression describes an integral of the propagator over the energy, p. Because of the Liouville interaction, there is no time translation symmetry, energy is not conserved and it has to be integrated over. If only macroscopic states contribute (p is real), and if the matter operator in the channel is massive, the integral over p converges and the contribution to the amplitude is finite. If, however, the operator is a tachyon (5.25) diverges. Shifting the integration contour to the complex p plane we find a finite answer with an imaginary part. Presumably, this imaginary part is an indication of the instability of the theory. Although it seems that one can perhaps make some sense of the correlation functions on the sphere, we see no way to justify a similar procedure at high genus. We conclude that the situation with these tachyonic theories is at best similar to the critical bosonic string with its tachyon instability.

5.6. Interesting theories without tachyons

It seems from the previous discussion that a theory with an infinite number of states necessarily has tachyons. We know one way around it [31]. Examining carefully the assumptions leading to the result in section 5.4, we see that we assumed that all the states of the two dimensional field theory contribute $+1$ to the partition function Z. This is clearly the case in any sensible (not necessarily unitary) field theory. By relaxing this assumption, we can evade the tachyon.

There are known examples of theories where this assumption is not satisfied. A typical one is a theory violating the spin-statistics relation. A more concrete example is a theory of a complex anticommuting scalar θ with the Lagrangian

$$L = \partial\theta\bar{\partial}\bar{\theta} \ . \tag{5.26}$$

The functional integral on the torus does not describe a trace over the states. The theory has a fermion number operator $(-1)^F$ classifying the states into bosons (F even) and fermions (F odd). The functional integral over the torus corresponds to the supertrace over the states, i.e. bosons contribute $+1$ and fermions contribute -1. This fermion number operator is not the two-dimensional fermion number. Using the terminology of string theory we can call it space-time fermion number. The fact that some states contribute -1 to the torus amplitude is interpreted as space-time statistics. We can then have a theory with an infinite number of space-time bosons, an infinite number of space-time fermions and no tachyon operators.

An interesting example of this kind is the theory of Marinari and Parisi [32]. Here the world-sheet theory is presumably

$$L = |\partial X + \bar{\theta}\partial\theta + \theta\partial\bar{\theta}|^2 \tag{5.27}$$

with one boson X and a complex spin zero fermion θ. It seems from the results of [32] that the theory defined by (5.27) has $c = 0$ and the lowest dimension operator is the identity which is not tachyonic.

If the Liouville field is interpreted as Euclidean time, this theory does not have space-time supersymmetry but only "space supersymmetry" – two supercharges do not anticommute to time translation. Space-time supersymmetry is explicitly broken when (5.27) is coupled to gravity. This breaking can presumably be thought of as spontaneous breaking in space-time.

Although the matrix model of the Marinari-Parisi theory is solvable, the two-dimensional theory (5.27) (the Green-Schwarz action), in the absence of gravity is not solvable. Another interesting example of a theory with an infinite number of states without tachyons is the following [31]. It is based on the free Lagrangian

$$L = \partial X^i \bar{\partial} X^i + \partial \theta^j \bar{\partial} \bar{\theta}^j \tag{5.28}$$

with $i = 1, ..., d$, $j = 1, ..., m$ which has Parisi-Sourlas supersymmetry. The central charge is $c = d - 2m$ and the lowest dimension operator is the identity which is not tachyonic. This theory is trivially solvable as a two-dimensional theory without gravity. Here the world-sheet is stable; in fact, by making d and m large we can have c large and negative and the gravitational part of the theory can be studied semi-classically. Unfortunately, unlike the Marinari-Parisi theory, the space-time theory does not satisfy the spin-statistics relation.

Acknowledgements

We would like to thank M. Douglas, J.-L. Gervais, D. Kutasov, E. Martinec, H. Neuberger, J. Polchinski and E. Witten, and especially, T. Banks, D. Friedan, G. Moore, S. Shenker, and A.B. Zamolodchikov, for valuable discussions. This research was supported in part by grants from the Department of Energy.

References

[1] A.M. Polyakov and A.B. Zamolodchikov, unpublished.
[2] A.M. Polyakov, Phys. Lett. **103B** (1981) 207, 211.
[3] H. Poincare, J. Math. Pure App. **5** se. 4 (1898) 157.
[4] T.L. Curtright and C.B. Thorn, Phys. Rev. Lett. **48** (1982) 1309; E. Braaten, T. Curtright and C. Thorn, Phys. Lett. **118B** (1982) 115; Ann. Phys. **147** (1983) 365; E. Braaten, T. Curtright, G. Ghandour and C. Thorn, Phys. Rev. Lett. **51** (1983) 19; Ann. Phys. **153** (1984) 147.
[5] D. Friedan, unpublished.
[6] J.-L. Gervais and A. Neveu, Nucl. Phys. **199** (1982) 59; **B209** (1982) 125; **B224** (1983) 329; **238** (1984) 125; 396; Phys. Lett. **151B** (1985) 271; J.-L. Gervais, LPTENS 89/14; 90/4.
[7] F. Smirnoff and L. Taktajan, Univ. of Colorado preprint (1990).
[8] J. Polchinski, Nucl. Phys. **B324** (1989) 123; UTTG-19-90.
[9] A. Belavin, A. Polyakov and A.B. Zamolodchikov, Nucl. Phys. **B241** (1984) 333.
[10] E. Witten, IASSNS-HEP-89/66; 90/37; R. Dijkgraaf and E. Witten, PUPT-1166.

[11] A.B. Zamolodchikov, Phys. Lett. **117B** (1982) 87.

[12] J. Liu and J. Polchinski, Phys. Lett. **203B** (1988) 39.

[13] E. D'Hoker and R. Jackiw, Phys. Rev. Lett. **50** (1983) 1719; Phys. Rev. **D26** (1982) 3517; E. D'Hoker, D. Freedman and R. Jackiw, Phys. Rev. **D28** (1983) 2583.

[14] J.B. Hartle and S.W. Hawking, Phys. Rev. **D28** (1983) 2960.

[15] J. Polchinski, UTTG-15-90.

[16] J. Ambjørn, B. Durhuus and J. Fröhlich, Nucl. Phys. **B257** [**FS14**](1985) 433; F. David, Nucl.Phys. **B257** [**FS14**] (1985) 45; V. A. Kazakov, I. K. Kostov and A. A. Migdal, Phys. Lett. **157B** (1985) 295; V. A. Kazakov, Phys. Lett. **119A** (1986) 140; Phys. Lett. **150B** (1985) 282; D. V. Boulatov and V. A. Kazakov, Phys. Lett. **B186** (1987) 379; V. A. Kazakov and A. Migdal, Nucl. Phys. **B311** (1988) 171.

[17] E. Brézin and V. Kazakov, Phys. Lett. **B236** (1990) 14.

[18] M. Douglas and S. Shenker, Nucl. Phys. **B335** (1990) 635.

[19] D. Gross and A. Migdal, Phys. Rev. Lett. **64** (1990) 127; PUPT-1159.

[20] T. Banks, M. Douglas, N. Seiberg and S. Shenker, Phys. Lett. **238B** (1990) 279.

[21] M. Douglas, Phys. Lett. **B238** (1990) 176.

[22] F. David, Mod. Phys. Lett. **A3** (1988) 1651; J. Distler and H. Kawai, Nucl. Phys. **B321** (1989) 509.

[23] V. Knizhnik, A. Polyakov and A. Zamolodchikov, Mod. Phys. Lett. **A3** (1988) 819.

[24] A. Polyakov, Mod. Phys. Lett. A **2** (1987) 893.

[25] E. Brezin, M. Douglas, V. Kazakov and S. Shenker, Phys. Lett. **B237** (1990) 43.

[26] J. Polchinski, Comm. Math. Phys. **104** (1986) 37.

[27] T. Banks, N. Seiberg and S. Shenker, unpublished.

[28] J. Distler, PUPT-1161; E. Verlinde and H. Verlinde, IASSNS-HEP-90/40; R. Dijkgraaf, E. Verlinde and H. Verlinde, to appear.

[29] C. Bachas and P.M.S. Petropoulos, CERN-TH.5714/90, CPTH-A964.0490; E. Witten, IASSNS-HEP-90/45.

[30] J. Cardy, Physica **140A** (1986) 219.

[31] T. Banks, N. Seiberg and S. Shenker, to appear.

[32] E. Marinari and G. Parisi, Phys. Lett. **240B** (1990) 375.